AF906178

SpringerWienNewYork

Hugues Duffau
Editor

Brain Mapping

From Neural Basis of Cognition to Surgical Applications

With a Foreword
by Marsel Mesulam

SpringerWienNewYork

Prof. Hugues Duffau, MD, PhD

Department of Neurosurgery, Gui de Chauliac Hospital,
and INSERM U1051, Team "Plasticity of Central Nervous
System, Human Stem Cells and Glial Tumors",
Institute for Neuroscience of Montpellier,
Montpellier University Medical Center,
Montpellier, France

SpringerWienNewYork is part of
Springer Science+Business Media
springer.at

Coverdesign: WMX Design GmbH, Heidelberg, Germany
Layout and Typesetting: Peter Sachartschenko, Vienna, Austria

Printed on acid-free and chlorine-free bleached paper
SPIN: 12791132

With 140 (mainly coloured) Figures

Library of Congress Control Number: 2011928715

ISBN 978-3-7091-0722-5 SpringerWienNewYork

Foreword

In order to achieve optimal precision and safety in the operating room, a surgeon needs to understand not only the shape and vasculature of an organ but also the function of its parts. Classic surgical textbooks offer exquisite and definitive detail of functional anatomy for most organ systems. This is not yet the case in neurosurgery, for the obvious reason that the human brain is the single most complex device in the known universe and that the function of its parts is far from being fully understood.

The human cerebral cortex alone contains 40 billion neurons crowded into 3 square meters of surface area. Each neuron makes thousands of synaptic contacts through which information rapidly flows from one neuron to another. The total number of neural contacts on the surface of the brain is in the order of 40 followed by 14 zeroes, a number that is as large as the number of all the stars in our galaxy.

This complexity is not without order. Consistent patterns of regional specializations give rise to a map where job descriptions can vary dramatically over the course of a few millimeters. The most obvious functional landmarks of this map, the primary sensory and motor areas, constitute only 10% of the cerebral cortex. The rest is subsumed by association cortex, a vast expanse of gray matter that mediates integrative processes known as cognition, emotion, and comportment.

Within association cortex, it is possible to identify a sensory-fugal hierarchy of synaptic relays that transmit impulses successively from primary to unimodal, heteromodal, paralimbic, and limbic cortices. This bottom-up synaptic cascade allows behavioral responses to be guided by extrapersonal events. Modern neuroanatomy has also identified a reciprocal top-down cascade that transmits impulses in the opposite direction, from limbic and association cortices toward sensory areas. This pathway introduces internally generated biases into the interpretation of sensory reality. The interaction of these two counter currents of neural transmission is somehow experienced as "consciousness."

The greater the synaptic distance from primary sensory-motor cortex, the more difficult it becomes to define the function of a cortical area. Even those functionalities that we believe to have identified seem to defy common sense. What kind of engineering logic would have made memory for recent events, a faculty essential for all aspects of behavior, critically dependent on a tiny part of the temporal lobe known as the hippocampus? Why is language, a faculty that permeates all aspects of thought, critically dependent on only one hemisphere? Why is the contralateral hemisphere so reluctant to take over some of this functionality in patients with left hemisphere damage?

The past 150 years have allowed us to accumulate mountains of facts on the functional cartography of the primate brain. The classic patient descriptions of the late 19th and early 20th centuries, the advent of new methods for tracing structural and chemical neuroanatomy, single cell recordings in behaving monkeys, and the modern revolution in neuroimaging are some of the engines that powered this growth. Neurosurgery has been the beneficiary of all these advances at the same time that it has contributed to them through intra-operative stimulation experiments and pivotal case reports such as those of H.M. and of patients will callosal resection.

An expanding body of evidence has led to the modern view that the brain does not have centers for "hearing words," "perceiving space," or "storing memories." The coordination of such cognitive and behavioral domains is now attributed to intersecting large-scale neural networks that contain interconnected cortical and subcortical components. The network approach to higher cerebral function has at least four implications of clinical relevance: (1) a single domain such as language or memory can be disrupted by damage to any one of several areas, as long as these areas belong to the same network; (2) damage confined to a single area can give rise to multiple deficits, involving the functions of all networks that intersect in that region; (3) damage to a network component may give rise to minimal or transient deficits if other parts of the network undergo compensatory reorganization; and (4) individual anatomic sites within a network display a relative (but not absolute) specialization for different behavioral aspects of the relevant function. Five anatomically defined large-scale networks are most relevant to clinical practice: a left-dominant perisylvian network for language; a right-dominant parietofrontal network for spatial cognition; an occipitotemporal network for face and object recognition; a limbic network for retentive memory; and a prefrontal network for attention and comportment. The inner organization of networks may vary from one person to another (as in the case of right versus left handers) and may therefore need to be ascertained individually when the goal is to guide surgical interventions.

What we have learned about the human brain is awe-inspiring. Nonetheless, we must also acknowledge that many aspects of the functional landscape remain to be filled. The next revolution in neuroscience will arise when existing facts are linked to explanatory theories of brain function, theories that can explain, in some principled way, how patterned synaptic activity can transform muscle contractions and sensory input into memories, words, feelings and purposeful actions. The scientist in each of us looks forward to such future revelations. As clinicians who care for sick patients, however, we do not have the luxury of waiting for all the details to become clarified before choosing a course of action at the bedside or operating room. This is why it is so essential for modern clinical neuroscience, and especially for neurosurgery, to incorporate developments in this field in as timely a fashion as possible, so that patient care becomes guided by the latest increments of relevant knowledge. I have no doubt that this comprehensive volume edited by Professor Duffau will serve this purpose with considerable distinction.

Marsel Mesulam, MD

Contents

Surgical applications

Perspectives

List of contributors

Céline Amiez
Neuropsychology/Cognitive Neuroscience Unit
Montreal Neurological Institute
Department of Neurology and Neurosurgery
McGill University,
Montreal, Quebec, Canada

Sylvain Baillet
Departments of Neurology and Biophysics
Medical College of Wisconsin
Milwaukee, USA

Paolo Bartolomeo
INSERM-UPMC UMR S 975
G.H. Pitié-Salpêtrière
Paris, France

Miriam H. Beauchamp
Centre de Recherche en Neuropsychologie et Cognition
Department of Psychology
University of Montreal
Quebec, Canada
Ste-Justine University Hospital Research Center
Montreal, Quebec, Canada

Pierre Bellec
UNF, Geriatric Institute,
University of Montreal
Montreal, Canada

John S. Bellerose
Centre de Recherche en Neuropsychologie et Cognition
Department of Psychology, University of Montreal
Quebec, Canada
Ste-Justine University Hospital Research Center
Montreal, Quebec, Canada

Lorenzo Bello
Neurochirurgia, Dipt. di Scienze Neurologiche
Università degli Studi di Milano
Istituto Clinico Humanitas
Milano, Italy

Alim Louis Benabid
Clinatec, CEA Grenoble
Grenoble, France

Habib Benali
INSERM U678, Paris, France
Faculté de médecine Pitié-Salpetrière
UPMC Université de Paris 06
Paris, France

Mitchel S. Berger
Departments of Neurological Surgery and Radiology
University of California
San Francisco, CA, USA

Ingeborg Bosma
Department of Neurology
VU University Medical Center
Amsterdam, The Netherlands

Christian Brogna
Department of Neuroscience-Neurosurgery
"Sapienza" University of Rome
Rome, Italy

Giuseppe Casaceli
Neurochirurgia, Dipt. di Scienze Neurologiche
Università degli Studi di Milano
Istituto Clinico Humanitas
Milano, Italy

Antonella Castellano
Neuroradiology, and CERMAC
Università Vita-Salute e Instituto Scientifico San Raffaele
Milano, Italy

Marco Catani
Natbrainlab, Department of Forensic and
Neurodevelopmental Sciences, Institute of Psychiatry
King's College London
London, UK

Anne Sophie Champod
Neuropsychology/Cognitive Neuroscience Unit
Montreal Neurological Institute
Department of Neurology and Neurosurgery
McGill University
Montreal, Quebec, Canada

Edward F. Chang
Departments of Neurological Surgery and Radiology
University of California
San Francisco, CA, USA

D. Louis Collins
Brain Imaging Centre
Montreal Neurological Institute
Department of Neurology and Neurosurgery
McGill University
Montreal, Quebec, Canada

Gérardo Conesa
Department of Neurosurgery
Hospital del Mar
Barcelona, Spain

David Coynel
Inserm U678, Paris, France
Faculté de médecine Pitié-Salpetrière
UPMC Université de Paris
Paris, France

Flavio Dell'Acqua
Natbrainlab, Department of Forensic and
Neurodevelopmental Sciences, Institute of Psychiatry
King's College London
London, UK

Rolando Del Maestro
Brain Tumour Research Centre
Montreal Neurological Institute
Department of Neurology and Neurosurgery
McGill University
Montreal, Quebec, Canada

Philip C. De Witt Hamer
Neurosurgical Center Amsterdam
VU University Medical Center
Amsterdam, The Netherlands

Anthony Steven Dick
Department of Psychology
Florida International University
Miami, FL, USA

Linda Douw
Department of Neurology
VU University Medical Center
Amsterdam, The Netherlands

Hugues Duffau
Department of Neurosurgery, Hôpital Gui de Chauliac
CHU Montpellier, Montpellier, France
Institute of Neuroscience of Montpellier
INSERM U1051, Plasticity of Central Nervous System,
Human Stem Cells and Glial Tumors, Hôpital Saint Eloi
CHU Montpellier
Montpellier, France

Josep M. Espadaler
Department of Neurosurgery, Hospital del Mar
Barcelona, Spain

Enrica Fava
Neurochirurgia, Dipt. di Scienze Neurologiche
Università degli Studi di Milano
Istituto Clinico Humanitas
Milano, Italy

Andrea Falini
Neuroradiology, and CERMAC
Università Vita-Salute e Instituto Scientifico San Raffaele
Milano, Italy

Emily Ferreira
Neuropsychology/Cognitive Neuroscience Unit
Montreal Neurological Institute
Department of Neurology and Neurosurgery
McGill University
Montreal, Quebec, Canada

Rodney A. Gabriel
Departments of Neurological Surgery and Radiology
University of California
San Francisco, CA, USA

Oliver Ganslandt
Neurochirurgische Klinik
Universität Erlangen-Nürnberg
Erlangen, Germany

Peter Grummich
Neurochirurgische Klinik
Universität Erlangen-Nürnberg
Erlangen, Germany

Jan J. Heimans
Department of Neurology
VU University Medical Center
Amsterdam, The Netherlands

Martin Klein
Department of Medical Psychology
VU University Medical Center
Amsterdam, The Netherlands

Penelope Kostopoulos
Neuropsychology/Cognitive Neuroscience Unit
Montreal Neurological Institute
Department of Neurology and Neurosurgery
McGill University
Montreal, Quebec, Canada

Alexandre Krainik
Clinique Universitaire de Neuroradiologie et IRM
CHU Grenoble
Grenoble, France

Matthew A. Lambon Ralph
Neuroscience and Aphasia Research Unit
School of Psychological Sciences
University of Manchester
Manchester, UK

Maryse Lassonde
Centre de Recherche en Neuropsychologie et Cognition
Department of Psychology
University of Montreal
Quebec, Canada
Ste-Justine University Hospital Research Center
Montreal, Quebec, Canada

Emmanuel Mandonnet
Department of Neurosurgery, Lariboisière Hospital
Paris, France
INSERM U678
Paris, France

Guillaume Marrelec
Inserm U678
Paris, France
Faculté de médecine Pitié-Salpetrière
UPMC Université de Paris
Paris, France

Juan Martino
Department of Neurological Surgery
Hospital Universitario Marqués de Valdecilla and
Instituto de Formación e Investigación Marqués de
Valdecilla (IFIMAV)
Santander, Cantabria, Spain

Arnaud Messé
Inserm U678
Paris, France
Faculté de médecine Pitié-Salpetrière
UPMC Université de Paris
Paris, France

Srikantan S. Nagarajan
Departments of Neurological Surgery and Radiology
University of California
San Francisco, CA, USA

Christopher Nimsky
Klinik für Neurochirurgie
Universitätsklinikum Giessen und Marburg GmbH
Marburg, Germany

George Ojemann
Department of Neurological Surgery
University of Washington School of Medicine
Seattle, WA, USA

Vincent Perlbarg
Inserm U678
Paris, France
Faculté de médecine Pitié-Salpetrière
UPMC Université de Paris
Paris, France

Michael Petrides
Neuropsychology/Cognitive Neuroscience Unit
Montreal Neurological Institute
Department of Neurology and Neurosurgery
McGill University
Montreal, Quebec, Canada

Thomas Picht
Department of Neurosurgery
Charité-Universitätsmedizin Berlin
Campus Benjamin Franklin
Berlin, Germany

Nick Ramsey
Department of Neurology and Neurosurgery
Rudolf Magnus Institute for Neuroscience
University Medical Center of Utrecht
Utrecht, The Netherlands

Jaap C. Reijneveld
Department of Neurology
VU University Medical Center
Amsterdam, The Netherlands
Department of Neurology
Academic Medical Center
Amsterdam, The Netherlands

Guilherme Carvalhal Ribas
Department of Surgery
University of São Paulo Medical School - LIM-02
Hospital Israelita Albert Einstein
São Paulo, Brazil

Marco Riva
Neurochirurgia, Dipt. di Scienze Neurologiche
Università degli Studi di Milano
Istituto Clinico Humanitas
Milano, Italy

Geert-Jan M. Rutten
Department of Neurology and Neurosurgery
Rudolf Magnus Institute for Neuroscience
University Medical Center of Utrecht
Utrecht, The Netherlands

Steven L. Small
Department of Neurology
The University of Chicago
Chicago, USA

Cornelis Jan Stam
Department of Clinical Neurophysiology
VU University Medical Center
Amsterdam, The Netherlands

PD Dr. Andrea Szelényi
Klinik und Poliklinik für Neurochirurgie
Klinikum der Johann Wolfgang Goethe Universität
Frankfurt a.M., Germany

Michel Thiebaut de Schotten
Natbrainlab, Department of Forensic and
Neurodevelopmental Sciences
Institute of Psychiatry
King's College London
London, UK
INSERM-UPMC UMR S 975
G.H. Pitié-Salpêtrière
Paris, France

Pascale Tremblay
Department of Neurology
The University of Chicago
Chicago, USA

Edwin van Dellen
Department of Neurology
VU University Medical Center
Amsterdam, The Netherlands

Peter A. Winkler
Department of Neurosurgery and Epilepsy Surgery
Schön-Klinik Vogtareuth
Vogtareuth, Germany

Christian Xerri
Neurobiologie Intégrative et Adaptative UMR 6149
Université de Provence/CNRS
Marseille, France

Introduction

From neurosciences to brain surgery and from brain surgery to neurosciences

Hugues Duffau

The goal of this book is to forge a link between fundamental research in the field of cognitive neurosciences, which now benefits from a better knowledge of the neural foundations of cerebral processing, and its clinical application, especially in neurosurgery – which itself is able to provide new insights into the organization of the brain.

First of all, the anatomical basis will be detailed. Indeed, a perfect knowledge of the anatomy of the central nervous system is crucial, both with regard to cortical (gyri and sulci) as well as subcortical (especially white matter pathways) structures. As stated by Broadmann in 1909 [1] "One thing must be stressed quite firmly: henceforth functional localization of the cerebral cortex without the lead of anatomy is utterly impossible in man as in animals ... So, first anatomy and then physiology; but if first physiology, then not without anatomy". A century later, this statement has lost none of its validity. Physiological understanding of the brain is not possible without a better knowledge of its anatomy. Yet, it is puzzling to note that very little data is available in the literature on the subcortical connectivity in humans since the seminal works of Déjerine [3] and Klinger [6]. Interestingly, recent dissection works have again begun studying the relationships between the different white matter tracts and their cortical terminations. This data is essential to understanding the neural foundations underlying distributed functional networks, as well as to performing surgery within the brain.

Nonetheless, due to the considerable interindividual anatomo-functional variability, neuroanatomy is not sufficient to predict eloquence and thus, in a surgical setting, to avoid the risks of postoperative permanent deficit. As a consequence, the second part this book will cover advances in and limitations of the different methods of functional cerebral mapping. In the past decades, dramatic technical developments have been made in the field of non-invasive (e.g., magnetoencephalography, functional MRI, diffusion tensor imaging) as well as invasive (e.g., extraoperative or intraoperative brain stimulation in conscious patients) human brain mapping. The aim is not to compare these different methods, as previously done in the literature, but rather to explain that they need to be combined in order to improve the reliability of the mapping, because their underlying principles are different. For instance, functional MRI shows activations based on neurovascular coupling, and therefore is unable to differentiate between crucial areas for the function and regions which could be compensated – but it is non-invasive and thus it can be repeated, opening the door to longitudinal studies (such as before and after brain surgery). Diffusion tensor imaging shows the anatomy of the white matter bundles, but not their function. On the contrary, direct brain electrostimulation inhibits functional networks both at the cortical and subcortical level, and makes it possible to identify the essential structures for the function (thanks to online anatomo-functional correlations) – but it is invasive and it can map only a part of the brain. Such knowledge of the respective pitfalls for each technique is very important for neuroscientists and physicians alike, in order to accurately interpret the results of the mapping, and on this basis, to propose new models of brain processing or to plan cerebral surgery.

In the book's third part, updated models of sensory-motor, visuo-spatial, language, memory, emotional and executive functions will be detailed. Advances in anatomical dissections and brain mapping methods, both in animals and humans, have allowed us to switch from a classical localisationist and fixed view of the organization of the central nervous system to a more dynamic view based on large-scale distributed and interactive networks [7]. It is worth noting that the elaboration of new models of cognition should definitely reflect the constraints of neuroanatomy, at least for their validation, rather than to be purely esoteric without any considerations of concrete aspects. In addition, beyond the jump from the modular to the integrated brain, relationships between different functions are now better understood. For instance, it is difficult to speak about language or memory without taking into account the executive functions as well as emotional aspects, leading to the concepts of theory of the mind and social cognition.

On this basis, new strategies for the surgical management of cerebral lesions will be proposed, with an optimization of the benefit-risk ratio of surgery. Functional neuroimaging and fiber tractrography may help to identify eloquent areas before surgery, and they can be integrated into a multimodal neuronavigational system during the surgical resection. However, it is important to keep in mind that, at least at the individual level in the case of brain disease, these techniques are not yet sufficently reliable, especially for methodological reasons (selection of tasks, choice of biomathematical model, neurovascular decoupling in cases of gliomas). Thus, intraoperative electrophysiological techniques (monitoring and electrostimulation mapping, particularly on conscious patients) are still the gold standard for cerebral surgery in eloquent structures. They allow the detection of functional cortical areas as well as subcortical connectivity, provided that a rigorous methodology is applied. Therefore, it is possible to tailor the resection according to individual functional boundaries in order (i) to extend the surgical indications within regions classically considered "inoperable" (such as Broca's area, the central area or the insula), (ii) to increase the extent of resection by avoiding leaving a margin around the crucial areas and thus increasing the impact on the natural progression of the disease (e.g., tumor or epilepsy), (iii) while decreasing the rate of permanent deficit (less than 2% in the recent literature) and even improving the quality of life (in particular thanks to seizure relief). To this end, beyond the classical neurological examination, longitudinal extensive neurocognitive assessment is now mandatory in cerebral surgery in order to objectively evaluate the impact of the diseases and treatments on health-related quality of life – which should be refined for each patient. Therefore, a better knowledge of models and the neural foundations of cognition is crucial to selecting the optimal tasks before, during and after surgical resection at the level of the individual patient. New specific programs of functional rehabilitation can also be created on this basis.

Lastly, translational studies involving serial (pre- and post-operative) functional neuroimaging and tractography, invasive intraoperative cortical and subcortical electrical mapping, as well as the biomathematical modeling of synchrony (especially based on the theory of graphs and small-world theory) opened the door to the new concepts of brain "hodotopy" [2,4] and plasticity [5], that is, a dynamic organization of the central nervous system constituted by parallel distributed networks that are interconnected and able to compensate for one another. This plastic potential, which implies that the subcortical connectivity must be preserved, now makes it possible to consider a multi-stage surgical approach in tumors involving eloquent areas, especially in slow-growing lesions such as low-grade gliomas. The principle is to perform a second (or even a third) surgery after glioma re-growth when the first resection was not complete for functional reasons, by improving the extent of tumor removal thanks to functional remapping over time – reshaping partly induced by adapted programs of rehabilitation. Finally, from a fundamental point of view, such methodological and conceptual developments also promote a deeper understanding of the neural foundations underlying brain functioning. Nevertheless, ethical aspects of (invasive) human brain mapping must not be forgotten, because the first goal of surgery remains that it be beneficial for the patient. To this end, longitudinal neuropsychological assessments, before and after each surgery, should be performed more systematically.

In summary, taking an integrative approach to the neurosciences and neurosurgery may improve both the understanding of the dynamic functional anatomy of the brain as well as the quality of life of patients bearing a cerebral lesion. Indeed, if the cognitive neurosciences are to be more systematically integrated in surgical strategy, brain surgery in itself also represents a unique opportunity to validate the hypotheses made on the basis of the findings provided by non-invasive cerebral mapping and biomathematical modeling. Furthermore, an improved knowledge of more sophisticated and thus more fragile circuits through the brain evolution may now open the door for a "preventive functional neurosurgery", i.e. for considering surgery before the disease has induced irreversible symptoms – or ultimately before any symptoms develop, especially in neurooncology, when growing tumors are incidentally discovered. Therefore, neuroanatomists, neuroradiologists, neurologists, neurophysiologists, neuropsychologists, speech therapists, neuroscientists and neurosurgeons should more regularly work together. The purpose of this book, which involves experts in these different fields, is to stimulate such conceptual and practical collaboration, in networks, mirroring the functioning of the nervous system itself.

References

[1] Broadmann K (1990) Vergleichende Lokalisationslehre der Grosshirnrinde. Verlag von Johann Ambrosius Barth, Leipzig

[2] Catani M (2007) From hodology to function. Brain 130: 602–605

[3] Déjerine JJ (1895) Anatomie des centres nerveux. Ruell, Paris

[4] Duffau H (2008) The anatomo-functional connectivity of language revisited: new insights provided by electrostimulation and tractography. Neuropsychologia 4: 927–934

[5] Duffau H (2005) Lessons from brain mapping in surgery for low-grade glioma: insights into associations between tumour and brain plasticity. Lancet Neurol 4: 476–486

[6] Klinger J (1935) Erleichterung der makroskopischen Präparation des Gehirns durch den Gefrierprozess. Schweiz Arch Neurol und Psychiatrie 36: 247–256

[7] Mesulam MM (1990) Large-scale neurocognitive networks and distributed processing for attention, language and memory. Ann Neurol 28: 597–613

Brain anatomy

The microneurosurgical anatomy of the cerebral cortex

Guilherme C. Ribas

In the relatively new field of microneurosurgery, the development and use of the transcisternal, transfissural, and transsulcal approaches [54, 57, 58] have established the sulci as fundamental landmarks on the brain surface. The well-known variability in cortical function [1, 2, 13, 29, 32, 50] calls for the aid of cortical mapping techniques to precisely identify specific sites related to cortical function. Nevertheless, detailed knowledge of the structure and form of the cerebral sulci and gyri continues to be mandatory for neuroimaging as well as intraoperative guidance. Once identified, the cerebral sulci can be used by the neurosurgeon either as microneurosurgical corridors or simply as cortical landmarks [41, 42]. On the other hand, historically, it is notable that despite the intense interest that humankind has always had in relation to the brain, it was only in the middle of the 19th century that the anatomical organization of the cerebral sulci and gyri was perceived and described [17], as detailed bellow.

General anatomical features

Given their phylogenetic [8, 28] and embryological [9, 28, 30, 53] development, especially the process of invagination of the surface of the brain, which effectively increases the cortical area without proportionally increasing the volume of the brain, [43, 53] the cerebral sulci, which delineate the respective gyri, can be considered natural extensions of the subarachnoid space. When they are deep and anatomically more constant, they are referred to as *fissures* [5, 6]. The gyri that are more rounded or quadrangular are usually referred to as *lobules*. The principal sulci have approximate depths ranging from 1 to 3 cm, and their walls harbor small gyri that face, adapt to, and connect with each other. Those gyri are generically designated the *transverse gyri*. The sulci that separate the transverse intrasulcal gyri vary in length and depth, and, at the surface of the brain, they become visible as incisures. The indentations caused by cortical arteries can have an appearance similar to that of the incisures.

It is noteworthy that the timing of their embryological development and their degree of variability [9, 28] define a true morphological hierarchy, at the top of which are the fissures and principal sulci (Table 1). It is equally notable that this structural hierarchy is directly correlated with the functional importance of the areas to which the sulci are related, the more anatomically constant sulci being those that are topographically related to areas that are more specialized [31, 50].

Table 1: Prenatal cerebral sulci development

Characteristic	Chi et al 1977	Nishikuni 2006
no. of fetuses	207	107
gestational age in wks	10–44	12–40
longitudinal cerebral fissure	10	12
superolat surface		
lateral sulcus	14	17
circular insular sulcus	18	17
central insular sulcus		29
central sulcus	20	21
precentral sulcus	24	26
superior frontal sulcus	25	25
inferior frontal sulcus	28	30
postcentral sulcus	25	29
intraparietal sulcus	26	29
transverse occipital sulcus		30
lunate sulcus		24
superior temporal sulcus	23	26
inferior temporal sulcus	30	31
transverse temporal sulcus	31	33
inferior surface		
olfactory sulcus	16	17
orbital sulcus		22
hippocampal sulcus	10	12
rhinal sulcus		25
collateral sulcus	23	29
occipitotemporal sulcus	30	33
medial surface		
callosal sulcus	14	12
cingulate sulcus	18	19
marginal ramus		33
paracentral sulcus		30
paraolfactory sulcus		29
subparietal sulcus		30
calcarine sulcus	16	17
parietooccipital sulcus	16	19
secondary sulcus	40	38

On the brain surface, the sulci can be long or short as well as continuous (sylvian fissure, callosal, calcarine, parietooccipital, collateral, and generally the central sulcus) or interrupted. Ono et al [30]. have described 4 main types of sulci: large primary sulci (for example, central, precentral, postcentral, and continuous sulci); short primary sulci (for example, rhinal, olfactory, lateral, and occipital sulci); short sulci composed of several branches (for example, orbital and subparietal sulci); and short, free supplementary sulci (for example, medial frontal and lunate sulci). Frequently, the sulci are composed of side branches that can be unconnected or connected (with end-to-side, end-to-end, or side-to-side connections that can also join 2 neighboring parallel sulci).

Since connections between sulci are common, the nomenclature varies widely, with different authors providing different interpretations [10, 30, 47]. The sulci can vary in size and shape from person to person and the cerebral gyri constitute a real continuum in that the surface presents a serpentine configuration because of the connections across the sulcal extremities and interruptions, and are continuous throughout the sulcal depths [56]. The gyral separation is only superficial and is defined by the continuity and depth of the adjacent sulci. Therefore, each gyrus should be understood as a region and not as a well-defined structure.

Because they result from an infolding process, the sulci of the superolateral and inferior surfaces of the brain are usually oriented toward the nearest ventricular cavity, although this feature does not apply to the medial surface of the cerebral hemisphere, where the sulci are particularly dependent on the development of the corpus callosum [30]. The single most common identifiable surface feature is the sylvian fissure, given its particular mechanism of development [53].

Their variations and irregularities give the sulci and gyri of the human brain a labyrinthine appearance. Nevertheless, they are arrayed in a particular configuration.

In rough terms, the human brain is organized as follows: the frontal and temporal regions of each hemisphere are each composed of 3 horizontal gyri; the central area is composed of 2 slightly oblique gyri; the parietal region is composed of 2 lobules, with a quadrangular superior lobule and an inferior lobule consisting of 2 semicircular gyri; the occipital region is composed of 3 irregular, less well-defined, predominantly longitudinal gyri that converge toward the occipital pole, the superior being vertical and the middle and inferior being horizontal; and the insula is composed of 4 – 5 diagonal gyri (Figs. 1 left and 2).

Medially, the external lateral gyri and lobules extend along the superior and inferolateral borders of each hemisphere. Together, these gyri constitute an outer medial ring, which surrounds a more well-defined C-shaped inner ring primarily composed of 2 continuous gyri. Inferiorly, the base of each hemisphere consists of 2 horizontal gyri longitudinally oriented between the lateral extended gyri (along the inferolateral and inferomedial borders) and the medial continuous gyri of the inner ring (Figs. 1 right and 3).

Regarding the artibrary division of the cerebral hemispheres into lobes, the International Anatomical Terminology published in 1998 [15] divided each brain hemisphere into 6 lobes: frontal, parietal, occipital, temporal, insular and limbic. Nevertheless, as proposed by Yasargil [56], the concept of a central lobe composed of the precentral and postcentral gyri is justified, since they compose a single morphological and functional unit, as has been suggested previously by Penfield and Rasmussen [31] and Rasmussen alone [33, 34, 35].

The sulci, gyri, and cerebral lobes

Given the visual evidence of the lateral cerebral sulcus, or sylvian fissure, together with the distinct, slightly oblique arrangement of the precentral and postcentral gyri as well as of the sulci that delineate them at approximately the center of the external brain surface, the character of the remaining gyri of the su-

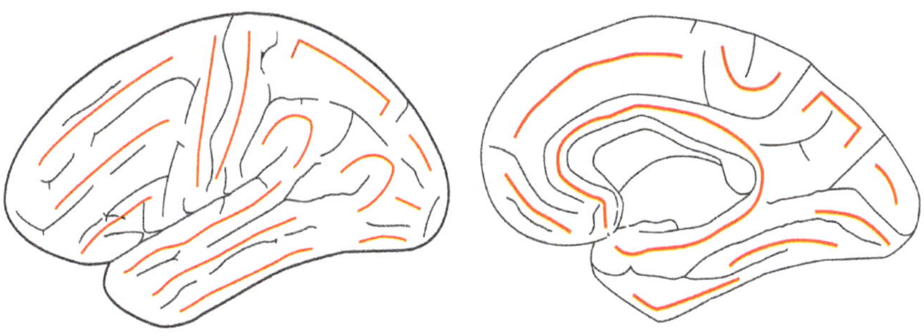

Fig. 1. Basic organization of the brain gyri: superolateral surface **(left)** and medial and basal surfaces **(right)**. *Red lines* indicate the constant arrangement of the brain gyri

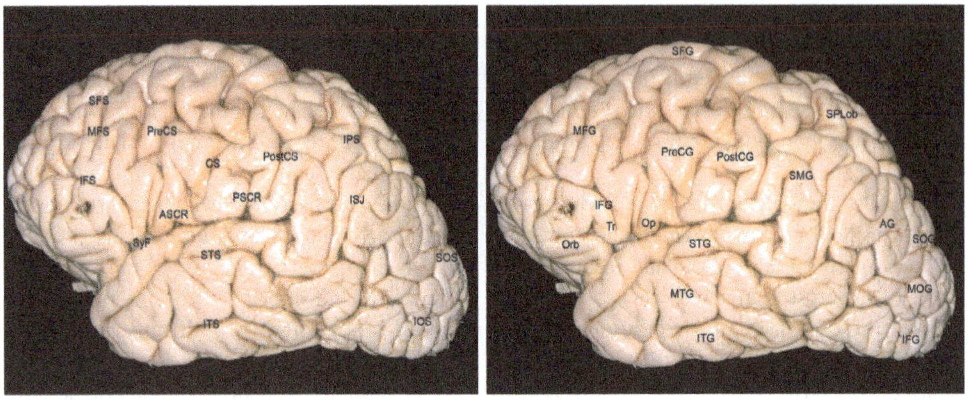

Fig. 2. The main sulci **(left)** and gyri **(right)** of the superolateral surface of the brain. AG = angular gyrus; ASCR = anterior subcentral ramus of sylvian fissure; CS = central sulcus; IFG = inferior frontal gyrus; IFS = inferior frontal sulcus; IOS = inferior occipital sulcus; IPS = intraparietal sulcus; ISJ = intermediary sulcus of Jensen; ITG = inferior temporal gyrus; ITS = inferior temporal sulcus; MFG = middle frontal gyrus; MFS = middle frontal sulcus; MOG = middle occipital gyrus; MTG = middle temporal gyrus; Op = opercular part of inferior frontal gyrus; Orb = orbital part of inferior frontal gyrus; PostCG = postcentral gyrus; PostCS = postcentral sulcus; PreCG = precentral gyrus; PreCS = precentral sulcus; PSCR = posterior subcentral ramus of sylvian fissure; SFG = superior frontal gyrus; SFS = superior frontal sulcus; SMG = supramarginal gyrus; SOG = superior occipital gyrus; SOS = superior occipital sulcus; SPLob = superior parietal lobe; STG = superior temporal gyrus; STS = superior temporal sulcus; SyF = sylvian fissure; Tr = triangular part of inferior frontal gyrus

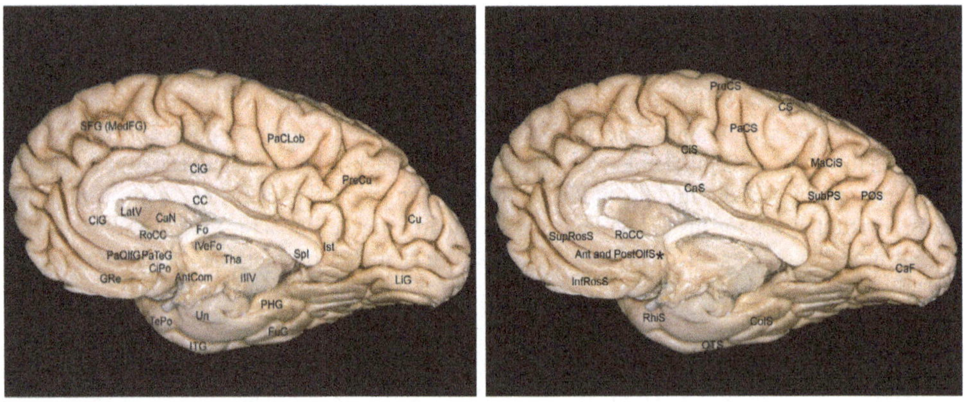

Fig. 3. The main sulci **(left)** and gyri **(right)** of the medial and basal temporooccipital surfaces. AntCom = anterior commissure; Ant and PostOlfS = anterior and posterior paraolfactory sulcus; CaF = calcarine fissure; CaN = caudate nucleus; CaS = callosal sulcus; CC = corpus callosum; CiG = cingulate gyrus; CiPo = cingulate pole; CiS = cingulate sulcus; ColS = collateral sulcus; CS = central sulcus; Cu = cuneus; Fo = fornix; FuG = fusiform gyrus; GRe = gyrus rectus; IIIV = third ventricle; InfRosS = inferior rostral sulcus; Ist = isthmus of cingulate gyrus; ITG = inferior temporal gyrus; IVeFo = interventricular foramen of Monro; LatV = lateral ventricle; LiG = lingual gyrus; MaCiS = marginal ramus of the cingulate sulcus; MedFG = medial frontal gyrus; OTS = occipitotemporal sulcus; PaCLob = paracentral lobule; PaCS = paracentral sulcus; PaOlfG = paraolfactory gyri; PaTeG = paraterminal gyrus; PHG = parahippocampal gyri; POS = parietooccipital sulcus; PreCS = precentral sulcus; PreCu = precuneus; RhiS = rhinal sulcus; RoCC = rostrum of the corpus callosum; SFG = superior frontal gyrus; Spl = splenium of corpus callosum; SubPS = subparietal sulcus; SupRosS = superior rostral sulcus; TePo = temporal pole; Tha = thalamus; Un = uncus

perolateral surface of each cerebral hemisphere can be more easily understood if we take as our starting point and basis those sulcal spaces and cerebral gyri. The macroscopic study of the sulci and gyri of each cerebral hemisphere should therefore begin with the identification of the sylvian fissure, which clearly separates the superolateral surfaces of the frontal, central, and parietal lobes from the temporal lobe, and should be followed by the identification of the precentral and postcentral gyri, which divide the portion of this surface that is superior and posterior to the sylvian fissure into its anterior and posterior halves. As above justified, we will describe the 7 lobes of each cerebral hemisphere as follows: frontal, central, parietal, occipital, temporal, insular, and limbic [56]. Since the connections of the limbic lobe are particularly complex, the text related to that lobe also addresses the adjacent and correlated areas.

The frontal lobe

The frontal lobe constitutes the largest, most anterior part of each hemisphere. Within the scheme adopted in this article, the frontal lobe is delineated posteriorly by the oblique precentral sulcus and is composed of the superior, middle, and inferior frontal gyri, which are oriented longitudinally and separated by the superior and inferior frontal sulci, also longitudinally oriented (Fig. 2). These gyri are often referred to as F1, F2, and F3, respectively.

On the cerebral surface posteriorly, the superior frontal gyrus is connected to the precentral gyrus by at least 1 fold, which more commonly lies medially along the interhemispheric fissure. Anteriorly, the superior frontal gyrus might be connected to the middle frontal gyrus, with the orbital gyri and the gyrus rectus. In general terms, the superior frontal gyrus is subdivided into 2 longitudinal portions by the so-called medial frontal sulcus, and its medial portion is at times designated the *medial frontal gyrus* [30]. Along the most medial portion of the superior frontal gyrus and immediately facing the precentral gyrus is the important region known as the *supplementary motor area*, which varies from person to person and has poorly defined borders [7, 53].

The middle frontal gyrus is typically the largest of the frontal gyri and often harbors a sulcus that is shallower than that of the other frontal sulci, usually running along its anterior two-thirds and known as the *middle* or *intermediate frontal sulcus* [30]. In most human brains, the middle frontal gyrus is superficially connected to the precentral gyrus by a prominent root that lies between the extremities of a marked interruption in the precentral sulcus. The frequent interruptions in the inferior frontal sulcus are attributable to connections between the middle and inferior frontal gyri.

The inferior frontal gyrus is irregular, crisscrossed by various small branches of the inferior frontal sulcus, which typically superiorly delineate the gyrus in an interrupted manner. Inferiorly, this gyrus is delineated and crisscrossed by rami of the sylvian fissure. Anteriorly, the inferior frontal gyrus terminates merging with the anterior portion of the middle frontal gyrus. Posteriorly, it is connected to the precentral gyrus. The inferior frontal gyrus is composed of, from anterior to posterior, its orbital, triangular, and opercular parts.

The emergence of the horizontal and ascending anterior rami from the same point in the sylvian fissure characterizes the triangular part of the inferior frontal gyrus, which is typically more retracted than the other 2 parts (Figs. 2 and 4). The orbital part is the most prominent of the 3, and the opercular part is consistently U-shaped. Given the usual retraction of the triangular part, the horizontal and ascending anterior branches of the sylvian fissure typically emerge from a small widening of the subarachnoid space, designated the *anterior sylvian point* [41, 59]. Therefore, the anterior sylvian point is situated inferior to the triangular part and anterior to the base of the opercular portion. The point, which is typically visible to the naked eye, divides the sylvian fissure into its anterior and posterior branches [41, 59].

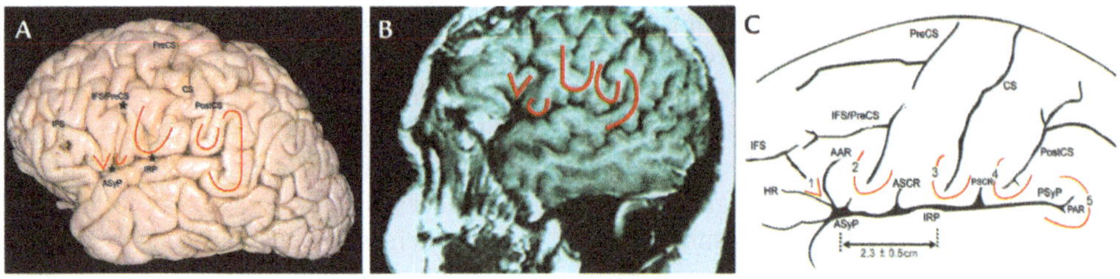

Fig. 4. The frontoparietal operculum: cadaveric specimen (**A**), MR image (**B**), and sketch of sulcal and gyri morphology (**C**). The frontoparietal operculum is characterized by a V-shaped convolution consisting of the triangular part of the inferior frontal gyrus (IFG) (1), located just superiorly to the anterior sylvian point (ASyP), and usually containing a descending branch of the inferior frontal sulcus (IFS); the 3 following U-shaped convolutions respectively composed by the opercular part of the IFG (2), which is always intersected by the inferior part of the precentral sulcus; the subcentral gyrus or rolandic operculum (3), composed of the inferior connection of the pre- and postcentral gyri enclosing the inferior part of the central sulcus (CS); the connection arm between the postcentral and supramarginal gyri (4) that contains the inferior part of the postcentral sulcus; and finally the Cshaped convolution (5) constituted by the connection of the supramarginal and superior temporal gyri that encircles the posterior end of the sylvian fissure. The bottoms of the U-shaped convolutions and their related sulcal extremities can be situated either superior to the fissure or inside its cleft. *Stars* designate the areas labeled. AAR = anterior ascending ramus of the sylvian fissure; ASCR = anterior subcentral ramus of sylvian fissure; HR = horizontal ramus of sylvian fissure; IFS/PreCS = IFS and precentral sulcus; IRP = inferior rolandic point, projection of the central sulcus in the sylvan fissure; PAR = posterior ascending ramus of sylvian fissure; PostCS = postcentral sulcus; PreCS = precentral sulcus; PSCR = posterior subcentral ramus of sylvian fissure; PSyP = posterior sylvian point

The triangular part is quite often divided superiorly by a small descending branch of the inferior frontal sulcus, and the inferiormost portion of the precentral sulcus is always harbored within the U of the opercular part [41]. Inferiorly and anteriorly delineated by the anterior sylvian point and posteriorly delineated by the anterior subcentral branch of the sylvian fissure, the posterior half of the U that characterizes the opercular part corresponds to the important connective fold that is always situated between this portion of the inferior frontal gyrus and the precentral gyrus. In some cases, the anterior basal portion of the opercular part is more developed and is divided by another branch of the sylvian fissure. That branch runs from front to back and is called the *diagonal sulcus of Eberstaller*. When the diagonal sulcus of Eberstaller is present, it divides the anterior portion of the opercular part into 2 triangular portions that are positioned inversely to each other.

In the dominant hemisphere, the opercular and triangular parts of the inferior gyrus correspond to the Broca area, which is responsible for the production of spoken language [3, 7, 18, 32, 53].

Inferiorly, although the orbital part continues with the lateral orbital gyrus, at times passing under a shallow sulcus known as the *fronto-orbital sulcus*, the triangular part is always superior to the sylvian fissure, and the base of the opercular part can be located either superiorly or within that same fissure [30, 39].

The triangular and opercular parts together with the subcentral gyrus, which connects the precentral and postcentral gyri, and the anteroinferior portion of the supramarginal gyrus cover the superior aspect of the insular surface and constitute the frontoparietal operculum. Therefore, the frontoparietal operculum is situated between the horizontal and posterior ascending branches of the sylvian fissure [41, 53].

Anteriorly, the frontal gyri are delineated by the appropriately named *frontomarginal sulcus*, which lies superior and parallel to the supraciliary margin, separating the superolateral and orbital frontal surfaces [30, 56].

On the frontobasal, or orbital, surface of each frontal lobe, the deep olfactory sulcus, which harbors the olfactory bulb and olfactory tract, is very deep and lies longitudinally in a paramedian position (Fig. 5). Posteriorly, the olfactory tract is divided into the medial and lateral striae, which delineate the anteriormost aspect of the anterior perforated substance.

Medial to the olfactory sulcus is the long and narrow gyrus rectus, which is continuous with the superior frontal gyrus along the medial surface of the hemisphere.

Lateral to the olfactory sulcus are the orbital gyri, which account for the greatest proportion of the frontobasal surface (Fig. 5). The H-shaped orbital sulcus (cruciform sulcus of Rolando) delineates the anterior, posterior, medial, and lateral orbital gyri. The posterior orbital gyrus is situated anterior to the anterior perforated substance and typically presents a configuration similar to a tricorner or "Napoleon" hat, which can facilitate its identification in anatomical specimens in which the H-shaped orbital sulcus presents variations [40].

The posterior orbital gyrus is connected medially to the medial orbital gyrus, characterizing the posteromedial orbital lobule situated posterior and along the olfactory tract and the lateral olfactory striae, [56] which is in turn connected to the anterior portion of the insula via the transverse insular gyrus. The remaining orbital gyri are connected to the superior, middle, and inferior frontal gyri, along the frontal pole.

The central lobe

The central lobe consists of the precentral (motor gyrus) and postcentral gyri (sensitive gyrus), which are oriented obliquely on the superolateral surface and are separated by the central sulcus, by their inferior (subcentral gyrus) and superior connections (paracentral gy-

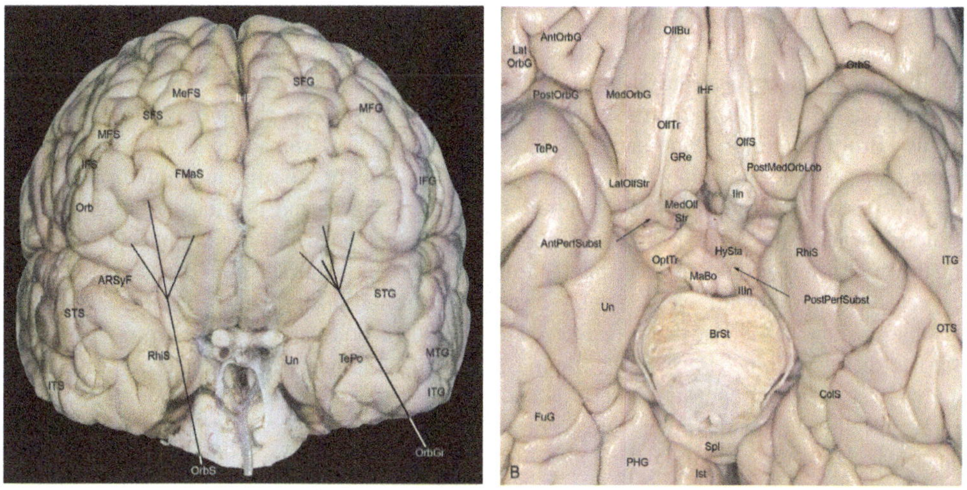

Fig. 5. Anterior view of cerebral hemispheres **(left)** and view of the basal frontotemporal surface **(right)**. AntOrbG = anterior orbital gyrus; AntPerfSubst = anterior perforated substance; ARSyF = anterior ramus of sylvian fissure; BrSt = brainstem (pons); ColS = collateral sulcus; FMaS = frontomarginal sulcus; FuG = fusiform gyrus; GRe = gyrus rectus; HySta = hypophyseal stalk; IFG = inferior frontal gyrus; IFS = inferior frontal sulcus; IHF = interhemispheric fissure; Ist = isthmus of cingulate gyrus; ITG = inferior temporal gyrus; ITS = inferior temporal sulcus; LatOlfStr = lateral olfactory striae; LatOrbG = lateral orbital gyrus; MaBo = mamillary body; MedOlfStr = medial olfactory striae; MedOrbG = medial orbital gyrus; MeFS = medial frontal sulcus; MFG = middle frontal gyrus; MFS = middle frontal sulcus; MTG = middle temporal gyrus; OlfBu = olfactory bulb; OlfS = olfactory sulcus; OlfTr = olfactory tract; OptTr = optic tract; Orb = orbital part of inferior frontal gyrus; OrbGi = orbital gyri; OrbS = orbital sulcus; OTS = occipitotemporal sulcus; PHG = parahippocampal gyri; PostMedOrbLob = posteromedial orbital lobule; PostOrbG = posterior orbital gyrus; PostPerfSubst = posterior perforated substance; RhiS = rhinal sulcus; SFG = superior frontal gyrus; SFS = superior frontal sulcus; Spl = splenium of corpus callosum; STG = superior temporal gyrus; STS = superior temporal sulcus; TePo = temporal pole; Un = uncus; IIn = optical nerve; IIIn = oculomotor nerve

rus, or lobule, located on the medial surface) and by the other related sulci [56].

Morphologically, the central lobe is situated obliquely over the sylvian fissure corresponding approximately to the median portion of the cerebral hemisphere.

On the superolateral surface, the central lobe is delineated anteriorly by the precentral and anterior subcentral sulci, and posteriorly by the postcentral and posterior subcentral sulci (Fig. 2). On the medial surface of the cerebral hemisphere (Fig. 3), the paracentral lobule is delineated anteriorly by the paracentral sulcus, and inferiorly and posteriorly by the ascending and distal part of the cingulate sulcus, which is known as the *marginal ramus of the cingulate sulcus* [30].

The precentral and postcentral gyri are situated obliquely in relation to the interhemispheric fissure, being less serpiginous than the other gyri of the cerebral convexity, and are connected to adjacent gyri via the usual interruptions in the precentral and postcentral sulci. The precentral and postcentral gyri are consistently united inferiorly by the subcentral gyrus (Broca's inferior frontoparietal *pli de passage*, or rolandic operculum) and superiorly by the paracentral lobule (Broca's superior frontoparietal *pli de passage*), which is located on the medial surface of each hemisphere. The precentral and postcentral gyri together resemble an elongated ellipse that is furrowed by the central sulcus, which is usually continuous, and are respectively delineated anteriorly and posteriorly by the precentral and postcentral sulci, which are typically discontinuous. This morphological unit, together with the functional interaction between motricity and sensitivity, justifies the characterization of these gyri as constituting a single lobe.

Inferiorly, the subcentral gyrus is delineated anteriorly and posteriorly by the anterior and posterior subcentral rami of the sylvian fissure, respectively. It can be situated either completely over the sylvian fissure or in part internal to the fissure, giving the false impression that the central sulcus is a branch of the sylvian fissure [39, 41]. The portion of the subcentral gyrus that corresponds to the base of the postcentral gyrus consistently lies over the transverse gyrus of Heschl, which is situated on the opercular surface of the temporal lobe [52] (Fig. 5).

Superiorly, and situated in the interhemispheric fissure, the paracentral lobule is delineated anteriorly by the paracentral sulcus and posteriorly by distal part of the cingulate sulcus, that is, the ascending marginal ramus of the cingulate sulcus.

The precentral gyrus typically presents 3 prominences known as *knees*: the superior and inferior knees are characterized by anterior convexities, and the middle knee is characterized by a posterior convexity. The precentral gyrus, in addition to its superior and inferior connections with the postcentral gyrus via the superior (paracentral lobule) and inferior frontoparietal fold (subcentral gyrus, or rolandic operculum), is usually also connected to the postcentral gyrus via a transverse gyrus that lies along the bottom of the central sulcus and constitutes the so-called Broca's middle frontoparietal *pli de passage* [2, 47]. This fold is situated at the level of the middle knee of the precentral gyrus, which is itself normally situated at the level of the superior frontal sulcus, corresponding to the portion of the gyrus that functionally harbors the motor representation of the contralateral hand. Therefore, the superior frontal sulcus tends to point the way to the middle frontoparietal *pli de passage*, as well as to the middle knee of the precentral gyrus, with its respective motor representation of the hand [2]. On axial MR images, this part of the precentral gyrus frequently presents a configuration that resembles the Greek letter ω [2].

Since they are disposed obliquely, the superior portions of the precentral and postcentral gyri that constitute the paracentral lobule on the medial surface of the cerebral hemisphere are topographically related to the ventricular atrium, which is situated posteriorly to the thalamus. In contrast, their inferior portions cover the posterior half of the insula and are topographically related to the body of the lateral ventricle, which is situated superior to the thalamus.

The parietal lobe

The parietal lobe anatomy is more complex in the sense that it is composed of gyri structurally less well defined and particularly serpiginous and curved. These gyri are also referred to as *lobules*.

On the superolateral surface, the parietal lobe is delineated anteriorly by the postcentral sulcus and posteriorly by the imaginary line running from the point from which the parietooccipital sulcus emerges (on the superomedial border) to the preoccipital notch, which is situated on the inferolateral border, ~ 5 cm anterior to the occipital pole [53]. The intraparietal sulcus, which originates from around the midpoint of the postcentral sulcus, is prominent along the parietal superolateral surface, in general runs almost parallel to the interhemispheric fissure, and posteriorly penetrates into the occipital lobe. The intraparietal sulcus divides the superolateral parietal surface into the superior and inferior parietal lobules (Fig. 2).

The superior parietal lobule is quadrangular and typically connected to the postcentral gyrus via a connection that transects the most superior portion of the postcentral sulcus and, occasionally, via another fold that more inferiorly interrupts the postcentral sulcus. The superior parietal lobule is delineated laterally by the intraparietal sulcus; medially, it is continuous with the precuneus gyrus along the superomedial border (Fig. 3); and, posteriorly, it continues to the superior occipital gyrus via the prominent and arched superior parietooccipital fold that surrounds the external perpendicular fissure, which represents the depth of the parietooccipital sulcus over the superolateral cerebral surface. In some brains, there is also a small sulcus, designated the *superior parietal sulcus*, emerging from the interhemispheric fissure, between the postcentral sulcus and the external perpendicular fissure, over the superior parietal lobule [30].

The inferior parietal lobule consists of, anteriorly, the supramarginal gyrus, which is a curved gyrus surrounding the distal portion of the sylvian fissure, and, posteriorly, the angular gyrus, which encircles the horizontal distal portion of the superior temporal sulcus. The supramarginal and angular gyri characterize the cranial parietal tuberosity, or bossa. Those 2 gyri are separated by the intermediate sulcus [50] (of Jensen), which is an inferior vertical branch of the intraparietal sulcus or a distal and superior vertical branch of the superior temporal sulcus, or both (Fig. 2).

Anteriorly, the supramarginal gyrus is connected to the postcentral gyrus via a fold that lies around the inferior portion of the postcentral sulcus; inferiorly, it consistently encircles the terminal portion of the sylvian fissure and continues to the superior temporal gyrus; and posteriorly, it occasionally rounds the inferior border of the intermediate sulcus, connecting to the angular gyrus (Fig. 2). In turn, the angular gyrus typically curves anteriorly around a distal horizontal branch of the superior temporal sulcus, also known as the *angular sulcus*, [30] continuing to the middle temporal gyrus, and posteriorly giving rise to a posterior fold that connects it to the middle occipital gyrus.

Therefore, the intraparietal sulcus delineates superiorly the supramarginal and angular gyri with a slightly arciform and inferiorly concave course, and, anteriorly, it typically continues to the inferior portion of the postcentral sulcus. Posteriorly, the continuation of the intraparietal sulcus becomes the intraoccipital sulcus, [10, 27] also known as the *superior occipital sulcus* [46] or *transverse occipital sulcus*, [30] which separates the (more vertical) superior occipital gyrus from its (more horizontal) middle counterpart [10, 30, 46, 47]. Along its length, the intraparietal sulcus typically gives rise to 2 vertical folds: a smaller, superior fold located anterior to the external perpendicular fissure, known as the *transverse parietal sulcus of Brissaud*, and another inferior, more developed fold that constitutes the previously mentioned intermediate sulcus of Jensen, which separates the supramarginal gyrus from the angular gyrus [46, 47, 50]. The superior parietal lobule, supramarginal gyrus, and angular gyrus are also known as P1, P2, and P3, respectively.

On the medial surface of each hemisphere, the precuneus gyrus is a medial extension of the superior parietal lobule along the superomedial border of the brain and corresponds to to the medial portion of the parietal lobe (Fig. 3). The precuneus is also quadrangular, delineated anteriorly by the marginal branch of the cingulate sulcus, posteriorly by the parietooccipital sulcus, and inferiorly by the subparietal sulcus. Inferiorly to the subparietal sulcus, the precuneus is connected to the isthmus of the cingulate gyrus.

The occipital lobe

On the superolateral cerebral surface, the occipital lobe is situated posterior to the imaginary line that connects the point of emergence of the parietooccipital sulcus (on the superomedial border of the cerebral hemisphere) with the preoccipital notch. The sulci and gyri of the occipital lobe have greater anatomical variation as compared with other lobes. Despite being less well defined and less anatomically constant than gyri in other dorsal cortical areas, the occipital gyri of the superolateral cerebral surface tend to consist of 3 gyri that are, for the most part, arrayed longitudinally in relation to the interhemispheric fissure and converge posteriorly to form the occipital pole of each hemisphere. As is the case for the frontal and temporal lobes, the occipital gyri of the superolateral surface are usually designated superior, middle, and inferior, [10, 46] or O1, O2, and O3, respectively. While the superior occipital gyrus is arranged more vertically along the interhemispheric fissure, the middle and inferior occipital gyri are arranged more horizontally and parallel to the inferior cerebral margin (Fig. 2).

On the medial surface, the occipital lobe is particularly well defined anatomically. It is separated from the parietal lobe by the parietooccipital sulcus and is composed of the cuneal and lingual gyri, which are separated by the calcarine fissure (Fig. 3). The basal, or inferior, surface of the occipital lobe, in turn, is contiguous with the basal surface of the temporal lobe.

On the superolateral surface, whereas the superior and middle occipital gyri are separated by the intraoccipital sulcus, [10, 27, 46] which is the continuation of the intraparietal sulcus and is also known as the *superior occipital sulcus* [10, 46] and *transverse occipital sulcus*, [30] the middle and inferior occipital gyri are separated by the less welldefined inferior occipital sulcus, [10, 46] also known as the *lateral occipital sulcus* [30]. When present, the so-called lunate sulcus is typically oriented vertically, immediately facing the occipital pole[10, 30]. Given the shallow depth and discontinuity as well as the (often) multiple branches of the 2 principal occipital sulci, the occipital gyri are not always well defined and are typically joined by various anastomotic folds, and thus constitute a cortical surface that is difficult to characterize morphologically.

Superiorly, the superior occipital gyrus extends along the superior border of the cerebral hemisphere, thus continuing along the medial surface to the cuneal gyrus. Inferiorly, the inferior occipital gyrus extends along the inferolateral margin, and its basal surface is lateral to the medial temporooccipital gyrus, also known as the *lingual gyrus*, as well as to the collateral sulcus that separates the two.

On the medial surface of the cerebral hemisphere, as previously mentioned, the occipital lobe is delineated and characterized by its well-defined and anatomically constant sulci and gyri (Fig. 3). Its principal sulcus is the calcarine fissure, which is located just above the inferomedial margin of the cerebral hemisphere. The calcarine fissure starts inferiorly to the splenium of the corpus callosum, delineating inferiorly the isthmus of the cingulate gyrus from the parahippocampal gyrus, and continues posteriorly as a gentle and superior convex curvature from whose apex emerges, superiorly, the parietooccipital sulcus, which in turn delineates anteriorly the occipital lobe on the medial surface of the cerebral hemisphere. Posteriorly, the calcarine fissure occasionally crosses the superomedial margin and extends along the occipital pole to the superolateral surface of the cerebral hemisphere.

The point of emergence of the parietooccipital sulcus divides the calcarine fissure into its proximal and distal parts. Superior to the proximal portion of the calcarine fissure and anterior to the parietooccipital sulcus is the precuneal gyrus, which is part of the parietal lobe. Superior to its distal part and posterior to the parietooccipital sulcus is the cuneus, or cuneal gyrus, so designated because of its wedge shape (Fig. 3).

Inferiorly and along the entire length of the calcarine fissure is the medial temporooccipital gyrus, or lingual gyrus, which anteriorly continues to the parahippocampal gyrus and constitutes the mediobasal portion of the occipital lobe, which is supported by the cerebellar tentorium.

The lingual gyrus is therefore delineated superiorly by the calcarine fissure and inferiorly by the collateral sulcus, which is a deep and generally continuous sulcus situated at the cerebral base, extending from the occipital pole to the temporal lobe and running parallel to the calcarine fissure.

The parietooccipital sulcus and calcarine fissure appear continuous on the surface. However, when their borders are retracted, it is obvious that they are separated by one or more small gyri. These gyri are composed of extensions of the cuneus and are known as the *cuneolingual gyri.*

The proximal part of the calcarine fissure creates a rise in the medial wall of the occipital horn of the lateral ventricle, designated the *calcar avis,* and the distal part runs along the visual cortex. Only the distal part includes the primary visual cortical areas, which are located on its superior (cuneal) and inferior (lingual) surfaces.

On the basal surface of the cerebral hemisphere, lateral to the lingual gyrus, is the medial temporooccipital gyrus or fusiform gyrus, situated between the collateral sulcus and the temporooccipital sulcus. The temporooccipital sulcus lies lateral and parallel to the collateral sulcus, rarely extends to the occipital pole, and in general is interrupted and divided into 2 or more parts. Anteriorly, the temporooccipital sulcus often bends medially and joins the collateral sulcus. The fusiform gyrus, in turn, extends to the basal surface of the temporal lobe, and lateral to its posterior portion lies the inferior occipital gyrus, whose lateral aspect constitutes the inferiormost portion of the lateral surface of the occipital lobe.

The temporal lobe

The temporal lobe is situated inferior to the sylvian fissure and delineated posteriorly by the imaginary line running from the superomedial portion of the parietooccipital sulcus to the preoccipital notch. Its lateral surface is composed of the superior, middle, and inferior temporal gyri – also respectively known as T1, T2, and T3 – which are separated by the superior and inferior temporal sulci, parallel to the lateral or sylvian fissure (Fig. 2). Anteriorly, the middle temporal gyrus is generally shorter, causing the superior and inferior temporal gyri to come together, and thereby forming the temporal pole.

The superior temporal sulcus is always a very well defined and deep sulcus and often presents as a continuous sulcus. The inferior temporal sulcus is usually discontinuous and composed of various parts. Both of the temporal sulci start at the proximal portion of the temporal pole and end posterior to its borders. Whereas the posterior portion of the sylvian fissure typically terminates as an ascending curve that penetrates the supramarginal gyrus, the superior temporal sulcus always terminates at a point posterior to the end of the sylvian fissure (posterior sylvian point). In general, the superior temporal sulcus then bifurcates into an ascending sulcal segment, which separates the supramarginal gyrus from the angular gyrus, and which corresponds to the intermediate sulcus of Jensen, and a distal and horizontal segment that penetrates the angular gyrus [30, 39, 42]. Given this configuration of the sulci, the superior temporal gyrus always continues posteriorly to the supramarginal gyrus, encircling the terminal portion of the sylvian fissure.

The middle temporal gyrus is often partially connected to the angular gyrus beneath the distal and horizontal portion of the superior temporal sulcus that penetrates the angular gyrus proper, and inferiorly is often connected to the inferior occipital gyrus. In turn, the inferior temporal gyrus continues to the inferior occipital gyrus, over the preoccipital notch that posteriorly delineates the temporal lobe. Inferiorly, the inferior temporal gyrus extends along the inferolateral margin of the cerebral hemisphere. Medially, its basal surface lies along the lateral temporooccipital gyrus, or fusiform gyrus, and along the temporooccipital sulcus that separates the 2 gyri.

The superior temporal gyrus constitutes the temporal operculum and covers the inferior aspect of the insular surface. Its superior, or opercular, surface (Fig. 6), which is within the sylvian fissure, is composed of various transverse gyri that emerge from the superior temporal gyrus and are directed obliquely toward the inferior part of the circular insular sulcus [48, 52].

Chief among these temporal gyri of the operculum is a much more voluminous transverse gyrus that originates in the posteriormost portions of the superior temporal gyrus and is oriented diagonally toward the posterior vertex of the floor of the sylvian fissure and toward the ventricular atrium. This gyrus is designated the

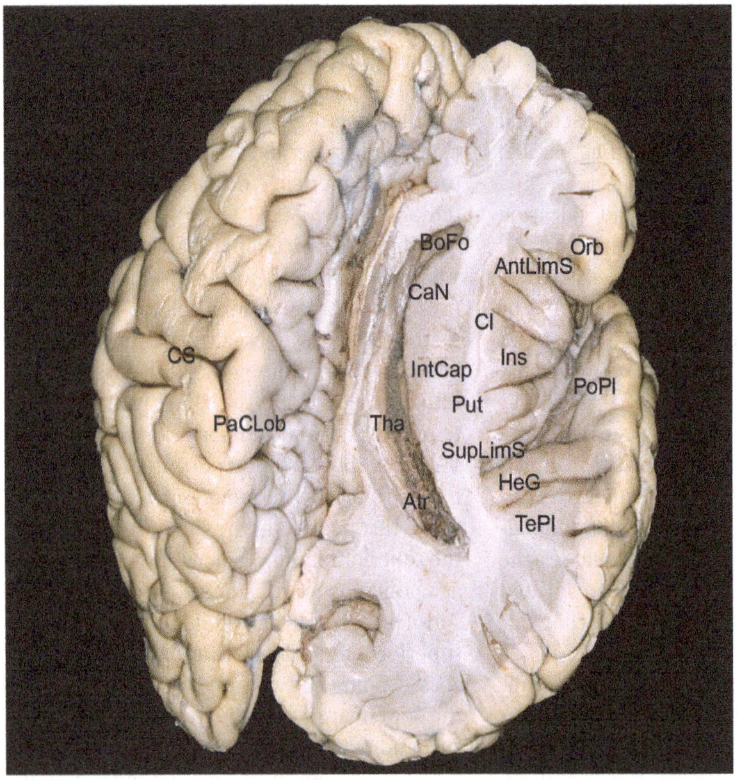

Fig. 6. The temporal opercular surface, the insula, and the central core of the brain. AntLimS = anterior limiting sulcus; Atr = atrium of lateral ventricle; BoFo = body of fornix; CaN = caudate nucleus; Cl = claustrum; CS = central sulcus; HeG = Heschl gyrus; Ins = insula; IntCap = internal capsule; Orb = orbital part of inferior frontal gyrus; PaCLob = paracentral lobule; PoPl = polar plane of the opercular temporal surface; Put = putamen; SupLimS = superior limiting sulcus of insula; TePl = temporal plane; Tha = thalamus

transverse gyrus of Heschl. In some brains, 1 or 2 sulci divide this gyrus; and in such cases, 2 or 3 gyri are also featured. Together with the posteriormost aspect of the superior temporal gyrus, the most anterior transverse gyrus of Heschl constitutes the primary auditory cortical area[48, 53]. This gyrus has particular topographical importance because it underlies the opercular surface of the postcentral gyrus, its longest axis is oriented toward the ventricular atrium, and it divides the temporal opercular surface into 2 planes: an anterior plane called the *polar plane* and a posterior plane known as the *temporal plane* (Fig. 6) [52].

The floor of the polar plane is composed of short transverse gyri at oblique angles, and the lower border of the plane is defined by the inferior portion of the circular insular sulcus that runs along the bottom of the sylvian fissure. The temporal plane is triangular with an internal vertex that exactly corresponds to the posterior vertex at the bottom of the sylvian fissure, where the superior part of the circular insular sulcus comes into contact with its inferior part. The temporal plane is oriented horizontally and faces the inferior surface of the supramarginal gyrus as if supporting its anteriormost portion. Whereas the sylvian fissure appears oblique in coronal slices taken in the polar plane, it appears horizontal in those taken in the temporal plane.

The basal surface of the temporal lobe is continuous with the basal surface of the occipital lobe and is situated over the floor of the middle cranial fossa, anterior to the petrous portion of the temporal bone, whereas the basal occipital surface lies over the superior surface of the cerebellar tentorium.

The base of the temporal lobe is composed laterally by the inferior surface of the inferior temporal gyrus, and by the anterior portion of the lateral temporooccipital gyrus, or fusiform gyrus, that lies laterally to the parahippocampal gyrus. The fusiform gyrus is then situated lateral to the parahippocampal and lingual gyri, between the collateral and temporooccipital sulci. Its temporal portion presents a slight basal prominence due to its adaptation to the concavity of the middle cranial fossa. Its anterior aspect is typically curved or pointed, because the anteriormost portion of the temporooccipital sulcus frequently presents a medial curvature toward the collateral sulcus. The anterior border of the fusiform gyrus, in general, corresponds medially to the level of the mesencephalic peduncle; as a whole, it constitutes the floor of both the atrium and the inferior horn of the lateral ventricle.

The parahippocampal gyrus belongs to the limbic lobe [15].

The insular lobe

On the publication of the fourth edition of the Paris *Nomina Anatomica* in 1975, (14) the insula came to be considered a cerebral lobe. The insular surface is composed of the so-called mesocortex, which is anatomically situated between the allocortex, which is older and topographically more medial (comprising the amygdala and hippocampus), and the isocortex, which is phylogenetically younger and topographically more lateral (comprising the neocortex of the cerebral hemispheres).

Embedded between the frontal and temporal lobes of each cerebral hemisphere and constituting the base of each sylvian cistern, the insula has an anterior surface and a lateral surface (Fig. 6) that are encased in their respective opercular convolutions [48]. The anterior surface of the insula is covered by the frontoorbital operculum (comprising the posterior portion of the posterior orbital gyrus and the orbital part of the inferior frontal gyrus), whereas its lateral surface is covered superiorly by the frontoparietal operculum (triangular and opercular parts of the inferior frontal gyrus, subcentral gyrus, and anterior and basal part of the supramarginal gyrus) and inferiorly by the temporal operculum (superior temporal gyrus) [48, 53, 60].

The lateral surface of the insula is characterized as a pyramid with a triangular base, whose anteroinferior vertex constitutes the limen insulae, and is divided by the oblique central sulcus of the insula into an anterior portion and a posterior portion, with the former having a larger area. The anterior portion is typically composed of 3 short gyri that originate at the apex of the insula, which corresponds to the most prominent aspect of the insular pyramid, with the anteriormost aspect extending over the anterior surface of the insula; and the posterior portion, in general, is composed of 2 long gyri not originating at the apex but oriented obliquely and in parallel. The transverse and accessory insular gyri, which together constitute the insular pole, also originate from the apex of the insula [48, 49]. The transverse insular gyrus, which is situated more inferiorly, runs along the limen insulae and is connected to the posteromedial orbital lobule, which is composed of the posterior portion of the medial orbital gyrus and the medial portion of the posterior orbital gyrus [48, 55] and is located anterior and along the lateral olfactory stria.

The insular surface is delineated peripherally by the circular sulcus of Reil, [10, 45] or periinsular sulcus, [47, 48] which is interrupted by the previously mentioned transverse insular gyrus. Given the triangular shape of the insula, its circular, or periinsular, sulcus is usually divided into 3 parts, that is, the anterior, superior, and inferior periinsular sulci, [48] also designated the *anterior*, *superior*, and *inferior limiting sulci of the insula* [36].

To understand the periinsular spaces more fully, one should remember that the insula has a lateral surface and an anterior surface. The superior and inferior limiting sulci are morphologically categorized as true sulci that delineate the respective transitions and deflections occurring among the lateral insular surface and the frontoparietal operculum, and the lateral insular surface and the temporal operculum. The so-called anterior limiting sulcus of the insula, on the other hand, is considerably deeper and morphologically characteristic of a true fissure or space that separates the anterior surface of the insula from the posterior surface of the posterior orbital gyrus.

The upper half of the fundus of the anterior limiting sulcus is separated from a true anterior ventricular recess at the head of the caudate nucleus only by the fibers of the thin anterior limb of the internal capsule, whereas the fundus of the lower half continues to the ventral striatopallidal region.

From a morphological and topographical perspective, the surface of the insula clearly represents the external shield of a true central core of the brain, [36] quite well defined anatomically. This central core of the brain comprises, in each cerebral hemisphere, the insula proper, the basal nuclei, the thalamus, and the internal capsule (Fig. 6). The anterior half of the lateral surface of the insula corresponds internally to the head of the caudate nucleus, whereas the posterior half corresponds to the thalamus and the body of the caudate nucleus. Each central core of each cerebral hemisphere, composed of all the structures mentioned above, is incorporated into the corresponding half of the mesencephalon, morphologically characterizing a brainstem with 2 heads that correspond to the 2 central cores.

The limbic lobe and correlated areas

The Limbic Lobe. The publication entitled *International Anatomical Terminology,* published in 1998 [15, 44] and replacing the previous *Nomina Anatomica,* introduced the limbic lobe as another of the cerebral lobes and described it as comprising the cingulate and parahippocampal gyri.

The term *limbic* was first used in the 19th century by Broca, [4] who observed that certain cerebral structures constituted a continuum arranged in the shape of a C surrounding the diencephalic region. Since then the term *limbic* – meaning border, ring, or surround [16] – came to be definitively established in the neuroanatomical literature. Subsequent studies introduced the notion that the limbic system is composed of telencephalic and diencephalic structures that, despite their anatomical and functional diversity, are particularly responsible for the physiology of emotions, memory, and learning [18, 22, 37, 38, 53].

Table 2: Principal limbic cortical structures

cingulate gyrus
parahippocampal gyrus
hippocampal formation
hippocampus (Ammon's horn)
subiculum
dentate gyrus
prehippocampal rudiment/indusium griseum
frontal mediobasal cortical area
paraterminal gyrus
paraolfactory gyri or subcallosal area
olfactory cortical areas

Within the medial surface of each cerebral hemisphere, one prominent feature is the cingulate gyrus, which wraps around the corpus callosum and continues posteriorly and inferiorly to the parahippocampal gyrus, forming the shape of a C around the diencephalon (Fig. 3). Broca originally described the cingulate and parahippocampal gyri as contiguous, jointly dubbing them the *greater limbic lobe*. He also considered the cingulate, subparietal, and collateral sulci to be segments of the sulcus he referred to as the *limbic sulcus* [4, 18, 19].

The cingulate gyrus is situated above the callosal sulcus and below the cingulate sulcus. It starts below the rostrum of the corpus callosum, and as it ascends around the knee of the corpus callosum, it typically presents a connection with the medial (or medial aspect of the superior) frontal gyrus. At the level of the trunk of the corpus callosum it is connected to the paracentral lobule, and more posteriorly it is connected to the precuneus.

Posterior to the splenium of the corpus callosum the cingulate gyrus consistently becomes narrower, at which point it is referred to as the *isthmus of the cingulate gyrus*, and continues to the parahippocampal gyrus. The site of transition between these 2 gyri is identified by the emergence of the anterior branch of the calcarine fissure, which originates beneath the isthmus of the cingulate gyrus.

As already mentioned, the terminal ascending branch of the cingulate sulcus delineates posteriorly the paracentral lobule and anteriorly the precuneus, whereas the subparietal sulcus is located inferior to the precuneus, separating it from the cingulate gyrus and appearing to be a posterior continuation of the cingulate sulcus after a short interruption of the latter. The connections between the cingulate and the precuneus gyri are anterior and posterior to the subparietal sulcus (Fig. 3).

The parahippocampal gyrus forms the lower half of the C that wraps around the diencephalic region. Posteriorly, it comprises the isthmus of the cingulate gyrus and is also the anterior continuation of the lingual gyrus, which lies under the calcarine fissure. The parahippocampal gyrus is situated lateral to the cerebral peduncle. Anteriorly, it folds back on itself medially, assuming the shape of a hook and constituting the uncus of the parahippocampal gyrus situated anterolateral to the cerebral peduncle, and harbors the uncal sulcus.

Medially, the parahippocampal surface curves superiorly and laterally, constituting the subiculum, characterized as a flat, superior surface running along the anteroposterior axis of the parahippocampal gyrus and under the pulvinar of the thalamus, so that these 2 surfaces constitute the portion of the transverse fissure of the brain that harbors the so-called wing of the ambient cistern. The hippocampus is situated lateral to the subiculum.

The hippocampus consists of Ammon's horn, which is characterized as an intraventricular prominence, and the small dentate gyri, which lie laterally along Ammon's horn. The small dentate gyri are separated from the subiculum by the hippocampal sulcus, which anteriorly terminates within the uncus. Given the greater magnitude of Ammon's horn, the term *hippocampus* is commonly used in reference to this structure. Within the ventricular cavity, Ammon's horn, or the hippocampus, is covered by the alveus, a thin layer of fibers that gives rise to the fimbria of fornix, the principal bundle of efferent fibers of the hippocampus. These structures are collectively known as the *hippocampal formation*. Between the dentate gyri and the fimbria of fornix is the fimbriodentate sulcus, which lies lateral and parallel to the hippocampal sulcus [11, 52].

The uncus of the parahippocampal gyrus is triangular with a medial vertex, such that its anteromedial surface faces the carotid cistern and its posteromedial surface faces and encircles the mesencephalic peduncle. Two small prominences known as the *semilunar gyrus* and *ambient gyrus*, which are separated by the semiannular sulcus, are evident in the anterior portion of the surface of the uncus. The ambient gyrus, which is more inferior, often presents with a depression caused by the pressure of the free edge of the tentorium cerebelli. The anterior half of the uncus includes the amygdala, whereas its posterior half includes the head of the hippocampus [52]. Superiorly, the amygdala runs toward the base of the globus pallidus so that, in a coronal slice, the base of the lentiform nucleus and the amygdala form a figure-eight or an hourglass shape [49, 52, 53].

Along the cerebral base, the parahippocampal gyrus is laterally delineated by the collateral sulcus, which separates it from the fusiform gyrus, and by the rhinal sulcus, which is occasionally continuous with the collateral sulcus. The rhinal sulcus, which is not always readily identifiable, is consistently the sulcus that separates the uncus from the rest of the temporal pole (Fig. 3).

The temporal stem. Laterally, the parahippocampal gyrus is contiguous with the fusiform gyrus and the remainder of the basal surface of the brain. Posteriorly, it continues along the cingulate gyrus. Medially, it runs under the thalamus along the natural space comprising the choroidal fissure. Anteriorly, its uncal portion is superiorly incorporated into the lateral-most aspect of the frontobasal region via a well-defined neural peduncle anterior to the inferior horn of the lateral ventricle.

Anteriorly and externally, this true temporal peduncle is composed of the cortex of the transverse insular gyrus, which crosses the limen insulae, connecting to the posteromedial orbital lobule [56]. Internally, it consists of the uncinate fascicle, which joins the frontal and temporal lobes; [12] fibers of the frontooccipital fascicle, which are arrayed immediately posterior to the uncinate fascicle; amygdalofugal fibers, which are composed of the ventral extensions of the amygdala – that project to the septothalamohypothalamic region [53] and the nucleus of the stria terminalis, situated under the head of the caudate nucleus [18, 19, 25] – and fibers of the anterior commissure. Medially, it includes the superior extension of the amygdala, which superiorly extends medially to the putamen and toward the globus pallidus [52].

In the literature, this set of structures has been given the generic and controversial name of *temporal stem*, equivalent to the *temporal axis*. According to Duvernoy, [10] "the temporal stem consists of only a thin layer of white matter situated between the ventricular cavity and the fundus of the superior temporal sulcus." In contrast, Wen et al [52] stated that the term refers "only to the connections between the temporal lobe and the insula, excluding the superior extension of the amygdala in the direction of the globus pallidus and the limen insulae." Türe et al [48] defined the temporal stem as "the portion of white matter that penetrates the temporal lobe between the anterior border of the insula and the inferior horn, [...] composed of the fronto-occipital fascicle and the anterior thalamic peduncle".

Posterolaterally, the temporal stem is contiguous with the layers of fibers called the *sagittal stratum,* [24, 49] which cover the inferior horn and ventricular atrium. All of these structures collectively join the temporal lobe to the remainder of the cerebral hemisphere beneath the insula. The so-called sagittal stratum consists of the frontooccipital fascicle, the inferior thalamic peduncle, or radiation – which encompasses the auditory and optic radiations – and the fibers that compose the anterior commissure and tapetum. The layer of callosal fibers known as the *tapetum* lies under the optic radiation and then constitutes the most inferior layer of the sagittal stratum. The sagittal stratum is situated inferiorly to the inferior limiting sulcus of the insula, forming the roof and lateral wall of the inferior horn and consituting the lateral wall of the ventricular atrium. Superficial to the sagittal stratum is the subcortical white matter of the entire neocortical portion of the temporal lobe.

From a topographical perspective, it is notable that the sagittal stratum corresponds to the set of fibers that cover the inferior horn and atrium of the ventricle, inferiorly and posteriorly to the insula, whereas the temporal stem is situated anterior to the inferior horn, connecting the anteromedial temporal portion to the basolateral frontal portion of the brain.

The basal forebrain and ventral striatopallidal region. The mediobasal frontal cortical area of each cerebral hemisphere, composed of the paraterminal gyrus and the paraolfactory gyri, is also considered a limbic cortical area. The paraterminal gyrus is situated on the medial wall of each cerebral hemisphere, immediately facing and quasi-continuous with the lamina terminalis, and is delineated anteriorly by a short, vertical sulcus known as the posterior olfactory sulcus. The small anterior curvature of the paraterminal gyrus is called the prehippocampal rudiment and extends superiorly, along the indusium griseum. Inferiorly, the paraterminal gyrus extends along the diagonal band of Broca and the lateral olfactory stria.

The posterior and anterior paraolfactory gyri, which are also vertical and separated by the anterior paraolfactory sulcus (the latter not always identifiable), are located anterior to the paraterminal gyrus. This area of the paraolfactory gyri is also known as the *subcallosal area.* Anterior to the subcallosal area is a fold that connects the basalmost portion of the cingulate gyrus with the gyrus rectus, encircles the posteriormost part of the superior rostral sulcus, and is called the *cingulate pole* [56].

The paraterminal gyri harbor the septal nuclei [53] and constitute the septal area, which corresponds to the socalled paraolfactory area of Broca [23]. Therefore, the septal region is situated on the medial surface of the cerebral hemisphere, immediately facing the anterior commissure.

The area known as the *anterior perforated substance* constitutes a particularly important topographical region of the basal forebrain (Fig. 5 right). Macroscopically, this area is delineated anteriorly by the olfactory trigone and the lateral and medial olfactory striae, whereas it is delineated posteriorly by the edges of the optic tracts, medially by the interhemispheric fissure, and laterally by the uncus of the parahippocampal gyrus and the limen insulae. Topographically, the anterior perforated substance is situated just above the bifurcation of the internal carotid artery, and thus forming the roof of the space that harbors the distal portion of the artery and the proximal segments of the anterior and middle cerebral arteries. The perforating branches that emerge from those arterial segments constitute the lenticulostriate arteries, and it is from the surface of the anterior perforated substance that they penetrate the frontobasal parenchyma.

The brain region currently known as the *ventral striatopallidal system,* [18, 21] or more simply the *ventral striatum,* corresponds in part to the substantia innominata, a name taken from the historical literature in German, and can be confused with the concept of the basal forebrain in current English literature. The ventral striatum refers to the basal forebrain region situated between the anterior perforated substance and the anterior commissure of each cerebral hemisphere. Superiorly, it is closely related to the most anterobasal portion of the anterior limb of the internal capsule, whereas laterally it is contiguous with the peduncle of the temporal stem, and medially it is particularly related to the septal region and the hypothalamus. The ventral striatopallidal region includes the nucleus accumbens, which corresponds to a basal connection of the head of the caudate nucleus with the most anterior and inferior portion of the putamen (hence the name *ventral striatum*), the globus pallidus, the magnocellular nucleus of the basal forebrain (nucleus basalis of Meynert), and the fibers that constitute the ventral extension of the amygdala and are directed toward the septal region, the hypothalamus, the thalamus, and the bed nucleus of the stria terminalis, located under the head of the caudate nucleus. The substantia innominata of Reichert corresponds most closely to the ventropallidal region [19, 23]. Because of its topography, the ventral striatum is crisscrossed by the perforating lenticulostriate arteries. Functionally, the ventral striatum is closely correlated with neuropsychiatric functions [19, 20, 22, 38].

It is interesting to emphasize that the mediobasal frontal cortical areas (paraterminal gyrus and subcallosal area), the olfactory cortical areas (anterior perforated substance and components of the piriform lobe), and the ventral pallidal-striatum region (with its subjacent nuclei) constitute a corticosubcortical continuum running along the ventral surface of the brain from the medial portion of the temporal pole to the posterior mediobasal portion of the frontal lobe, with its posterior border being delineated by the anterior commissure.

In parallel with these observations, Mesulam [26] proposed that because of their particularly superficial presentation, the most medial portions of the amygdaloid complex (within the uncus), the substantia innominata (within the ventral striatopallidal region), and the septal nuclei (within in the paraterminal gyri), which together constitute the basal forebrain, should be considered constituents of the cerebral mantle.

In conclusion, it should be noted that although the cortical areas mentioned compose the cortical portion of the limbic lobe, the concept of the limbic system as a functional unit also involves the participation of deep structures and is controversial in terms of its conception and composition [18, 19, 20, 38]. From a morphological perspective, however, the structures that make up the limbic system present as a series of C-shaped curves centered around the thalamus and hypothalamus in each cerebral hemisphere.

Final remarks

To understand and correctly identify the sulci, and consequently the cerebral gyri, it is fundamental to consider the notion that the characterization of a given sulcus does not necessarily imply that it is composed of a single continuous space. A sulcus can consist of one or more parts, which in some cases can be oriented in different directions. Those parts can be long or short and can be isolated or connected to other sulci [30].

The principal sulci can vary in form and size from person to person and the surface of the brain constitutes a true continuum, presenting a serpentine configuration given its various forms of connections surrounding the sulcal extremities and running under the fundus of the sulci [56]. The separation between neighboring and adjacent gyri is therefore only superficial and is structurally defined by the continuity and the fundi of the sulci that surround them. The interruption of a sulcus or the presence of a free sulcal extremity necessarily indicates the presence of a fold that connects different gyri or different sectors of the same gyrus. Each cerebral gyrus should therefore be understood as a region of the brain surface and not as an individual, anatomically well-defined neural structure.

For microneurosurgical applications, is also notable that given the mechanism of invagination of the surface of the brain throughout its evolution and embryological development, [43] the sulci of the superolateral and inferior surfaces of the brain are consistently oriented toward the nearest ventricular cavity, which is especially evident in coronal slices of MR imaging studies. This disposition of the sulci is not seen on the medial surface of the hemispheres because the development of the sulci on this surface is directly related to that of the corpus callosum, and these sulci therefore tend to be arranged in parallel with this commissure [30].

References

[1] Berger MS, Cohen WA, Ojemann GA (1990) Correlation of motor cortex brain mapping data with magnetic resonance imaging. J Neurosurg 72: 383–387

[2] Boling W, Olivier A, Bittar RG, Reutens D (1999) Localization of hand motor activation in Broca's pli de passage moyen. J Neurosurg 91: 903–910

[3] Broca P (1861) Remarques sur le siège de la facultè du langage articulé: suiries d'une observation d'aphémie (perte de la perole). Bull Soc d'Anth (Paris) 6: 330–357, 397–407; cited in Finger S (1994) Origins of neuroscience. Oxford University Press, New York

[4] Broca P (1877) Sur la cinconvolution limbique et al scissure limbique. Bull Soc d'Anth 12: 646–657; cited in Finger S (1994) Origins of neuroscience. Oxford University Press, New York

[5] Broca P (1876) Sur la topographie crânie-cerébrale ou sur les rapports anatomiques du crâne et du cerveau. Rev d' Anthrop 5: 193–248

[6] Broca P (1861) Sur les rapports anatomiques des divers points de la surface du crâne et des diverses parties des hémisfères cérébraux. Bull Soc d'Anth 2: 340; cited in Gusmão S, Silveira RL, Cabral G (2000) Broca and the birth of modern surgery: Arq Neuropsiquiatr 58: 1149–1152 (in Portuguese)

[7] Brodal A (1981) Neurological anatomy in relation to clinical medicine, 3rd edn. Oxford University Press, New York

[8] Butler AB, Hodos W (2005) Comparative vertebrate neuroanatomy: Evolution and adaptation, 2nd edn. John Wiley & Sons, Hoboken

[9] Chi JG, Dooling EC, Gilles FH (1977) Gyral development of the human brain. Ann Neurol 1: 86–93

[10] Duvernoy HM (1991) The human brain. Springer, Wien

[11] Duvernoy MH (1998) The human hippocampus, 2nd edn. Springer, Berlin

[12] Ebeling U, von Cramon D (1992) Topography of the uncinate fascicle and adjacent temporal fiber tracts. Acta Neurochir (Wien) 115: 143–148

[13] Ebeling U, Eisner W, Gutbrod K, Ilmberger I, Schmid UD, Reulen HJ (1992) Intraoperative speech mapping during resection of tumors in the posterior dominant temporal lobe. J Neurol 369: 104

[14] Excerpta Medica Foundation (1975) Nomina anatomica, 4th edn. Excerpta Medica, Amsterdam

[15] Federative Committee on anatomical terminology (1998) International anatomical terminology. Thieme, Stuttgart

[16] Ferreira AG (1966) [Latin-Portuguese Dictionary.] Porto Editora, Porto (in Portuguese)

[17] Gratiolet LP (1854) Memoire sur les plis cerébraux de l'homme et des primates. Paris: Bertrand, cited in Tamraz JC, Comair YG (2000) Atlas of regional anatomy of the brain using MRI. Springer, Berlin

[18] Heimer L (1995) The human brain and spinal cord: Functional neuroanatomy and dissection guide, 2nd edn. Springer, New York

[19] Heimer L (2003) A new anatomical framework for neuropsychiatric disorders and drug abuse. Am J Psychiatry 160: 1726–1739

[20] Heimer L, Van Hoesen GW (2006) The limbic lobe and its output channels: implications for emotional functions and adaptive behavior. Neurosci Biobehav Rev 30: 126–147

[21] Heimer L, Switzer RC, Van Hoesen GV (1982) Ventral striatum and ventral pallidum. Trends Neurosci 5: 83–87

[22] Heimer L, Van Hoesen GW, Trimble M, Zahm DS (2008) Anatomy of neuropsychiatry: The new anatomy of the basal forebrain and its implications for neuropsychiatric illness. Elsevier, Amsterdam

[23] Lockard I (1977) Desk reference for neuroanatomy. A guide to essential terms. Springer, New York

[24] Ludwig E, Klinger J (1956) Atlas Cerebri Humani. S. Karger, Basel

[25] McGinty JF (ed) (1999) Advancing from the ventral striatum to the extended amygdala: Implications for neuropsychiatry and drug abuse. Ann NY Acad Sci 877. The New York Academy of Sciences, New York

[26] Mesulam MM (1987) Patterns in behavioral neuroanatomy: association areas, the limbic system, and hemispheric specialization. In: Davis FA (ed) Principles of behavioral neurosurgery. Philadelphia, pp 1–70

[27] Naidich TP, Valavanis AG, Kubik S (1995) Anatomic relationships along the low-middle convexity: Part I – Normal specimens and magnetic resonance imaging. Neurosurgery 36: 517–532

[28] Nishikuni K (2006) Study of the fetal and post-natal morphological development of the sulci of the brain (thesis). Faculdade de Medicina, Universidade de São Paulo, São Paulo (in Portuguese)

[29] Ojemann G, Ojemann J, Lettich E, Berger M (1989) Cortical language localization in left, dominant hemisphere. An electrical stimulation mapping investigation in 117 patients. J Neurosurg 71: 316–326

[30] Ono M, Kubik S, Abernathey CD (1990) Atlas of cerebral sulci. Thieme, Stuttgart

[31] Penfield W, Rasmussen T (1952) The cerebral cortex of man. Macmillan, New York

[32] Quiñones-Hinojosa A, Ojemann SG, Sanai N, Dillon WP, Berger MS (2003) Preoperative correlation of intraoperative cortical mapping with magnetic resonance imaging landmarks to predict localization of the broca area. J Neurosurg 99: 311–318

[33] Rasmussen T (1979) Cortical resection for medically refractory focal epilepsy: results, lessons and questions. In: Rasmussen T, Marino R Jr (ed) Functional neurosurgery. Raven Press, New York, pp 253–269

[34] Rasmussen T (1991) Surgery for central, parietal and occipital epilepsy. Can J Neurol Sci 18 (Suppl 4): 611–616
[35] Rasmussen T (1991) Tailoring of cortical excisions for frontal lobe epilepsy. Can J Neurol Sci 18 (Suppl 4): 606–610
[36] Rhoton AL Jr (2003) Cranial anatomy and surgical approaches. Neurosurgery 53: 1–746
[37] Ribas GC (2006) Considerations about the nervous system phylogenetic evolution, behavior, and the emergence of consciousness. Rev Bras Psiquiatr 28: 326–338 (in Portuguese)
[38] Ribas GC (2007) Neuroanatomical basis of behavior: history and recent contributions. Rev Bras Psiquiatr 29: 63–71 (in Portuguese)
[39] Ribas GC (2005) Surgical anatomy of microneurosurgical sulcal key-points (thesis). Faculdade de Medicina, Universidade de São Paulo, São Paulo (in Portuguese)
[40] Ribas GC (2010) The cerebral sulci and gyri. Neurosurg Focus 28(2): E2
[41] Ribas GC, Ribas EC, Rodrigues CJ (2005) The anterior sylvian point and the suprasylvian operculum. Neurosurg Focus 18(6): E2
[42] Ribas GC, Yasuda A, Ribas EC, Nishikuni K, Rodrigues Junior AJ (2006) Surgical anatomy of microneurosurgical sulcal key-points. Neurosurgery 59 (ONS Suppl 4): ONS177–ONS209
[43] Sarnat HB, Netsky MG (1981) Evolution of the nervous system, 2nd edn. Oxford University Press, New York
[44] Sociedade Brasileira de Anatomia (2001) Anatomical Terminology] Manole, São Paulo (in Portuguese)
[45] Taveras JM, Wood EH (1976) Diagnostic neuroradiology, Vol 1, 2nd edn. Williams and Wilkins, Baltimore
[46] Testut L, Jacob O (1932) Topographic anatomy textbook, 5th edn. Salvat, Barcelona (in Spain)
[47] Testut L, Latarjet A (1932) Human anatomy textbook, 8th edn. Salvat, Barcelona (in Spain)
[48] Türe U, Yaşargil DCH, Al-Mefty O, Yaşargil MG (1999) Topographic anatomy of the insular region. J Neurosurg 90: 730–733
[49] Türe U, Yaşargil MG, Friedman AH, Al-Mefty O (2000) Fiber dissection technique: lateral aspect of the brain. Neurosurgery 47: 417–427
[50] Uematsu S, Lesser R, Fisher RS, Gordon B, Hara K, Krauss GL et al (1992) Motor and sensory cortex in humans: topography studied with chronic subdural stimulation. Neurosurgery 31: 59–72
[51] Von Economo C (2009) Cellular structure of the human cerebral cortex (transl. by Triarhou LC). Karger, Basel
[52] Wen HT, Rhoton AL Jr, de Oliveira E, Cardoso ACC, Tedeschi H, Baccanelli M et al (1999) Microsurgical anatomy of the temporal lobe: part 1: mesial temporal lobe anatomy and its vascular relationships as applied to amygdalohippocampectomy. Neurosurgery 45: 549–592
[53] Williams PL, Warwick R (ed) (1980) Gray's anatomy, 36th edn. Saunders, Philadelphia
[54] Yaşargil MG (1999) A legacy of microneurosurgery: memoirs, lessons, and axioms. Neurosurgery 45: 1025–1092
[55] Yaşargil MG (1984) Microneurosurgery, vol 1. Georg Thieme, Stuttgart
[56] Yaşargil MG (1994) Microneurosurgery, vol 4a. Georg Thieme, Stuttgart
[57] Yaşargil MG, Cravens GF, Roth P (1988) Surgical approaches to "inaccessible" brain tumors. Clin Neurosurg 34: 42–110
[58] Yaşargil MG, Kasdaglis K, Jain KK, Weber HP (1976) Anatomical observations of the subarachnoid cisterns of the brain during surgery. J Neurosurg 44: 298–302
[59] Yaşargil MG, Krisht AF, Türe U, Al-Mefty O, Yaşargil DCH (2002) Microsurgery of insular gliomas: Part I: surgical anatomy of the Sylvian cistern. Contemp Neurosurg 24: 1–8
[60] Yaşargil MG, Teddy PJ, Roth P (1985) Selective amygdalo-hippocampectomy. Operative anatomy and surgical technique. Adv Tech Stand Neurosurg 12: 93–123

Anatomy of the white-matter pathways

Juan Martino and Christian Brogna

Introduction

The white matter of the brain consists of myelinated bundles of nerve fibers, known as fascicles or fiber tracts. These fibers form a complex three-dimensional architecture within the hemispheres and the brainstem (Fig. 1). Neurosurgical approaches to different compartments of the brain cannot be performed without an in-depth understanding of the systems of the white fibers. As a consequence, a detailed knowledge of the architectural anatomy of the white-matter tracts is paramount when dealing with intrinsic neoplastic and vascular lesions within the brain. In addition, the precise knowledge of the trajectories and cortical terminations of the different white-matter bundles is of great value for neuroscience research, as it opens a new door to understand brain functioning.

Neuroanatomic laboratory training is the best way to learn and appreciate the anatomy of white-matter fibers, in conjunction with vascular, cysternal, and gyral anatomy [44, 45]. In particular, the fiber dissection technique has allowed neurosurgeons to discover the three-dimensional anatomy of association, commissural, and projection fibers.

It is of considerable importance that, in the history of neuroanatomy, the fiber dissection technique was one of the first solutions that gave the anatomists the invaluable opportunity to describe the internal structure of the brain. Türe et al [40] made a great contribution by tracing the history of the fiber dissection technique. Andreas Vesalius (1514–1564) differentiated for the first time the softer and yellowish cerebrum from the harder and whiter deeper substance below it and provided the first description of the corpus callosum. Raymond Vieussens (1641–1715), who was one of the first anatomists to disclose the internal structure of the brain, introduced the term "centrum ovale" and could demonstrate the continuity of the corona radiata, internal capsule, and pyramidal tracts within the brainstem. The crossing of the pyramidal fibers below the pons was described by Domenico Mistichelli (1675–1715). Johann Christian Reil (1759–1813), who was the first to introduce the method of alcohol fixation for preservation of the brain, illustrated the tapetum fibers and the optic radiation. The Italian anatomist Luigi Rolando (1773–1831) described the continuity of fibers starting within the medial olfactory stria and proceeding through the subcallosal area and cingulate and parahippocampal gyri, forming a nearly complete circle, and ending in the uncus. Bartholomeo Panizza (1785–1867) demonstrated the visual pathway in 1855 and Louis Pierre Gratiolet (1815–1865) described the optic radiation from the lateral geniculate body to the occipital cortex.

In the first two decades of the 20th century, due to the fact that microtome and histological techniques were more widely used, the fiber dissection technique was abandoned. Nevertheless, white-matter tracts could not be followed within histological specimens. Therefore, in 1935 Joseph Klingler (1888–1963), in Basel, Switzerland, recognized the importance of studying white matter by the fiber dissection technique. He revolutionized the technique by developing a new method of brain fixation which consisted of freezing the already formalin-fixed brains before dissection [20].

In the 1950s, Yasargil was the first neurosurgeon to apply the knowledge of fiber systems acquired by the fiber dissection technique directly during neurosurgical procedures [46]. In the 1990s, Türe et al made highly detailed white-matter dissection studies with the intention to spread the use of this technique in neuroanatomical and neurosurgical training [38–40].

Klingler'st echnique

The main concept of Klingler's technique is to freeze formalin-fixed brains before dissecting them with wooden spatulas. Freezing allows the formalin between fibers to crystallize: by this process the fibers get expanded and separated, allowing to easily follow them during the dissection.

In the preparation of anatomic specimens the following steps must be followed. The human cerebral hemispheres are first fixed in a 10% formalin solution for at least 3 weeks. Then the arachnoidal and vascular structures are carefully removed by use of surgical magnification. Afterward, the specimens are frozen at −16 °C for 3 weeks, and finally the dissection is performed with fine custom-shaped wooden spatulas. A good point to start with the dissection is to remove the gray matter within the superior temporal sulcus.

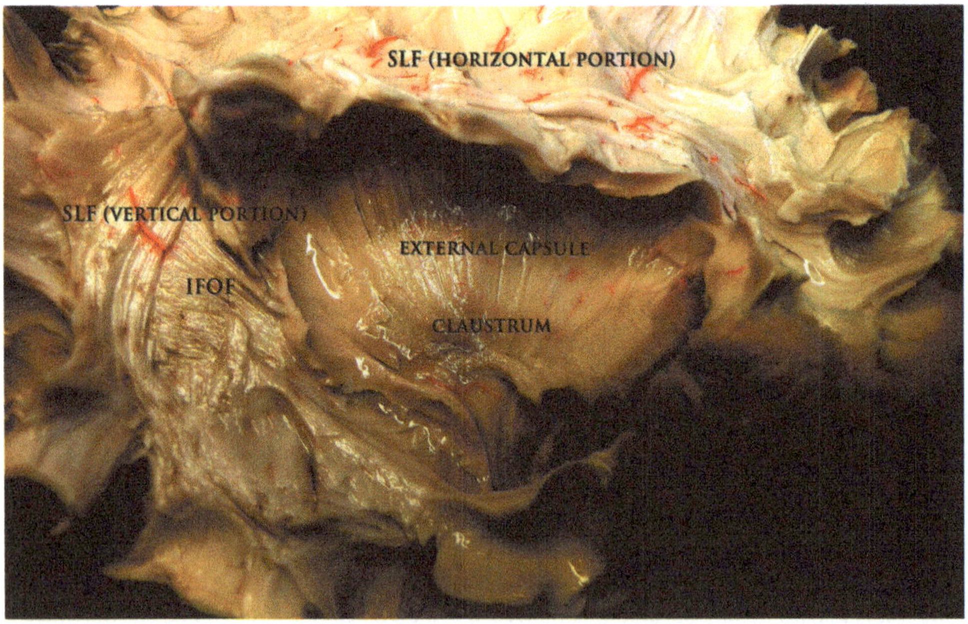

Fig. 1. Complex three-dimensional architecture of the white-matter fibers of the brain. The superior longitudinal fasciculus (*SLF*) is running around the insula and crosses above the inferior fronto-occipital fasciculus (*IFOF*). The claustrum and the external capsule are located deep to the insula and the extreme capsule and above the IFOF

White-matter pathways

White-matter pathways within the brain are distinguished into association, commissural, and projection fibers. Short association fibers, also called U fibers, connect adjacent gyri running just below the deepest portion of sulci, while long association fibers connect different lobes within the same hemisphere. The main long association fibers are the superior longitudinal fasciculus (SLF), the inferior fronto-occipital fasciculus (IFOF), the uncinate fasciculus (UF), the inferior longitudinal fasciculus (ILF), and the cingulum. The commissural fibers, which cross the midline interconnecting regions of the two hemispheres, include the corpus callosum, the anterior commissure, and the hippocampal commissure. Projection fibers connect the cerebral cortex with the brainstem and the spinal cord and form the corona radiata and the internal capsule.

Magnetic resonance imaging (MRI) studies of the developing human brain showed that the internal capsule, the optic radiation, and the corpus callosum are among the first to myelinate. In contrast, long association fibers such as the SLF and the IFOF myelinate relatively late [29].

Superior longitudinal fasciculus

Reil and Autenrieth (1809, 1812) were the first to describe the SLF as a group of fibers located in the white matter of the temporal, parietal, and frontal regions and around the Sylvian fissure. Burdach (1819–1826) and subsequently Dejerine (1895), following this initial identification of the tract, described in detail this fiber system as a tract that arches around the Sylvian fissure and connects the posterior temporal lobe with the frontal lobe. They named this tract "fasciculus arcuatus" (arcuate fasciculus, AF) and considered it as part of the SLF, using interchangeably the terms "superior longitudinal fasciculus" or "fasciculus arcuatus" in their descriptions. In contrast to these classical descriptions, recent diffusion tensor imaging (DTI) tractography studies demonstrated that the AF is a subdivision of the SLF. Moreover, neuroimaging experiments with nonhuman primates and humans have revealed the SLF to be a complex brain association fiber system composed of three different portions [5, 8, 16, 21] (Fig. 1) (and Fig. 6 in online version): (i) fronto-parietal or horizontal segment, (ii) temporo-parietal or vertical segment, and (iii) temporo-frontal segment or arcuate fasciculus.

The fronto-parietal or horizontal segment of the SLF originates in the inferior parietal lobe, at the level of the angular and supramarginal gyrus. Then it runs within the white matter of the frontal and parietal operculum lateral to the AF. Finally, it terminates at the posterior and inferior frontal lobe, at the level of the precentral gyrus and the posterior portion of the inferior frontal gyrus (Broca's territory) [8, 16, 21]. Isotope studies of non-human primate brains [30, 35, 36] as well as DTI analyses of human of brains [23] revealed that this fronto-parietal segment is not one single fiber tract but can be divided into three dorsal to ventral components in the white matter of the parietal and frontal lobes: the SLF I, II, and III. The SLF I originates from the dorsal superior parietal lobe and the medial parietal lobe (precuneus), runs through the white matter of the superior parietal and frontal regions and terminates at the premotor and prefrontal cortex (dorsal parts of areas 6, 8, and 9 and the supplementary motor area). The SLF II originates from the posterior portion of the inferior parietal lobe (angular gyrus), runs through the central core of the white matter above the superior limiting sulcus of the insula and terminates in the dorsal premotor and prefrontal regions. The SLF III originates at the anterior portion of the inferior parietal lobe (supramarginal gyrus), runs through the opercular white matter of the parietal and frontal lobes and terminates at the ventral premotor and prefrontal cortex (Broca's territory). This third portion of the SLF seems to correspond to the horizontal segment previously described.

The temporo-parietal or vertical segment of the SLF originates in the posterior portion of the superior and middle temporal gyrus (Wernicke's territory). The fibers turn vertically, running parallel and lateral to the AF, and terminate at the inferior parietal lobe [8, 16].

The temporo-frontal segment corresponds to the classical AF and is a long white-matter tract that directly connects the posterior temporal lobe with the posterior frontal lobe. The posterior projection of the AF is not limited to a well-defined anatomical territory with precise landmarks. It mostly encompasses the medial and posterior portions of the superior, middle, and inferior temporal gyri [2, 3, 18, 32, 34]. Then the fibers converge in a single tract that arches around the caudal end of the Sylvian fissure, running within the white matter of the parietal and frontal operculi. This fascicle runs parallel and medial to the two superficial tracts previously described and lateral to the corticospinal tract. Controversy exists about the frontal termination of the AF. Some authors suggested that it terminates at the precentral gyrus, posterior portion of the inferior frontal gyrus (pars opercularis and pars triangularis), and the middle frontal gyrus [18, 32, 34], while others have described it as mainly connected to the precentral gyrus and not the inferior frontal gyrus [2]. Interestingly, Catani et al [8], on the basis of tractographic evidence, have described a rostrocaudal organization of the terminations of the SLF within Broca's territory, with the AF reaching the anterior portion and the fronto-parietal segment projecting to the posterior portion.

In summary, such recent data suggest two parallel pathways connecting temporal and frontal regions [8, 16]: (i) a direct pathway corresponding to the classical AF, (ii) an indirect pathway running parallel and lateral to the direct pathway and consisting of two segments – an anterior or horizontal segment linking Broca's territory with the inferior parietal lobe; a posterior or vertical segment linking the inferior parietal lobe with Wernicke's territory. The SLF has been central to the neurobiological interpretation of higher brain functions, specially language functions, and language disorders [19, 43]. The present subdivision of the SLF into direct and indirect segments highlights the importance of the inferior parietal cortex as a separate primary language area with dense connections to the classical language areas [8].

Middle longitudinal fasciculus

Experimental and imaging studies of macaque monkey brains described a fascicle originating from the caudal part of the inferior parietal lobe and extending into the white matter of the superior temporal gyrus [36]. This tract was named the middle longitudinal fasciculus. Interestingly, this fascicle has been recently identified in human brains. Makris et al [24] delineated the trajectory and terminations of this tract in 4 human subjects by MRI tractography. They described the middle longitudinal fasciculus as a thin tract that is located medial and caudal to the classical AF and connects the angular gyrus with the temporal pole. Further studies are necessary to confirm this initial observation and clarify the fascicle's functional role.

Inferior longitudinal fasciculus

The ILF, first described by Burdach in 1822, runs in the inferior temporal lobe with an anterior-posterior orientation and connects the anterior part of the temporal lobe to the occipital lobe. Recent DTI studies demonstrated that the ILF is composed of a direct and an indirect pathway [7]. The indirect pathway, i.e., the occipitotemporal projection system, is constituted by U-shaped fibers that connect adjacent gyri at the inferior temporal and occipital convexity. The direct pathway is composed of long association fibers located medial to the short fibers.

Despite numerous DTI studies analyzing the course of the direct segment of the ILF, controversy exists about the anterior cortical termination of this fascicle. There have been described connections with the anterior portion of the superior, middle, and inferior temporal gyri on the lateral surface of the temporal lobe, the fusiform gyrus, parahippocampal gyrus, the amygdale, and the hippocampus [7, 15, 35]. These fibers, at the anterior portion of the temporal horn of the lateral ventricle, gather in a single bundle running laterally and inferiorly to the lateral wall of the temporal horn. At this level, the ILF is located lateral and below the optic pathways, whereas the IFOF runs medial and above the optic radiations. Thus, the roof of the ventricle is a good anatomical landmark to distinguish the ILF (below) from the IFOF (above) [25]. At the atrium of the lateral ventricle, the ILF is lateral to the sagittal stratum (IFOF, optic radiations, and tapetum) and medial to the AF and the temporo-parietal segment of the SLF. Posteriorly, the ILF terminates on the convexity surface of the occipital pole, posterior lingual gyrus, posterior fusiform gyrus, and the cuneus [7].

Inferior fronto-occipital fasciculus

The IFOF is a ventral associative bundle that connects the frontal lobe with the occipital and parietal lobes via the temporal lobe and insula. In 1909, Curran first described this fascicle by postmortem fiber dissection. Since then many other authors used white-matter dissection to elucidate the anatomical course of this fascicle [26, 27, 31, 40, 41]. Recently, DTI studies and studies combining fiber dissection with DTI described the main course of the IFOF at the level of the limen insulae, the roof of the temporal horn, and within the anterior and middle temporal lobe [6, 5, 16].

The exact frontal cortical terminations of the IFOF remain unclear, due to their strong intersection with the terminal branches of other long association fascicles (mainly the superior longitudinal fasciculus). Recent DTI tractography studies have described frontal connections of the IFOF with the dorso-lateral prefrontal and orbito-frontal cortex [5, 16]. At the insula, the IFOF runs parallel to the UF, crossing the anteroinferior portion of the external capsule and claustrum. Then, both fascicles cross the temporal stem. The anatomical course of the IFOF at the temporal stem was recently analyzed by fiber dissection of postmortem human hemispheres [27]. It was observed that the UF crosses the anterior one-third of the temporal stem, passing through the region of the limen insulae and a few millimeters of the inferior limiting sulcus of the insula, while the IFOF crosses the posterior two-thirds of the temporal stem, in the region between the posterior limit of the UF and the lateral geniculate body (Fig. 2) (and Fig. 7 in online version). At the lateral portion of the temporal stem, the fibers of the UF curve in an anterior direction to reach the anterior temporal lobe, whereas the fibers of the IFOF turn in a posterior direction coursing above the roof of the temporal horn, superior and medial to the optic radiations. The auditory radiations, the claustro-opercular and insulo-opercular fibers of the external and extreme capsules pass through the temporal stem above the IFOF, whereas the optic radiations pass below (Fig. 2). The UF crosses the anterior portion of the temporal stem in the same plane as the IFOF. The anterior commissure and the inferior thalamic peduncle cross the temporal stem below the UF. Therefore, the IFOF separates the origin of the auditory and optic radiations at the medial and lateral geniculate bodies: the initial segment of the auditory radiations coming from the medial geniculate body passes above the IFOF at the temporal stem, whereas the initial segment of the optic radiations coming from the lateral geniculate body passes below the IFOF.

At the level of the middle temporal region, the IFOF runs in the roof of the temporal horn, superior and lateral to the optic radiations and medial to the ILF. In the posterior portion of the temporal stem the fibers of the IFOF turn medially to join the sagittal stratum. At this level the IFOF is located medial to the fibers of the AF and lateral to the optic radiations in the lateral surface of the atrium of the ventricle.

Important controversy exists about the posterior cortical terminations of the IFOF. Indeed, some authors consider that the IFOF terminates at the level of the ventral occipital lobe [5], while for others it terminates into the middle and posterior temporal lobe [16], and still others consider that it terminates at the level of both the temporal and occipital lobes (i.e., into the middle and inferior temporal gyri and into the lingual and fusiform gyri) [28]. In a recent publication, fourteen postmortem human hemispheres were dissected and the posterior cortical terminations of the IFOF were analyzed [26]. Two different subcomponents of this fascicle were described: a superficial and dorsal portion and a deeper and ventral portion, both at the ventral part of the external capsule. The superficial and dorsal portion (Fig. 3) has an inferior and posterior orientation. It crosses the anterior portion of the temporal stem, passing above the anterior part of the roof of the temporal horn, and then it turns superiorly, passing underneath the posterior insula. Finally, it joins the sagittal stratum in the superior part of the lateral surface of the atrium, to reach the convexity surface of the parietal and occipital lobes, at the superior parietal lobe and the posterior portion of the superior and medial occipital gyri. The deeper and ventral portion of the IFOF (Fig. 8 in online version) is situated between the superficial portion and the deeper claustrum. At this level, it has an inferior and posterior orientation, crossing the anterior part of the temporal isthmus. It runs above the roof of the temporal

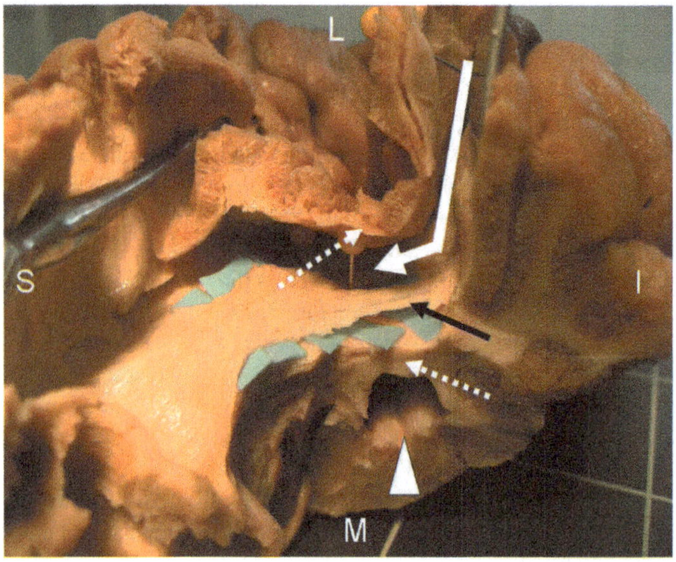

Fig. 2. Dissection of IFOF at insula and temporal stem of left hemisphere. A coronal cut of the temporal lobe 8 mm posterior to the limen insulae has been performed; the temporal horn is opened (arrowhead). At the insula and temporal stem, the IFOF (black arrow) has been completely dissected and separated from the superficial and deeper structures. Small pieces of blue paper were placed between the IFOF and the deeper structures to demonstrate that the fibers of the fascicle were completely dissected and isolated from the surrounding fiber tracts. The temporal operculum, the dorsal portion of the insula, claustrum, and extreme and external capsules have been separated from the IFOF and lifted superiorly. The claustro-opercular and insulo-opercular fibers of the extreme and external capsules (upper dotted arrow) are U-fibers that connect the claustrum and insular cortex with the temporal operculum. These fibers cross the temporal stem passing through the narrow space between the superior surface of the IFOF and the inferior limiting sulcus of the insula. The optic radiations (lower dotted arrow) pass through the temporal stem below the IFOF. *L* Lateral, *M* medial, *S* superior, *I* inferior

horn, superficial to the optic radiations, and then it proceeds posteriorly along the lateral surface of the floor of the atrium and occipital horn. Finally, it terminates at the posterior portion of the inferior occipital gyrus, posterior portion of the fusiform gyrus, temporo-occipital sulcus, and basal surface of the inferior temporal gyrus.

Extreme capsule, external capsule, uncinate fasciculus

The extreme capsule is a fiber system located immediately deep to the insular cortex. It is mainly composed of short association fibers connecting the insular gyri with each other and with the frontal, parietal, and temporal operculae [15].

The external capsule and the claustrum are intimately related as most of the fibers of the external capsule originate in the claustrum (Fig. 1). The anatomy of the claustrum and external capsule was recently reviewed by Fernandez-Miranda et al [15], who described in detail the claustrum and its projection system by DTI tractography and fiber dissection. The claustrum is a thin collection of gray matter located deep to the extreme capsule and the insula. Both the claustrum and the external capsule have two parts: dorsal (or posterosuperior) and ventral (or anteroinferior). The

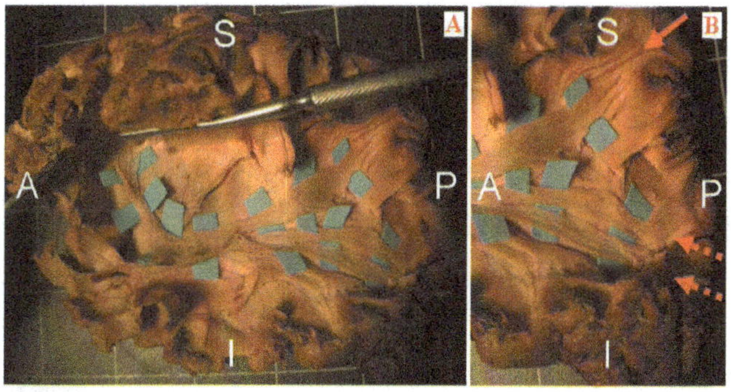

Fig. 3. **(A)** Dissection of the superficial and dorsal subcomponents of IFOF. Small pieces of blue paper have been placed between the superficial portion of the IFOF and the deeper structures, to demonstrate that the fibers of the fascicle were completely dissected and isolated from the surrounding fiber tracts. This superficial portion of the IFOF, at the ventral part of the external capsule, has an inferior and posterior orientation. It crosses the anterior portion of the temporal stem, passing above the anterior portion of the roof of the temporal horn, and then turns superiorly passing underneath the posterior insula. Finally it joins the sagittal stratum in the superior portion of the lateral surface of the atrium to reach the parietal and occipital lobes. **(B)** Enlarged view of cortical terminations of the superficial and dorsal portion of IFOF. It is connected with the cortex of the superior parietal lobe (solid arrow) and with the posterior portion of the superior and middle occipital gyri (dotted arrows). *A* Anterior, *P* posterior, *S* superior, *I* inferior

dorsal or posterosuperior portion of the claustrum and external capsule are located beneath the posterior short and the anterior and posterior long gyri of the insula. The dorsal external capsule is composed predominantly of claustrocortical fibers that connect the claustrum with the superior frontal, precentral, postcentral, superior parietal, and parieto-occipital regions. Furthermore, a topographical organization of the dorsal portion was described where posterior cortical areas project into the posterior part of the dorsal claustrum and more anterior cortical areas converge in the anterior part. The ventral or anteroinferior portion of the claustrum and external capsule are located deep with respect to the anterior and middle short insular gyri. The ventral portion of the external capsule is formed by the UF and the IFOF. The ventral claustrum consists of a group of diffuse or "island-like" gray masses that are separated and fragmented by the UF and the IFOF.

The UF (Fig. 4) is a ventral associative hooklike-shaped bundle that connects the anterior temporal lobe with the medial and lateral orbitofrontal cortex [6]. The precise cortical connection of the UF at the temporal lobe is unknown and connections with the temporal pole, amygdala, hippocampal formation, and superior and middle temporal gyri were described [12, 36, 37]. The fibers converge in a single tract that curves in a superior direction to cross the temporal stem. As it was previously described, the UF crosses the anterior one-third of the temporal stem, passing through the region of the limen insulae and a few millimeters of the inferior limiting sulcus of the insula (Fig. 7 in on-line version) [27]. Subsequently, the bundle hooks around the limen insulae at the level of the ventral portion of the external

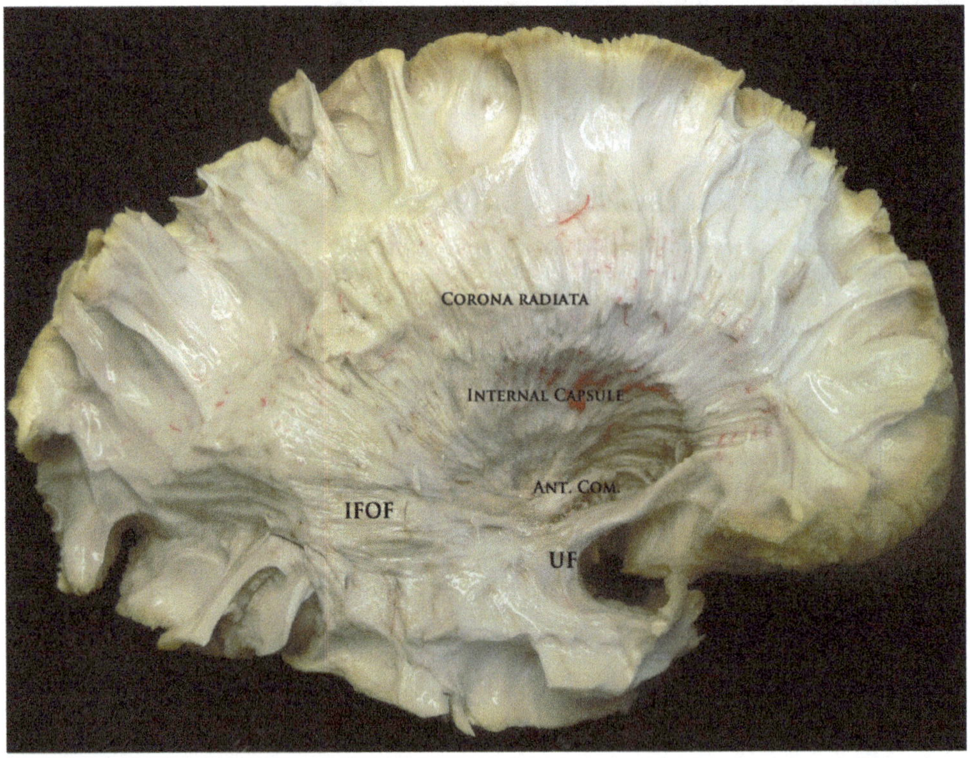

Fig. 4. Anatomical relationships between corona radiata, internal capsule, and anterior commissure (*Ant. Com.*). *IFOF* Inferior fronto-occipital fasciculus, *UF* uncinate fasciculus

capsule. Finally, it fans out into the frontal lobe to terminate in the orbital gyri, area subcallosa, gyrus rectus, frontal pole, and inferior frontal gyrus [12, 31, 36, 41].

Subcallosal fasciculus

The subcallosal fasciculus in human brains was first described by Onufrowicz in 1887. This bundle originates at the fronto-mesial precentral structures (the supplementary motor area and the cingulum). Then, it runs in a vertical direction passing through the white matter surrounding the lateral angle of the frontal horn of the ventricle. It terminates at the head of the caudate nucleus [9, 24]. It has been argued that the lateral portion of the subcallosal fasciculus is in fact the superior fronto-occipital fasciculus [6, 42].

Periventricular white matter

Intraoperative electrical stimulation of the fibers beneath the ventral premotor cortex and the lateral part of the putamen elicits dysarthria or anarthria with high reproducibility [10, 11, 17]. Dysarthria and anarthria may be due to a contraction of the muscles that participate in the speech process. On the basis of these observations, it was hypothesized that there exists a descending subcortical white bundle that may be the final pathway of speech production [9]. This bundle may originate from the grey structures specifically involved in speech production, including the primary sensorimotor area of the mouth, the ventral premotor cortex, the anterior insula, and the lentiform nucleus.

However, neither fiber dissection nor tractography studies have identified this connec-

Fig. 5. Commissural fibers of corpus callosum and their complex relationships with the other structures belonging to the walls of the lateral ventricle

tion; therefore, further anatomical studies are needed to demonstrate the existence of this fascicle and to define its exact trajectory and projections.

Corona radiata and internal capsule

The corona radiata and the internal capsule contain projection fibers with a rostrocaudal extension (Fig. 4). The corona radiata continues caudally with the more compact internal capsule, the latter joining the cerebral peduncles. To expose the corona radiata, the SLF, the IFOF, the ILF, and the optic radiations should be removed. In particular, removal of the optic radiations exposes the connection between the internal capsule and the cerebral peduncles.

The internal capsule is medial and caudal to the corona radiata (Fig. 4). The extreme capsule, claustrum, external capsule, putamen, and globus pallidus should be all removed in a lateromedial fashion to reveal the internal capsule. During dissection, by its being more fibrous the globus pallidus can be distinguished from the putamen. Nevertheless, the globus pallidus should be removed carefully to avoid damaging the anterior commissure just below it. The caudate nucleus and the thalamus are located just medial to the internal capsule.

The internal capsule has an anterior and a posterior limb, a genu and retrolenticular and sublenticular portions [16, 40, 47]. The anterior limb of the internal capsule descends between the head of the caudate nucleus and the lentiform nucleus, and the posterior limb passes between the lentiform nucleus and the thalamus. It is notable that the anterior limb is located above the anterior perforated substance, while the genu reaches the ventricular surface just lateral to the foramen of Monro.

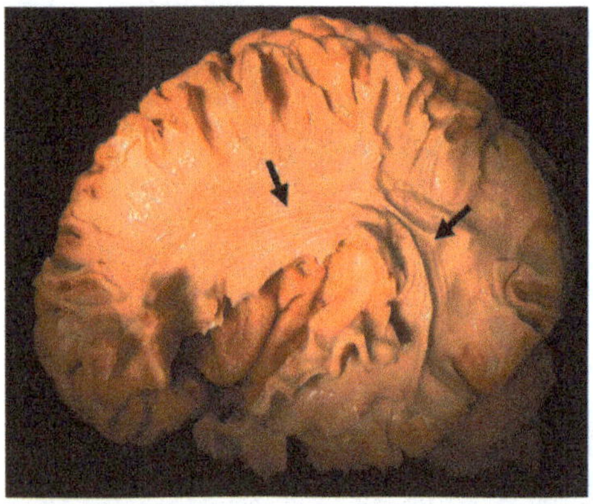

Fig. 6. (on-line version). Dissection of temporo-frontal segment of superior longitudinal fasciculus or arcuate fasciculus (black arrows)

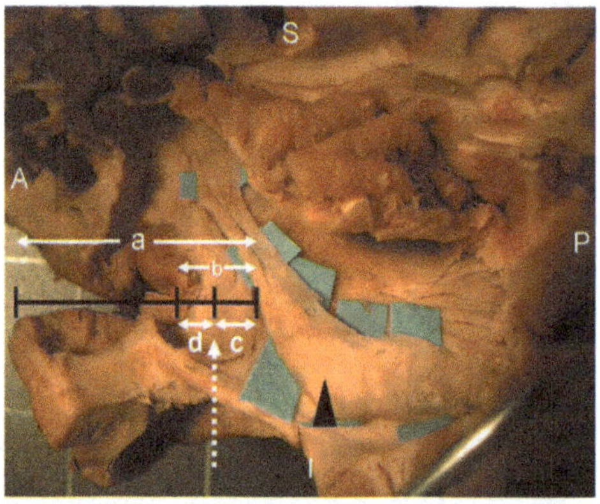

Fig. 7. (on-line version). Dissection of IFOF (arrowhead) at insula, temporal stem, and temporal lobe of left hemisphere. Dotted arrow indicates the superior projection of the tip of the temporal horn in the temporal stem. Distances measured at the temporal stem, at the level of the inferior limiting sulcus: a anteroposterior distance between the temporal pole and the anterior edge of the IFOF; b anteroposterior distance between the limen insulae and the anterior edge of the IFOF; c anteroposterior distance between the anterior tip of the temporal horn and the anterior edge of the IFOF; d anteroposterior distance between the limen insulae and the anterior tip of the temporal horn

Fibers connecting the anterior and medial thalamus and the pontine nuclei to the frontal lobe pass through the anterior limb of the internal capsule, while the genu, in addition to the corticothalamic and thalamocortical fibers, contains corticobulbar fibers to the motor nuclei of the cranial nerves. Fibers interconnecting the thalamus and cortex and the corticospinal fibers to the motor nuclei of the upper and the lower extremity and the trunk belong to the posterior limb of the internal capsule. The fibers responsible for superior limb movements are more anterior and nearer to the genu than those responsible for inferior limb movements. In a projection of the internal capsule to the lateral surface of the hemisphere, the precentral gyrus is just rostrally located to the posterior limb of the internal capsule [47]. The corticospinal tract, which originates from Brodmann's areas 4, 4a, and 4p [13], runs through the posterior limb of the internal capsule to reach the brainstem. Corticospinal tract fibers are in close topographical relationship with the SLF, crossing each other at an angle of 90° (Fig. 9 in on-line version).

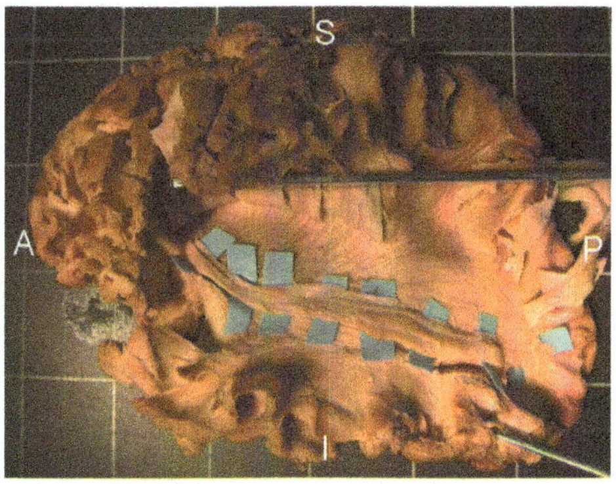

Fig. 8. (on-line version). The fibers of the superficial portion of the IFOF and the extreme capsule have been lifted to expose the deep and ventral subcomponent of the IFOF. Small pieces of blue paper have been placed between the deep portion of the IFOF and the deeper structures, to demonstrate that the fibers of the fascicle were completely dissected and isolated from the surrounding fiber tracts. This deep portion of the IFOF, at the ventral part of the external capsule, is situated between the superficial portion of the IFOF and the deeper claustrum. At this level, it has an inferior and posterior orientation, crossing the anterior portion of the temporal stem. It courses above the roof of the temporal horn superficially to the optic radiations and then proceeds posteriorly along the lateral surface of the floor of the atrium and occipital horn. Finally, it terminates at the occipital lobe and at the basal temporal lobe. *A* Anterior, *P* posterior, *S* superior, *I* inferior

The retrolenticular portion encases all the fibers belonging to the internal capsule curving around the posterior edge of the lentiform nucleus, while the sublenticular portion contains fibers passing below the lentiform nucleus. The inferior thalamic peduncle and the temporopontine fibers are composed of the sublentiform portion of the internal capsule [40]. Optic radiation fibers in their majority pass through the sublenticular part, whereas only a few belong to the retrolenticular portion. Moreover, the sublenticular portion hosts the auditory radiation fibers which originate in the lateral geniculate body and terminate in the transverse temporal gyrus (or *Heschl's gyrus – Brodmann's area 41*) and superior temporal gyrus.

Optic radiations

Fibers of the optic radiations take origin from the lateral geniculate body, which belongs to the metathalamus, located lateroventrally to the pulvinar [1]. To expose the optic radiations, removal of the short and long association pathways, extreme capsule, claustrum, external capsule, and lenticular nucleus is required. This exposes the sublenticular and retrolenticular portions of the internal capsule: at this level the optic radiations come into view and they can be dissected and followed to the occipital cortex, where they fan out to reach the calcarine fissure.

To dissect the optic radiations, the floor of the temporal horn must be removed, since optic fibers run on the roof and lateral wall of the temporal horn. These fibers directly overlie the ependyma of the temporal horn and are separated from the atrium by only a thin layer of tapetal fibers. At this stage a detailed knowledge of the sagittal stratum is paramount to achieve a good dissection of the optic fibers.

The sagittal stratum is composed of fibers of the IFOF, posterior thalamic peduncle, and optic radiations [40]. As its name suggests, the sagittal stratum is oriented in the sagittal plane from the posterior temporal to the occipital lobe. Within the sagittal stratum, even if the IFOF and posterior thalamic peduncle are intermingled and difficult to separate, the optic radiations can be carefully dissected and well appreciated.

Optic radiations become thicker as they extend more posteriorly, due to the fact that more optic radiation fibers are picked up by the sagittal stratum, which, as a consequence, becomes thicker itself within the occipital lobe, gaining also more fibers from the IFOF, anterior commissure, and posterior thalamic peduncle [22].

Three groups of optic fibers can be distinguished. Fibers of the first group, also called posterior bundle, run straight from the lateral geniculate body within the sagittal stratum to the occipital cortex. These fibers pass above the atrium and occipital horn to reach the upper lip of the calcarine fissure. It is from this perspective that the optic radiations can be seen to be part of the posterior thalamic peduncle. The second group, also called central bundle, makes an anterior curve but does not extend toward the anterior tip of the temporal horn. These fibers course along the lateral wall of the atrium and occipital horn. The third group, also called anterior bundle or Meyer's loop, passes entirely around the lateral half of the tip of the temporal horn, before coursing within the sagittal stratum, passes below the atrium and occipital horn and terminates posteriorly in the lower lip of the calcarine fissure. The anterior loop of the radiation always reaches the uncal recess and is usually located about 25 mm behind the temporal pole [22, 40].

In a coronal section, the shape of the optic radiations changes as we move posteriorly from Meyer's loop to the occipital cortex. While anteriorly the optic radiations appear flat, posteriorly they have the shape of a comma. Moreover, the latero-inferior edge of the optic radiations is never below the inferior temporal sulcus.

In summary, all surgical approaches tailored to remove intrinsic lesions within the

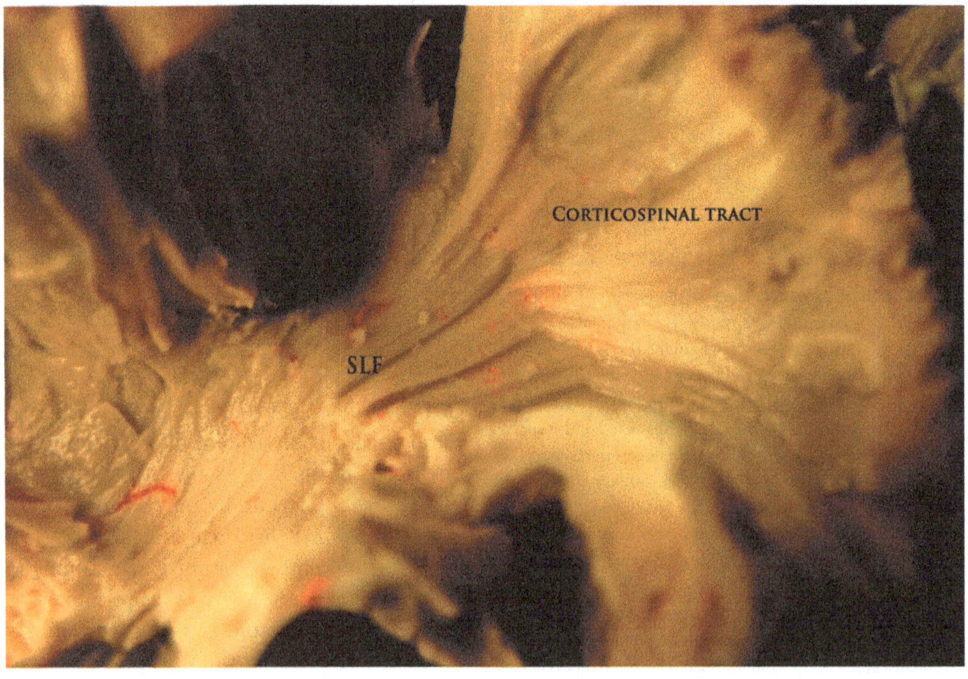

Fig. 9 (on-line version). Fibers belonging to the spinal tract cross at an angle of 90° the fibers of the SLF

temporal lobe should consider the following anatomical landmarks. The optic radiation covers the lateral half of the tip of the lateral wall and the roof of the temporal horn. On the other hand, the medial wall of the temporal horn is free from optic radiations except at the level of the lateral geniculate body, where the optic radiations arise and ascend over the roof of the temporal horn.

Fornix and cingulum

The fornix is the main efferent system of the hippocampus, starting as the posterior extension of the fimbria. It is divided into three portions: crura, body, and columns [33]. The crura of the fornix run in a rostromedial direction just below the splenium of corpus callosum, forming the anterior wall of the lateral ventricle. At the level of the body, both fornices run together in the midline along the superomedial border of the thalami, in the medial wall of the body of the lateral ventricle. Fornices fuse below the body of the corpus callosum in the commissure of the fornix.

The columns of fornix constitute the anterior wall of the foramen of Monro. The anterior columns each bifurcate into a postcommissural portion directed toward the mammillary body and a precommissural portion ending in the septal region. The mammillothalamic tract interconnects the mammillary bodies with the anterior nucleus of the thalamus.

To expose the cingulum, the cortex, the subcortex, and short fibers of the cyngulate gyrus must be removed. The cingulum is a longitudinal compact fasciculus running above and parallel to the corpus callosum, connecting the prefrontal lobes with the posterior cortices and the hippocampus [40]. The cingulum receives fibers from the anterior thalamic nucleus, superior frontal gyrus, paracentral lobule, and precuneus. Fibers coming from the precuneus greatly contribute to the enlargement of the cingulum.

Rostrally the cingulum curves anteriorly in front of the genu of the corpus callosum and ends in the subcallosal gyrus, also known as paraolfactory area of Broca [40] and paraterminal gyrus. Caudally the cingulum crosses the back of the fibers of the forceps major, covers the inferior lip of the anterior portion of the calcarine sulcus, and continues toward the anterior parahippocampal region, ending in the presubiculum and entorhinal cortex [4].

Anterior commissure

The anterior commissure is a commissural system of fibers located below the globus pallidus and above the stria terminalis. Its fibers cross the midline perpendicular to the optic tract and medial to the UF (Fig. 4). The lateral extension of the anterior commissure can be followed into the temporal lobe. While a few fibers of the anterior commissure merge with the UF, most are directed latero-posteriorly merging with the IFOF and sagittal stratum [40].

Corpus callosum

Corpus callosum (Fig. 5) is the main commissural system of fibers connecting the two hemispheres. Rostrum, genu, body, and splenium of the corpus callosum come into view after opening the interhemispheric fissure and removing the cingulate gyrus. If dissection is conducted in a lateromedial fashion, callosal fibers can be followed crossing the midline at an angle of nearly 90° relative to the fibers of corona radiata. At the level of the genu of the corpus callosum, the fibers take an anterior oblique direction, forming the forceps minor, which interconnects the prefrontal and orbitofrontal regions. At the level of the splenium, the fibers take a posterior oblique direction, forming the forceps major, which interconnects the parieto-occipital and calcarine regions.

The tapetum represents a subgroup of callosal fibers in the splenial region. To expose the tapetum, the roof and the lateral wall of the temporal horn must be dissected away [40]. The tapetum forms the roof and the lateral wall of the atrial portion of the lateral ventricle. It curves anteriorly into the temporal lobe, extending almost to the tip of the temporal horn just lateral to the tail of the caudate nucleus, and runs below the temporal horn, separating the posterior thalamic peduncle from the temporal horn.

Conclusions

Despite the development in recent years of different accurate surgical guides, such as neuronavigation, intraoperative ultrasounds or intraoperative MRI, in our opinion, none of them can substitute the comprehensive understanding of the three-dimensional fiber pathways organization acquired in the microsurgical laboratory by the fiber dissection technique. This is essential for planning strategically the surgical approach to a wide spectrum of intrinsic brain lesions and for preserving the subcortical structures in order to avoid postsurgical permanent deficits. It is worth noting that intraoperative subcortical mapping requires a precise anatomical knowledge of the spatial relationships of the different white-matter bundles.

A new era in the understanding of white-matter organization has begun with the development of MRI tractography, which allows to visualize the white-matter fiber orientation in the living human brain. Nevertheless, neurosurgeons dealing with intrinsic neoplastic and vascular lesions should be trained in the fiber dissection technique in order to acquire their own view of the three-dimensional complex anatomy of the white-matter pathways.

References

[1] Archambault L (1906) Le faisceau longitudinal inférieur et le faisceau optique central: quelques considérations sur les fibres d'association du cerveau. Rev Neurol (Paris) 4: 1206

[2] Bernal B, Altman N (2010) The connectivity of the superior longitudinal fasciculus: a tractography DTI study. Magn Reson Imaging 28: 217–225

[3] Bernal B, Ardila A (2009) The role of the arcuate fasciculus in conduction aphasia. Brain 132: 2309–2316

[4] Carpenter MB, Sutin J (1983) Human neuroanatomy. Williams & Wilkins, Baltimore

[5] Catani M, Thiebaut de Schotten M (2008) A diffusion tensor imaging tractography atlas for virtual in vivo dissections. Cortex 44: 1105–1132

[6] Catani M, Howard RJ, Pajevic S, Jones DK (2002) Virtual in vivo interactive dissection of white matter fasciculi in the human brain. Neuroimage 17: 77–94

[7] Catani M, Jones DK, Donato R, ffytche DH (2003) Occipito-temporal connections in the human brain. Brain 129: 2093–2107

[8] Catani M, Jones DK, ffytche DH (2005) Perisylvian language networks of the human brain. Ann Neurol 57: 8–16

[9] Duffau H (2008) The anatomo-functional connectivity of language revisited. New insights provided by electrostimulation and tractography. Neuropsychologia 46: 927–934

[10] Duffau H, Capelle L, Sichez N, Denvil D, Lopes M, Sichez JP, Bitar A, Fohanno D (2002) Intraoperative mapping of the subcortical language pathways using direct stimulations. An anatomo-functional study. Brain 125: 199–214

[11] Duffau H, Capelle L, Denvil D, Gatignol P, Sichez N, Lopes M, Sichez JP, Van Effenterre R (2003) The role of dominant premotor cortex in language: a study using intraoperative functional mapping in awake patients. Neuroimage 20: 1903–1914

[12] Ebeling U, von Cramon D (1992) Topography of the uncinate fascicle and adjacent temporal fiber tracts. Acta Neurochir (Wien) 115: 143–148

[13] Economo C von, Koskinas GN (1925) Die Cytoarchitektonik der Hirnrinde des erwachsenen Menschen. Springer, Wien Berlin

[14] Epelbaum S, Pinel P, Gaillard R, Delmaire C, Perrin M, Dupont S, Dehaene S, Cohen L (2008) Pure alexia as a disconnection syndrome: new diffusion imaging evidence for an old concept. Cortex 44: 962–974

[15] Fernandez-Miranda JC, Rhoton AL Jr, Kakizawa Y, Choi C, Alvarez-Linera J (2008) The claustrum and its projection system in the human brain: a microsurgical and tractographic anatomical study. J Neurosurg 108: 764–774

[16] Fernandez-Miranda JC, Rhoton AL Jr, Alvarez-Linera J, Kakizawa Y, Choi C, de Oliveira EP (2008) Three-dimensional microsurgical and tractographic anatomy of the white matter of the human brain. Neurosurgery 62 (Suppl 3): 989–1026

[17] Gil Robles S, Gatignol P, Capelle L, Mitchell MC, Duf-

fau H (2005) The role of dominant striatum in language: a study using intraoperative electrical stimulations. J Neurol Neurosurg Psychiatry 76: 940–946
[18] Glasser MF, Rilling JK (2008) DTI tractography of the human brain's language pathways. Cereb Cortex 18: 2471–2482
[19] Goodglass H, Wingfield A (1993) Selective preservation of a lexical category in aphasia: dissociations in comprehension of body parts and geographical place names following focal brain lesion. Memory 1: 313–328
[20] Klingler J (1935) Erleichterung der makroskopischen Präparation des Gehirns durch den Gefrierprozess. Schweiz Arch Neurol Psychiatr 36: 247–256
[21] Lawes IN, Barrick TR, Murugam AB, Spierings N, Evans DR, Song M, Clark CA (2008) Atlas-based segmentation of white matter tracts of the human brain using diffusion tensor tractography and comparison with classical dissection. Neuroimage 39: 62–79
[22] Mahaney KB, Abdulrauf SI (2008) Anatomic relationship of the optic radiations to the atrium of the lateral ventricle: description of a novel entry point to the trigone. Neurosurgery 63 (Suppl 2): 195–202
[23] Makris N, Kennedy DN, McInerney S, Sorensen AG, Wang R, Caviness VS Jr, Pandya DN (2005) Segmentation of subcomponents within the superior longitudinal fascicle in humans: a quantitative, in vivo, DT-MRI study. Cereb Cortex 15: 854–869
[24] Makris N, Papadimitriou GM, Kaiser JR, Sorg S, Kennedy DN, Pandya DN (2009) Delineation of the middle longitudinal fascicle in humans: a quantitative, in vivo, DT-MRI study. Cereb Cortex 19: 777–785
[25] Mandonnet E, Nouet A, Gatignol P, Capelle L, Duffau H (2007) Does the left inferior longitudinal fasciculus play a role in language? A brain stimulation study. Brain 130: 623–629
[26] Martino J, Brogna C, Gil Robles S, Vergani F, Duffau H (2010) Anatomic dissection of the inferior fronto-occipital fasciculus revisited in the lights of brain stimulation data. Cortex 46: 691–699
[27] Martino J, Vergani F, Gil Robles S, Duffau H (2010) New insights into the anatomic dissection of the temporal stem with special emphasis on the inferior fronto-occipital fasciculus: implications in surgical approach to left mesiotemporal and temporoinsular structures. Neurosurgery 66(Suppl 1): 4–12
[28] Nieuwenhuys R, Voogd J, van Huijzen C (1988) The human central nervous system. Springer, Berlin
[29] Paus T, Collins DL, Evans AC, Leonard G, Pike B, Zijdenbos A (2001) Maturation of white matter in the human brain: a review of magnetic resonance studies. Brain Res Bull 54: 255–266
[30] Petrides M, Pandya DN (1984) Projections to the frontal cortex from the posterior parietal region in the rhesus monkey. J Comp Neurol 228: 105–116
[31] Peuskens D, van Loon J, Van Calenbergh F, van den Bergh R, Goffin J, Plets C (2004) Anatomy of the anterior temporal lobe and the frontotemporal region demonstrated by fiber dissection. Neurosurgery 55: 1174–1184
[32] Powell HW, Parker GJ, Alexander DC, Symms MR, Boulby PA, Wheeler-Kingshott CA, Barker GJ, Noppeney U, Koepp MJ, Duncan JS (2006) Hemispheric asymmetries in language-related pathways: a combined functional MRI and tractography study. Neuroimage 32: 388–399
[33] Rhoton AL Jr (2002) The cerebrum. Neurosurgery 51(Suppl 1): 1–51
[34] Rilling JK, Glasser MF, Preuss TM, Ma X, Zhao T, Hu X, Behrens TE (2008) The evolution of the arcuate fasciculus revealed with comparative DTI. Nat Neurosci 11: 426–428
[35] Schmahmann JD, Pandya DN (2006) Fiber pathways of the brain. Oxford Univerity Press, New York
[36] Schmahmann JD, Pandya DN, Wang R, Dai G, D'Arceuil HE, de Crespigny AJ, Wedeen VJ (2007) Association fibre pathways of the brain: parallel observations from diffusion spectrum imaging and autoradiography. Brain 130: 630–653
[37] Sincoff EH, Tan Y, Abdulrauf SI (2004) White matter fiber dissection of the optic radiations of the temporal lobe and implications for surgical approaches to the temporal horn. J Neurosurg 101:739–746
[38] Türe U, Yasargil MG, Pait TG (1997) Is there a superior occipitofrontal fasciculus? A microsurgical anatomic study. Neurosurgery 40: 1226–1232
[39] Türe U, Yasargil DC, Al-Mefty O, Yasargil MG (1999) Topographic anatomy of the insular region. J Neurosurg 90: 720–733
[40] Türe U, Yasargil MG, Friedman AH, Al-Mefty O (2000) Fiber dissection technique: lateral aspect of the brain. Neurosurgery 47: 417–426
[41] Wang F, Sun T, Li X-G, Liu N-J (2008) Diffusion tensor tractography of the temporal stem on the inferior limiting sulcus. J Neurosurg 108: 775–781
[42] Yakovlev PI, Locke S (1961) Corticocortical connections of the anterior cingulate gyrus, the cingulum and subcallosal bundle. Trans Am Neurol Assoc 86: 252–256
[43] Yamada K, Nagakane Y, Mizuno T, Hosomi A, Nakagawa M, Nishimura T (2007) MR tractography depicting damage to the arcuate fasciculus in a patient with conduction aphasia. Neurology 68: 789
[44] Yasargil MG (1984) Microneurosurgery, 1: microsurgical anatomy of the basal cisterns and vessels of the brain, diagnostic studies, general operative techniques and pathological considerations of the intracranial aneurysms. Thieme, Stuttgart
[45] Yasargil MG (1994) Microneurosurgery, 4A: CNS tumors: surgical anatomy, neuropathology, neuroradiology, neurophysiology, clinical considerations, operability, treatment options. Thieme, Stuttgart
[46] Yasargil MG (1996) Microneurosurgery, 4B: microneurosurgery of CNS tumors. Thieme, Stuttgart
[47] Young P (1997) Basic clinical neuroanatomy. Williams & Wilkins, Philadelphia

Methods of brain mapping: advances and limitations

Functional MRI

Alexandre Krainik

Introduction

Since blood oxygenation level dependent (BOLD) signal changes have been observed using MRI [45] and modulated using neuronal stimuli [4, 46], functional MRI (fMRI) has quickly become the most popular functional neuroimaging technique in clinical practice and cognitive neuroscience. Indeed, high-field MR scanners and BOLD sensitive sequences are now widely accessible in both clinical and research settings. Moreover, fMRI is non-invasive. BOLD signal that relies on deoxyhemoglin (deoxyHb) concentration is detectable without injection of external contrast media. The colorful activation maps combined with three-dimensional brain anatomy may have also made this imaging method as much attractive as controversial [44, 56].

Numerous promising applications of fMRI have been suggested in medicine [43]. After a fifteen-year long history in clinical practice and thousands of scientific papers even in prestigious journals, the role of fMRI remains mostly dedicated to map eloquent cortex before a neurosurgical procedure. In fact, BOLD fMRI is challenging because the relationship between the neuronal response to a stimulus and the activation blobs relies on neurovascular coupling, hemodynamic response, MR signal detection, and complex time-series analyses [40, 8].

Besides an obvious and partially elucidated complexity, and several concerns on the interpretation of experimental paradigms in cognitive neuroscience, fMRI is based on a robust physiological and physical framework (see recent and outstanding references for details [40, 8]. BOLD signal is reproducible across subjects and MR scanners [61]. fMRI requires a rigorous methodology to acquire and analyze data, an advanced knowledge in sulcogyral and functional neuroanatomy to estimate spatial displacement and reorganization in patient with focal lesion, and a solid experience in BOLD imaging to distinguish artifacts and potential confounds from appropriate results [22]. In order to better advocate for fMRI in clinical practice, here is a brief review of the principles of fMRI and the key points in fMRI interpretation.

Principles of fMRI

Biophysical framework of the BOLD signal

Most fMRI experiments in humans are conducted using blood oxygenation level dependent (BOLD) signal. This task-related signal is thought to be a consequence of presynaptic neurotransmitter release that reflects local signaling, depending on relative inhibitory and excitatory input [40]. However the BOLD signal is not a direct marker of the neuronal activity, it relies on changes in blood oxygenation. In the arterial blood, the hemoglobin carries oxygen, the oxyhemoglobin is diamagnetic. In the capillary compartment, the oxygen is delivered to the parenchyma for aerobic metabolism. The deoxyhemoglobin (deoxyHb) is paramagnetic due to subtle changes in the spatial conformation of the heme. This effect, the magnetic susceptibility, increases the dephasing of the spins of the hydrogen nuclei, producing magnetic field gradients in both intravascular and extravascular compartments. At 1.5 T, about 50% of the BOLD contrast is due to the intravascular signal although blood volume represents about 4% of the volume of tissular voxel [5]. Additionally, the difference between the internal and external susceptibilities at the surface of the vessels is correlated to the spatial extent of the dephasing. The effects of deoxyHb on the magnetic field is particularly detected in adjacent extravascular tissue is larger in tissue adjacent to larger blood vessels, such as veins. The diffusion of water molecules contributes also to the extravascular magnetic susceptibility in increasing phase dispersion, especially around larger veins. Thus, the increased dephasing of spins due to magnetic susceptibility shortens the T2*; and largest BOLD changes are identified near veins draining the activated area. These changes are better detected using gradient recalled echo (GRE) T2*-weighted images (WI). Spin echo (SE) T2-weighted images are less sensitive to BOLD signal because of the weakness of the extravascular effect but SE T2-WI are more specific to BOLD signal originating from capillaries, especially at higher static fields [8].

As BOLD imaging is based on deoxyHb concentration as an endogenous contrast agent, at least both neuronal and vascular basal characteristics and task-related responses are involved. Indeed, change in cerebral metabolic rate of O_2 ($CMRO_2$), cerebral blood flow (CBF), and cerebral blood volume (CBV) modify the local deoxyHb content. Moreover, the functional properties of the vasculature such as the neurovascular coupling are critical. In fact, the nature of the BOLD signal is complex because physiological parameters have opposite effects. The neuronal activity increases the $CMRO_2$ and the deoxyHb concentration which decreases the BOLD signal. Although inconstant and controversial, this early effect has been called the "initial dip" or "fast response" and would rather reflect the early change of $CMRO_2$ than the venular response [13]. Then, the neurovascular coupling increases the CBF dramatically. While the CBF increases, the oxygen extraction fraction (OEF) decreases, the deoxyHb is washed-out and its concentration decreases. After 1–2 seconds, BOLD signal increases along a 5–8 s ramp to reach a plateau when then neuronal activity is sustained with a magnitude of 0.5–5% [2]. Thus, the enhanced BOLD signal is due to an increased delivery of oxygenated blood out of proportion to the utilization of oxygen by neural cells (Fig. 1). The moderate increase of the CBV and the intravoxel content of deoxyHb do not initially compensate the increase of the BOLD signal. However, a persistent dilatation of the venous compartment has been proposed to explain a "poststimulus undershoot" which lasts several seconds after the signal drops below the baseline [59]. Again, the contributions of neuronal, vascular and metabolic components in the "poststimulus undershoot" are controversial [8].

A calibrated-BOLD method has been proposed to better estimate $CMRO_2$ changes. It consists in measuring simultaneously BOLD and CBF with a combined BOLD and arterial spin labeling (ASL) technique. The BOLD calibration is performed under hypercapnia [15, 25] or hyperoxia [10]. Thus, the calibration relies on the vasoreactivity to mild inhalation challenges without change of $CMRO_2$. For a given CBF, neuronal-related BOLD signal is smaller than capnic-related BOLD signal. This difference corresponds to the signal drop due to the $CMRO_2$ (Davis 1998). This approach to measure $CMRO_2$ using MRI has also been called quantitative fMRI. Other MR techniques have been proposed to perform neuronal fMRI using dynamic perfusion ASL [6], and diffusion imaging [35]. Although attractive, these methods are, still today, technically challenging and remain out of the clinical practice.

According to this biophysical framework, BOLD signal appears to be: (1) a combination of neuronal and vascular changes that may rather give information on the neurovascular unit, the vasculature and its properties than on single neural events, (2) a relative measurement that requires different experimental conditions to obtain a BOLD contrast.

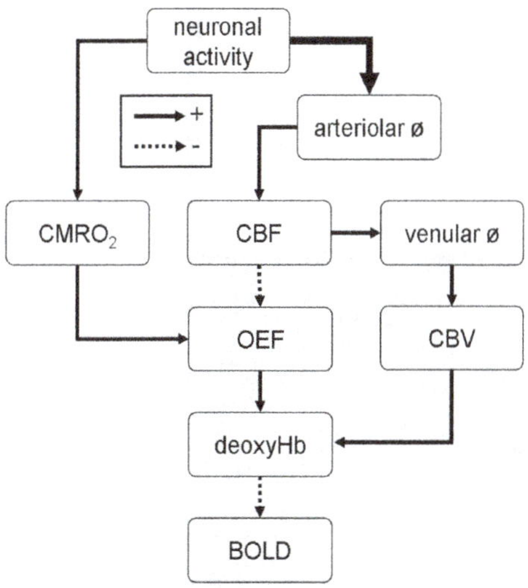

Fig. 1. BOLD response to neuronal activity
Plain and dotted lines represent positive and negative relationships, respectively. The thickness of the line is related to the importance of the effect.
The time-course of the BOLD response is still debated but the following scenario could be proposed. First, the neuronal activity increases the $CMRO_2$ that produces deoxyHb which decreases the BOLD signal (initial dip). Second, the neurovascular coupling allows a fractional increase of CBF 2-3 times larger than the fractional increase $CMRO_2$. This conducts to a paradoxical decrease of the OEF that leads to an increase of the BOLD signal (plateau). Third, the rapid growth of the blood flow would be transiently stored in the venular compartment. At the voxel level and despite of the OEF decrease, the increase of CBV and deoxyHb due the neuronal activity and $CMRO_2$ would decrease the BOLD signal (post-stimulus undershoot)

Experimental design

To investigate neural basis in cognitive neuroscience, most of experimental paradigms rely on the cognitive subtraction approach. The validity of this method is debated because of the assumption of pure insertion, which asserts that cognitive processes could remain additive and independent; an assumption that is often not tenable [40]. In time-series, BOLD signal is measured under alternating conditions and BOLD contrasts between these conditions allow mapping eloquent cortico-subcortical networks. Because MR signals are noisy and the amplitude of the BOLD signal changes is weak, repeated measures have to be performed. The temporal distribution of the alternating stimuli could be set according to a block design or an event-related design.

Block design requires for each condition to be maintained during a sufficient duration that allows the BOLD signal to reach a plateau. For each condition, the number of stimuli, the duration of the blocks, and the number of blocks have to be determined. Paradigms using block designs have more power than event related designs when detecting the magnitude of the BOLD response [7].

Event-related design intends to explore single events scattered over time. Design optimization is required to obtain sufficient BOLD contrast [38]. Event-related design is less powerful in detecting response magnitude but more efficient at estimating the shape of the hemodynamic response [38, 7]. Moreover, the necessity to increase the number of stimuli and the interstimulus interval remain an important limitation, especially in patients with increased risks of movements, fatigue, and poor performance.

When combined with EEG recordings, EEG-fMRI uses the time-course of interictal spikes as regressor in an event-related fashion to identify epileptic networks [19].

Contrary to the task-related fMRI, resting state fMRI has a single rest condition [15]. Statistical analyses identify temporal correlation of BOLD signals across sets of brain regions, unveiling potential resting state networks. Now, this technique still remains out of the clinical applications.

Workflow in fMRI

Functional MRI is a reliable technique when the whole procedure is well-known and promptly conducted by the investigators. Hesitations during data acquisition waist time, and may increase fatigue, anxiety, discomfort, movements, and poor performances. Thus, the fMRI workflow has to be tested during preliminary examinations, learned, and practiced to be fluent in order to be able to face potential technical disorders. In practice, an MRI

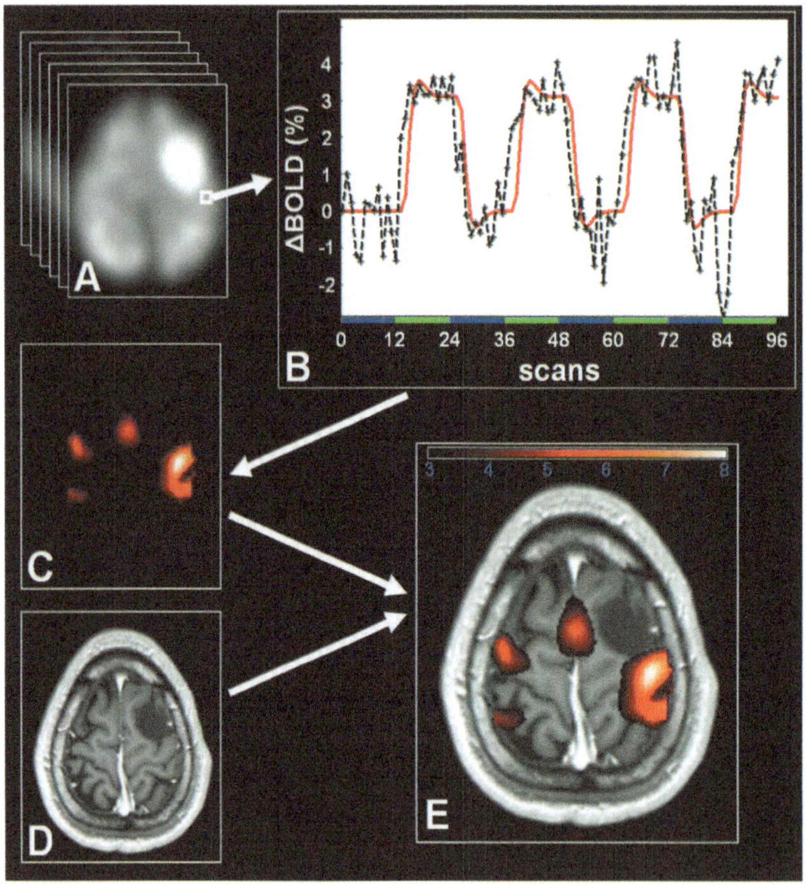

Fig. 2. Data processing
Functional scans were obtained using T2*-WI GRE-EPI (**A**). In this example, each scan covering the whole brain was acquired repeatedly every 3 secs, 96 times for a total duration of 3x96=288 secs (**B**). During this time, the subject performed a block-design paradigm alternating 4 periods of rest (blue line) and right fingers movement (green line). Each block lasts 12 scans. After preprocessing, the signal time-course of each voxel (dotted line) was plotted against a theoretical signal change (plain line), given by the convolution of the paradigm with a canonical hemodynamic response function. A voxelwise statistical analysis, here a linear regression study, tested the relationship between the theoretical and the observed signal changes. A statistical map was generated using the statistical value of the test for each voxel according to a color scale (**C**). An anatomical volume with a millimetric resolution was acquired during the same examination (**D**). Finally, the statistical map was overlaid onto the anatomical image for display (**E**)

examination should not last over one hour. Data processing also requires preparation, especially in clinical settings, to allow the results to be provided and considered in due time for the patient care.

Before the examination, extensive explanations are necessary to obtain full cooperation. Instructions are clearly explained, and need several re-explanations during the examination. A preliminary training is also needed.

The installation has to be comfortable with the head maintained within the coil. Specific devices are placed according to the stimuli (headsets, goggles or mirror, correcting glasses if necessary...) and the expected responses (buttons, joystick...).

Image acquisitions include both anatomical and functional volumes. Anatomical images consist usually in a 3D T_1-weighted gradient-echo sequence covering the whole brain using a millimetric spatial resolution to coregister functional data. Functional images consist usually in single-shot GRE echo-planar-imaging (EPI) T_2*-WI covering the whole brain using voxels with a lower spatial resolution of 3–5 mm in each dimension. The echo time is adapted to the field strength. Such volume is acquired in 2–5 seconds depending on the number of planes scanned. The acquisition of this functional volume is repeated over time at a temporal resolution given by the time of repetition. Thus, this procedure provides a time-course of signal change for each voxel (Fig. 2).

During the functional acquisition, the subject has to follow the cognitive paradigm, and absolutely avoid head motion. In clinical practice, cortical mapping of sensorimotor functions, visual areas, and hemispheric dominance for language are routinely performed (Fig. 3). Additional measurements could be also achieved such as perfusion or BOLD calibration if necessary [7].

A popular approach to analyze the data is based on the general linear model, which assumes the resulting BOLD signals as a linear combination of regressors. Given to the experimental design, the stimuli presentation is convolved by a canonical hemodynamic response function to obtain a theoretical BOLD time-course for each condition. Thus, the regression analyses are conducted in order to estimate the causal relationships between theoretical and observed BOLD time-courses for each voxel. Again, the validity of the canonical hemodynamic response function is questionable for all voxels and between healthy subjects and patients [11, 48, 61, 18]. This issue might also change the results dramatically, even in EEG-fMRI [20]. Now, such analyses could be computed in a real-time mode. These recent advances widely distributed by manufacturers allow estimating the quality of the examination.

Key points in fMRI interpretation

Despite recent works using advanced imaging techniques; numerous limitations maintain BOLD fMRI far from being able to map directly neural activity in clinical practice. Besides reserves on the physiological basis of the BOLD signal and the cognitive subtraction approach, imaging resolutions are out of the neuronal range. Indeed, a standard voxel of 55 mm^3 (3–5 mm in each dimension) contains 5.5 million neurons, 2.2–5.5 × 10^{10} synapses, 22km of dendrites, and 220 km of axons Additionally, a spatial smoothing twice larger than the voxel size is commonly applied [40]. The temporal resolution of 2–5 s is also far from having the neuronal millisecond scale, and multiple repetitions have to be performed and averaged over time to reach a sufficient contrast between cognitive conditions.

Besides obvious methodological challenges, fMRI has been validated in humans at the individual level by confrontation to others mapping techniques such as PET [47], magnetoencephalography [52], preoperative electrical stimulations [36], functional impairment induced by Wada test [24] and lesions [30, 31, 32]. Because of the vascular and mostly venous nature of the BOLD signal, the activations do rather delineate a region involved in the execution of the tasks than the spatial extent of the activated cortex. Thus, an interval of confidence of 1 cm around the center of mass of the activation is commonly admitted [3].

Before addressing the most common issues in clinical fMRI interpretation, expected results and potential methodological and physiological confounds have to be clear in mind. The whole procedure of data acquisition and analysis has to be tested and validated prior to start clinical practice.

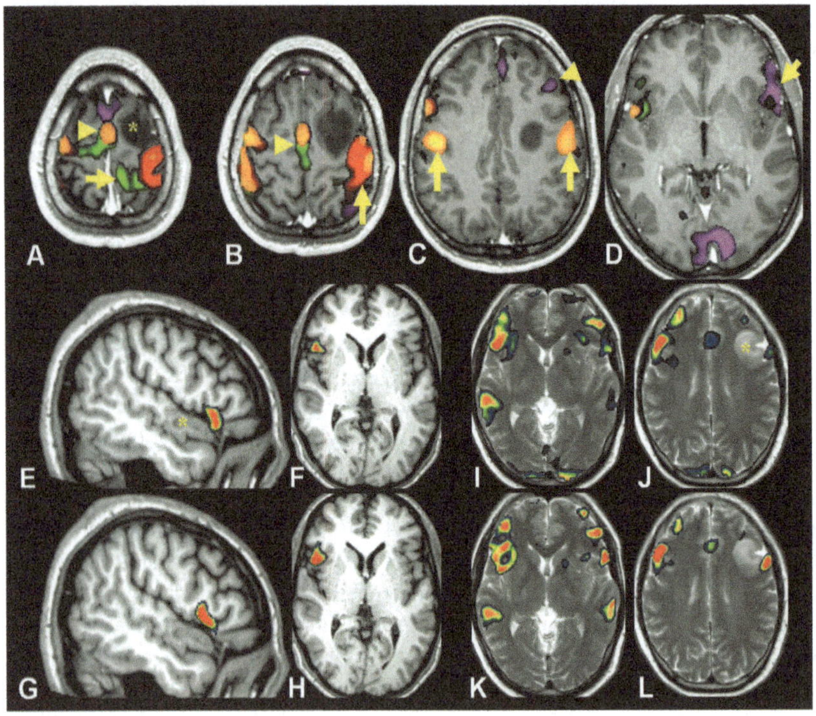

Fig. 3. Preoperative fMRI

(A–D): Preoperative fMRI in a right-handed patient with a low grade glioma of the left frontal lobe (asterisks). Primary sensorimotor activations were detected along the central sulcus, posteriorly to the lesion during movements of the right toes (green, arrow on **A**), the right fingers (red, arrow on **B**), the lips (yellow, arrows on **C**). Covert sentences generation showed a dominance of the left hemisphere with an activation of the Broca area (arrow on **D**) and the left middle gyrus (arrowhead on **C**). Activations of the supplementary and pre-supplementary motor areas were superimposed and visible on **A** and **B** (arrowhead).

(E–H): Preoperative fMRI in a left-handed patient with dysplasia of the right superior temporal gyrus (asterisks). The covert sentences generation task was repeated twice in the same examination for testing reproducibility. Both sessions (**E–F** and **G–H**) showed a reproducible activation of the Broca area in the right frontal hemisphere. Speech arrest was evoked during surgery using awake peroperative cortical electrical stimulations.

(I–L): Preoperative fMRI in a right-handed patient with a low grade glioma of the left middle frontal gyrus (asterisks). The covert sentences generation task was repeated twice (**I–J** and **K–L**) during the same examination. Most activations were detected in the right hemisphere, despite activation adjacent to the tumor. In such cases, confrontation with peroperative electrical stimulations was necessary to better assess the cortical reorganization that might have improved the postoperative outcome

Image quality

First of all, fMRI is an imaging technique. Checking the image quality sounds trivial. However and contrary to conventional MRI, checking for the quality of the whole dataset of an fMRI is hard to perform in clinical practice. Indeed, a standard fMRI examination may for instance include 4 tasks acquired in 4 separate sessions of 100 volumes containing 40 slices, giving a total of 16000 images to visually scan for artifacts.

In practice, magnetic susceptibility artifacts due to hemorrhage, calcifications, metal (implanted materials), and previous surgical procedures are common in neuro-oncology and vascular malformations [29]. Beside these lesion-related artifacts, constitutional magnetic susceptibility artifacts due to important signal changes between air, bone, and tissue are detected in the vicinity of the skull base. Thus, fMRI of orbito-frontal, temporo-polar, and temporo-basal regions is particularly difficult to perform using GRE-T2*. Such artifacts could be reduced using spin-echo T2-weighted images [60].

When present, these artifacts might be responsible of both false negative and false positive results. On the one hand, no BOLD signal change related to the activation paradigm can be extracted within a steady and homogeneous "black hole". On the other hand, false activations are commonly observed on the margins of artifacts. Such activations are mostly due to subtle movements, identified as significant because of the important range of artifactual signal changes, synchronous to the paradigm (Fig. 4).

Recent advances in MR processing with real-time fMRI packages offer online reconstruction, to display native images and to superimpose activation maps. In such situation, it provides a better comprehension of the results during the examination, instead of suggesting another cause of abnormal results such as abnormal performance or movement that would require performing the task again.

Image position

In clinical practice, statistical maps calculated using BOLD images are usually overlaid onto an anatomical image to better depict the spatial relationships of the activations with the surrounding sulci or a lesion. To be valid, the superimposition has to be performed between datasets with similar geometric parameters or using software able to interpret potential change of position, angle and dimensions across volumes, appropriately. In fact, many image formats have been used in the last two decades according to the performance of the image reconstructors, informatics networks, storage media etc. Exhaustive geometric information was not always saved or handled by processing software. This point is absolutely critical in order to avoid spatial offsets across datasets. Coregistration can be easily checked

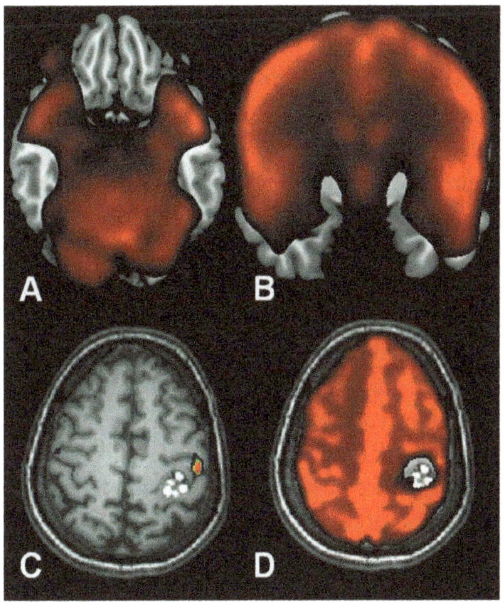

Fig. 4. Susceptibility artifacts
(A–B): Native axial T2*GRE-EPI in red overlaid onto an anatomical template showing common signal loss due to susceptibility artifacts in orbito-frontal and temporo-basal regions. **(C–D):** Cavernoma of the "hand knob" of the left precentral gyrus, responsible a pathological susceptibility artifiact. Right hand movements elicited a primary sensorimotor activation, lateral to the artifact that might have partially masked eloquent cortex

by superimposing BOLD native images onto the anatomical image. Right-left orientation has also to be carefully checked. When necessary, coregistration has to be conducted.

Since postprocessing can now be performed on DICOM images by MR manufacturers, these steps are handled properly. Other image formats might also interpret appropriately the position of images volumes across series. However, one must always remain critical on image geometrics, especially when data are exported to or imported from an external source. Looking at datasets and checking for their registration is a basic step before postprocessing.

Despite extensive care for spatial coregistration, a brain shift might occur during surgery and the removal of a space-occupying lesion. Thus beside shift simulation implemented within the navigation system, peroperative fMRI has been proposed [17].

Individual performance

As other functional neuroimaging techniques, poor performance is a trivial cause of poor results. Indeed, the assessment of task-related neuronal activity relies on the appropriate perception of the stimuli and the execution of the task when required. The attention has also to be controlled across conditions and sessions. Indeed attention and fatigue modulate BOLD contrast [40]. Overly long paradigm should be avoided. In patients, experimental setup and design have to be simple and feasible, limiting fatigue, task difficulty and speed. Thus, preliminary tests are necessary to estimate the appropriateness of the task.

Before fMRI, a full explanation of the time course of the examination and the tasks has to be given and sufficient training has to be performed. During the examination, instructions ought to be re-explained and repeated before each run. A dedicated software is useful to send stimuli at specific onsets. The task execution has to be monitored, at least visually for motor tasks. However, changes in force, amplitude, frequency, or preparation of the movement influence the results. In cognitive tasks, behavioral data can be recorded, using dedicated device to obtain specific answers. To study the hemispheric dominance for language, production tasks are robust and simple. Covert tasks are usually conducted because overt paradigms increase task-related movements and their control "non-language" condition is difficult to perform overtly. However, monitoring a covert task is particularly difficult. Again, real-time analyses are useful to estimate results, to repeat a task when necessary, and for positive reinforcement. To better estimate the relationship between the task execution and signal changes, recorded performances can be tested as a statistical regressor.

Movements

Because of the brief duration of each image acquisition (<100ms), movements artifacts within the image are unlikely. However in awake subjects, subtle movements are always identified across images along time. The motion amplitude and its spatial coherence and synchrony with the paradigm may be responsible of significant signal changes [21]. In case of movements, most important signal changes are commonly detected along the parenchymal borders in voxels containing both brain tissue and cerebrospinal fluid. When these changes are synchronous with the task execution, long strips of false-positive activation are detected along the margins of the brain and ventricles. The borders of foci of magnetic susceptibility, such as hemorrhage or calcification with voxels containing both deep hypointensity and parenchymal hyperintensity, might also be associated with peripheral false-positive activation.

To minimize subject motion, the installation has to be comfortable with the head maintained within the coil by adapted cushions. Paradigms have to be tested to avoid excessive task-related movements and the overall duration of fMRI has to be limited. Again, preliminary evaluation, explanation, training, dialog, and real-time analyses are precious to avoid excessive motion. After image acquisition, motion correction could be performed using spatial realignment. Statistical analyses might also estimate the confounding effect of head motion.

Physiological confounds

As previously mentioned, BOLD contrast relies on task-related changes in neuronal activity, perfusion, and blood oxygenation. This complex mechanism is modulated by basal conditions and physiological properties such as resting neuronal activity, oxygenation, neurovascular coupling, perfusion, and vasomotricity. Age, medication, pathology, capnia, nicotine or caffeine influence BOLD contrast [11, 18, 7, 8, 22]. These confounds need to be controlled in comparative studies across populations. Individually, brain lesions may influence locally BOLD contrast and may lead to inappropriate interpretation [26, 37, 53, 16, 54, 33, 39, 27, 41, 9, 58, 28]. For instance, the vicinity of a tumor decreases adjacent activation even for a distance greater than 10 mm [39, 27, 41, 58, 28]. In preoperative fMRI, BOLD signal is impaired in patients with higher grade gliomas and meningiomas [26, 54, 39, 41, 9, 28]. However, this impairment is not detected beside low grade gliomas [28], suggesting a higher reliability of fMRI in such population which could expect a greater benefit of complete tumoral resection [14].

As BOLD contrast impairment might underestimate the local neuronal activity, discordance in language lateralization has been reported when compared to the Wada test [37, 53, 58]. In fact, interhemispheric comparisons are not necessary accurate to assess cortical reorganization when unilateral lesion or vascular impairment is detected. In such cases, BOLD contrast should be at least tested (Fig. 5) [23, 33, 28].

Impaired BOLD contrast may reveal change in oxygenation as suggested by combined fMRI-NIRS studies [50]. During motor tasks in patients with primary brain neoplasms or stroke, an unexpected deoxyHb increase has been observed in the ipsilesional eloquent cortex [16]. Main hypotheses are: (1) a decreased oxygen delivery, (2) an increased oxygen extraction that may due to impaired hemodynamics including an increased blood transit time, and (3) changes in venous oxygenation and blood volume.

Besides oxygenation disorders in the vicinity of brain lesions, impaired vasomotor responses have been previously proposed to explain BOLD discrepancies [11, 18]. Schematically, pathophysiological alterations might be secondary to changes in: (1) the functional mechanisms that link a specific stimulus and a vasomotor response (neurovascular coupling in response to neural activity, vasoreactivity to circulating gases, and autoregulation to perfusion pressure); (2) the quality of the hemodynamic responses that might be affected by loco-regional changes in basal perfusion and structural abnormalities of the vasculature.

A functional hypothesis would rely on a selective dysfunction of the physiological properties of the brain vessels. Although partially elucidated, neurovascular coupling and vasoreactivity that are mediated by common metabolic changes, such as NO and H^+, rely on the integrity of the brain blood barrier (BBB) [18, 34]. Thus, BBB disruption may cause BOLD contrast impairment as previously suggested in stroke [11, 49, 33, 26, 9, 28].

A hemodynamic hypothesis would be supported by regional perfusion changes either due to the lesional vascularization, or to structural alterations of the surrounding brain vessels. In line with experimental data in healthy subjects showing that BOLD signal may decrease as CBV increases [8], local changes in basal perfusion have been advocated to explain BOLD contrast variations in patients with tumor [27, 41, 28], stroke [1], and arteriovenous malformation [42] (Fig. 5).

The regional effects of the hypervascularization of meningiomas and of neo-angiogenesis in HGG have been previously reported to explain decreased fMRI activations [26, 27, 41, 9]. During hemodynamic responses to vasomotor stimuli, the functional hyperperfusion would be partially absorbed by the surrounding hyperperfusion of the tumor, mimicking a

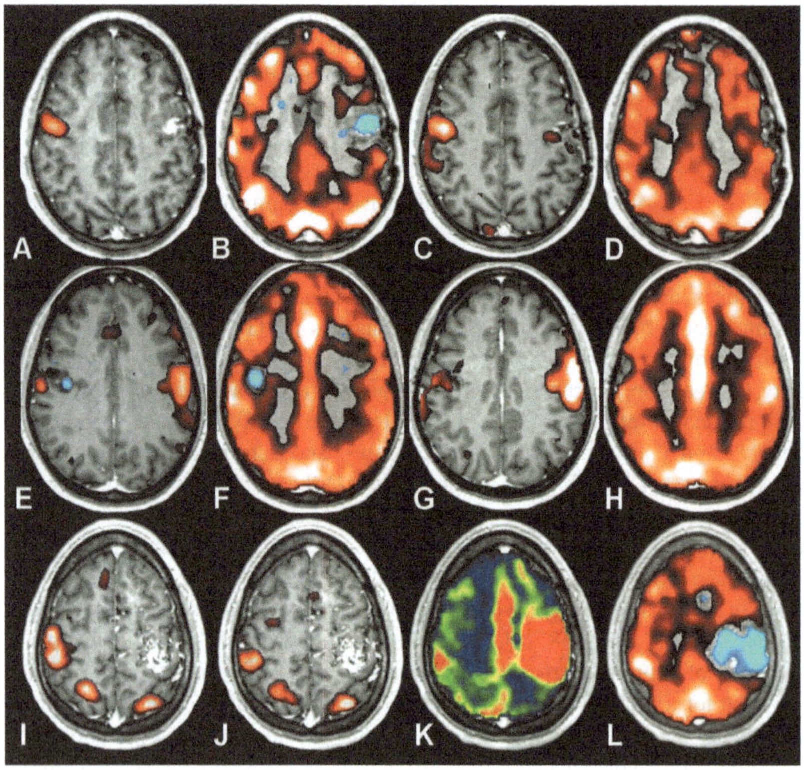

Fig. 5. "False negative" preoperative fMRI

During fMRI, positive changes of the BOLD contrast are expected close to the eloquent cortex. When absent, a perilesional BOLD contrast impairment has to be suggested. A negative activation could to be detected. Perfusion and vasoreactivity imaging would be helpful to better characterize these disorders.

(A–D): Patient with a recurrence of a xantho-astrocytoma of the left precentral gyrus, revealed by seizures of the right hemiface. No preoperative facial palsy was detected. Before surgery (**A–B**), lips movements did not elicit primary sensorimotor activation along the central sulcus posteriorly to the tumor, contrary to the contratumoral hemisphere (**A**). The vasoreactivity map using BOLD contrast during a capnic challenge showed a broad cortical response in red, expected for an inverted peritumoral response in blue covering the eloquent cortex (**B**). Postoperative fMRI (**C–D**) showed a "recovery" of the primary sensorimotor activation adjacent to the partial resection (**C**), in line with the recovery of the peritumoral vasoreactivity (**D**).

(E–H): Patient with a ganglioglioma of the right precentral gyrus without facial palsy. Before surgery (**E–F**) and beside the tumor, lips movements elicited postcentral sensory activation with a negative precentral motor activation (**E**). The vasoreactivity map showed a focal negative response over the eloquent cortex, confirmed by peroperative electrical stimulations (**F**). Postoperative fMRI (**G–H**) showed a "recovery" of both sensorimotor activation (**G**) and vasoreactivity (**H**) along the central sulcus, adjacent to the resection.

In both cases, perfusion of the eloquent cortex was normal. A perilesional vasomotor disorder was likely in the vicinity of BBB disruption.

(I–L): Patient with a left precentral arteriovenous malformation (AVM). Movements of the left fingers elicited right primary sensorimotor activation (**I**) whereas right fingers did not elicited activation along the left central sulcus posterior the AVM (**J**). Perfusion imaging showed a major increase of the CBF (**K**) and vasoreactivity map showed a broad perilesional impairment of the BOLD contrast with a negative response to the hypercapnic challenge (**L**) functional steal phenomenon. Although inconstant, local increase of CBV has also been reported adjacent to the tumor [49, 9]. This hyperperfusion could be due to the tumoral hypervascularisation or a compensating autoregulation which maintains perfusion pressure by compensating the decrease of the mean arterial pressure due to a steal phenomenon or the local mass effect. Autoregulation could maintain basal perfusion but the vasodilatation would be partially exhausted, limiting the range of both potential perfusion increase and evoked BOLD signals. Besides arteriolar disorders, impaired venous drainage and stasis may also impair BOLD contrast. Finally, increased MTT could also be responsible of a decrease in BOLD signal secondary to an increase in oxygen extraction and deoxyHb concentration [50].

Structural alterations of brain vessels might also impair morphological changes elicited by functional stimuli. In such case, one may expect that all functional properties could be affected whatever the functional stimuli used. These alterations could be due to an increase in vascular stiffness that might lead to a chronic hypoperfusion. It has been previously suggested in patients with cerebrovascular disorders including stroke, diabetes mellitus, chronic hypertension and ageing [11, 23, 49, 33, 18], and even in patients with neurodegenerative disorders such as Alzheimer disease [18,051]. In patients with tumors, structural abnormalities of the microvessels were described in case of brain edema [57].

In brain-lesioned patients, quantitative fMRI using the calibrated-BOLD method seems to be a promising technique to better estimate the underlying neuronal activity. However, this method remains difficult to conduct routinely in patients. An alternative could be to test the quality of the BOLD contrast using a mild and simple respiratory challenge such as hyperventilation [33], CO_2 or carbogen inhalation (a gas mixture containing 5–7% CO_2 in 93–95% O_2) [55, 28]. Indeed, imaging of cerebral vasoreactivity (CVR) using BOLD signal to carbogen and CO_2 inhalation has been tested, especially in patients with vascular disorders and false negative fMRI results [23, 28]. This approach provides BOLD maps that overlap 95% of the functional activation. In patients with stroke and tumors, regional asymmetries in eloquent areas detected on CVR maps were the best predictors for impaired motor activation [33, 28].

As BOLD fMRI aggregates evoked blood oxygenation and functional changes in brain perfusion, fMRI interpretation might remain equivoqual, especially in case of focal lesions that modify these parameters. Multimodal advanced imaging, including DTI to better detect peritumoral infiltration and edema, perfusion study with permeability and vessel size imaging, functional imaging of the perfusion using arterial spin labeling during vasomotor challenge, oxygenation imaging using MRI or NIRS, could be proposed to better interpret fMRI data in patients, and to better understand structural and functional changes in the vicinity of brain lesions.

Conclusion

BOLD fMRI relies on a solid but partially understood biophysical framework, which is sensitive to general and regional pathophysiological changes Since BOLD fMRI is widely available, brain mapping is performed daily in patients before surgery, and in humans or animals to better depict cortical changes during cognitive challenges. To face a large demand, manufacturers and research institutions have broadly provided efficient tools to easily produce activation maps. However fMRI remains a complex technical and physiological challenge. Indeed, data acquisition and interpretation impose a rigorous practice.

To remain a reliable technique for brain mapping, one must remember that fMRI is an indirect functional neuroimaging technique that implies 4 trivial but major constraints to keep in mind to interpret results.

First and like any other imaging technique, the native images have to be watched, especially in brain-lesioned patients and when a fascinating neuropsychological or ethical question is addressed. Definitely, instantaneous colorful maps tend to "hide" the native images. Forgetting to carefully watch images first is probably the most common source of misinterpretation.

Second, this neuroimaging technique with a millimetric spatial resolution requires a solid knowledge in neuroanatomy in order to be able to precisely locate brain activation at the individual level, especially beside brain lesion. Stereotactic atlases may be useful to provide a general picture of the results, but such approach is not sufficient and even dangerous at the individual level.

Third and because of its functional purpose, fMRI relies on the appropriate perception of the stimuli and execution of the task. This survey is necessary during the data acquisition.

Now, most manufacturers provide simple and useful tools to perform "real-time" fMRI, in order to monitor images, and provide initial results for block-design paradigms during the data acquisition.

Fourth, BOLD fMRI provides indirect pictures of the neuronal activity by measuring at a much larger spatiotemporal scale changes in blood oxygenation and brain perfusion. Additional measurements of the perfusion and mapping BOLD contrast independently of the neuronal activity could be necessary; especially in patients.

Thereafter, EEG-fMRI, resting state fMRI, calibrated-BOLD fMRI, perfusion fMRI, and diffusion fMRI are promising but challenging techniques. Today, their validations by confrontation with invasive procedures remain under investigation.

Abbreviations

ASL:	arterial spin labeling
BOLD:	blood oxygenation level dependent
CBF:	cerebral blood flow
CBV:	cerebral blood volume
CMRO2:	change in cerebral metabolic rate of O_2
CVR:	cerebral vasoreactivity
deoxyHb:	deoxyhemoglobin
DICOM:	digital imaging and communication in medicine
EEG-fMRI:	electroencephalography combined with fMRI
EPI:	echo planar imaging
fMRI:	functional magnetic resonance imaging
GRE:	gradient recalled echo
MTT:	mean transit time
NIRS:	near infrared spectroscopy
OEF:	oxygen extraction fraction
SE:	spin echo
WI:	weighted image
PET:	positron emission tomography

References

[1] Altamura C, Reinhard M, Vry MS, Kaller CP, Hamzei F, Vernieri F, Rossini PM, Hetzel A, Weiller C, Saur D (2009) The longitudinal changes of BOLD response and cerebral hemodynamics from acute to subacute stroke. A fMRI and TCD study. BMC Neurosci 10: 151

[2] Bandettini PA, Jesmanowicz A, Wong EC, Hyde JS (1993) Processing strategies for time-course data sets in functional MRI of the human brain. Magn Reson Med 30: 161-173

[3] Bartos R, Jech R, Vymazal J, Petrovicky P, Vachata P, Hejcl A, Zolal A, Sames M (2009) Validity of primary motor area localization with fMRI versus electric cortical stimulation: a comparative study. Acta Neurochir (Wien) 151: 1071-1080

[4] Belliveau JW, Kwong KK, Kennedy DN, Baker JR, Stern CE, Benson R, Chesler DA, Weisskoff RM, Cohen MS, Tootell RB et al (1992) Magnetic resonance imaging mapping of brain function. Human visual cortex. Invest Radiol 27 (Suppl) 2: S59-65

[5] Boxerman JL, Bandettini PA, Kwong KK, Baker JR, Davis TL, Rosen BR, Weisskoff RM (1995) The intravascular contribution to fMRI signal change: Monte Carlo modeling and diffusion-weighted studies in vivo. Magn Reson Med 34: 4-10

[6] Brown GG, Eyler Zorrilla LT, Georgy B, Kindermann SS, Wong EC, Buxton RB (2003) BOLD and perfusion response to finger-thumb apposition after acetazolamide administration: differential relationship to global perfusion. J Cereb Blood Flow Metab 23: 829-837

[7] Brown GG, Perthen JE, Liu TT, Buxton RB (2007) A primer on functional magnetic resonance imaging. Neuropsychol Rev 17: 107-125

[8] Buxton RB (2009) Introduction to functional magnetic resonance imaging. 2nd edn. Cambridge University Press, Cambridge

[9] Chen CM, Hou BL, Holodny AI (2008) Effect of age and tumor grade on BOLD functional MR imaging in preoperative assessment of patients with glioma. Radiology 248: 971–978

[10] Chiarelli PA, Bulte DP, Wise R, Gallichan D, Jezzard P (2007) A calibration method for quantitative BOLD fMRI based on hyperoxia. Neuroimage 37: 808–820

[11] D'Esposito M, Deouell LY, Gazzaley A (2003) Alterations in the BOLD fMRI signal with ageing and disease: a challenge for neuroimaging. Nat Rev Neurosci 4: 863–872

[12] Davis TL, Kwong KK, Weisskoff RM, Rosen BR (1998) Calibrated functional MRI: mapping the dynamics of oxidative metabolism. Proc Natl Acad Sci USA 95: 1834–1839

[13] Devor A, Dunn AK, Andermann ML, Ulbert I, Boas DA, Dale AM (2003) Coupling of total hemoglobin concentration, oxygenation, and neural activity in rat somatosensory cortex. Neuron 39: 353–359

[14] Duffau H, Lopes M, Arthuis F, Bitar A, Sichez JP, Van Effenterre R, Capelle L (2005) Contribution of intraoperative electrical stimulations in surgery of low grade gliomas: a comparative study between two series without (1985–96) and with (1996–2003) functional mapping in the same institution. J Neurol Neurosurg Psychiatry 76: 845–851

[15] Fox MD, Raichle ME (2007) Spontaneous fluctuations in brain activity observed with functional magnetic resonance imaging. Nat Rev Neurosci 8: 700–711

[16] Fujiwara N, Sakatani K, Katayama Y, Murata Y, Hoshino T, Fukaya C, Yamamoto T (2004) Evoked-cerebral blood oxygenation changes in false-negative activations in BOLD contrast functional MRI of patients with brain tumors. Neuroimage 21: 1464–1471

[17] Gasser T, Sandalcioglu E, Schoch B, Gizewski E, Forsting M, Stolke D, Wiedemayer H (2005) Functional magnetic resonance imaging in anesthetized patients: a relevant step toward real-time intraoperative functional neuroimaging. Neurosurgery 57: 94–99; discussion 94–99

[18] Girouard H, Iadecola C (2006) Neurovascular coupling in the normal brain and in hypertension, stroke, and Alzheimer disease. J Appl Physiol 100: 328–335

[19] Gotman J (2008) Epileptic networks studied with EEG-fMRI. Epilepsia 49 (Suppl) 3: 42–51

[20] Grouiller F, Vercueil L, Krainik A, Segebarth C, Kahane P, David O (2010) Characterization of the hemodynamic modes associated with interictal epileptic activity using a deformable model-based analysis of combined EEG and functional MRI recordings. Hum Brain Mapp (in press)

[21] Hajnal JV, Myers R, Oatridge A, Schwieso JE, Young IR, Bydder GM (1994) Artifacts due to stimulus correlated motion in functional imaging of the brain. Magn Reson Med 31: 283–291

[22] Haller S, Bartsch AJ (2009) Pitfalls in FMRI. Eur Radiol 19: 2689–2706

[23] Hamzei F, Knab R, Weiller C, Rother J (2003) The influence of extra- and intracranial artery disease on the BOLD signal in FMRI. Neuroimage 20: 1393–1399

[24] Hertz-Pannier L, Gaillard WD, Mott SH, Cuenod CA, Bookheimer SY, Weinstein S, Conry J, Papero PH, Schiff SJ, Le Bihan D, Theodore WH (1997) Noninvasive assessment of language dominance in children and adolescents with functional MRI: a preliminary study. Neurology 48: 1003–1012

[25] Hoge RD, Atkinson J, Gill B, Crelier GR, Marrett S, Pike GB (1999) Linear coupling between cerebral blood flow and oxygen consumption in activated human cortex. Proc Natl Acad Sci USA 96: 9403–9408

[26] Holodny AI, Schulder M, Liu WC, Wolko J, Maldjian JA, Kalnin AJ (2000) The effect of brain tumors on BOLD functional MR imaging activation in the adjacent motor cortex: implications for image-guided neurosurgery. Am J Neuroradiol 21: 1415–1422

[27] Hou BL, Bradbury M, Peck KK, Petrovich NM, Gutin PH, Holodny AI (2006) Effect of brain tumor neovasculature defined by rCBV on BOLD fMRI activation volume in the primary motor cortex. Neuroimage 32: 489–497

[28] Jiang Z, Krainik A, David O, Salon C, Tropres I, Hoffmann D, Pannetier N, Barbier EL, Bombin ER, Warnking J, Pasteris C, Chabardes S, Berger F, Grand S, Segebarth C, Gay E, Le Bas JF (2010) Impaired fMRI activation in patients with primary brain tumors. Neuroimage 52: 538–548

[29] Kim MJ, Holodny AI, Hou BL, Peck KK, Moskowitz CS, Bogomolny DL, Gutin PH (2005) The effect of prior surgery on blood oxygen level-dependent functional MR imaging in the preoperative assessment of brain tumors. Am J Neuroradiol 26: 1980–1985

[30] Krainik A, Lehericy S, Duffau H, Vlaicu M, Poupon F, Capelle L, Cornu P, Clemenceau S, Sahel M, Valery CA, Boch AL, Mangin JF, Bihan DL, Marsault C (2001) Role of the supplementary motor area in motor deficit following medial frontal lobe surgery. Neurology 57: 871–878

[31] Krainik A, Lehericy S, Duffau H, Capelle L, Chainay H, Cornu P, Cohen L, Boch AL, Mangin JF, Le Bihan D, Marsault C (2003) Postoperative speech disorder after medial frontal surgery: role of the supplementary motor area. Neurology 60: 587–594

[32] Krainik A, Duffau H, Capelle L, Cornu P, Boch AL, Mangin JF, Le Bihan D, Marsault C, Chiras J, Lehericy S (2004) Role of the healthy hemisphere in recovery after resection of the supplementary motor area. Neurology 62: 1323–1332

[33] Krainik A, Hund-Georgiadis M, Zysset S, von Cramon DY (2005) Regional impairment of cerebrovascular reactivity and BOLD signal in adults after stroke. Stroke 36: 1146–1152

[34] Lavi S, Gaitini D, Milloul V, Jacob G (2006) Impaired cerebral CO2 vasoreactivity: association with endothelial dysfunction. Am J Physiol Heart Circ Physiol 291: H1856–1861

[35] Le Bihan D (2007) The "wet mind": water and functional neuroimaging. Phys Med Biol 52: R57–90

[36] Lehericy S, Duffau H, Cornu P, Capelle L, Pidoux B, Carpentier A, Auliac S, Clemenceau S, Sichez JP, Bitar A, Valery CA, Van Effenterre R, Faillot T, Srour A, Fohanno D, Philippon J, Le Bihan D, Marsault C (2000) Correspondence between functional magnetic resonance imaging somatotopy and individual brain anatomy of the central region: comparison with intraoperative stimulation in patients with brain tumors. J Neurosurg 92: 589–598

[37] Lehericy S, Biondi A, Sourour N, Vlaicu M, du Montcel ST, Cohen L, Vivas E, Capelle L, Faillot T, Casasco A, Le Bihan D, Marsault C (2002) Arteriovenous brain malformations: is functional MR imaging reliable for studying language reorganization in patients? Initial observations. Radiology 223: 672–682

[38] Liu TT, Frank LR, Wong EC, Buxton RB (2001) Detection power, estimation efficiency, and predictability in event-related fMRI. Neuroimage 13: 759–773

[39] Liu WC, Feldman SC, Schulder M, Kalnin AJ, Holodny AI, Zimmerman A, Sinensky R, Rao S (2005) The effect of tumour type and distance on activation in the motor cortex. Neuroradiology 47: 813–819

[40] Logothetis NK (2008) What we can do and what we cannot do with fMRI. Nature 453: 869–878

[41] Ludemann L, Forschler A, Grieger W, Zimmer C (2006) BOLD signal in the motor cortex shows a correlation with the blood volume of brain tumors. J Magn Reson Imaging 23: 435–443

[42] Mangia S, Di Salle F, Garreffa G, Esposito F, Giove F, Cirillo S, Scarabino T, Morrone R, Maraviglia B (2004) Perfusion- and BOLD-based fMRI in the study of a human pathological model for task-related flow reductions. Brain Res Bull 63: 1–5

[43] Matthews PM, Honey GD, Bullmore ET (2006) Applications of fMRI in translational medicine and clinical practice. Nat Rev Neurosci 7: 732–744

[44] Miller G (2008) Neuroimaging. Growing pains for fMRI. Science 320: 1412–1414

[45] Ogawa S, Lee TM, Kay AR, Tank DW (1990) Brain magnetic resonance imaging with contrast dependent on blood oxygenation. Proc Natl Acad Sci USA 87: 9868–9872

[46] Ogawa S, Tank DW, Menon R, Ellermann JM, Kim SG, Merkle H, Ugurbil K (1992) Intrinsic signal changes accompanying sensory stimulation: functional brain mapping with magnetic resonance imaging. Proc Natl Acad Sci USA 89: 5951–5955

[47] Ramsey NF, Kirkby BS, Van Gelderen P, Berman KF, Duyn JH, Frank JA, Mattay VS, Van Horn JD, Esposito G, Moonen CT, Weinberger DR (1996) Functional mapping of human sensorimotor cortex with 3D BOLD fMRI correlates highly with H2(15)O PET rCBF. J Cereb Blood Flow Metab 16: 755–764

[48] Rombouts SA, Goekoop R, Stam CJ, Barkhof F, Scheltens P (2005) Delayed rather than decreased BOLD response as a marker for early Alzheimer's disease. Neuroimage 26: 1078–1085

[49] Rossini PM, Altamura C, Ferretti A, Vernieri F, Zappasodi F, Caulo M, Pizzella V, Del Gratta C, Romani GL, Tecchio F (2004) Does cerebrovascular disease affect the coupling between neuronal activity and local haemodynamics? Brain 127: 99–110

[50] Sakatani K, Murata Y, Fujiwara N, Hoshino T, Nakamura S, Kano T, Katayama Y (2007) Comparison of blood-oxygen-level-dependent functional magnetic resonance imaging and near-infrared spectroscopy recording during functional brain activation in patients with stroke and brain tumors. J Biomed Opt 12: 062110

[51] Silvestrini M, Pasqualetti P, Baruffaldi R, Bartolini M, Handouk Y, Matteis M, Moffa F, Provinciali L, Vernieri F (2006) Cerebrovascular reactivity and cognitive de-

cline in patients with Alzheimer disease. Stroke 37: 1010–1015

[52] Stippich C, Freitag P, Kassubek J, Soros P, Kamada K, Kober H, Scheffler K, Hopfengartner R, Bilecen D, Radu EW, Vieth JB (1998) Motor, somatosensory and auditory cortex localization by fMRI and MEG. Neuroreport 9: 1953–1957

[53] Ulmer JL, Krouwer HG, Mueller WM, Ugurel MS, Kocak M, Mark LP (2003) Pseudo-reorganization of language cortical function at fMR imaging: a consequence of tumor-Induced neurovascular uncoupling. Am J Neuroradiol 24: 213–217

[54] Ulmer JL, Hacein-Bey L, Mathews VP, Mueller WM, DeYoe EA, Prost RW, Meyer GA, Krouwer HG, Schmainda KM (2004) Lesion-induced pseudo-dominance at functional magnetic resonance imaging: implications for preoperative assessments. Neurosurgery 55: 569–579; discussion 580–561

[55] van der Zande FH, Hofman PA, Backes WH (2005) Mapping hypercapnia-induced cerebrovascular reactivity using BOLD MRI. Neuroradiology 47: 114–120

[56] Van Horn JD, Poldrack RA (2009) Functional MRI at the crossroads. Int J Psychophysiol 73: 3–9

[57] Vaz R, Borges N, Sarmento A, Azevedo I (1996) Reversion of phenotype of endothelial cells in brain tissue around glioblastomas. J Neurooncol 27: 127–132

[58] Wellmer J, Weber B, Urbach H, Reul J, Fernandez G, Elger CE (2009) Cerebral lesions can impair fMRI-based language lateralization. Epilepsia 50: 2213–2224

[59] Yacoub E, Ugurbil K, Harel N (2006) The spatial dependence of the poststimulus undershoot as revealed by high-resolution BOLD- and CBV-weighted fMRI. J Cereb Blood Flow Metab 26: 634–644

[60] Ye Y, Zhuo Y, Xue R, Zhou XJ (2010) BOLD fMRI using a modified HASTE sequence. Neuroimage 49: 457–466

[61] Zou P, Mulhern RK, Butler RW, Li CS, Langston JW, Ogg RJ (2005) BOLD responses to visual stimulation in survivors of childhood cancer. Neuroimage 24: 61–69

Mapping white matter pathways with diffusion imaging tractography: focus on neurosurgical applications

Marco Catani and Flavio Dell'Acqua

Introduction to diffusion magnetic resonance imaging

Diffusion magnetic resonance imaging (MRI) tractography is a method that can be used for the *in vivo* quantification of tissue microstructural integrity and the virtual reconstruction of white matter pathways [3]. The diffusion-weighted MRI pulse sequences are sensitive to the displacement of water molecules within biological tissues [2]. Generally the displacement of water molecules follows Einstein's equation, where the mean squared displacement $\langle r^2 \rangle$ is directly proportional to the observation time (t) according to:

$$\langle r^2 \rangle = 6Dt.$$

Given a body temperature of 37 °C, the diffusion coefficient of free water (D) is about 3×10^{-3} mm^2s^{-1}. Thus, within a voxel containing, for example, cerebrospinal fluid, where D is very close to the coefficient of free water, the water molecules will displace randomly for an average distance of 20 μm in all directions in about 20 ms. This is true only if the water molecules are allowed to move freely, which is not the case for the water contained in the nervous tissue, where the presence of cell membranes, proteins, myelin, intracellular filaments, organelles, and so forth, hinders the water displacement. It follows that the displacement and, therefore, the diffusion coefficient in brain tissue are smaller than those in free water. This is why the term "Apparent Diffusion Coefficient" (ADC) was coined, to reflect in part the fact that in the tissue the diffusion of water is hindered by several biological barriers. The hindering effect of these biological barriers can change in brain disorders (Le Bihan et al, 1986) [38a]. Pathological changes that occur, for example, early in ischemic tissue affect the diffusivity of water, leading to a general reduction of the ADC. In this case diffusion MRI offers the possibility of visualizing stroke-related changes within an earlier time window compared to other structural MRI sequences [49].

In diffusion MRI, the signal is usually sensitized to the displacement of water molecules along a selected direction. The ADC measurement is therefore strictly dependent on the chosen direction. If the composition of the tissue is *isotropic* (i.e., its physical properties are identical in all directions), the diffusion of water molecules is reduced equally along all orientations. This is the case, for example, of the grey subcortical nuclei or the cortex (Fig. 1 left).

Other tissues, like the white matter in the cerebrum and the spinal cord are *anisotropic*. Here, the fibres are parallelly oriented and the axonal membranes, together with the myelin sheets, represent the greatest biological barrier to the diffusivity of water. Hence, the ADC measured along a direction perpendicular to the fibers is always lower compared to the ADC measured along the direction of the fibres. Diffusion in anisotropic white matter tissue is therefore characterized by having a preferential direction, which varies according to the main orientation of the fibers (Fig. 1 right). In other words the measurement of the ADC inside the white matter is orientationally variant (Fig. 2 top).

The property of orientational variance of the ADC represents a problem for the exact interpretation of regional changes in the diffusion signal, especially for those lesions localized within specific tracts where the decrease of ADC could simply be related to a different orientation of the fibers [39, 50]. As we will see in the next paragraph this problem was resolved mathematically using the diffusion tensor to model the water diffusion inside each brain voxel.

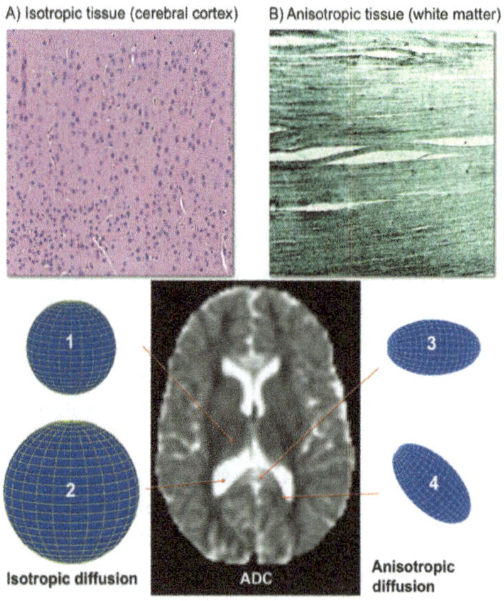

Fig. 1. *Top*: Histology of the cerebral cortex **(A)** and white-matter fibers **(B)**. The two tissues differ for the structural composition and architectural organization of their biological constituents. *Bottom*: Axial ADC map of a human brain and visualization of the corresponding three-dimensional displacement of water molecules in different brain regions: (1) low isotropic diffusivity in the thalamus; (2) high isotropic diffusivity in the cerebrospinal fluid of the lateral ventricles; (3) horizontal anisotropic diffusivity along the midsagittal fibers of the splenium; (4) oblique anisotropic diffusivity along the lateral fibers of the splenium. Note that 3 and 4 have the same isotropic characteristic but the ADC signal is different due to a different orientation of the underlying fibres

Diffusion tensor imaging

In 1994, Peter Basser, James Mattiello, and Denis Le Bihan published a seminal paper on diffusion imaging in which they showed that if diffusion is measured along at least six different directions, it is possible to obtain a mathematical description of the overall displacement of water molecules, the diffusion tensor (DT) [2] (Fig. 2 bottom). The DT provides a synthetic description of the water diffusion in the three-dimensional space and can be visualised as a "diffusion" ellipsoid. This representation describes the geometrical profile of the water displacement and it is defined only by three diffusion coefficients or eigenvalues (λ_1, λ_2, λ_3) and the orientations or eigenvectors (v_1, v_2, v_3) of its three principal axes (Fig. 2 bottom). The DT can be used to extract quantitative indexes that are rotationally invariant (i.e., independent from the orientation of the measurement).

Mean diffusivity (MD) describes the average mobility of water molecules. MD is the average of the three eigenvalues of the tensor:

$$MD = \frac{\lambda_1 + \lambda_2 + \lambda_3}{3}.$$

In the brain, normal values range from more than 2.0 x10^{-3} mm^2/s (in the cerebro-spinal) to 0.6 and 0.9 x10^{-3} mm^2/s for grey and white matter, respectively. MD reduces with age within the early years of life and increases in those disorders characterized by demyelination, inflammation, axonal injury and edema.

Fractional anisotropy (FA) varies from 0 to 1 and represents a quantitative index of the degree of anisotropy. FA is also a rotational invariant like MD and is calculated according to the formula:

$$FA\frac{\sqrt{3\left((\lambda_1-\bar{\lambda})^2(\lambda_2-\bar{\lambda})^2(\lambda_3-\bar{\lambda})^2\right)}}{\sqrt{2\left(\lambda_1^2+\lambda_2^2+\lambda_3^2\right)}},$$

(where $\bar{\lambda} = MD$).

FA gives information about the organization of the microstructure and the integrity of the white matter (e.g., the level of the myelination or the density of white matter fibres).

In white matter voxels with a single fiber orientation the axial diffusivity is defined as the diffusivity along the principal direction of the DT and therefore of the underlying fibre bundle. This measure is sensitive to intraxonal changes and can be defined as:

$$ADC\| = \lambda_1.$$

The radial diffusivity is the diffusivity perpendicular to the direction of λ_1 and therefore reflects the hinderance of the axonal membrane and myelin sheet. The radial diffusivity is defined as:

$$ADC_{\perp} = \frac{\lambda_2+\lambda_3}{2}.$$

The radial diffusivity is particularly sensitive to pathological changes that affect the degree of myelination of fibres or axonal integrity.

The diffusivity indexes described above provide complementary information about the tissue microstructural organization.

A detailed analysis of the ellipsoids can give

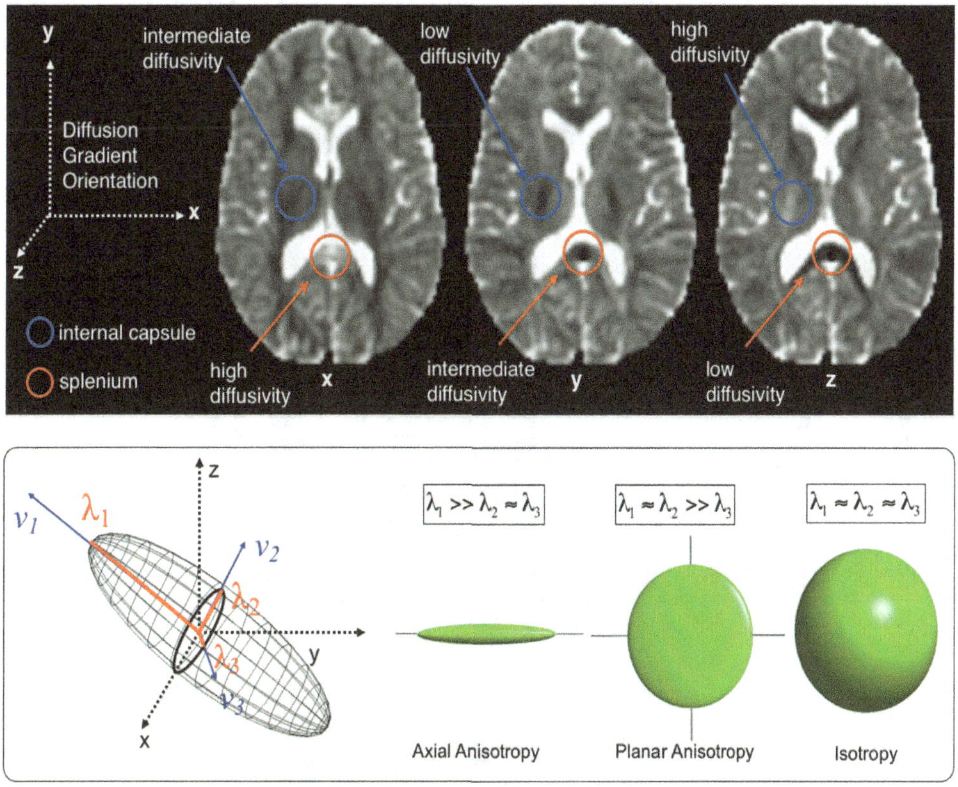

Fig. 2. *Top:* Monodirectional ADC maps where the signal is sensitized to the displacement of water molecules along the three orthogonal planes (x, latero-lateral direction; y, antero-posterior; z, superior-inferior). The fibers of the internal capsule and the splenium have different orientations, therefore their ADC values change according to the direction of the measured diffusivity. *Bottom:* Representation of the diffusion tensor as a diffusion ellipsoid. The size and the shape are completely defined by the three eigenvalues (red), whereas the spatial orientation is described by the three eigenvectors (blue)

precise information about not only the average water molecular displacement (i.e., MD) within a voxel but also the degree of tissue anisotropy (i.e., FA) and the main orientation of the underlying white matter fibers (principal eigenvector or orientation of maximum diffusivity). Thus, for example, two regions like the thalamus and the mid-splenium with similar MD may have different FA (Fig. 3 top). This is well described by their respective tensor ellipsoids that have different shapes (i.e., different FA) but the same average "size" (i.e., same MD) (Fig. 1 and Fig. 3, top).

Conversely, two white matter regions with similar MD and FA like the mid-splenium and lateral splenium, can have different direction of maximum diffusivity (i.e., different fibre orientation) as shown in the principal eigenvector maps or the color-coded maps (Fig. 3 bottom) [59].

Virtual reconstruction of white matter pathways

Compared to established methods for tracing fiber pathways, such as those using axonal tracers, diffusion tensor tractography offers the advantage of being a completely non-invasive technique and therefore it can be applied to study connections of the living human brain. Furthermore, the data required by the tractography process can be readily obtained on standard clinical MRI systems with acquisition times ranging from 5 to 20 minutes. The main assumption underpinning DT tractography is that the diffusion of water molecules inside the brain can be described mathematically by the diffusion tensor and that the principal axis of this tensor aligns with the predominant fiber orientation within each voxel [39]. Tractography algorithms use this information

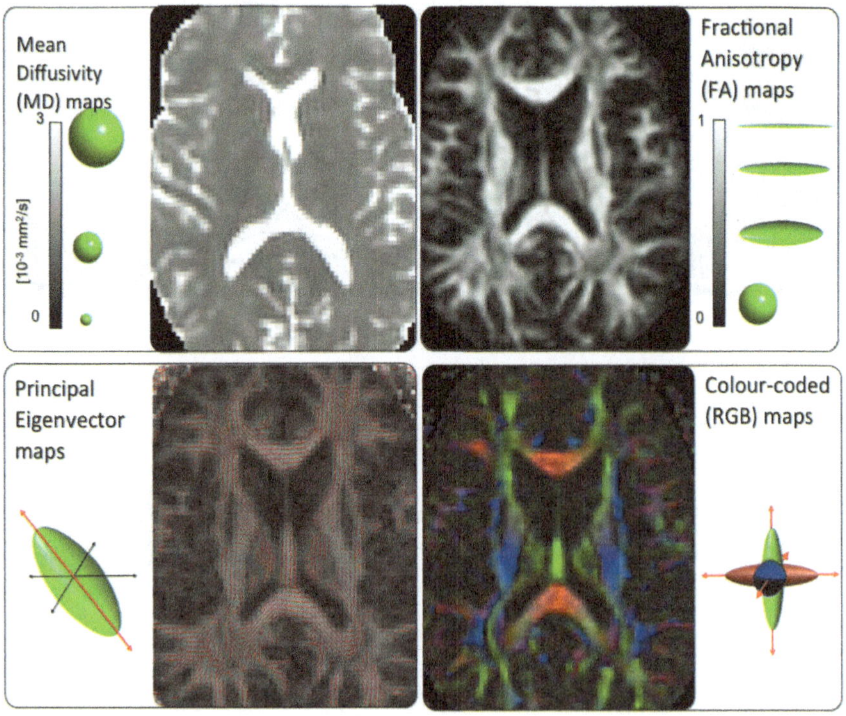

Fig. 3. The diffusion tensor allows to extract quantitative indexes and reconstruct 2D maps that provide information about the microstructural properties of the biological tissues and their organization. Numbers indicate (1) thalamus, (2) mid-splenium and (3) lateral splenium

to track white matter pathways by inferring continuity of fibers from voxel to voxel. This process is achieved using tractography softwares that reconstruct continuous streamlines by following the direction of maximum diffusion from a given voxel into a neighbouring voxel (Fig. 4) [3, 39, 46].

Most of the tractography algorithms use "tracking" and "stopping" rules that reduce errors in the virtual reconstruction of pathways. The most common of these rules are the adoption of angular and anisotropy thresholds to avoid unrealistic fiber bending or tracking outside white-matter regions [3, 47]. Tractography offers the possibility to study *in vivo* the trajectories of white-matter pathways [10] and parcellate the cortex according to its pattern of connectivity [4]. By extracting quantitative diffusion indexes along the dissected tract, it is possible to obtain tract-specific measurements indicative of the microstructural organization, composition, and integrity of the tract of interest. The most used indexes are FA, MD, axial and radial diffusivities. The number of streamlines is also commonly used as a surrogate measure of tract volume, although assuming a direct correspondence with the anatomy is incorrect (e.g., the number of streamlines do

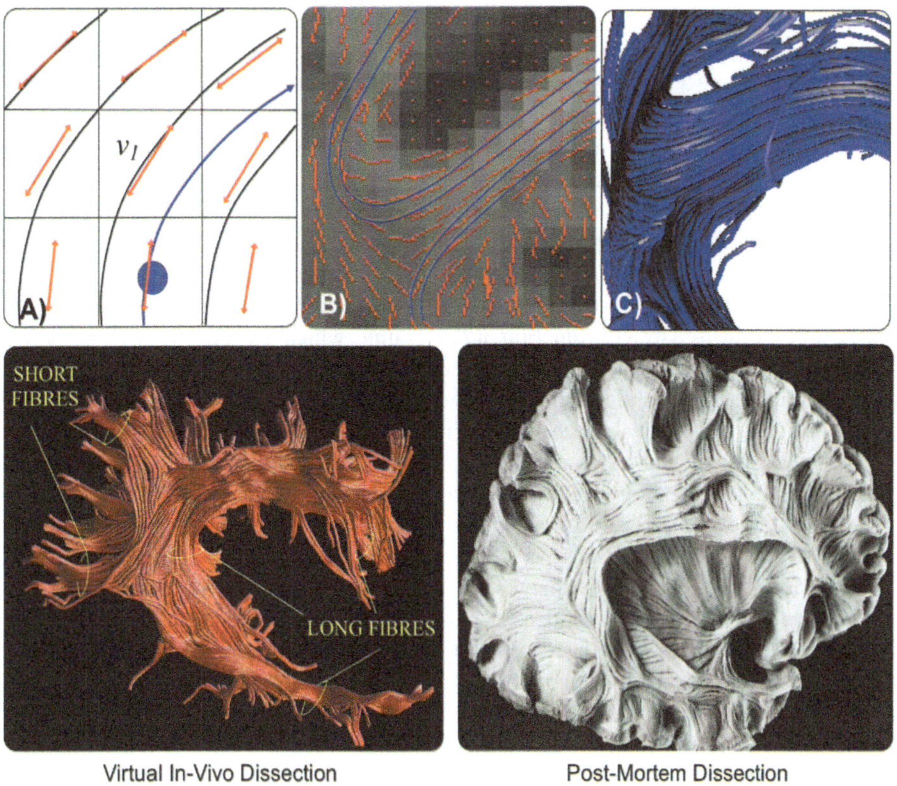

Fig. 4. *Top:* Tracking continuous pathways with diffusion tensor imaging. (**A**) Streamline tractography is based on the assumption that in each white-matter voxel the principal eigenvector (red arrow) is tangent to the main trajectory of the underlying fibers (black lines). Starting from a seed voxel (blue circle) the tractography algorithm propagates, voxel by voxel, a streamline (blue) parallel to the white matter fibers. (**B**) Axial section of the eigenvector map and streamlines (blue) passing through the splenium of the corpus callosum. (**C**) Tractography reconstruction of the splenial streamlines visualized as three-dimensional streamtubes. *Bottom:* Comparison between the virtual in vivo reconstruction (left) of the arcuate fasciculus [10] and the corresponding post-mortem dissection (right) from an atlas of human brain connections [25]

not correspond to the number of axonal fibres). These indexes have been applied in vivo to measure white-matter changes in neurosurgical patients with epilepsy and brain tumor (Fig. 5).

Diffusion signal changes in epilepsy and brain tumor

Biological changes of white-matter tissue in epilepsy and brain tumor are often detectable as changes in the diffusion signal. These changes are associated with three main pathological mechanisms: cytotoxic edema, vasogenic edema and cellularity.

Cytotoxic edema is related to an arrested cellular metabolism due to the reduced availability of energy substrates. The reduction of the energy substrates results in an impaired regulation of the cellular water content with subsequent increase of the intracellular fluid and decrease of the extracellular space. In voxels containing tissue with cytotoxic edema, the water diffusion is most frequently decreased (i.e. reduced MD). In contrast, the vasogenic edema is characterized by a dysfunction of the blood–brain barrier with plasma proteins and other macromolecules passing freely into the perivascular and extracellular space with a consequent increase of the extracellular fluid. Vasogenic edema is therefore frequently associated with increased water diffusion (i.e., increased MD and radial diffusivity and reduced FA). An increase in cellularity is characteristic of brain neoplasms and is associated, especially in more advanced stages, with decreased diffusion secondary to reduced water mobility in the extracellular space. In the early stage of growth of some forms of tumors (e.g., gliomas) the increase in cellularity is associated with water accumulation in the extracellular space leading to increased diffusion of water.

In patients with epilepsy, a characteristic pattern of modifications of the diffusion signal is observed, with decreased diffusivity in the early postictal phase, followed by normalization and then transient or chronic elevation [63]. The early reduction in diffusivity is related to the presence of cytotoxic edema [72], whilst the delayed increase is probably secondary to the vasogenic edema [55, 62]. The intensity of the diffusivity changes is related to the severity of the seizure activity [55, 67, 72], with prolonged seizures potentially causing permanent alterations [77]. These changes are to be considered as a possible consequence of cellular death [73] and subsequent membrane cytolysis, which causes an increase in the extracellular space and therefore in diffusivity [28, 66].

In patients with brain tumors, the peritumoral edema is predominantly vasogenic in nature and is related to a dysfunction of the blood–brain barrier. The vasogenic edema determines an increase in diffusivity and decrease in FA [51]. The cause of vasogenic edema in peritumoral tissue is not well understood, but it is probably related to the leakiness of the blood–brain barrier due to a defect of the inter-endothelial tight junctions. Another possible mechanism acting in more aggressive tumors is related to the increase of proteins in the extracellular space due to necrosis and gliosis. This results in the increase of the extracellular volume, which may offer low resistance and, thus, facilitate the passage of fluid from the intracapillary to the extracellular space [57]. Once in the extracellular space inside the tumor, the fluid has to cross a rim of gliotic tissue around the tumor to reach the perilesional parenchyma. The gliotic tissue presents a significant barrier to the bulk flow motion of the extracellular fluid and it is likely to determine an increase of the extracellullar pressure inside the tumor. Eventually, the edema fluid creates paths for its movement through the gliotic rim into the peritumoral parenchyma and the vasogenic edema spreads by bulk flow through the white-matter fibers [68].

It is likely that such changes in diffusivity are not homogeneous across the pathological tissue due to the presence of other factors affecting the distribution of the edema. For example, compared to low-grade tumors (e.g., diffuse astrocytomas), high-grade gliomas have lower diffusivity probably due to increased cellarity [15, 26]. These findings suggest that diffusion imaging could potentially be used to differentiate between tumor grades and indirectly quantify the degree of infiltration. Recent data from both animal models and human studies suggest that diffusion imaging may be sensitive to predict the response to therapy [14], but further studies are needed to evaluate the clinical potentials of the technique.

Finally it is important to highlight that modifications of the tumoral tissue and surrounding white matter affect the diffusion signal and therefore the ability to reconstruct pathways. In other words in pathological tissue the orientation of the DT may not reflect the orientation of the underlying fibres. This aspect suggests a possible dissociation between virtual reconstructions of pathways and the real anatomy of fibres in neurosurgical patients.

Clinical applications in neurosurgical patients with brain tumours

The use of tractography for mapping eloquent white-matter pathways for neurosurgical planning and neuronavigation was one of the first clinical applications of the technique [16, 32, 48]. Early attempts to utilize tractography in patients with tumors undergoing surgery focused on the demonstration of motor pathway distortion or disruption as a result of mass effect or infiltration [16]. One of the limitations of the use of tractography for the reconstruction of the motor pathways is the inability to visualise tracts originating from the lateral motor cortex. This problem limits the application of tractography only to patients where the most medial component of the cortico-spinal tract (CST) connecting to the leg and trunk area is affected.

Direct comparison of tractography of the motor pathways with intraoperative cortical electrostimulation shows that a number of factors can affect the reconstruction of the CST and generate pathways that have no direct correspondence with the exact location of the functional fibres [48]. Berman et al [8] demonstrated that stimulation sites for electro-cortical mapping could be used to seed tractography of the motor pathway, successfully reaching the cerebral peduncle in 16 out of 27 stimulation site locations in patients with glioma. Edema resulted in incomplete or diverted trajectories for 5 sites, suggesting as possible mechanism the expansion of the extracellular space and reduced anisotropy. This is therefore an important confounder that needs to be considered, as previously suggested by Clark et al

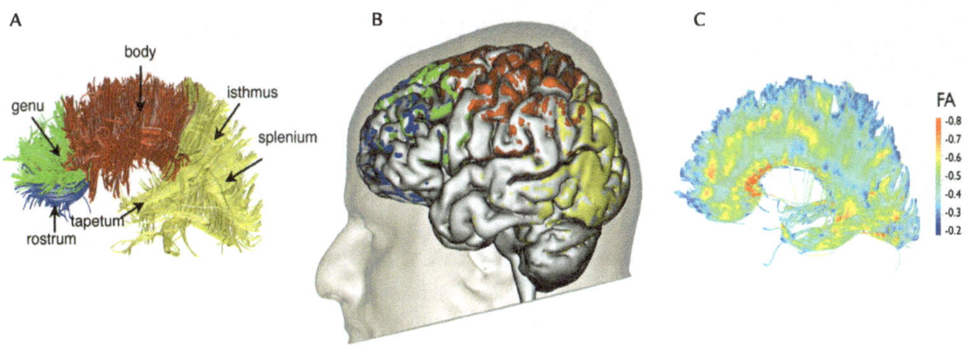

Fig. 5. Possible applications of DT tractography to the study of the white-matter anatomy and microstructural organization of tracts. The corpus callosum is used as an example. **(A)** Virtual dissections allow to reconstruct and study the three-dimensional trajectories of the major divisions of the corpus callosum. **(B)** Cortical regions can be segmented according to the projections of the different subcomponents of the corpus callosum. **(C)** Tract-specific measurements (in this case of FA) along the pathways can be used to obtain quantitative measurements and perform case-control comparisons in pathological conditions affecting white matter anatomy (Second image courtesy of Dr. Michel Thiebaut de Schotten)

[16]. Figure 6 shows an example of the CST deviation in a patient with brain tumor on the left hemisphere. Kinoshita et al [37] used a relatively high FA threshold of 0.3, which may have partly contributed to their finding that tractography underestimates the size of the motor pathways.

Berman et al [9] found that the mean distance between subcortical stimulation sites and the tractography-derived motor pathway was 8.7 ± 3.1 mm for 16 stimulation sites in nine patients with gliomas. Mikuni et al [45] compared cortical electrostimulation with tractography for 40 patients undergoing surgery for treatment of brain tumors located near the motor pathways. In 18 of 20 patients, motor evoked potentials were elicited from the subcortex within 1 cm of the reconstructed motor pathways, and in the remaining 20 patients the distance between the motor pathway and the stimulated subcortex was more than 1 cm, but with motor evoked potentials detected in only 3 of the patients. Mikuni et al emphasized that tractography and intraoperative cortical electrostimulation are complementary techniques [45] and can lead to better outcomes compared to those obtained with cortical electrostimulation [13, 22, 35] or tractography alone [7, 76]. Furthermore, these authors suggest that tractography can be used to identify initial sites for cortical electrostimulation, allowing a more rapid localization of eloquent cortex during surgery.

Tractography can be integrated into neuronavigation systems (Fig. 6) [31, 53, 54, 56]. Nimsky et al demonstrated that the intraoperative shifting of white-matter tracts, such as the internal capsule, can range from −8 mm to +15 mm, patients undergoing glioma surgery. These findings emphasize the importance of an intraoperative update of navigation systems during resection of deep-seated tumor por-

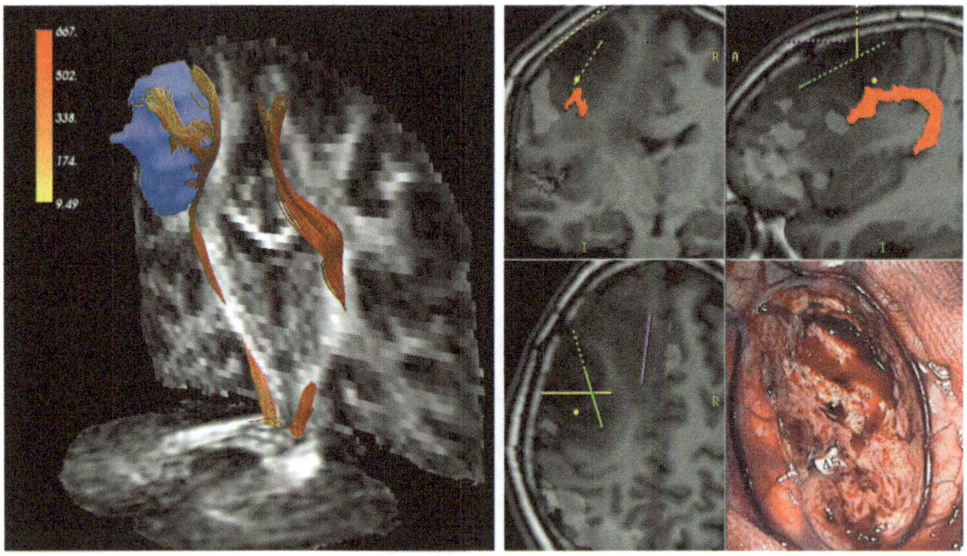

Fig. 6. *Left*: Deviation of the tractography reconstruction of the left corticospinal tract in the proximity of a glioblastoma multiforme (in light blue) in the left perirolandic area. Note that the anisotropy along the left corticospinal tract is reduced in the segment close to the tumor. The two corticospinal tracts are overlayed on coronal and axial FA maps. *Right*: Tractography reconstruction of the arcuate fasciculus (red) exported into the neuronavigator system to assist the neurosurgeon during the resection of a fibrillary astrocytoma infiltrating the left dorsolateroprefrontal cortex of a 27-year-old subject. Validation of tractography results was made with intraoperative subcortical electrostimulation: stimulation in the floor of the operative cavity generated arrest of speech during a verb generation task. The coordinates of the site of stimulation (tag no. 45) with a bipolar electrode is reported as a yellow dot on the neuronavigator's three-dimensional display (Courtesy of Drs. Carlo Marras and Alberto Bizzi, Fondazione IRCCS – Istituto Neurologico "Carlo Besta", Milan, Italy)

tions near eloquent brain areas [53]. In their study of the motor pathways, Nimsky et al found an intra- and interobserver variability of up to 1 mm and 2–3 mm respectively with regard to the rendered volume of the tract [54]. These values were comparable with the target registration error, which was 1 mm.

Tractography has been used also to map language and visual pathways. The arcuate fascicle connecting frontal, parietal and temporal regions is implicated in language function and the anatomy and organization of this structure has been studied using tractography [11, 12]. In a series of 22 patients, Kamada et al [34] described tractography of the arcuate fascicle, generated by fMRI activations with a verb generation task, and magnetoencephalography with a reading task. For 2 of the 22 cases studied, imaging data were imported into a neuronavigation system, and the location of the arcuate fascicle was compared with cortical electrostimulation mapping of the area of activation identified by fMRI; mapping tags were found to be within 6 mm of the arcuate fascicle. Henry et al [30] used tractography in a 40-year-old male subject to investigate pathways initiated from sites of intraoperative cortical stimulation associated with speech and naming. Disruption by glioma of language pathways, notably the arcuate fascicle, as well as the inferior fronto-occipital fascicle (IFOF) and uncinate fascicle, has recently been described [6].

The optic radiations can be visualized with tractography for the neurosurgical planning of epileptic patients undergoing anterior temporal lobectomy. In a series of 10 patients with arteriovenous malformation, Kikuta et al [36] found that incomplete reconstruction of the optic radiation was associated with visual field loss.

Tractography of the optic radiations has also been integrated retrospectively into a gamma knife radiosurgery system, illustrating the principle of modified treatment planning with the aim of reducing radiation damage to the optic radiation [42].

Clinical applications to neurosurgical patients with epilepsy

Tractography has been employed in patients with epilepsy [75] undergoing surgery to evaluate the extension of structural changes following surgical procedures and to predict outcomes in patients with epilepsy [75]. The white-matter changes in the optic radiations and limbic structures induced by surgery have been widely investigated. Concha et al [18] assessed the fornix and cingulum in patients with temporal lobe epilepsy and unilateral mesial temporal sclerosis before and 1 year after temporal lobectomy. Patients showed preoperative diffusion abnormalities (i.e., reduced FA and increased MD), which became more evident after surgery, suggesting Wallerian degeneration of the white-matter fibers affected by the surgery [18]. Even the contralateral tracts, which were not directly affected by the surgery, did not normalize their diffusion indices in seizure-free patients, indicating the presence of irreversible white-matter abnormalities. The same group performed a longitudinal analysis of the diffusion indices of the genu and body of the corpus callosum in patients who underwent corpus callosotomy [17]. In the three patients studied, the tractography measures along the transected portion of callosal fibres showed diffusion changes indicative of wallerian degeneration. These findings suggest that an exciting application of diffusion tractography is the in vivo assessment of the microscopic structural changes that follow axonal injury.

Another interesting application of tractography is to predict the complications of surgery for refractory epilepsy. In future, this will lead to tractography being used to aid preoperative planning and prevent damage to eloquent cortical functions. For example, Powell et al [60] reported that probabilistic tractography can be used to assess the extension and location of the Flechsig-Meyer's loop and to predict the contralateral superior quadrantanopia, which often follows a resection of the anterior temporal lobe. The authors showed that the trajectory of the optic radiation was disrupted in a patient with a visual-field defect, whilst another patient with no disruption to the trajectory suffered no deficit. More recently, Nilsson et al [52] extended these results by using a deterministic tractography to reconstruct the optic radiation and measuring the distance between the most anterior part of the Flechsig-Meyer's loop and both the temporal pole and horn. The authors demonstrated a disruption of the Flechsig-Meyer's loop in a patient with postoperative quadrantanopia after temporal lobe resection, while the Flechsig-Meyer's loop was reported to be intact in a patient without postoperative visual-field defect. Interestingly, the difference in the Flechsig-Meyer's loops between the two patients was their distance from the tip of the temporal horn, which was shorter by about 5 mm in the patient with postoperative quadrantanopia, suggesting that the relationship between these two structures may be useful to predict the risk of developing a visual-field defect after temporal lobe resection for therapy-resistant epilepsy [52]. Other investigators combined fMRI and probabilistic tractography to study patients undergoing a resection of the anterior temporal lobe of the language-dominant hemisphere [61]. Greater lateralization of tracts to the dominant hemisphere prior to the surgery was associated with greater postoperative decline in naming function, suggesting that this technique can be extended to predict postoperative language deficits.

From these reports, an attractive, future application of tractography to patients with epilepsy relates to intraoperative MRI, which has the potential to reduce the risks of surgical complications and eventually visualize the white-matter connections that need to be transected in order to disconnect the seizure focus. In this respect, a recent paper has shown that combining tractography with EEG and fMRI can delineate the pathways of propagation of epileptic activity [27]. Although this was a case report, and tractography cannot distinguish between afferent and efferent pathways, this paper reported on a promising technique for identifying the functional interaction of regions involved in the interictal epileptiform discharges and its structural correlates. Future studies will clarify whether factors such as age on onset, severity of epilepsy, and handedness are associated with the lateralization of the tracts and functional plasticity [15a].

Limitations and future directions of diffusion tractography

Whilst injected tracers are able to follow single neuronal terminations, tractography follows the principal axes of the DT, which is obtained by averaging the MRI signal within a voxel. Typically the voxel resolution is too low to identify small fibre bundles. Also the levels of noise in the diffusion data and the intrinsic MR artefacts [40] constitute important factors that affect the precision and accuracy of the diffusion measurements and therefore the tractography reconstruction. It should be recognized that the result that we obtain from tractography is also dependent on a number of factors under the control of the experimenter, such as the angular and anisotropy threshold and the choice of the tractography algorithm itself. Finally, DT tractography methods assume that fibers in each voxel are well described by a single orientation estimate, and so these techniques perform poorly in regions where more than one population of fibers exist and where fibers are crossing, kissing, merging, or diverging. More recent tractography developments based on HARDI (high angular resolution diffusion imaging) methods [24] and appropriate processing techniques are able to accomodate the crossing problem [1, 5, 69, 70, 74]. Preliminary work suggests that it is possible to combine tractography with a spherical-deconvolution algorithm to perform dissections of white-matter pathways in regions with multiple fiber crossing, like in the corpus callosum (Fig. 7) [20]. Spherical deconvolution datasets can be acquired using clinically feasible protocols.

Compared to classical axonal tracing studies, tractography is unable to differentiate anterograde and retrograde connections, detect the presence of synapses, or determine whether a pathway is functional. All these limitations may lead to tracking pathways that do not exist (false positive) or ineffectively track those that do exist (false negative), and therefore the interpretation of tractography results requires experience and a priori anatomical knowledge.

Furthermore, in the diseased brain, alteration and anatomic distortion due to the presence of pathological processes, such as brain edema, bleeding, and compression generates tissue changes likely to lead to artefactual tract reconstructions.

A few studies have so far dealt with the issue of validating the tractography results with neuronal tracers [19, 23] or performing reproducibility analysis on human subjects [29, 38, 71]. It should also be noted that numerous algorithms have been described in the literature and there

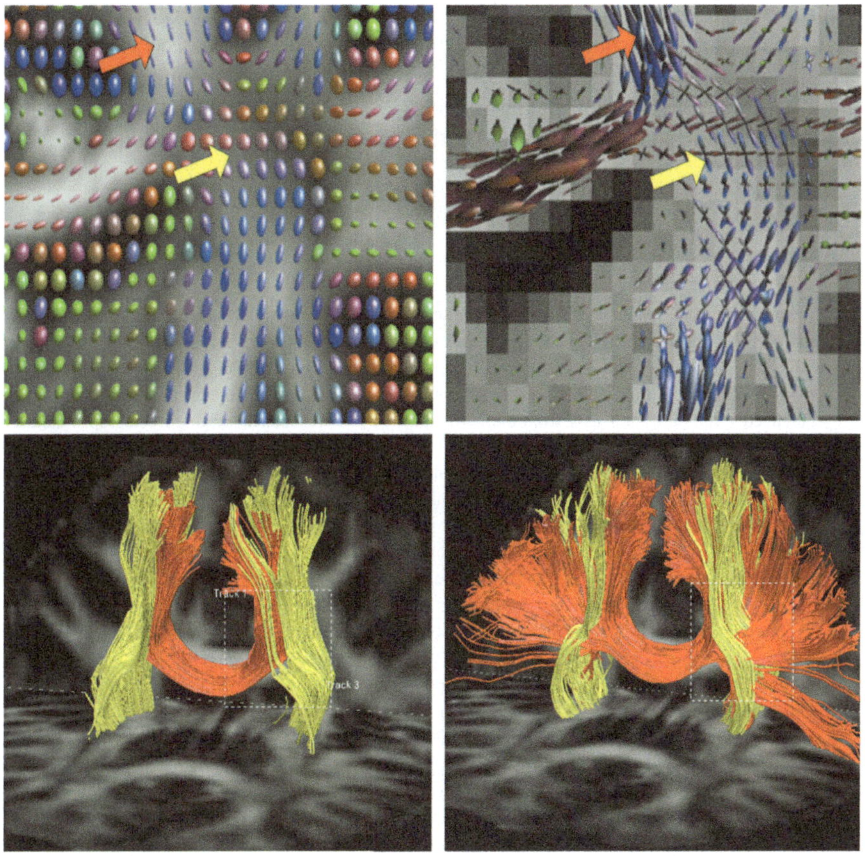

Fig. 7. *Top:* Visualization of the white matter organization based on the tensor model (left) and spherical deconvolution (right) [21] inside the corona radiata. For voxels with one fiber population such as the corpus callosum (red arrow), both models describe orientations that are consistent with the known anatomy. However, in regions with more than one population of crossing fibers (yellow arrow) the tensor model gives an average representation of the water diffusion, whereas the spherical-deconvolution model separates different fiber components and describes their individual orientations. *Bottom:* Diffusion tensor-based and spherical-deconvolution tractography. The virtual dissections of the corpus callosum based on diffusion tensor tractography (left) reconstruct only the most central part of the corpus callosum (red), while spherical-deconvolution tractography shows several streamlines of the corpus callosum crossing the streamlines of the corticospinal tract (yellow) and reaching the lateral cortex

is no consensus on which approach is the most valid.

In summary, tractography is growing in importance in the field of neurosurgical planning and is increasingly being requested by neurosurgical departments worldwide. However, its current use is largely restricted to specialist institutions with the necessary infrastructure (local technical support from physicists and imaging specialists) required to guarantee a reliable and clinically feasible methodology. With improvements in diffusion acquisition (e.g. higher spatial resolution and higher signal to noise ratio), processing (e.g., complex fibre orientation modelization, tractography algorithms) and a better characterization of the relationship between diffusion parameters and pathological changes, tractography methods are likely to become established as routine clinical investigations in the years to come.

Acknowledgments

In-house developed software and Trackvis were used to produce the figures. The members of the Natbrainlab are funded by the Wellcome Trust, Guy's & St Thomas' Charity, Autism Speaks, Marie Curie, the Medical Research Council (UK – the AIMS network) and the NIHR Specialist Biomedical Research Centre for Mental Health at the South London and Maudsley NHS Foundation Trust and Institute of Psychiatry, King's College London, London, UK.

References

[1] Alexander DC (2005) Multiple-fiber reconstruction algorithms for diffusion MRI. Ann NY Acad Sci 1064: 113–133

[2] Basser PJ, Mattiello J, LeBihan D (1994) Estimation of the effective self-diffusion tensor from the NMR spin echo. J Magn Reson B 103: 247–254

[3] Basser PJ, Pajevic S, Pierpaoli C, Duda J, Aldroubi A (2000) In vivo fiber tractography using DT-MRI data. Magn Reson Med 44: 625–632

[4] Behrens TE, Johansen-Berg H, Woolrich MW, Smith SM, Wheeler-Kingshott CA et al (2003) Non-invasive mapping of connections between human thalamus and cortex using diffusion imaging. Nat Neurosci 6: 750–757

[5] Behrens TE, Woolrich MW, Jenkinson M, Johansen-Berg H, Nunes RG et al (2003) Characterization and propagation of uncertainty in diffusion-weighted MR imaging. Magn Reson Med 50: 1077–1088

[6] Bello L, Gambini A, Castellano A, Carrabba G, Acerbi F et al (2008) Motor and language DTI fiber tracking combined with intraoperative subcortical mapping for surgical removal of gliomas. Neuroimage 39: 369–382

[7] Beppu T, Inoue T, Kuzu Y, Ogasawara K, Ogawa A, Sasaki M (2005) Utility of three-dimensional anisotropy contrast magnetic resonance axonography for determining condition of the pyramidal tract in glioblastoma patients with hemiparesis. J Neurooncol 73: 137–144

[8] Berman JI, Berger MS, Mukherjee P, Henry RG (2004) Diffusion-tensor imaging-guided tracking of fibers of the pyramidal tract combined with intraoperative cortical stimulation mapping in patients with gliomas. J Neurosurg 101: 66–72

[9] Berman JI, Berger MS, Chung SW, Nagarajan SS, Henry RG (2007) Accuracy of diffusion tensor magnetic resonance imaging tractography assessed using intraoperative subcortical stimulation mapping and magnetic source imaging. J Neurosurg 107: 488–494

[10] Catani M, Howard RJ, Pajevic S, Jones DK (2002) Virtual in vivo interactive dissection of white matter fasciculi in the human brain. Neuroimage 17: 77–94

[11] Catani M, Jones DK, Donato R, Ffytche DH (2003) Occipito-temporal connections in the human brain. Brain 126: 2093–2107

[12] Catani M, Jones DK, Ffytche DH (2005) Perisylvian language networks of the human brain. Ann Neurol 57: 8–16

[13] Cedzich C, Taniguchi M, Schäfer S, Schramm J (1996) Somatosensory evoked potential phase reversal and direct motor cortex stimulation during surgery in and around the central region. Neurosurgery 38: 962–970

[14] Chenevert TL, McKeever PE, Ross BD (1997) Monitoring early response of experimental brain tumors to therapy using diffusion magnetic resonance imaging. Clin Cancer Res 3: 1457–1466

[15] Chenevert TL, Stegman LD, Taylor JM, Robertson PL, Greenberg HS et al (2000) Diffusion magnetic resonance imaging: an early surrogate marker of therapeutic efficacy in brain tumors. J Natl Cancer Inst 92: 2029–2036

[15a] Ciccarelli O, Catani M, Johansen-Berg H, Clark C, Thompson A (2008) Diffusion-based tractography in neurological disorders: concepts, applications, and future developments. Lancet Neurology 7: 715–727

[16] Clark CA, Barrick TR, Murphy MM, Bell BA (2003) White matter fiber tracking in patients with space-occupying lesions of the brain: a new technique for neurosurgical planning? NeuroImage 20: 1601–1608

[17] Concha L, Gross DW, Wheatley BM, Beaulieu C (2006) Diffusion tensor imaging of time-dependent axonal and myelin degradation after corpus callosotomy in epilepsy patients. Neuroimage 32: 1090–1099

[18] Concha L, Beaulieu C, Wheatley BM, Gross DW (2007) Bilateral white matter diffusion changes persist after epilepsy surgery. Epilepsia 48: 931–940

[19] Dauguet J, Peled S, Berezovskii V, Delzescaux T, Warfield SK et al (2007) Comparison of fiber tracts derived from in-vivo DTI tractography with 3D histological neural tract tracer reconstruction on a macaque brain. Neuroimage 37: 530–538

[20] Dell'Acqua F, Scifo P, Catani M, Scotti G, Fazio F (2008) Combining spherical deconvolution with streamline tractography: preliminary results. Proc 16th Annual Meeting Int Society of Magnetic Resonance in Medicine, 2008, Toronto, Ontario, Canada

[21] Dell'Acqua F, Scifo P, Rizzo G, Catani M, Simmons A, Scotti G, Fazio F (2010) A modified damped Richardson-Lucy algorithm to reduce isotropic background effects in spherical deconvolution. Neuroimage 49: 1446–1458

[22] Duffau H (2005) Lessons from brain mapping in surgery for low-grade glioma: insights into associations between tumour and brain plasticity. Lancet Neurol 4: 476–486

[23] Dyrby TB, Søgaard LV, Parker GJ, Alexander DC, Lind NM et al (2007) Validation of in vitro probabilistic tractography. Neuroimage 37: 1267–1277

[24] Frank LR (2001) Anisotropy in high angular resolution diffusion-weighted MRI. Magn Reson Med 45: 935–939

[25] Gluhbegovic N, Williams TH (1980) The human brain. Harper and Row, Hagerstown, MD

[26] Guo AC, Cummings TJ, Dash RC, Provenzale JM (2002) Lymphomas and high-grade astrocytomas: comparison of water diffusibility and histologic characteristics. Radiology 224: 177–183

[27] Hamandi K, Powell HW, Laufs H, Symms MR, Barker GJ et al (2008) Combined EEG-fMRI and tractography to visualise propagation of epileptic activity. J Neurol Neurosurg Psychiatry 79: 594–597

[28] Hasegawa D, Orima H, Fujita M, Nakamura S, Takahashi K et al (2003) Diffusion-weighted imaging in kainic acid-induced complex partial status epilepticus in dogs. Brain Res 983: 115–127

[29] Heiervang E, Behrens TE, Mackay CE, Robson MD, Johansen-Berg H (2006) Between session reproducibility and between subject variability of diffusion MR and tractography measures. Neuroimage 33: 867–877

[30] Henry RG, Berman JI, Nagarajan SS, Mukherjee P, Berger MS (2004) Subcortical pathways serving cortical language sites: initial experience with diffusion tensor imaging fiber tracking combined with intraoperative language mapping. Neuroimage 21: 616–622

[31] Hlatky R, Jackson EF, Weinberg JS, McCutcheon IE (2005) Intraoperative neuronavigation using diffusion tensor MR tractography for the resection of a deep tumor adjacent to the corticospinal tract. Stereotact Funct Neurosurg 83: 228–232

[32] Holodny AI, Ollenschleger MD, Liu WC, Schulder M, Kalnin AJ (2001) Identification of the corticospinal tracts achieved using blood-oxygen-level-dependent and diffusion functional MR imaging in patients with brain tumors. AJNR Am J Neuroradiol 22: 83–88

[33] Kamada K, Todo T, Masutani Y, Aoki S, Ino K et al (2005) Combined use of tractography-integrated functional neuronavigation and direct fiber stimulation. J Neurosurg 102: 664–672

[34] Kamada K, Todo T, Masutani Y, Aoki S, Ino K et al (2007) Visualization of the frontotemporal language fibers by tractography combined with functional magnetic resonance imaging and magnetoencephalography. J Neurosurg 106: 90–98

[35] Keles GE, Lundin DA, Lamborn KR, Chang EF, Ojemann G, Berger MS (2004) Intraoperative subcortical stimulation mapping for hemispherical perirolandic gliomas located within or adjacent to the descending motor pathways: evaluation of morbidity and assessment of functional outcome in 294 patients. J Neurosurg 100: 369–375

[36] Kikuta K, Takagi Y, Nozaki K, Hanakawa T, Okada T et al (2006) Early experience with 3-T magnetic resonance tractography in the surgery of cerebral arteriovenous malformations in and around the visual pathway. Neurosurgery 58: 331–337

[37] Kinoshita M, Yamada K, Hashimoto N, Kato A, Izumoto S et al (2005) Fiber-tracking does not accurately estimate size of fiber bundle in pathological condition: initial neurosurgical experience using neuronavigation and subcortical white matter stimulation. NeuroImage 25: 424–429

[38] Lawes IN, Barrick TR, Murugam V, Spierings N, Evans DR et al (2008) Atlas-based segmentation of white matter tracts of the human brain using diffusion tensor tractography and comparison with classical dissection. Neuroimage 39: 62–79

[38a] Le Bihan D, Breton E, Lallemand D, Grenier P, Cabanis E, Laval-Jeantet M (1986) MR imaging of intravoxel incoherent motions: application to diffusion and perfusion in neurologic disorders. Radiology 161: 401–407

[39] Le Bihan D, Mangin JF, Poupon C, Clark CA, Pappata S et al (2001) Diffusion tensor imaging: concepts and applications. J Magn Reson Imaging 13: 534–546
[40] Le Bihan D, Poupon C, Amadon A, Lethimonnier F (2006) Artifacts and pitfalls in diffusion MRI. J Magn Reson Imaging 24: 478–488
[41] Lo CY, Chao YP, Chou KH, Guo WY, Su JL, Lin CP (2007) DTI-based virtual reality system for neurosurgery. Conf Proc IEEE Eng Med Biol Soc 2007: 1326–1329
[42] Maruyama K, Kamada K, Shin M, Itoh D, Masutani Y et al (2007) Optic radiation tractography integrated into simulated treatment planning for Gamma Knife surgery. J Neurosurg 107: 721–726
[43] Mikuni N, Okada T, Enatsu R, Miki Y, Hanakawa T et al (2007) Clinical impact of integrated functional neuronavigation and subcortical electrical stimulation to preserve motor function during resection of brain tumors. J Neurosurg 106: 593–598
[44] Mikuni N, Okada T, Enatsu R, Miki Y, Urayama S et al (2007) Clinical significance of preoperative fibre-tracking to preserve the affected pyramidal tracts during resection of brain tumours in patients with preoperative motor weakness. J Neurol Neurosurg Psychiatry 78: 716–721
[45] Mikuni N, Okada T, Nishida N, Taki J, Enatsu R et al (2007) Comparison between motor evoked potential recording and fiber tracking for estimating pyramidal tracts near brain tumors. J Neurosurg 106: 128–133
[46] Mori S, Van Zijl PC (2002) Fiber tracking: principles and strategies – a technical review. NMR Biomed 15: 468–480
[47] Mori S, Crain BJ, Chacko VP, van Zijl PC (1999) Three-dimensional tracking of axonal projections in the brain by magnetic resonance imaging. Ann Neurol 45: 265–269
[48] Mori S, Frederiksen K, van Zijl PC, Stieltjes B, Kraut MA et al (2002) Brain white matter anatomy of tumor patients evaluated with diffusion tensor imaging. Ann Neurol 51: 377–380
[49] Moseley ME, Cohen Y, Mintorovitch J, Chileuitt L, Shimizu H et al (1990) Early detection of regional cerebral ischemia in cats: comparison of diffusion-and T2-weighted MRI and spectroscopy. Magn Reson Med 14: 330–346
[50] Moseley ME, Kucharczyk J, Asgari HS, Norman D (1991) Anisotropy in diffusion-weighted MRI. Magn Reson Med 19: 321–326
[51] Nedelcu J, Klein MA, Aguzzi A, Boesiger P, Martin E (1999) Biphasic edema after hypoxic-ischemic brain injury in neonatal rats reflects early neuronal and late glial damage. Pediatr Res 46: 297–304
[52] Nilsson D, Starck G, Ljungberg M, Ribbelin S, Jönsson L et al (2007) Intersubject variability in the anterior extent of the optic radiation assessed by tractography. Epilepsy Res 77: 11–16
[53] Nimsky C, Ganslandt O, Hastreiter P, Wang R, Benner T et al (2005) Preoperative and intraoperative diffusion tensor imaging-based fiber tracking in glioma surgery. Neurosurgery 56: 130–137; discussion 138
[54] Nimsky C, Ganslandt O, Merhof D, Sorensen AG, Fahlbusch R (2006) Intraoperative visualization of the pyramidal tract by diffusion-tensor-imaging-based fiber tracking. Neuroimage 30: 1219–1229
[55] Nitsch C, Klatzo I (1983) Regional patterns of blood-brain barrier breakdown during epileptiform seizures induced by various convulsive agents. J Neurol Sci 59: 305–322
[56] Okada T, Mikuni N, Miki Y, Kikuta K, Urayama S et al (2006) Corticospinal tract localization: integration of diffusion-tensor tractography at 3-T MR imaging with intraoperative white matter stimulation mapping – preliminary results. Radiology 240: 849–857
[57] Papadopoulos MC, Saadoun S, Binder DK, Manley GT, Krishna S, Verkman AS (2004) Molecular mechanisms of brain tumor edema. Neuroscience 129: 1009–1018
[58] Parmar H, Sitoh YY, Yeo TT (2004) Combined magnetic resonance tractography and functional magnetic resonance imaging in evaluation of brain tumors involving the motor system. J Comput Assist Tomogr 28: 551–556
[59] Pierpaoli C, Jezzard P, Basser PJ, Barnett A, Di Chiro G (1996) Diffusion tensor MR imaging of the human brain. Radiology 201: 637–648
[60] Powell HW, Parker GJ, Alexander DC, Symms MR, Boulby PA et al (2005) MR tractography predicts visual field defects following temporal lobe resection. Neurology 65: 596–599
[61] Powell HW, Parker GJ, Alexander DC, Symms MR, Boulby PA et al (2008) Imaging language pathways predicts postoperative naming deficits. J Neurol Neurosurg Psychiatry 79: 327–330
[62] Prichard JW, Zhong J, Petroff OA, Gore JC (1995) Diffusion-weighted NMR imaging changes caused by electrical activation of the brain. NMR Biomed 8: 359–364
[63] Righini A, Pierpaoli C, Alger JR, Di Chiro G (1994) Brain parenchyma apparent diffusion coefficient alterations associated with experimental complex partial status epilepticus. Magn Reson Imaging 12: 865–871
[64] Schonberg T, Pianka P, Hendler T, Pasternak O, Assaf Y (2006) Characterization of displaced white matter by brain tumors using combined DTI and fMRI. Neuroimage 30: 1100–1111
[65] Staempfli P, Reischauer C, Jaermann T, Valavanis A, Kollias S, Boesiger P (2008) Combining fMRI and DTI: a framework for exploring the limits of fMRI-guided DTI fiber tracking and for verifying DTI-based fiber tractography results. Neuroimage 39: 119–126
[66] Sundgren PC, Dong Q, Gómez-Hassan D, Mukherji SK, Maly P, Welsh R (2004) Diffusion tensor imaging of

the brain: review of clinical applications. Neuroradiology 46: 339–350

[67] Tanaka T, Tanaka S, Fujita T, Takano K, Fukuda H et al (1992) Experimental complex partial seizures induced by a microinjection of kainic acid into limbic structures. Prog Neurobiol 38: 317–334

[68] Thapar KR (1995) Brain edema, increased intracranial pressure, vascular effects and other epiphenomena of human brain tumors. In: Kaye AH and Laws ER (eds) Brain tumors: an encyclopedic approach. Churchchill Livingstone, Edinburgh, pp 163–189

[69] Tournier JD, Calamante F, Gadian DG, Connelly A (2004) Direct estimation of the fiber orientation density function from diffusion-weighted MRI data using spherical deconvolution. Neuroimage 23: 1176–1185

[70] Tuch DS (2004) Q-ball imaging. Magn Reson Med 52: 1358–1372

[71] Wakana S, Caprihan A, Panzenboeck MM, Fallon JH, Perry M et al (2007) Reproducibility of quantitative tractography methods applied to cerebral white matter. Neuroimage 36: 630–644

[72] Wang Y, Majors A, Najm I, Xue M, Comair Y et al (1996) Postictal alteration of sodium content and apparent diffusion coefficient in epileptic rat brain induced by kainic acid. Epilepsia 37: 1000–1006

[73] Wasterlain CG, Fujikawa DG, Penix L, Sankar R (1993) Pathophysiological mechanisms of brain damage from status epilepticus. Epilepsia 34 (Suppl 1): S37–53

[74] Wedeen VJ, Hagmann P, Tseng WY, Reese TG, Weisskoff RM (2005) Mapping complex tissue architecture with diffusion spectrum magnetic resonance imaging. Magn Reson Med 54: 1377–1386

[75] Yogarajah M, Duncan JS (2008) Diffusion-based magnetic resonance imaging and tractography in epilepsy. Epilepsia 49: 189–200

[76] Yu CS, Li KC, Xuan Y, Ji XM, Qin W (2005) Diffusion tensor tractography in patients with cerebral tumors: a helpful technique for neurosurgical planning and postoperative assessment. Eur J Radiol 56: 197–204

[77] Zhong J, Petroff OA, Pleban LA, Gore JC, Prichard JW (1997) Reversible, reproducible reduction of brain water apparent diffusion coefficient by cortical electroshocks. Magn Reson Med 37: 1–6

Magnetoencephalography

Sylvain Baillet

Introduction

A well-documented and thoroughly-discussed limitation of hemodynamic and metabolic neuroimaging techniques is their lack of temporal resolution. This basic limitation has become salient with the emergence of new questions in neuroscience to investigate the brain as an ensemble of complex networks that form, reshape, and flush information dynamically [41, 42].

This chapter briefly reviews how magnetoencephalography (MEG) is able to offer time-resolved image sequences of brain functions and dysfunctions.

MEG principles and instrumentation

Physiological sources of electromagnetic fields

The brain sustains ionic current flows within and across cell assemblies, with neurons as the strongest elementary generators. A simple model of intracellular current pathways is a small, straight electrical dipole conducting ionic currents from a source to a sink (Fig. 1). Intracellular current sources are twofold in a neuron: (1) fast discharges of action potentials along axons and (2) slower (from a few tens to hundreds of milliseconds) excitatory and inhibitory postsynaptic potentials, which typically create an electrical imbalance between the basal and the apical dendritic tree or the cell soma. Both of these current sources generate electromagnetic fields, although it is very unlikely that action potentials sufficiently overlap in time and space to add up to a massive current flow. A great share of neural cells is shaped along a longitudinal geometry –the pyramidal cells in neocortical layers II/III and V – which is favorable to yield a large net current. Also, neurons are grouped into assemblies of tightly interconnected cells. Therefore it is likely that postsynaptic potentials be identically distributed across a given assembly, with the immediate benefit that they build up efficiently to drive higher levels of currents, which in turn generate electromagnetic fields that are strong enough to be detected outside the head (Fig. 1).

A typical source strength of 10 nA·m generates magnetic fields that are sufficiently large in amplitude to be detected at the head surface by MEG. This may correspond to the activity of as few as 50,000 pyramidal cells with an individual current density of 0.2 pA·m [31].

Instrumentation

While the neural generators of MEG and electroencephalography (EEG) are equivalent, the respective technologies involved differ significantly because MEG brain signals reach only up to about 10×10^{-15} to 50×10^{-15} $T/Hz^{1/2}$. Very

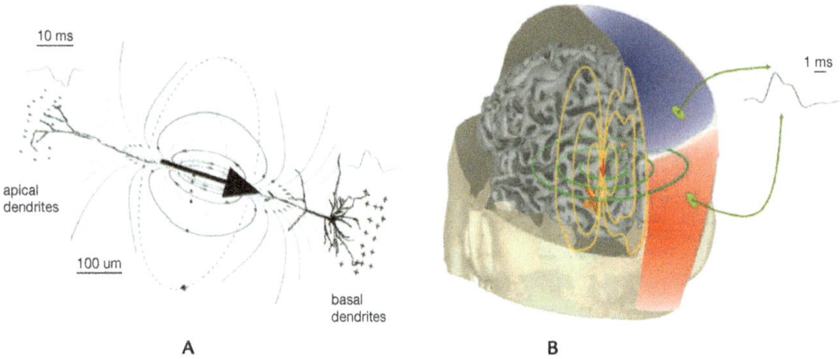

Fig. 1. Basic electrophysiological principles of MEG and EEG. **(A)** Large neural cells – just like this pyramidal neuron from cortex layer V – drive ionic electrical currents. These latter are essentially impressed by the difference in electrical potentials between the basal and apical dendrites or the cell body, which is due to a mixture of excitatory and inhibitory postsynaptic potentials, which are slow (>10 ms) relatively to action potentials firing and therefore add up efficiently at the scale of synchronized neural ensembles. These primary currents can be modeled using an equivalent current dipole, here represented by a large black arrow. The electrical circuit of currents is closed within the entire head volume by secondary, volume currents shown with the dark plain lines. Additionally, magnetic fields are generated by the primary and secondary currents. The magnetic field lines induced by the primary currents are shown using dashed lines arranged in circles about the dipole source. **(B)** At a larger spatial scale, the mass effect of currents due to neural cells sustaining similar mixtures of postsynaptic potentials add up locally and behave also as a current dipole (shown in red). This primary generator induces secondary currents (shown in yellow) that travel through the head tissues. They eventually reach the scalp surface, where they can be detected with pairs of electrodes by EEG. Magnetic fields (in green) travel more freely within tissues and are less distorted than current flows. They can be captured with arrays of magnetometers by MEG. The distribution of blue and red colors on the scalp illustrates the continuum of magnetic and electric fields and potentials distributed at the surface of the head

sensitive magnetometers are therefore necessary. For MEG, a basic magnetometer consists of a pickup coil paired with a superconducting current detector (called superconducting quantum interference device). The MEG instrument itself consists of a rigid whole-head helmet containing up to 300 sensors (Fig. 2): single or pairs of magnetometers, called gradiometers, which are less sensitive to far-field sources such as road traffic, elevators, and heartbeats. EEG and other ancillary electrophysiological recordings (such as electro-oculogram, myogram, and cardiogram) can be recorded simultaneously with MEG, which completes the electromagnetic signature of neural currents. The sampling rate can reach up to 5 kHz on all channels. The superconducting sensing technology necessary to keep instrumental noise levels at less than a few femtotesla per square root hertz requires cooling at –269 °C with liquid helium.

Working with ultrasensitive sensors is also problematic as they are very good at picking up the nuisances and electromagnetic perturbations generated by external sources. A magnetically shielded room made of layers of metal alloys (and possibly complemented by active shielding solutions) attenuates external magnetic fields and makes MEG recordings possible.

The technology involved in MEG sensing, the weekly helium refills, and the materials constituting the magnetically shielded room, make MEG a costly piece of equipment. Exciting recent developments – e.g., in high-temperature magnetometers – however contribute to constant progress in cost-effectiveness, practicality and the future of MEG sensing science.

MEG has indeed substantial benefits compared with EEG: (1) EEG is strongly degraded by the heterogeneity in conductivity within head tissues (e.g., insulating skull versus conducting scalp) – this effect is extremely limited in MEG, resulting in greater spatial discrimination of neural contributions; (2) subject preparation time is reduced considerably; (3) measures are absolute, i.e., they are not dependent on the choice of a reference; (4) subject's comfort is improved as there is no direct contact of the sensors on the skin. The installation of new MEG systems is presently steadily growing within research and clinical centers (about 200 worldwide).

Scenarios of most typical MEG and EEG sessions

A successful MEG or EEG study is a combination of quality instrumentation, careful paradigm design, and well-understood preprocessing and analysis methods integrated in efficient software tools.

Subject preparation

Precautions should be taken as any magnetic material carried by the subject would cause major MEG artifacts. Recording leads for electrooculography, electrocardiography, and electromyography are recommended for artifact monitoring and subsequent correction. Head positioning coils are taped to the subject's head to detect its position with respect to the sensor array. Some MEG systems feature the possibility for continuous head position monitoring during the very recording and off-line head movement compensation (see Fig. 3 for more details and the digitization of additional anatomical fiducial points).

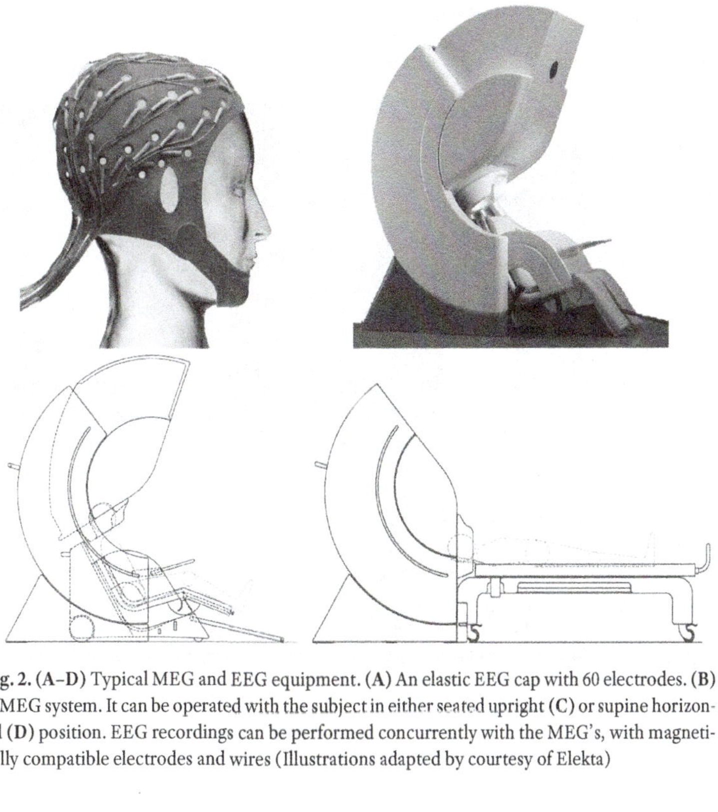

Fig. 2. (A–D) Typical MEG and EEG equipment. **(A)** An elastic EEG cap with 60 electrodes. **(B)** A MEG system. It can be operated with the subject in either seated upright **(C)** or supine horizontal **(D)** position. EEG recordings can be performed concurrently with the MEG's, with magnetically compatible electrodes and wires (Illustrations adapted by courtesy of Elekta)

Paradigm design

The time dimension accessible to MEG and EEG offers some considerable variety in the design of experimental paradigms. In a nutshell, an MEG and EEG experimental design is conditioned to the type of event-related brain responses of foremost interest to the investigator: evoked, induced, or sustained. The by far most common experimental design is the interleaved presentation of transient stimuli representing multiple conditions to be tested. The durations of interstimulus intervals are typically much shorter than in fMRI paradigms and range from a few tens of milliseconds to a few seconds. Each presentation is considered as an experimental event; hence such paradigms are called event-related.

The second category of experimental designs utilizes sustained stimulus presentations with specific temporal encoding (e.g., visual pattern reversals or sound modulations at a well-defined frequency) to trigger steady-state brain responses locked to the stimulus presentation rate or its harmonics. This approach is sometimes called frequency tagging (of brain responses) [35].

Data acquisition

A typical MEG and EEG session usually consists of several runs. A run is a series of experimental trials. A trial is an experimental event whereby a stimulus is presented to a subject or the subject performs a predefined action, within a certain condition of the paradigm.

The vast majority of studies target brain responses that are evoked by stimulation and revealed after trial averaging. Most of these responses have a typical half-cycle of about 20 ms or greater, hence a characteristic frequency of 100 Hz or less. A sampling rate of 300 to 600 Hz would therefore be sufficient. High-frequency (600–900 Hz) oscillatory components have however been evidenced in the somatosensory cortex or in interictal epilepsy [5], hence the necessity of faster sampling rates (3–5 kHz) in such situations.

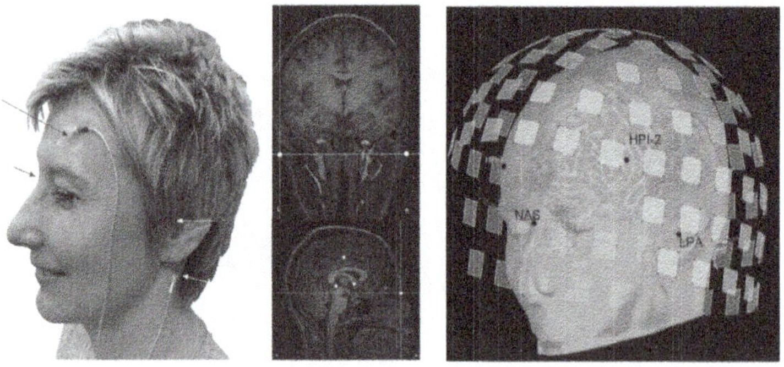

Fig. 3. (A–C) Multimodal MEG-MRI geometrical registration. **(A)** 3 to 5 head-positioning indicators are taped onto the subject's scalp. Their positions, together with 3 additional anatomical fiducials (nasion, left and right periauricular points) are digitized by a magnetic pen digitizer. **(B)** The anatomical fiducials need to be detected and marked (white dots) in the subject's anatomical MRI volume data, together with 3 optional, additional points defining the anterior and posterior commissures and the interhemispheric space, for the definition of Talairach coordinates. **(C)** These anatomical landmarks henceforth define a geometrical referential in which the MEG sensor locations and the surface envelopes of the head tissues (e.g., the scalp and brain surface, segmented from the MRI volume) are coregistered. MEG sensors are shown as squares positioned about the head. The anatomical fiducials (*NAS*, *LPA*) and locations of the head position indicators (*HPI*) are marked with black dots

Data preprocessing

The purpose of data preprocessing is to enhance the levels of signals of interest, while attenuating nuisances or rejecting artifact episodes in the recordings. Artifacts include physiological sources such as the eyes, heart, muscles, electromagnetic perturbations from other ancillary experimental devices and leaking power line contamination, which are not all well-handled by classical band-pass filtering. For perturbations which are thought to be independent of the brain processes of interest, empirical statistics obtained from a series of representative events (e.g., eye blinks, heartbeats, muscle contractions) are likely to properly capture the nuisance they systematically generate in the MEG recordings. Approaches like principal or independent component analysis have proven to be effective in that respect [9]. Signal space separation and its variants have also been proposed for MEG [39]: They basically consist in designing software spatial filters that attenuate sources of nuisance originating from outside a virtual spherical volume designed to contain the subject's head within the MEG helmet.

Evoked responses across trials in signal extraction via epoch averaging

An enduring tradition of MEG and EEG signal analysis consists in enhancing brain responses that are evoked by a stimulus or an action, by averaging the data about each event – thereby defining an epoch – across trials. The underlying assumption is that there exist some consistent brain responses that repeat systematically and which are time-locked with no phase variations ("phase-locked") with respect to the occurrence of an event. Extraction of these responses is enhanced by averaging epochs across trials, under the assumption that the rest of the data is inconsistent in time or phase with respect to the event of interest. This simple practice has generated a vast amount of contributions to the field of event-related potentials (in EEG, ERP) and fields (in MEG, ERF) [15]. Averaging epochs across trials can be conducted for each experimental condition at the individual and the group level ("grand-averaging"). Measurements are taken on ERF components, which are defined as waveform elements that emerge from the baseline of the recordings. They may be characterized in terms of, e.g., relative latency, topography, amplitude, and duration with respect to baseline or a specific test condition.

Induced responses across trials in signal extraction via epoch averaging

Massive event-related cell synchronization is not guaranteed to take place with consistent temporal phase with respect to the onset of the event. Therefore averaging trials when phase jitters occur across event repetitions might lead to decreased effect sensitivity. This assumption can be further elaborated in the theoretical and experimental framework of distributed, synchronized oscillatory cell assemblies during perception and cognition [37, 41], especially in the gamma range (40–60 Hz and greater [20]). Further, it has been recently discussed that these stimulus-induced responses tend to colocalize with the sites of hemodynamics fluctuations in fMRI [32].

Time-frequency decomposition (using, e.g., wavelets [28]) is a methodology of choice to detect induced components: it proceeds to the estimation of instantaneous power in the time-frequency domain of time series, which is insensitive to variations of the signal phase.

New trends and methods: connectivity and complexity analysis

The analysis of brain connectivity is a rapidly evolving field of neuroscience. The time resolution of MEG offers a unique perspective on the mechanisms of rapid neural connectivity engaging cell assemblies at multiple temporal and spatial scales. Two approaches have developed somewhat distinctly in the recent years, though we might predict they will ultimately converge with forthcoming research efforts.

The first strategy assumes that cell synchronization is a central feature of neural communication. Hence, signal analysis techniques dedicated to the estimation of signal interdependencies in the broad sense have been largely applied in MEG. Coherence measures, for instance, are sensitive to simultaneous variations of power that are specific to each frequency bin of the signal spectrum [33]. There is however a concurrent assumption that neural signals may synchronize their phases, without the necessity of simultaneous, increased power modulation [41]. Connectivity analysis has also been recently studied through the concept of causality, whereby some neural regions influence others in a nonsymmetric, directed fashion [13]. The investigation of directed influence between not only pairs but larger sets of time series (i.e., MEG sensors or brain regions) is usually ruled by parametric models. These latter may be related either to the definition of the time series (i.e., through autoregressive modeling for Granger causality assessment [27]), or to models of connectivity between neural assemblies (e.g., structural equation modeling [1] or dynamic causal modeling [23]).

The second approach to connectivity analysis pertains to the emergence of complex-network studies and associated methodology. Complex-network science is a recent branch of applied mathematics that provides quantitative tools to identify and characterize patterns of organization among large interconnected networks. The concept of the brain "connectome" has recently emerged in this context and aims at capturing the characteristics of spatially distributed dynamical neural processes at multiple spatial and temporal scales [36]. The new science of brain "connectomics" is contributing both to theoretical and computational models of the brain as a complex system [19] and to experimental methodology by suggesting new indices and metrics – such as nodes, hubs, efficiency, and modularity – to characterize and scale the functional organization of the healthy and diseased brain [4].

Electromagnetic source imaging

A direct assessment of the anatomical origins of the effects detected at the sensor level may be required. Electromagnetic source imaging addresses this issue by elaborating a model for the generators responsible for the signals in the data.

MEG-EEG source estimation as a modeling problem

From a methodological standpoint, MEG source modeling is referred to as an inverse problem, a ubiquitous concept well known to physicists in a wide variety of scientific fields, from medical imaging to particle physics [38]. The inverse problem framework helps conceptualize and formalize the fact that, in experimental sciences, models are related to observations to draw scientific conclusions and/or estimate some model parameters that were originally unknown. Parameters are quantities that might be changed without fundamentally violating and thereby invalidating the theoretical model. Predicting observations from a model with a given set of parameters is called solving the forward modeling problem. The reciprocal situation, where observations are used to estimate the values of some model parameters, is the inverse modeling problem.

MEG forward modeling consists in predicting the electromagnetic fields generated by any arbitrary source model, for any location, orientation, and amplitude parameter values of the neural currents. In general, MEG forward modeling considers that some parameters are known and fixed: the geometry of the head, conductivity of tissues, sensor locations, etc. MEG inverse modeling consists in relating the data to the forward model, so that parameters of source activity can be estimated.

A fundamental principle is that, whereas the forward problem has a unique solution in classical physics, the inverse problem may admit multiple solutions, which are models that equivalently predict the observations. This issue of nonuniqueness is not specific to MEG and can be rigorously addressed with the mathematics of ill-posedness and inverse modeling, which formalize the necessity of bringing additional contextual information to complement a basic theoretical model. Today, a reasonable degree of technical maturity has been reached by electromagnetic brain imaging using MEG. Methods reduce to only a handful of classes of approaches, which are now well identified.

Modeling the electromagnetics of head tissues

Models of neural generators
MEG forward modeling requires two basic models that are bound to work together in a complementary manner: a physical model of neural sources and a model that predicts how these sources generate electromagnetic fields outside the head. The canonical source model of the net primary intracellular currents within a neural assembly is a simple, equivalent current dipole (ECD) model. ECDs can be distributed in the entire brain volume or on the cortical surface, thereby forming a dense grid of elementary sites of activity, whose intensity distribution is determined from the data.

Modeling head tissues
Predicting the electromagnetic fields produced by a source model is called head modeling. The physics of MEG are ruled by the theory of electrodynamics [10], reduced to Maxwell's equations under quasistatic assumptions. These latter consider that the propagation delay of the electromagnetic waves from brain sources to the MEG sensors is negligible [14]. This is a very important, simplifying assumption which has immediate consequences on the computational aspects of MEG head modeling.

Indeed, the equations of magnetostatics determine that there exist analytical, closed-form solutions to MEG head modeling when the head geometry is considered as spherical. Hence, the simplest and consequently by far most popular model of head geometry in MEG is a single-layer sphere. Indeed, MEG is insensitive to the number of spherical shells with different conductivities and it can be shown that only the location of the center of the sphere matters. This relative sensitivity to tissue conductivity values is therefore a general, important difference between EEG and MEG.

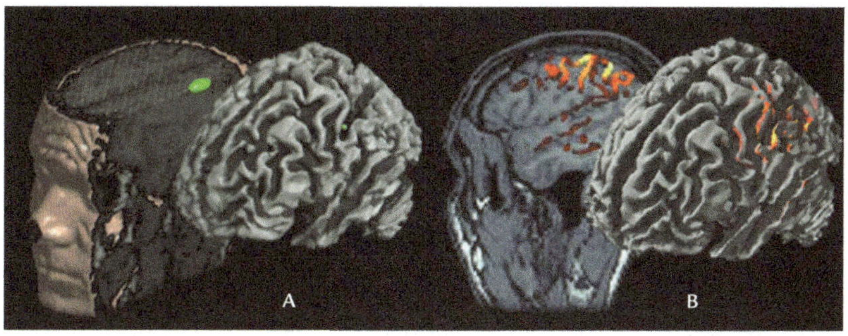

Fig. 4. Inverse modeling: localization (**A**) versus imaging (**B**) approaches. Source modeling through localization consists in decomposing the MEG/EEG generators in a handful of elementary source contributions; the simplest source model in this situation being the ECD. This is illustrated here with experimental data from testing the somatotopic organization of primary cortical representations of hand fingers. The parameters of the single ECD have been adjusted on the 20–40 ms time window following stimulus onset (right median nerve stimulation). The ECD was found to localize along the contralateral central sulcus as revealed by the three-dimensional rendering obtained after the source location has been registered to the individual anatomy. In the imaging approach, the source model is spatially distributed on a large number of ECDs. Here, a surface model of MEG-EEG generators was constrained to the individual brain surface extracted from T1-weighted MR images. Elemental source amplitudes are interpolated onto the cortex, which yields an imagelike distribution of the amplitudes of cortical currents

Another remarkable consequence of the spherical symmetry is that radially oriented brain currents produce no magnetic field outside a spherically symmetric volume conductor. For this reason, MEG signals from currents generated within the gyral crest or sulcal depth are attenuated, with respect to those generated by currents flowing perpendicularly to the sulcal walls. This is another important difference between MEG and EEG's respective sensitivity to source orientation [17].

Finally, the amplitude of magnetic fields decreases faster than that of electrical potentials with the distance from the generators to the sensors. Hence it has been argued that MEG is less sensitive to mesial and subcortical brain structures than is EEG. Experimental and modeling efforts have shown however that MEG can detect neural activity from deeper brain regions [2].

Although spherical head models are convenient, they are poor approximations of the human head shape, which has some influence on the accuracy of MEG source estimation [12]. More realistic head geometries have been investigated and all require solving Maxwell's equations by numerical methods. Boundary element and finite element methods are generic numerical approaches to the resolution of continuous equations in discrete domains. Geometric tessellations of the different envelopes forming the head tissues need to be extracted from the individual MRI volume data to yield a realistic approximation of their geometry. This is still considered a relatively complex task, although efficient software exists. Computation times for boundary element and particularly for finite element methods remain cumbersome but both algorithmic and pragmatic solutions to this problem are available in academic and commercial software packages.

MEG source modeling

Source localization versus source imaging

The localization approach to MEG source estimation considers that brain activity at any time instant is generated by a relatively small number (a handful, at most) of brain regions. Each source is therefore represented by an elementary model, such as an ECD, that captures local distributions of neural currents.

The alternative imaging approaches were originally inspired by the plethoric research in image restoration and reconstruction. Dense grids of current dipoles are defined over the entire brain volume or limited to the cortical gray-matter surface. These dipoles are fixed in location and, generally, orientation and are homologous to pixels in a digital image. The imaging procedure estimates the amplitudes of all these elementary currents at once. Hence contrarily to the localization model, there is no intrinsic sense of distinct, active source regions per se. Explicit identification of regional activity usually necessitates empirical or inference-driven amplitude thresholding. In that respect, MEG source images are very similar in essence to the activation maps obtained in fMRI, with the additional benefit of time resolution (Fig. 6).

The early MEG literature is abundant in studies reporting on single-dipole source models. The somatotopic, tonotopic auditory, and retinotopic visual responses are examples where the single-dipole model contributed to the better temporal characterization of primary brain responses.

Later components of evoked fields usually necessitate that more elementary sources be adjusted. This is however detrimental to the numerical stability and robustness of the inverse model. The number of elementary sources in the model is often qualitatively assessed by expert users, which may question the reproducibility of the analyses. Hence, special care should be brought to the evaluation of the stability and robustness of the estimated source models. With all that in mind, source localization techniques have proven to be effective, even on complex experimental paradigms [16].

Signal classification and spatial filtering techniques are efficient alternative approaches in that respect. They have gained considerable momentum in the MEG community in the recent years.

Scanning techniques: spatial filters, beamformers and signal classifiers

Rather than attempting to identify discrete sets of sources by adjusting their location parameters, scanning techniques have emerged and proceed by systematically sifting through the brain space to evaluate how a predetermined elementary source model would fit the data at every voxel of the brain volume. For this local model evaluation to be specific, contributions from other sources in the brain volume need to be blocked. Hence, these techniques are known as spatial filters and beamformers (creating a virtual beam directed and "listening" exclusively at some brain region [18]).

Beamforming is therefore a convenient method to translate the source localization problem into a signal detection issue. Initial technical shortcomings of beamforming have long been thoroughly investigated and techniques such as multiple signal classification were suggested as a more robust alternative for MEG (for a review, see [30]).

In summary, spatial filters, beamformers, and signal classification approaches bring us closer to a distributed representation of the brain electrical activity. As a caveat, the results generated by these techniques are not an estimation of the distribution of neural currents. They represent a score map of a source model (generally a single current dipole) evaluated on a predefined spatial lattice, which sometimes leads to misinterpretations.

Distributed source imaging

Imaging source models consist of distributions of elementary sources, generally with fixed locations and orientations, whose amplitudes are estimated at once. MEG source images represent estimations of the global neural current intensity maps, distributed within the entire brain volume or constrained at the cortical surface. The free parameters of the imaging model are the amplitudes of the elementary source currents distributed on the brain's geometry.

A reasonable spatial sampling of the image space requires several thousands (typically about 10,000) elementary sources. Consequently, the imaging inverse problem is dramatically underdetermined and imaging models need to be complemented by *a priori* information [3], which may take multiple faces: promote current distributions with high spatial and temporal smoothness, penalize models with currents of unrealistic, physiologically implausible amplitudes, favor the match with an fMRI activation map, or prefer source image models made of piecewise homogeneous active regions, etc. An appealing benefit from well-chosen priors is that they may ensure the uniqueness of the optimal solution to the imaging inverse problem.It is important to understand that the numerous source imaging techniques available usually have the same technical background. Also, the selection of image priors can be seen also as arbitrary and subjective. Comprehensive solutions for model selection are now emerging and suggest that the data can help decide on the best general class of models that would properly account for the data [8]. It is likely however that the critical ill-posedness of the source modeling problem be detrimental to the efficiency of establishing tight bounds on the admissible model parameters. Further, these techniques are still extremely demanding in terms of computational resources.

A widely used prior in the field of image reconstruction considers that the expected source amplitudes be as small as possible on average. This is the well-described minimum norm model and its multiple variations [25, 26], for which quantitative and qualitative empirical evidence demonstrates spatial resolution at the centimeter scale [6] (see Fig. 4).

Appraisal of MEG source models

MEG imaging is modeling which implies dealing with uncertainty: Data are complex and contaminated with various nuisances, source models are simplistic, head models have approximated geometries and conductivity properties, etc. It is therefore necessary to assess how sensitive source estimates are to modeling errors and biases. This can be approached through the estimation of confidence intervals about the estimated values of a source model by parametric or nonparametric statistics [7, 29].

In turn, hypothesis testing can be addressed by techniques of statistical inference. In neuroimaging, the population samples that will support the inference are either trials or subjects, for hypothesis testing at the individual and group levels, respectively. As in the case of the estimation of confidence intervals, there exist both parametric and nonparametric approaches to statistical inference. Parametric models have been extensively studied for fMRI and positron emission tomography before being recently adapted to EEG and MEG [22], and popularized with software such as Statistical Parametric Mapping [11]. Alternatively, nonparametric approaches such as permutation tests have emerged for statistical inference applied to neuroimaging data. Rather than applying transformations to the data to secure the assumption of normally distributed measures, nonparametric statistical tests take the data to be robust to departures from normal distributions [34]. Both parametric and nonparametric techniques properly handle the problem of multiple hypotheses testing, whereby the same inference is conducted at multiple instances through a vast source domain, which leads to possible detection errors.

The emergence of statistical inference solu-

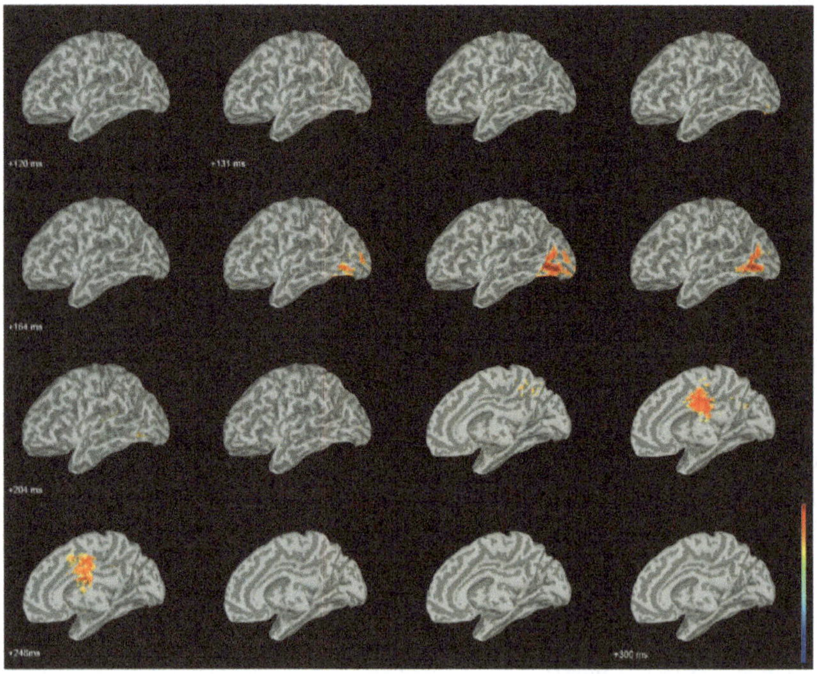

Fig. 5. Distributed source imaging of the 120–300 ms time interval following the presentation of the target face object in a rapid serial visual presentation in an oddball paradigm. The experiment consists of the detection of a visual "oddball". Pictures of faces are presented very rapidly to the participants every 100 ms, for a duration of 50 ms and an interstimulus interval of 50 ms. In about 15% of the trials, a face known to the participant is presented. This is the target stimulus and the participant needs to count the number of times he or she has seen the target individual among the unknown, distracting faces. Here, the experiment consisted of 4 runs of about 200 trials, hence resulting in a total of 120 target presentations. The images show a slightly smoothed version of one participant's cortical surface. Colors encode the contrast of MEG source amplitudes between responses to target versus control faces. Visual responses are detected by 120 ms and rapidly propagate anteriorly. From 250 ms onwards, strong anterior midline responses are detected in the cingular cortex. These latter are the main contributors of the brain response to target detection

tions adapted to MEG has brought electromagnetic source imaging to a considerable degree of maturity that is quite comparable to other neuroimaging techniques (Fig. 5 and 6).

Conclusions: a pragmatic point of view

Throughout this chapter, I have attempted to provide a pragmatic point of view on the technical difficulties of MEG. It is indeed quite striking that despite multiple shortcomings, MEG source analysis is able to reveal exquisite relative spatial resolution when localization approaches are used appropriately, and how imaging models help investigators tell a story on the cascade of brain events that have been occurring in controlled experimental conditions. An increasing number of groups from other neuroimaging modalities have come to realize that beyond mere cartography, temporal and oscillatory brain responses are essential keys to the understanding and interpretation of the basic mechanisms ruling information processing amongst neural assemblies. The growing number of EEG systems installed in MR magnets and the steady increase in MEG equipments demonstrate an active and dynamic scientific community, with exciting perspectives for the future of multidisciplinary brain research.

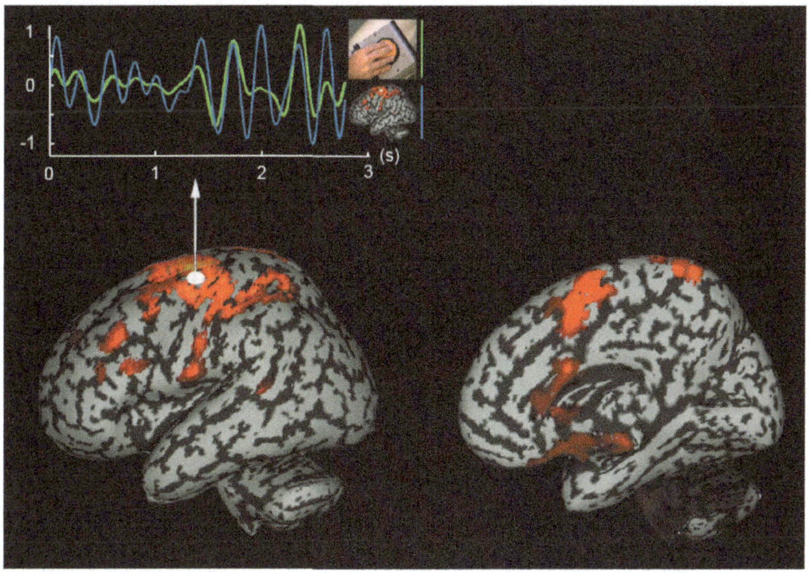

Fig. 6. MEG functional connectivity and statistical inference at the group level. Jerbi et al [21] have revealed a cortical functional network involved in hand movement coordination at low frequency (4 Hz). The statistical inference at group level first consisted in fitting for each trial in the experiment a distributed source model constrained to the individual anatomy of each of the 14 subjects involved. The brain area with maximum coherent activation with instantaneous hand speed was identified within the contralateral sensorimotor area (white dot). The traces at the top illustrate excellent coherence in the 3–5 Hz range between these measurements (hand speed in green and M1 motor activity in blue). Secondly, the search for brain areas with activity in significant coherence with the M1 revealed a larger distributed network of regions. All subjects were coregistered to a brain surface template in Talairach normalized space with the corresponding activations interpolated onto the template surface. A nonparametric t-test contrast was completed by permutations between rest and task conditions ($p < 0.01$)

References

[1] Astolfi L, Cincotti F, Babiloni C, Carducci F, Basilisco A, Rossini PM, Salinari S, Mattia D, Cerutti S, Dayan DB, Lei Ding, Ying Ni, Bin He, Babiloni F (2005) Estimation of the cortical connectivity by high-resolution EEG and structural equation modeling: simulations and application to finger tapping data. IEEE Trans Biomed Eng 52: 757–768

[2] Attal Y, Bhattacharjee M, Yelnik J, Cottereau B, Lefèvre J, Okada Y, Bardinet E, Chupin M, Baillet S (2009) Modelling and detecting deep brain activity with MEG and EEG. IRBM Ing Rech Biomed 30: 133–138

[3] Baillet S, Mosher J, Leahy R (2001) Electromagnetic brain mapping. IEEE Signal Process Mag 18(6): 14–30

[4] Bassett DS, Bullmore ET (2009) Human brain networks in health and disease. Curr Opin Neurol 22: 340–347

[5] Cimatti Z, Schwartz DP, Bourdain F, Meunier S, Bleton JP., Vidailhet M, Renault B, Garnero L (2007) Time-frequency analysis reveals decreased high-frequency oscillations in writer's cramp. Brain 130: 198–205

[6] Darvas F, Pantazis D, Kucukaltun-Yildirim E, Leahy RM (2004) Mapping human brain function with MEG and EEG: methods and validation. Neuroimage 23 (Suppl 1): S289–S299

[7] Darvas F, Rautiainen M, Pantazis D, Baillet S, Benali H, Mosher JC, Garnero L, Leahy RM (2005) Investigations of dipole localization accuracy in MEG using the bootstrap. Neuroimage 25: 355–368

[8] Daunizeau J, Mattout J, Clonda D, Goulard B, Benali H, Lina J-M (2006) Bayesian spatio-temporal approach for EEG source reconstruction: conciliating ECD and distributed models. IEEE Trans Biomed Eng 53: 503–516

[9] Delorme A, Sejnowski T, Makeig S (2007) Enhanced detection of artifacts in EEG data using higher-order statistics and independent component analysis. Neuroimage 34: 1443–1449

[10] Feynman RP, Leighton RB, Sands ML (1964) The Feynman lectures on physics, vol 2. Addison-Wesley, Reading, MA

[11] Friston K, Ashburner J, Kiebel S, Nichols T, Penny W (eds) (2007) Statistical parametric mapping: the analysis of functional brain images. Academic Press, London

[12] Fuchs M, Drenckhahn R, Wischmann H, Wagner M (1998) An improved boundary element method for realistic volume-conductor modeling. IEEE Trans Biomed Eng 45: 980–997

[13] Gourévitch B, Le Bouquin Jeannès R, Faucon G (2006) Linear and nonlinear causality between signals: methods, examples and neurophysiological applications. Biol Cybern 95: 349–369

[14] Hämäläinen M, Hari R, Ilmoniemi R, Knuutila J, Lounasmaa O (1993). Magnetoencephalography – theory, instrumentation and applications to noninvasive studies of the working human brain. Rev Mod Phys 65: 413–497

[15] Handy TC (ed) (2004) Event-related potentials: a methods handbook. MIT Press, Cambridge, MA

[16] Helenius P, Parviainen T, Paetau R, Salmelin R (2009) Neural processing of spoken words in specific language impairment and dyslexia. Brain 132: 1918–1927

[17] Hillebrand A, Barnes GR (2002) A quantitative assessment of the sensitivity of whole-head MEG to activity in the adult human cortex. Neuroimage 16: 638–650

[18] Hillebrand A, Singh KD, Holliday IE, Furlong PL, Barnes GR (2005) A new approach to neuroimaging with magnetoencephalography. Hum Brain Mapp 25: 199–211

[19] Honey CJ, Kötter R, Breakspear M, Sporns O (2007) Network structure of cerebral cortex shapes functional connectivity on multiple time scales. Proc Natl Acad Sci USA 104: 10240–10245

[20] Hoogenboom N, Schoffelen JM, Oostenveld R, Parkes LM, Fries P (2006) Localizing human visual gamma-band activity in frequency, time and space. Neuroimage 29: 764–773

[21] Jerbi K, Lachaux J, N'Diaye K, Pantazis D, Leahy R, Garnero L, Baillet S (2007) Coherent neural representation of hand speed in humans revealed by MEG imaging. Proc Natl Acad Sci USA 104: 7676–7681

[22] Kiebel SJ, Tallon-Baudry C, Friston KJ (2005) Parametric analysis of oscillatory activity as measured with EEG/MEG. Hum Brain Mapp 26: 170–177

[23] Kiebel SJ, Garrido MI, Moran RJ, Friston KJ (2008) Dynamic causal modelling for EEG and MEG. Cogn Neurodyn 2: 121–136

[24] Leahy RM, Mosher JC, Spencer ME, Huang MX, Lewine JD (1998) A study of dipole localization accuracy for MEG and EEG using a human skull phantom. Electroencephalogr Clin Neurophysiol 107: 159–173

[25] Lin FH, Belliveau JW, Dale AM, Hamalainen MS (2006) Distributed current estimates using cortical orientation constraints. Hum Brain Mapp 27: 1–13

[26] Lin FH, Witzel T, Ahlfors SP, Stufflebeam SM, Belliveau JW, Hämäläinen MS (2006) Assessing and improving the spatial accuracy in MEG source localization by depth-weighted minimum-norm estimates. Neuroimage 31: 160–171

[27] Lin FH, Hara K, Solo V, Vangel M, Belliveau JW, Stufflebeam SM, Hämäläinen MS (2009) Dynamic Granger-Geweke causality modeling with application to interictal spike propagation. Hum Brain Mapp 30: 1877–1886

[28] Mallat S (1998) A wavelet tour of signal processing. Academic Press, San Diego, CA

[29] Mosher JC, Spencer ME, Leahy RM, Lewis PS (1993) Error bounds for EEG and MEG dipole source localization. Electroencephalogr Clin Neurophysiol 86: 303–321

[30] Mosher JC, Baillet S, Leahy RM (1999) EEG source localization and imaging using multiple signal classification approaches. J Clin Neurophysiol 16: 225–238

[31] Murakami S, Okada Y (2006) Contributions of principal neocortical neurons to magnetoencephalography and electroencephalography signals. J Physiol 575: 925–936
[32] Niessing J, Ebisch B, Schmidt KE, Niessing M, Singer W, Galuske RA (2005) Hemodynamic signals correlate tightly with synchronized gamma oscillations. Science 309: 948–951
[33] Nunez PL, Srinivasan R, Westdorp AF, Wijesinghe RS, Tucker DM, Silberstein RB, Cadusch PJ (1997) EEG coherency. I: Statistics, reference electrode, volume conduction, Laplacians, cortical imaging, and interpretation at multiple scales. Electroencephalogr Clin Neurophysiol 103: 499–515
[34] Pantazis D, Nichols TE, Baillet S, Leahy RM (2005) A comparison of random field theory and permutation methods for the statistical analysis of MEG data. Neuroimage 25: 383–394
[35] Parkkonen L, Andersson J, Hämäläinen M, Hari R (2008) Early visual brain areas reflect the percept of an ambiguous scene. Proc Natl Acad Sci USA 105: 20500–20504
[36] Sporns O, Tononi G, Kötter R (2005) The human connectome: a structural description of the human brain. PLoS Comput Biol 1: e42
[37] Tallon-Baudry C (2009) The roles of gamma-band oscillatory synchrony in human visual cognition. Front Biosci 14: 321–332
[38] Tarantola A (2004) Inverse problem theory and methods for model parameter estimation. Society for Industrial and Applied Mathematics, Philadelphia, PA
[39] Taulu S, Kajola M, Simola J (2004) Suppression of interference and artifacts by the signal space separation method. Brain Topogr 16: 269–275
[40] Tikhonov AN, Arsenin VYa (1977) Solutions of ill-posed problems. Winston, Washington, DC
[41] Varela F, Lachaux JP, Rodriguez E, Martinerie J (2001) The brainweb: phase synchronization and large-scale integration. Nat Rev Neurosci 2: 229–239
[42] Werner G (2007). Brain dynamics across levels of organization. J Physiol Paris 101: 273–279

Extraoperative electrical mapping

Peter A. Winkler

Introduction

The main goal for all preoperative investigation is to define the epileptogenic zone and/or tumor area and its location to functionally important cortex, as well as to evaluate patient's psychological condition, in order to assure that epilepsy surgery will not worsen patient's life of quality.

One of the main groups of investigations during pre-surgical evaluation is invasive investigation and our work mainly focuses on describing and comparing two methods of invasive investigation: intraoperative and extraoperative cortex stimulation [3], [6]. Both of these methods are essential for defining an epileptogenic zone as well as functional cortex. In our work, we try to describe and compare these two methods and to present an analysis of epilepsy patients who have undergone cortex stimulation.

Evaluation of cortical function and cortical stimulation

There are 3 physiological tests for evaluation of cortical function:

- Recording of spontaneous activity [13], [34], [39];
- Recording of cortical evoked potentials [2], [5], [16], [31];
- Electrical stimulation, [9], [12], [24], [40];

Regarding spontaneous activity, unfortunately these physiological rhythms are frequently not sufficiently well defined to be used as reliable markers of cortical function [30]. Regarding different types of evoked potentials, only the somatosensory evoked potentials provide useful information when trying to define cortical function [30]. But the most powerful technique for localisation of cortical function is direct electrical stimulation of the cortex. This not only identifies the extent of the sensorimotor areas, but also permits excellent localisation of language areas and other higher cortical func-

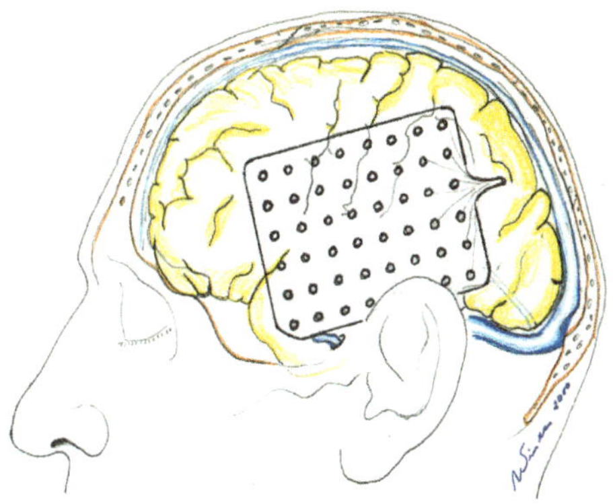

Fig. 1. Authors drawing to illustrate a grid implantation for covering the left fronto-temporo-parietal region. Please note the importance of preservation of the inferior anastomotic vein, i.e., the vein of Labbé

tions [30]. Cortical stimulation involves passage of a small electrical current through individual electrodes, with close observation for symptoms of interference with cortical function [28], [38]. It is performed either by electrodes that are placed in subdural or intracerebral space (extra-operative stimulation), or by direct cortical stimulation during operation (intra-operative stimulation). The meaning of cortical stimulation process is to define functionally significant cortical regions that should be preserved in epilepsy surgery. Both of these methods also are used for cortical functional mapping. Comparison of the effects of electrical stimulation (without afterdischarges) and the effects produced by afterdischarges limited to the same electrode that had been stimulated electrically indicates that the physiological result of electrical stimulation and result of activation by an epileptiform discharge are extremely similar if not identical [48]. In other words, electrical stimulation is the ideal experimental setting to evaluate the symptomatogenic areas of the human cortex [26]. Nowadays it is an integral part of pre-surgical evaluation process for epilepsy surgery, although its significance has reduced following introduction of several non-invasive pre-surgical investigation methods (like positron emission tomography (PET), single photon emission computed tomography (SPECT), functional MRI, etc). It is usually indicated when non-invasive pre-surgical investigations give diverse results about the possible location of epileptogenic zone and when we need more information to proceed with epilepsy surgery. Thus the questions to be answered with intracranial electrodes are shaped by the results of the non-invasive evaluation, including extracranial EEG [32], [33], [41].

Invasive electroencephalography and Video-EEG-monitoring

Invasive electroencephalography (EEG) is performed through the same intracranial electrodes, which are used for extra-operative cortical stimulation. The need for intracranial recordings may arise if an ictal extra-cranial EEG and other pre-operative studies do not give sufficient evidence for the same localisation of epileptogenic zone, if their results are discordant, or if the suspected seizure onset is in or very near critical eloquent cortical areas [5]. Invasive EEG might also be offered to a patient with no underlying structural pathology identified on neuroimaging, but in whom other investigations have generated a plausible hypothesis as to location of the epileptogenic region [46]. The use of intracranial EEG necessitates careful and individual prejudgement of the hypotheses and goals for every planned recording [45]. Main indications for invasive video-EEG monitoring can be divided into three overlapping groups: defining extent and distribution of the epileptogenic zone (irritative and pacemaker zones), defining epileptogenic zone versus structural lesion and defining epileptogenic zone versus eloquent cortex [17].

The following are examples that may require invasive intracranial monitoring:

- Seizures are lateralized but not localized (e.g., a left-sided, widespread frontal-temporal onset);
- Seizures are localized but not lateralized (e.g., ictal EEG patterns that appear maximally over both temporal lobes);
- Seizures are neither localized nor lateralized (e.g., stereotyped complex partial seizures with diffuse ictal changes or initial changes obscured by artifact);
- Seizure localization is discordant with other data (e.g., EEG ictal scalp data discordant with neuroimaging [MRI, PET, SPECT] or neuropsychological data);
- Relationship of seizure onset to functional tissue must be determined (e.g., seizures with early involvement of language or motor function);
- Relationship of seizure onset to lesion must be determined (e.g., dual pathology or multiple intracranial lesions);
- If seizures are clinically suspected, but video-EEG is inadequate for defining them (e.g., simple partial seizures with no detectable scalp EEG ictal discharge or suspected

epileptic seizures with unusual semiology that suggests psychogenic seizures [pseudo-pseudo seizures]).

By knowing functional map and epileptogenic zone, we can see the overlap of these two zones and decide whether and in what extent should we precede with resective surgery. We can also assess if there is a need for any other investigations and additional invasive electroencephalography or cortical stimulation (repeated extra-operative or additional intra-operative). Today, due to the improvement in various diagnostic methods, the need for semi-invasive (i.e., use of epidural and/or foramen ovale electrodes) or invasive EEG monitoring (i.e., subdural and/or intracerebral depth electrodes) in precise delineation of the epileptogenic zone is much smaller than it used to be. In video-EEG monitoring we can determine the exact type of seizures of a given patient. This investigation also provides information for localisation of seizure origin [19].

Effect of cortical stimulation

Electrical stimulation of the human cortex is the best experimental model to reproduce the effect of activation of the cortex by an epileptiform discharge [26]. The electrical stimulus, which consists of alternating-polarity 0.3-msec duration pulses, produces synchronised depolarisation and hyperpolarisation of the neurons underlying the electrode [30]. Depolarisation and hyperpolarisation of cell membranes produces action potentials, which results as positive or negative clinical effect. Positive or negative clinical effects are dependant on stimulus intensity and stimulated area [22], [25], [30]. Also we know both from kindling models and cortical stimulation experience that not all impulses trigger response, even if we stimulate functionally significant zone. Therefore it is very important to use certain physical parameters. By using these parameters, we enlarge the possibility to elicit response.

Description of different electrodes used for cortical stimulation and/or eeg recording, indications for their use

Subdural grid- and strip-electrodes

Subdural strip electrodes are made from a single row of electrode contacts. The material is biologically inert, flexible (Silastic, Teflon, etc) and commonly has stainless steel electrode contacts. Some manufacturers have also made strips with platinum disk electrodes as well as electrodes that are compatible for magnetoencephalographic (MEG) imaging [54]. Common strip electrode lengths are 5 to 9 cm with four to eight electrode contacts in each strip, but can be up to 16 cm in length. Electrode contacts have been made in diameters generally varying from 2 to 5 mm, with center-to-center distances of 1 to 2 cm between electrodes [23]. Closer distances than 1 cm between electrodes are produces on request, mainly for investigation of highly eloquent areas.

Indications or use

The indications for placement of subdural strip electrodes are exploration and confirmation of epileptogenic foci. Exploration usually assumes that the patient is thought likely to have an epileptogenic focus, but the exact location and lateralisation of the focus is unknown. Confirmatory recordings will often arise in cases where a discrepancy exists between various data accumulated during the presurgical evaluation, and so the localisation of the focus is not sufficient to proceed with surgery [54]. In difference from strip electrodes, grid electrodes are made from parallel rows of multiple electrode contacts. They are made from the same materials as strip electrodes and are generally composed of from two to eight electrodes and rectangular or square arrays each containing up to 64 electrodes. Here, electrode contacts have been made in diameters generally varying from 2 to 5 mm with center-to-center distances of 1 to 2 cm between electrodes [23].

Subdural grid electrodes can be broadly utilized in two circumstances. The first relates to the establishment of the location of the seizure focus and the second relates to the localisation of functional areas that one might not wish to remove during a resective procedure [23]. More particularly, subdural grids are method of choice for situations where we suspect epileptogenic activity from cortical regions and situations when we must perform a cortical stimulation procedure to localise eloquent cortex.

Depth electrodes

Depth electrodes are multiple-contact "needles" of polyurethane or other material that typically are inserted into the brain by way of twist-drill skull holes under stereotactic guidance [56].

Indications for use

In total the indications for depth electrodes are the same as for subdural strip electrodes - exploration and confirmation of epileptogenic foci. More particularly, the most common indication is definition of the side of seizure onset in patients with suspected bitemporal epilepsy. Depth electrodes can also be used to resolve some discrete localisation issues, like mesial temporal versus orbitofrontal or cingulated seizure onset. Deep electrodes are considered to be a "gold standard" for patients with bitemporal epilepsy [56]. We recently started a project for the additional use of depth electrodes together with grid- and strip-electrodes (see Fig. 3b).

The additional usefulness of 3-D reconstructed images of the human cerebral cortex for localization of subdural electrodes in epilepsy surgery was stressed by a work of our Imaging Group at the Ludwig Maximilians University of Munich [53].

Extra-operative cortical stimulation

Extra-operative cortical stimulation is performed mainly by grid electrode, which is placed during the first operation on brain surface (subdural space) in the suspected epileptogenic zone. Subdural strip electrodes are implanted mainly for registering intracranial EEG. The type, number and position of the electrodes are determined by the location of the suspected epileptogenic zone in each patient, according to finding from clinical history, neuroimaging, neuropsychology, scalp electroencephalographic recordings and videotelemetry [1]. After operation patient is brought to the epilepsy intensive station for further observation, recording of intracranial EEG and stimulation. In the intensive station, after gradual reduction of antiepileptic medication, patient is monitored for epileptic seizures, and cortical stimulation of different electrodes is performed, with reaction observed and cortical map drown. As we know, it is very important to use certain stimulation parameters that can provoke some physiological effect. At the beginning of investigation, each 2 electrodes on the strip/grid are stimulated and reaction observed. Initially we start with current strength 1 mA for 5 sec. and proceed until some response or maximal current strength (15 mA) is reached. During stimulation, patient has to perform several tasks, varying from the stimulated zone. The main tasks are moving his arms and fingers in different directions (up-down); naming several objects, presented to him, also months of the year; reading in loud a book or journal, etc. If we observe any changes in this action or patient reports some uncustomary feelings, we imply more detailed tasks to clarify this response. Symptoms during stimulation may include positive motor phenomena (tonic or clonic contraction of the muscle group), negative motor phenomena (inhibition of voluntary movements of the tongue, fingers or toes), somatosensory phenomena (tingling, tightness, or numbness of a part of the body), or language impairment (speech hesitation or arrest, anomia, or repetitive difficulties) [4].

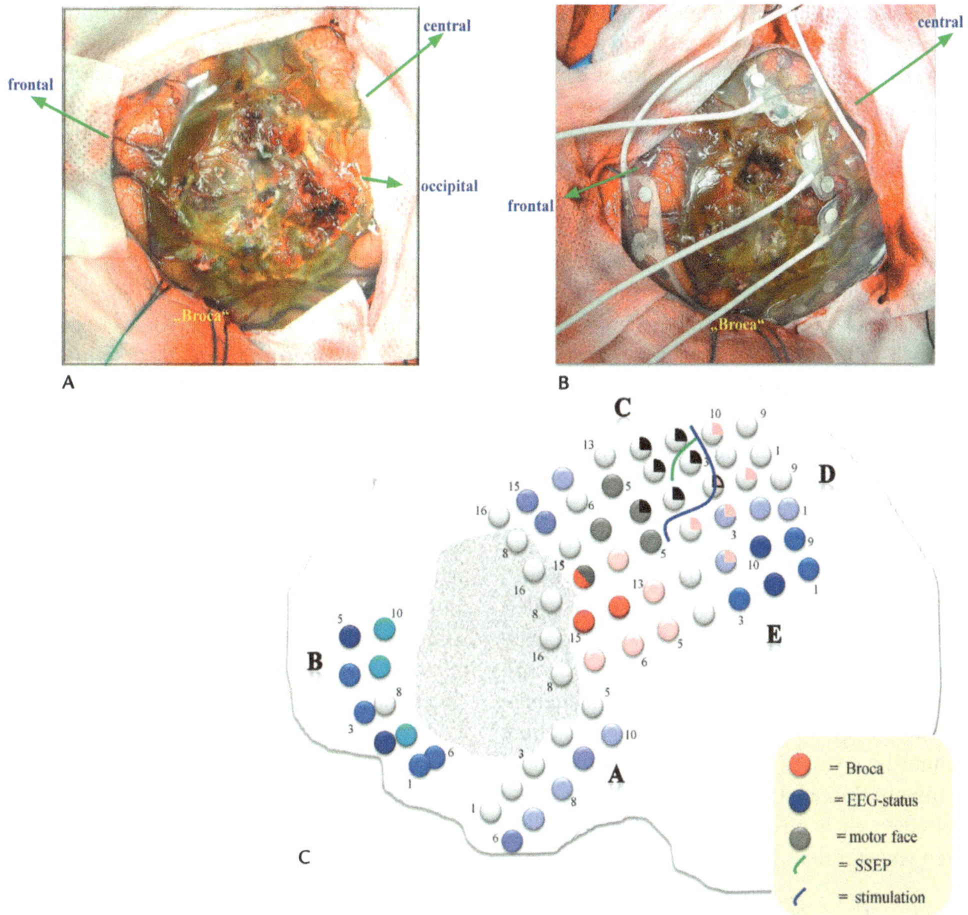

Fig. 2. Case I: 12 years old girl with severe pharmacoresistant epilepsy and an extensive cavernous hemangioma in the left frontobasal area involving the middle and inferior frontal gyrus (pars orbitalis, pars opercularis and pars triangularis) and extending until the precentral area. No speech disturbance after surgery, no neurological deficit was ascertained. The patient is seizure-free since surgery.

(A) Exposure of the cavernous hemangioma.

(B) Covering of the surroundings of the lesion, especially Broca-area and basal central region using 5x2 grid-electrodes.

(C) Mapping of motor speech function in the Broca-area, motor function of face and EEG-status and SSEP-stimulation results

Significant response is considered to be any response during stimulation, which we can observe or which is noted by the patient. "No response" is when we see no reaction during stimulation and also patient reports no changes in his body. In the second phase of stimulation, 1 reference electrode is chosen in the site where no first phase response was gained. Then we repeat stimulation between reference electrode and all the other strip/ grid electrodes. Usually stimulation is performed for 4 to 5 hours in 5 to 7 days. Gradually obtained results are gathered and functional cortical map is drawn. Later patient's data are discussed between epileptologist, neuropsychiatrist, neurophysiologist and epilepsy surgeon. In case of positive decision regarding resective epilepsy surgery, in average 8 to 14 days after 1st operation, second operation with excision of epileptogenic lesion is performed. Used electrodes are disposed in order to prevent transmission of Creutzfeld – Jacob disease. In infants and young children, cortical stimulation studies are more challenging. Sensory, negative motor and language function cannot be assessed reliably during stimulation in infants. Special stimulation paradigms are required to elicit positive motor effects in children younger than 3 or 4 years [20], [35].

Combined use of both stimulation methods

Usually combined use of both stimulation methods is needed in cases when ictal onset zone is close to or overlies critical motor or speech areas. In these situations extra-operative stimulation has been already carried out and intra-operatively we stimulate cortex only in functionally significant points, which lay close to or on the area of planned resection. In our work these were the cases when planned resection was close to speech areas detected in extra-operative cortical mapping. To assess language zones during operation, we performed awake surgeries with direct intra-operative stimulation and consecutive resection of functionally non-significant epileptogenic zone. The main idea of combined cortical mapping is to ensure maximally precise stimulation results and perform the maximal resection of epileptogenic tissue in areas adjacent to functionally signicifant cortex. In such cases, the epileptic zone may be resected up to the pial margin without disturbing function, as long as the vascular structures within the pia are preserved [8].

Afterdischarges

One of the most common problems in cortical stimulation is afterdischarge. By definition it is "the portion of the response to stimulation in a nerve which persists after the stimulus has ceased and consists of rhythmic, high-voltage, high frequency spikes, sharp waves, or spike-wave complexes which occur at the region stimulated and are distinctly different from background activity"[7]. It can be triggered only in certain circumstances when electrical stimulation is done in sufficient intensity, with repetitive rate and some duration. Initially afterdischarges tend to be limited to the stimulating electrode but not infrequently spread to adjacent electrodes producing activation of extensive cortical areas [14], [15]. The symptomatology elicited when afterdischarges are triggered is not only an expression of the area directly stimulated electrically but also of the whole region activated by the afterdischarges. Therefore in these cases we cannot be sure if the answer we get at the electrode site where afterdischarges are elicited is from our stimulation, or it is produced by afterdischarge [42], [43]. This limits us to take into account only those symptoms and signs elicited by stimuli that do not produce afterdischarges.

One way to handle this problem is to stimulate each cortical site with stepwise increasing stimulus intensities until we either elicit signs or symptoms, or afterdischarges are recorded. The drawback of this methodology is that there are great varieties of intensity levels at which afterdischarges can be produced. Therefore, if we test a given cortical site and get no signs or symptoms at stimulus intensity immediately subtreshold for afterdischarges, we cannot necessarily conclude that the tested site is "silent" [29].

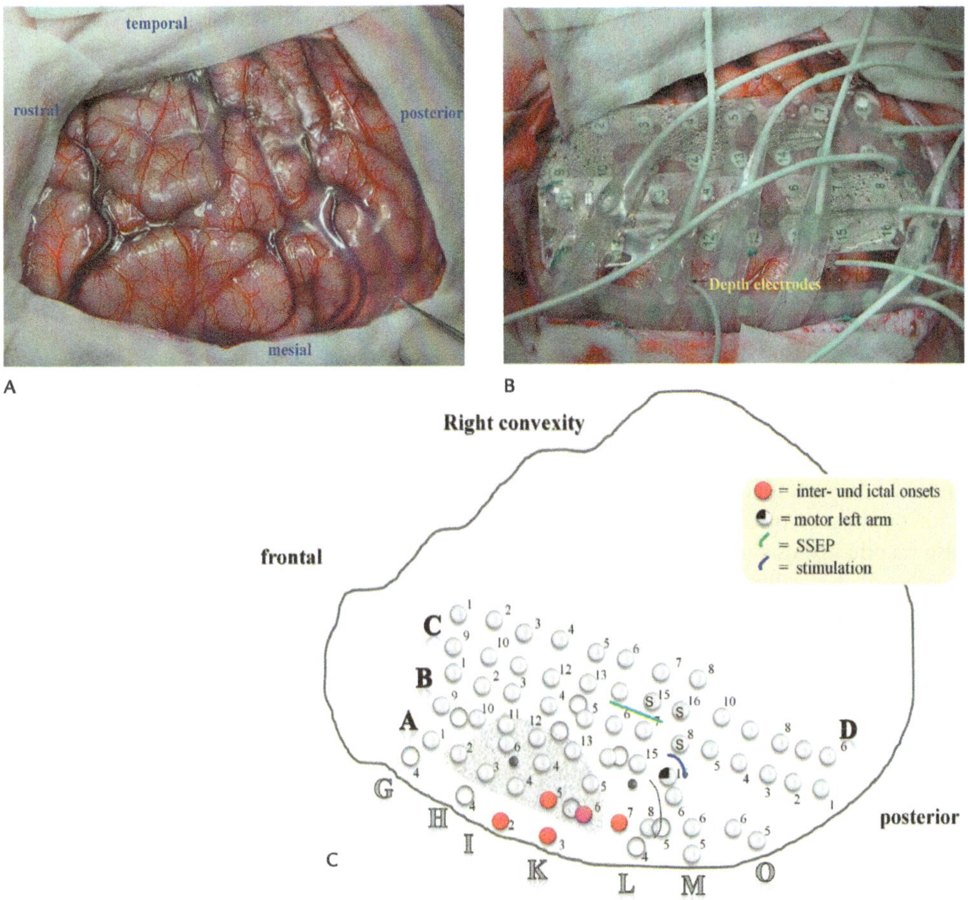

A B C

D

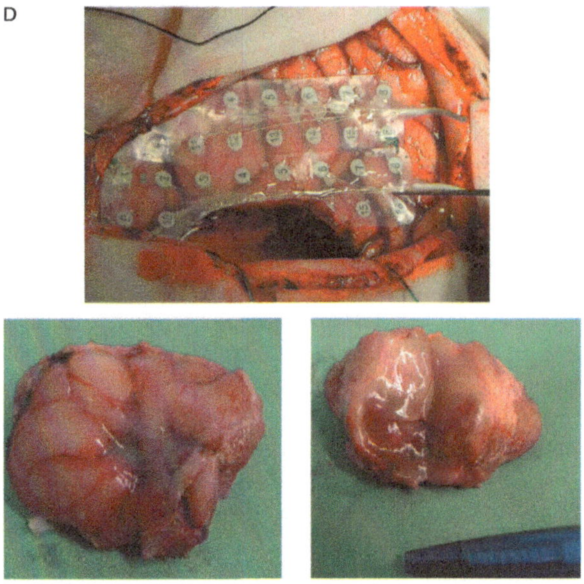

Fig. 3. Case II: 2 years old male suffering from M. Pringle-Bourneville-tuberous sclerosis with a devastating epilepsy and high seizure frequency. A large tuber close to the central and precentral area was diagnosed, involving partly the supplementary sensorimotor area (SSMA). No neurological deficits and seizure freedom since surgery were ascertained.

(A) Exposure of the large tuber in the right precentral area, involving a great portion of the SSMA.

(B) Implantation of many grid- and strip-electrodes. Additionally two depth electrodes in the frontal and parietal border of the tuber were implanted (see arrows).

(C) Mapping of ictal and interictal onsets, motor function of left arm, SSEP- and stimulation-results

(D) Situs after complete resection of the tuber and specimens

Surgical complications after extra-operative stimulation

The likelihood of complications from intracranial recordings is roughly proportional to the number of electrodes implanted [1]. The main complication for extra-operative stimulation is CSF fistula and the most frequent cause for it is cable exit site [44]. In most cases CSF fistula leads to infection which is the most common source of potential morbidity associated with subdural electrode placement, with reported incidence rates of 7% in grid and 0.85% in strip investigations [55]. If meningitis occurs, the electrodes must be removed and infection treated. Other complication is wound infection with or without osteomyelitis. There can be also aseptic necrosis of bone flap. When large number of electrodes is implanted, brain swelling may be extent. Cerebral edema is more likely to occur with an increase in the duration of the monitoring session or with a greater number of intracranial electrodes [52]. Also some thickness of subdural hematomas occurs in all patients. Therefore this diagnosis should always be considered in any patient incurring a new or increasing deficit at any time the electrodes are in place. Transmission of Creutzfeldt-Jacob disease has been reported in the past but can be avoided by disposing of used electrodes.

Concluding remarks

It is interesting to mark that when electrical stimulation elicits some cortical area with functional significance, this finding does not mean that the area stimulated is the only one that can generate that function or even that this area is crucial for that function. In some cases it is possible to resect cortical regions with functional significance without evident postoperative functional deficits [49], [50]. It has also been observed that when an electrode is stimulated continuously for a long period of time or repeatedly with only brief separations between periods of stimulation, the deficit elicited may, in some cases, no longer be found after a period of stimulation. This is analogous to the findings that lesions caused by strokes may cause acute but not chronic deficits in functioning and that after not only central, but also peripheral lesions, cortex may apparently reorganize its functional-anatomical representations [10], [11], [21]. It can be considered that these changes are not due to the growth of new synaptic connections, but instead represent strengthening of existing synaptic connections and relationships due, presumably, to alterations in the inhibitory and excitatory connections between the involved regions [18], [34], [47]. We know that the basic assumption of functional mapping using electrical stimulation is that local electrical stimulation of cortex generates localized responses. However, the extent of affected cortical surface by electrical stimulation is controversial [7]. Ojemann and associates found that bipolar stimulation produced a localized effect between electrode contacts both with monkey occipitocortical stimulation and in human motor as well as association cortices [37]. On the other hand, Van Buren and co-workers noted NADH florescence changes away from electrode contacts. Especially larger currents may generate local afterdischarges which may spread to distant areas [51]. It has also been noted that due to fact that by bipolar stimulation, we stimulate smaller cortical region, higher current levels are needed to elicit response [7]. Therefore, functional mappings should be done very carefully and punctiliously. It has been claimed that recent advances in MRI technology will eventually abolish the need for chronic intracranial recordings. This has not been our experience in recent years. As with any diagnostic test, improvements in MRI sensitivity will necessarily be associated with increments in true and false positive detections. In some patients multiple lesions are detected, raising the question of whether there is a single focal seizure onset and which lesion generates seizures. In other patients, mild non-specific unclear abnormalities can be seen, raising doubt about whether they are indeed epileptogenic [1]. The final decision on whether intracranial recordings and cortical stimulation is needed relies on the specific clinical and personal circumstances of each patient.

Acknowledgments

The kind help of Dipl. Med. Tom Pieper, Department of Neuropediatrics – Epileptology and Dr. Manfred Kudernatsch, Department of Neurosurgery and Epilepsy Surgery at the Schoen Klinik Vogtareuth in providing the intraoperative photographs and of Dr. Daniel Martin from the Department of Neurosurgery at the University of Dresden for reviewing the manuscript and the reference list is highly appreciated.

References

[1] Alarcon G, Valentin A, Watt C, Selway RP et al (2006) Is it worth pursuing surgery for epilepsy in patients with normal neuroimaging? J Neurol Neurosurg Psychiatry 77: 474–480
[2] Allison T, Goff WR, Williamson PD, VanGilder JC (1980) On the neural origin of early components of the human somatosensory evoked potential. In: Desmedt JE (ed) Clinical uses of cerebral, brain stem, and spinal somatosensory evoked potentials. Progress in Clinical Neurophysiology, vol 7. Karger, pp 51–68
[3] Awad IA, Rosenfeld J, Ahl J, Hahn JF, Lüders H (1990) Intractable epilepsy and structural lesions of the brain: mapping, resection strategies and seizure outcome. Epilepsia 32(2): 179–186
[4] Benbadis S, Wyllie E, Bingman WE (2001) Intracranial electroencephalography and localisation studies. In: Wyllie E (ed) The treatment of epilepsy, 3rd edn. Lippincott Williams & Wilkins
[5] Broughton RJ (1967) Somatosensory evoked potentials in man: cortical and scalp recordings. Thesis, McGill University
[6] Carreno M, Lüders HO (2001) General principles of presurgical evaluation. Epilepsy Surg 20: 185–199
[7] Cataltepe O, Comair YG (2000) Intrasurgical cortical electrical stimulation. In: Lüders HO and Noachtar S (eds) Epileptic seizures pathophysiology and clinical semiology. Elsevier LTD, Oxford, pp 172–186
[8] Cohen-Gadol AA, Britton JW, Collignon FP et al (2003) Nonlesional central lobule seizures: use of awake cortical mapping and subdural grid monitoring for resection of seizure focus. J Neurosurg 98(6): 1255-1262
[9] Cushing H (1909) A note upon the faradic stimulation of the postcentral gyrus in conscious patients. Brain 32: 44–54
[10] Donoghue JP, Sanes JN (1988) Organization of adult motor cortex representation patterns following neonatal forelimb nerve injury in rats. J Neurosci 8: 3221–3232
[11] Donoghue JP, Sunner S, Sanes JN (1990) Dynamic organisation of primary motor cortex output to target muscles in adult rats, II. Rapid reorganization following motor cortex nerve lesions. Exp Brain Res 79: 492–503
[12] Foerster O (1936) Sensible cortical field. In: Bumke O and Foerster O (eds) Handbuch der Neurologie. Springer, Berlin, pp 358–448
[13] Gastaut H (1952): Etude electrocorticographique de la reactivite des rhytmes rolandiques. Rev Neurol 87: 176–182
[14] Goddard GV (1982) The kindling model of epilepsy Trends Neurosci 6: 275–279
[15] Goddard GV, McIntyre DC, Leech CK (1969) A permanent change in brain function resulting from daily electrical stimulation. Exp Neurol 25: 295–330
[16] Golding S, Gregorie EM (1984) Surgical management of epilepsy using epidural recordings to localise the seizure focus. Review of 100 cases. J Neurosurg 60: 451–466
[17] Hamer HM, Morris HH (2001) Indications for invasive video-electroencephalographic monitoring. In: Lüders HO and Comair YG (eds) Epilepsy surgery. Lippincott Williams & Wilkins, Philadelphia, pp 559–566
[18] Jacobs KM, Donoghue JP (1991) Reshaping the cortical motor map by unmasking latent intracortical connections. Science 252: 94
[19] Jayakar P, Duchowny M, Reschnik TJ, Alvarez LA (1991) Localization of seizure foci: pitfalls and caveats. J Clin Neurophysiol 8: 414–431
[20] Jayakar P, Alvarez LA, Duchowny MS et al (1992) A safe and effective paradigm to functionally map the cortex in childhood. J Clin Neurophysiol 9: 288–293
[21] Jenkins WM, Merzenich MM (1987) Reorganization of neocortical representations after brain injury: A neurophysiological model of the bases of recovery from stroke. Prog Brain Res 71: 249–266
[22] Krnjevic K, Randic M, Strnoghan DW (1966) An inhibitory process in the cerebral cortex. J Physiol (Lond) 184: 16–48
[23] Lesser RP, Gordon B, Fisher R, Hart J, Uematsu S (1991) Subdural grid electrodes in surgery of epilepsy. Epilepsy Surg 46: 399–408
[24] Lesser RP, Gordon B (2000) Methodologic considerations in cortical electrical stimulation in adults. In: Lüders HO and Noachtar S (eds) Epileptic seizures pathophysiology and clinical semiology. Elsevier LTD, Oxford, pp 153–165

[25] Li CL, Chou SN (1962) Cortical intracellular synaptic potentials and direct cortical stimulation. J Cell Comp Physiol 60: 1–66
[26] Lüders HO, Noachtar S (1995) Atlas und Video epileptischer Anfälle und Syndrome. Wehr, Ciba-Geigy Verlag
[27] Lüders HO, Awad I (1991) Conceptual considerations. Epilepsy Surg 7: 51–62
[28] Lüders H, Lesser RP, Dinner DS, Morris HH et al (1987) Chronic intracranial recording and stimulation with subdural electrodes. In: Engel J Jr (ed) Surgical treatment of the epilepsies. Raven Press, New York, pp 297–321
[29] Lüders HO (2000) Symptomatogenic areas and electrical cortical stimulation. In: Lüders HO and Noachtar S (eds) Epileptic seizures pathophysiology and clinical semiology. Elsevier LTD, Oxford, pp 131–140
[30] Lüders H, Lesser RP, Dinner D, Morris HH, Hahn JF, Friedman L, Skipper G, Wyllie E, Friedman D (1987) Commentary: Chronic intracranial recording and stimulation with subdural electrodes. In: Engel J Jr (ed) Surgical treatment of the epilepsies. Raven Press, New York
[31] Lüders H, Lesser RP, Hahn J, Dinner DS, Kliem G (1983) Cortical somatosensory evoked potentials in response to hand stimulation. J Neurosurg 58: 885–894
[32] Lüders HO, Dinner DS, Morris HH et al (1989) EEG evaluation for epilepsy surgery in children. Cleve Clin J Med 56: S53–S61
[33] Lüders HO, Engel J Jr, Munari C (1993) Noninvasive preoperative evaluation: general principles. In: Engel J Jr (ed) Surgical treatment of epilepsy, 2nd edn. Raven Press, New York, pp 137–155
[34] Morris HH, Lueders H (1985) Electrodes, long-term monitoring in epilepsy. Elsevier Science Publishers, Amsterdam, pp 3–25
[35] Nespeca M, Wyllie E, Lüders HO et al (1990) Subdural electrodes in infants and young children. J Epilepsy 3: 107–124
[36] Noachtar S, Winkler P, Lüders HO (2003) Surgical therapy of epilepsy. Neurological disorders: course and treatment, 2nd edn. Elsevier Science Publishers, Amsterdam
[37] Ojemann GA, Sutherling WW, Lesser RP et al (1993) Cortical stimulation. In: Engel J Jr (ed) Surgical treatment of the epilepsies, 2nd edn. Raven Press, New York, p 339
[38] Ojemann GA, Sutherling WW, Lesser RP et al (1993) Cortical stimulation. Engel J Jr (ed) Surgical treatment of the epilepsies, 2nd edn. Raven Press, New York, pp 399–414
[39] Penfield W, Jasper HH (1954) Epilepsy and the functional anatomy of the human brain. Little Brown, Boston
[40] Penfield W (1958) The excitable cortex in conscious man. Liverpool University Press, Liverpool
[41] Quesney LF, Risinger MW, Shewmon DA (1993) Extracranial EEG evaluation. In: Engel J Jr (ed) Surgical treatment of epilepsy, 2nd edn. Raven Press, New York, pp 173–196
[42] Racine RJ (1972) Modification of seizure activity by electrical stimulation. II. Motor seizure. Electroencephalogr Clin Neurophisiol 32: 281–294
[43] Racine RJ (1972) Modification of seizure activity by electrical stimulation. I. Afterdischarge treshold. Electroencephalogr Clin Neurophysiol 32: 269–279
[44] Roper SN (1995) Implantation of grid and strip electrodes. Tech Neurosurg 1: 5–10
[45] Rosenow F, Lueders H (2001) Presurgical evaluation of epilepsy. Brain 124: 1683–1700
[46] Smith SJM (2005) EEG in the diagnosis, classification, and management of patients with epilepsy. J Neurol Neurosurg Psychiatry 76 (Suppl 2): ii2–ii7
[47] Smits E, Gordon DC, Witte S et al (1991) Synaptic potentials evoked by convergent somatosensory and corticocortical inputs in raccoon motor cortex: Substrates of plasticity. J Neurophysiol 66: 688–695
[48] Tailarach J, Bancaud J (1974) Stereotactic exploration and therapy in epilepsy. Clin Neurol 15: 758–782
[49] Uematsu S, Lesser RP, Gordon B (1992) Localization of sensorimotor cortex: The influence of Sherrington and Cushing on the modern concept. Neurosurgery 30: 904–913
[50] Uematsu S, Lesser RP, Fisher RP et al (1992) Motor and sensory cortex in humans: Topography studied with chronic subdural stimulation. Neurosurgery 31: 59–72
[51] Van Buren J, Lews D, Sceuttew W et al (1978) Fluorometric monitoring of NADH levels in cerebral cortex. Neurosurgery 2: 114
[52] Wiggins GC, Elisevich K, Smith BJ (1999) Morbidity and infection in combined subdural grid and strip electrode investigation for intractable epilepsy. Epilepsy Res Oct 37(1): 73–80
[53] Winkler PA, Vollmar C, Krishnan KG, Pfluger T, Brückmann H, Noachtar S (2000) Usefulness of 3-D reconstructed images of the human cerebral cortex for localization of subdural electrodes in epilepsy surgery. Epilepsy Res 41: 169–178
[54] Wyler AR (1991) Subdural strip electrodes in surgery of epilepsy. In: Lüders H (ed) Epilepsy surgery. Raven Press, New York, pp 395–398
[55] Wyler AR, Walker G, Somes G (1991) The morbidity of long-term seizure monitoring using subdural strip electrodes. J Neurosurg 74: 734–737
[56] Wyllie E, Awad I (1993) Intracranial EEG and localization studies. In: Wyllie E (ed) The treatment of epilepsy: principles and practices. Lea & Febiger, Philadelphia, pp 1023–1037

Intraoperative electrical mapping: advances, limitations and perspectives

Emmanuel Mandonnet

The use of direct electrical stimulation (DES) to perform intraoperative brain mapping in order to preserve eloquent areas during epilepsy surgery has been established as a reliable routine procedure by Penfield in his seminal paper in 1937 [42]. The technique has been further refined by Ojemann in the seventies, by introducing a biphasic current-constant pulse and by optimizing the intraoperative testing tasks [52]. In the nineties, Berger applied the method for mapping eloquent cortical areas during oncological neurosurgery [5], and Duffau then extended the indication to the preservation of axonal pathways [10]. In the last decade, popularized by the reported successes [9, 46], the technique spread all over the world [24].

The aim of this chapter is to review what is our current understanding of the interaction between brain and Ojemann Stimulation (OS: 60 Hz 1ms biphasic current-constant bipolar stimulation) and to analyze its sensitivity and specificity in predicting the clinical consequences of a focal surgical removal.

Focal electrophysiological effects of DES: insights from experimental and modeling studies

We will now introduce some basic results on the local biophysics of DES. We will first review what is known about individual neuronal behaviour (i.e., membrane dynamics properties) from *intracellular* stimulation experiments and models, before focusing on the more complex situation of *extracellular* stimulation.

Studies of individual neuronal properties have been initiated in 1952 by Hodgkin and Huxley, in the series of papers describing both experimentally and mathematically the response of the transmembrane potential to an intracellularly controlled current [18–22]. This pioneering work has been since refined with the introduction of intracellular patch-clamped techniques (see [38] and refs. 10, 20, 28, 33 in [44]), leading to the experimental determination of membrane characteristics for different kinds of neurons. Part of the original equation is based on a bioelectrical model of the membrane (considered as an equivalent of multiple parallel RC circuits for each ionic currents), while part of it has been derived purely phenomenologically, to fit at best the experimental data. The fundamental notion is that an *outward* current going through the membrane towards the outside of the neuron will depolarize the transmembrane potential (i.e., shifting from negative towards positive values the difference between the intracellular potential and the extracellular potential, which is negative at rest). They proposed a phenomenological law for the variation of the resistances of the sodium and potassium channels with the transmembrane potential. As a consequence of the non-linear coupled set of differential equations, if a certain threshold of depolarization is reached, an action potential is generated, without the need of further support from electrical stimulation. As stated in the conclusion of the theoretical paper [18], the model was in good agreement with experiments regarding:

- the threshold to generate an action potential, (as well as the waveform of this potential)
- the velocity of propagation of this potential,
- the refractory period,
- the subthreshold response, including oscillations,
- the accommodation phenomenon,
- the generation of an action potential after stopping abruptly a hyperpolarization (the so called anodal break response).

Three separate systems have then to be considered when modeling and analyzing the effects of any *extracellular* DES: the individual neuronal unit, the surrounding tissue and the stimulator. As proposed by McNeal in [36] and reminded by Warman et al in [50], modeling the effects of electrical *extracellular* stimulation is a two-steps process. The first one is the determination of the extracellular potentials induced by the stimulator in the surrounding tissue. The second step involves the solving of the non-linear equations governing the evolution of the transmembrane potential of the target neuron, with transmembrane currents derived from extracellular potentials calculated in the first step. Ultimately, as stated by Rattay in [44], "the high density of excitable elements in CNS stimulation which become active at the same time cause rather strong membrane currents that may significantly influence the extracellular potential", so that the first step should be recomputed after the first events of excitation.

Before exploring in more details each of the two steps, we will briefly present an historical perspective on the first experimental and modeling studies of *extracellular* stimulations. Indeed, long before the work of Hodgkin and Huxley, Lapicque proposed in 1907 to model the neuronal membrane of a peripheral nerve (in frogs) as a simple constant capacitor in parallel with a leaky resistance [6, 25, 26]. In this model, the membrane is thus fully characterized by a time-constant (product of capacitance by resistance). This hypothesis allowed him to derive the mathematical law of the strength-duration curve (that is the experimental curve representing the threshold intensity of the extracellular current needed to generate an excitation as a function of the single pulse duration). Two values can be defined on this strength-duration curve: the rheobasis (minimal amount of current needed to generate an excitation with a very long duration pulse) and the chronaxie (pulse duration for a stimulating threshold intensity at twice the rheobasis). It can be shown easily that the chronaxie is directly proportional to the time-constant of the membrane. Thus, measuring the chronaxie of a tissue is the simplest way to quantify tissue excitability and to estimate the intrinsic membrane properties of the stimulated elements, "independently on the conditions in which the excitation is given" [26]. This is why chronaxie values are still in use 100 years later, especially in cardiac tissue stimulation, although, as explained later, modern computational modeling renders this notion somehow inaccurate and old-fashioned for the case of central nervous system electrical stimulation (see, for example [33, 48]). Moreover, within this model, it is elementary to show that the chronaxie corresponds to the point on the strength-duration curve where the energy required to generate an action potential is minimal, a notion of utmost importance for reducing the risk of damaging the stimulated tissue and optimizing the longevity of the battery in chronically stimulating devices.

The first step of computing the extracellular potentials produced by a given stimulator is quite straightforward if the tissue is considered as isotropic and non-limited, as an explicit analytical formula is tractable [33, 44]. This is no more possible when considering the exact gyral and sulcal anatomy of the cortex: solving the problem then requires the use of numerical methods [29, 30]. To our knowledge there are no studies on the influence of the highly anisotropic nature of fibers on the determination of extracellular potentials when stimulating an axonal pathway. One of the important parameter characterizing a tissue is its resistivity. It is useful to know that the resistivity of gray matter is 4–6 times greater than cerebrospinal fluid (meaning that CSF may shunt the current and be the cause of a false negative stimulation) and about 2–3 times less than the resistivity of white matter [43]. The value of 800 Ω cm found in an orthogonal direction of the fibers is about ten times higher than parallel to the fibers [39].

The second step of solving the non-linear equations of the neuronal unit under the influence of the previously computed extracellular potentials might be computationally intensive. However, solving these equations confirmed the experimental data evidencing that a wide range of phenomena arise from the specificity of stimulating *extracellularly* rather than *intracellularly*. We refer the reader to previous work for extensive review of these phenomena [3, 12, and 43]. It has been shown that computing the *activating function*, namely the spatial derivative of the component of the electric field along the axis of the axon, is an approximate but very efficient way to understand intuitively these phenomena. These include, but are not confined to:

- the axon is the site where an action potential is induced [33, 40, 41, 44],
- the larger the diameter of the axon, the greater its excitability [3, 12],
- a cathode may act on both distant sides of the stimulation center as a virtual anode, leading to the observation of surround anodal block, and to the paradox that close to the cathode, small diameter axons may be stimulated and larger ones may not be [3, 12, 43],
- similarly, an anode may act on both distant sides of the stimulation center as a virtual cathode,
- an anode will stimulate preferentially axons perpendicular to the cortical surface, while a cathode will stimulate axons longitudinally aligned with the cortical surface [29],
- it is possible to design specifically an electrode to obtain an unidirectional propagation stimulation [49] or a diameter selective stimulation [27].

All in all, these effects illustrate the complexity of *extracellular* cortical electrical stimulation, and the question raised by Ranck in 1975[43] "Which elements are excited in electrical stimulation of mammalian central nervous system" still holds, although a more appropriate formulation would be "which elements are excited or inhibited or just perturbed (by sub-threshold effects)?". Moreover, it should be kept in mind that most of the aforementioned knowledge, although probably relevant for the case of human brain mapping with OS as well, has been derived in other fields of DES, essentially with monophonic pulse and monopolar electrodes and, for some of them, in the peripheral nervous system. Actually, to fill the gap between these studies and clinical situations, more realistic computational models have been recently developed in the context of cortical stimulation for pain [29, 30] and deep brain stimulation for movement disorders [33, 35], leading to a better understanding of the local stimulation effects for both cases. To our knowledge, such studies have never been performed in the case of OS for brain mapping. Only one experimental study has shown by optical imaging that the intrinsic signal was spatially highly localized around the electrode tip with the lowest impedance [16].

Effects of a focal lesion[1] on a non-locally functioning brain

There is no doubt that high-order cognitive processes are sustained by large-scale networks distributed all over the brain. In other words, it is not possible to attribute a specific function to a focal area. But we do not know how to interact in a non-local way with an entity like a large-scale network. All we can do at the present time is to analyze the effects of a local perturbation, applied on a subpart of the system. However, any focal lesion will induce a perturbation at least over areas mono-synaptically linked to that area and likely over pluri-synap-

1 In this paragraph, the term lesion will be used in a generic way, including focal surgical removal.

tically connected areas (but not all over the brain, as there exists a certain degree of spatial segregation between separate functions). This means that a whole large-scale network (including peri-lesional, intra- and inter- hemispheric areas, controlateral areas, subcortical and cerebellar areas) is indeed disturbed. Thus, part of the whole brain is forced to passively reorganize in response to this external lesion, meaning that all the dynamics of all cognitive processes will be affected and redistributed over the remaining unaffected networks. Eventually, the re-shaping might be not effective enough to restore a fully normally functioning brain, leading to an observable clinical deficit.

It seems intuitive to infer from the observed deficit what the function of an area was prior it has been damaged. However, one cannot exclude that this function could have been successfully redistributed over the non-affected brain, but as a collateral effect of this redistribution, another function could have been impaired (thus leading to a deficit completely unrelated to the function normally supported by the destroyed area). In other words, localizationism, a powerful theory when applied to primary sensori-motor areas, could be a misleading theory when applied to high-order cognitive processes. The lack of one-to-one relationship between the function of an area and the observed deficit after its lesion puts a fundamental limitation for any "non-interfering" brain mapping method in predicting the cognitive consequences of a surgical removal. It explains, at least in part, the non-optimal sensitivity and specificity of fMRI compared to OS for language networks identification [45].

In most instances, the whole brain reshaping is immediate, but in some situations, nonlinear complex dynamics may require a timescale of several minutes to hours before reaching a new steady state, leading to a *delayed onset of the deficit*. The fact that a clinical deficit might become apparent only after half an hour after a resection has been already reported for the case of the supplementary motor area [11].

The same process of fast dynamical re-arrangement of brain networks may also explain observations that an initial deficit after a resection might recover very shortly, over half an hour [37]. This phenomenon could be called *dynamical short-term plasticity*.

Finally, after an initial lesion causing a deficit, long-term plasticity mechanisms will take place in an attempt to compensate, over a timescale ranging from days to years. Biological modifications of individual neuronal properties will support this second rewiring of the brain, usually facilitated by rehabilitation strategies. This corresponds to a *biological long-term plasticity*.

Comparison between a transient DES lesion and a definitive surgical removal

DES does introduce an artificial non-physiologic signal into the brain, whereas a surgical removal, while certainly disturbing brain networks, does not. As a consequence, a surgical removal always tends to damage some function, whereas the effect of OS is more complex. Indeed, for sensori-motor functions, Ojemann stimulation seems to create what is sometimes called a "positive" effect, that is mimicking some kind of sensori-motor behavior [7, 13, 42], whereas for higher order cognitive processes like language, arithmetics, or spatial awareness, the same stimulation technique seems to block the function, as if the physiological signal was jammed by the 60 Hz stimulation (a theory introduced to explain the therapeutic effect of high frequency stimulation in the context of movement disorders [4]). One can always argue that the movement elicited by Ojemann stimulation might be considered as well as a dysfunction of motor networks, as the patient did not intent to move. However, this would still not explain why Ojemann stimulation does not seem to force the patient to pronounce some words. This rises the possibility that depending on the signal engineering (high versus low frequency, periodic versus random, waveform of single pulse), some functions would be somehow simulated, whereas some others would be inhibited by OS. Thus, one has to admit that with the current technique, OS either mimicks the deficit one would obtain with surgical removal, or activates a roughly simplified version of normal sensori-motor behavior, indicating in both cases the need to preserve the responsive area.

Moreover, this non-physiologic signal might interfere with other network that the one linked to the stimulated area. For example, let us consider a pathway projecting unidirectional from a network A towards another network B in physiologic condition (for example, network A of auditory comprehension projecting toward network B of autobiographical verbal memory). An axonal spike generated by OS in the linking fasciculus might propagate antidromically up to network A, causing a deficit, which would have not been observed by the removal of this pathway.

Finally, the fact that OS is applied for short period of 3 s precludes observing any of the aforementioned phenomena whose occurrence is time-dependent, including:

- a delayed onset of a deficit,
- a fast recovery by dynamical short-term plasticity,
- a late recovery by biological long-term plasticity.

Analysis of false negative and false positive of OS

A false negative site is by definition an area whose removal led to a *permanent* postoperative diagnosis, despite its OS did not evoke any functional impairment during intraoperative mapping. Some methodological errors have been already identified and discussed in [28] as the main source of false negative, including:

- sub-threshold stimulation (either by low intensity or short period or CSF shunt),
- stimulation during the refractory period following an after discharge,
- inappropriate selection of the testing task.

Last but not least, the synchronization between stimulation and the course of the tested cognitive process has been recently shown to generate false negatives, when mapping is performed with short-train pulses technique [2]. Although the relatively long period of 3s stimulation with OS technique should limit the risk of false negative, the approximate synchronization performed by the surgeon (applying stimulation in correlation with the bip of a newly presented picture) could restrict the reproducibility of the errors: a perfectly synchronized stimulation could cause a complete anomia, whereas a poorly synchronized stimulation would only generate a semantic paraphasia.

However, even assuming a rigorous application of the technique, a high rate of patients do present *immediate transient* post-operative deficits [14], whose cause is not clearly identified. We acknowledge that inflammatory processes and/or small venous infarcts are likely to contribute to the deficit. Alternatively, the aforementioned phenomena of *delayed onset of deficit* could be invoked: the perturbation induced by OS lasting 3s is initially silent, as it would require maintaining the perturbation for a longer time before the deficit could be apparent. A somehow similar phenomenon has been observed after turning off high frequency stimulation for movement disorders: functional impairment is not immediate, but appears progressively on a time-scale of a few hours [15, 47].

We have already investigated in [28] the potential causes of false positive – i.e., the fact that an area has been preserved on the ground of stimulation-induced functional response, whereas no long term deficit would have been observed after its removal. Whereas some methodological causes (like patient tiredness or after discharge occurrence) can be easily fixed, some others definitely limit intrinsically the specificity of OS:

- the fact that the transient deficit is not related to the stimulated network but to a distant (albeit axonally linked) network, as a result of an anti-dromic propagation [23],
- a post-operative deficit (caused by the removal of an intraoperatively responsive area) could actually resolved later in time, either

by dynamical short term plasticity [37] or biological long-term plasticity [8].

Conclusion and future works

The fundamental concept that, for high-order cognitive processes, there is no reciprocal relation between a function and a focal area (as assumed by localizationist theory) explains why any pre-operative non-interfering brain mapping techniques will remain of limited value for predicting the effect of a focal surgical removal. The only way to take advantage of these studies would be to enter these data in a biomathematical model of large-scale networks (in order to personalize general frameworks to an individual patient) and then simulate in-silico the effect of a focal removal on the whole network dynamics. Considerable efforts are devoted to this end, leading to some promising results [1]. But even if this methology should supplant invasive brain mapping in the long term future, it is currently in its infancy and OS still provide in 2010 the best tool to tailor resection according to functional boundaries.

While OS are certainly the gold standard in brain mapping, the present chapter aimed to illustrate some of their limitations. Most of them could be overcome by improving our understanding of both local and non-local effects of OS on cognitive processes. To this intent, the case of axonal stimulation rather than cortical stimulation seems to be the best candidate. Indeed, given the simple anatomical configuration of axonal pathways (compared to the complex organization of the cortical layers), a realistic computational modeling of local effects of OS on the fibers could be achieved. Compared to previous similar work performed in the context of cortical stimulation for pain [29, 30] and deep brain stimulation for movement disorders [35], the novelty would be to analyze the specificity generated by the geometric configuration of the 5 mm spaced bipolar probe, the biphasic nature of the pulse, and the highly anisotropic nature of the tissue. The influence of several parameters (including the distance and the orientation between the fibers as well as various waveforms of pulse) could be theoretically determined. Model predictions should be experimentally verified in animals, at the local level of the stimulated fascicle, for example by means of optical imaging [17].

Once the OS effects on local fibers would have been accurately described, one could focus on the study of non-local effects of OS. The concept that a whole large-scale network is the system to be modeled has emerged in several clinical settings of direct electrical stimulation, such as brain mapping [28], deep brain stimulation [34], or epilepsy [51]. This would allow to determine the role of signal engineering in generating a functional response, by simulating the effects of short-train pulses versus repetitive pulses [2], of periodically versus randomly distributed pulses, and of low versus high frequency signals [53]. Results of these in-silico simulations could then be compared with intraoperative cortical electrophysiological recordings during axonal stimulation, modifying the cortico-cortical-evoked potential methodology [31, 32] to an axono-cortical-evoked potential set-up [28].

We are confident that this combination of computational modeling and experimental studies will allow to propose new stimulating methods, with a better specificity, resulting in a better understanding of neural correlates of brain function, and ultimately, in better surgical outcomes for patients.

References

[1] Alstott J, Breakspear M, Hagmann P, Cammoun L, Sporns O (2009) Modeling the impact of lesions in the human brain. PLoS Comput Biol 5: e1000408

[2] Axelson HW, Hesselager G, Flink R (2009) Successful localization of the Broca area with short-train pulses instead of "Penfield" stimulation. Seizure 18: 374–375

[3] Basser PJ, Roth BJ (2000) New currents in electrical stimulation of excitable tissues. Annu Rev Biomed Eng 2: 377–397

[4] Benabid AL, Pollak P, Gao D, Hoffmann D, Limousin P, Gay E, Payen I, Benazzouz A (1996) Chronic electrical stimulation of the ventralis intermedius nucleus of the thalamus as a treatment of movement disorders. J Neurosurg 84: 203–214

[5] Berger MS, Ojemann GA (1992) Intraoperative brain mapping techniques in neuro-oncology. Stereotact Funct Neurosurg 58: 153–161

[6] Brunel N, van Rossum MC (2007) Lapicque's 1907 paper from frogs to integrate-and-fire. Biol Cybern 97: 337–339

[7] Desmurget M, Reilly KT, Richard N, Szathmari A, Mottolese C, Sirigu A (2009) Movement intention after parietal cortex stimulation in humans. Science 324: 811–813

[8] Duffau H (2009) Does post-lesional subcortical plasticity exist in the human brain? Neurosci Res 65: 131–135

[9] Duffau H (2009) Surgery of low-grade gliomas: towards a "functional neurooncology". Curr Opin Oncol 21: 543–549

[10] Duffau H, Capelle L, Sichez N, Denvil D, Lopes M, Sichez JP, Bitar A, Fohanno D (2002) Intraoperative mapping of the subcortical language pathways using direct stimulations. An anatomo-functional study. Brain 125: 199–214

[11] Duffau H, Lopes M, Denvil D, Capelle L (2001) Delayed onset of the supplementary motor area syndrome after surgical resection of the mesial frontal lobe: a time course study using intraoperative mapping in an awake patient. Stereotact Funct Neurosurg 76: 74–82

[12] Durand DM (2000) Electric stimulation of excitable tissue. In: Bronzino JD (ed) The biomedical engineering handbook, 2nd edn. Boca Raton, CRC Press

[13] Fried I, Katz A, McCarthy G, Sass KJ, Williamson P, Spencer SS, Spencer DD (1991) Functional organization of human supplementary motor cortex studied by electrical stimulation. J Neurosci 11: 3656–3666

[14] Gil-Robles S, Duffau H (2010) Surgical management of World Health Organization grade II gliomas in eloquent areas: the necessity of preserving a margin around functional structures. Neurosurg Focus 28: E8

[15] Grips E, Blahak C, Capelle HH, Bazner H, Weigel R, Sedlaczek O, Krauss JK, Wohrle JC (2007) Patterns of reoccurrence of segmental dystonia after discontinuation of deep brain stimulation. J Neurol Neurosurg Psychiatry 78: 318–320

[16] Haglund MM, Ojemann GA, Blasdel GG (1993) Optical imaging of bipolar cortical stimulation. J Neurosurg 78: 785–793

[17] Histed MH, Bonin V, Reid RC (2009) Direct activation of sparse, distributed populations of cortical neurons by electrical microstimulation. Neuron 63: 508–522

[18] Hodgkin AL, Huxley AF (1952) A quantitative description of membrane current and its application to conduction and excitation in nerve. J Physiol 117: 500–544

[19] Hodgkin AL, Huxley AF (1952) Currents carried by sodium and potassium ions through the membrane of the giant axon of Loligo. J Physiol 116: 449–472

[20] Hodgkin AL, Huxley AF (1952) The components of membrane conductance in the giant axon of Loligo. J Physiol 116: 473–496

[21] Hodgkin AL, Huxley AF (1952) The dual effect of membrane potential on sodium conductance in the giant axon of Loligo. J Physiol 116: 497–506

[22] Hodgkin AL, Huxley AF, Katz B (1952) Measurement of current-voltage relations in the membrane of the giant axon of Loligo. J Physiol 116: 424–448

[23] Ishitobi M, Nakasato N, Suzuki K, Nagamatsu K, Shamoto H, Yoshimoto T (2000) Remote discharges in the posterior language area during basal temporal stimulation. Neuroreport 11: 2997–3000

[24] July J, Manninen P, Lai J, Yao Z, Bernstein M (2009) The history of awake craniotomy for brain tumor and its spread into Asia. Surg Neurol 71: 621–624; discussion 624–625

[25] Lapicque L (1907) Recherches quantitatives sur l'excitation électrique des nerfs traitée comme une polarisation. J Physiol Pathol Générale 9: 620–635

[26] Lapicque L (2007) Quantitative investigations of electrical nerve excitation treated as polarization. 1907. Biol Cybern 97: 341–349

[27] Lertmanorat Z, Gustafson KJ, Durand DM (2006) Electrode array for reversing the recruitment order of peripheral nerve stimulation: experimental studies. Ann Biomed Eng 34: 152–160

[28] Mandonnet E, Winkler PA, Duffau H (2010) Direct electrical stimulation as an input gate into brain functional networks: principles, advantages and limitations. Acta Neurochir (Wien) 152: 185–193

[29] Manola L, Holsheimer J, Veltink P, Buitenweg JR (2007) Anodal vs cathodal stimulation of motor cortex: a modeling study. Clin Neurophysiol 118: 464–474

[30] Manola L, Roelofsen BH, Holsheimer J, Marani E, Geelen J (2005) Modelling motor cortex stimulation for chronic pain control: electrical potential field, activating functions and responses of simple nerve fibre models. Med Biol Eng Comput 43: 335–343

[31] Matsumoto R, Nair DR, LaPresto E, Bingaman W, Shibasaki H, Lüders HO (2007) Functional connectivity in

human cortical motor system: a cortico-cortical evoked potential study. Brain 130: 181–197

[32] Matsumoto R, Nair DR, LaPresto E, Najm I, Bingaman W, Shibasaki H, Lüders HO: Functional connectivity in the human language system (2004) a cortico-cortical evoked potential study. Brain 127: 2316–2330

[33] McIntyre CC, Grill WM (1999) Excitation of central nervous system neurons by nonuniform electric fields. Biophys J 76: 878–888

[34] McIntyre CC, Hahn PJ (2010) Network perspectives on the mechanisms of deep brain stimulation. Neurobiol Dis 38: 329–337

[35] McIntyre CC, Miocinovic S, Butson CR (2007) Computational analysis of deep brain stimulation. Expert Rev Med Devices 4: 615–622

[36] McNeal DR (1976) Analysis of a model for excitation of myelinated nerve. IEEE Trans Biomed Eng 23: 329–337

[37] Mikuni N, Ohara S, Ikeda A, Hayashi N, Nishida N, Taki J, Enatsu R, Matsumoto R, Shibasaki H, Hashimoto N (2006) Evidence for a wide distribution of negative motor areas in the perirolandic cortex. Clin Neurophysiol 117: 33–40

[38] Neher E, Sakmann B (1976) Single-channel currents recorded from membrane of denervated frog muscle fibres. Nature 260: 799–802

[39] Nicholson PW (1965) Specific impedance of cerebral white matter. Exp Neurol 13: 386–401

[40] Nowak LG, Bullier J (1998) Axons, but not cell bodies, are activated by electrical stimulation in cortical gray matter. I. Evidence from chronaxie measurements. Exp Brain Res 118: 477–488

[41] Nowak LG, Bullier J (1998) Axons, but not cell bodies, are activated by electrical stimulation in cortical gray matter. II. Evidence from selective inactivation of cell bodies and axon initial segments. Exp Brain Res 118: 489–500

[42] Penfield W, Boldrey E (1937) Somatic motor and sensory representation in the cerebral cortex of man as studied by electrical stimulation. Brain 60: 389–443

[43] Ranck JB, Jr. (1975) Which elements are excited in electrical stimulation of mammalian central nervous system: a review. Brain Res 98: 417–440

[44] Rattay F (1999) The basic mechanism for the electrical stimulation of the nervous system. Neuroscience 89: 335–346

[45] Roux FE, Boulanouar K, Lotterie JA, Mejdoubi M, LeSage JP, Berry I (2003) Language functional magnetic resonance imaging in preoperative assessment of language areas: correlation with direct cortical stimulation. Neurosurgery 52: 1335–1345; discussion 1345–1337

[46] Sanai N, Berger MS (2010) Intraoperative stimulation techniques for functional pathway preservation and glioma resection. Neurosurg Focus 28: E1

[47] Temperli P, Ghika J, Villemure JG, Burkhard PR, Bogousslavsky J, Vingerhoets FJ (2003) How do Parkinsonian signs return after discontinuation of subthalamic DBS? Neurology 60: 78–81

[48] Testerman RL (2005) Comments on "accuracy limitations of chronaxie values". IEEE Trans Biomed Eng 52: 750

[49] van den Honert C, Mortimer JT (1979) Generation of unidirectionally propagated action potentials in a peripheral nerve by brief stimuli. Science 206: 1311–1312

[50] Warman EN, Grill WM, Durand D (1992) Modeling the effects of electric fields on nerve fibers: determination of excitation thresholds. IEEE Trans Biomed Eng 39: 1244–1254

[51] Wendling F (2008) Computational models of epileptic activity: a bridge between observation and pathophysiological interpretation. Expert Rev Neurother 8: 889–896

[52] Whitaker HA, Ojemann GA (1977) Graded localisation of naming from electrical stimulation mapping of left cerebral cortex. Nature 270: 50–51

[53] Zangaladze A, Sharan A, Evans J, Wyeth DH, Wyeth EG, Tracy JI, Chervoneva I, Sperling MR (2008) The effectiveness of low-frequency stimulation for mapping cortical function. Epilepsia 49: 481–487

New insights into neuro-cognition provided by brain mapping

Experience-dependent reorganization of somatosensory and motor cortical areas: towards a neurobiology of rehabilitation

Christian Xerri

Introduction

The cerebral cortex is a dynamic construct of highly interconnected and spatially distributed neuronal networks whose morphological and functional connectivity is continuously modified by experience-dependent plasticity mechanisms. Use-dependent activity of neuronal networks has been shown to reshape neuronal circuits by promoting changes in synapse strength and the formation and elimination of synapses. Both basic and clinical research over the past decades have clearly established that salient experience and intensive training lead to widespread organizational changes within the subcortical and cortical representations underlying sensory perception, motor integration, and memory formation, thereby paving the route for the acquisition of new sensorimotor and cognitive skills. Conversely, disuse and social isolation have long been shown to exert adverse effects on sensorimotor and cognitive abilities through deleterious morphological and biochemical alterations of neuronal circuits. Recent findings have disclosed powerful top-down influences on the interplay of cortical and thalamic processes that allow attention, expectation, and motivation to reshape cortical representations in a behaviorally relevant way. In addition, disruption of neuronal networks as a result of damage to the peripheral or central nervous system triggers large-scale reorganizations that evolve over time and are capable of mediating functional recovery by engaging experience-dependent neuroplasticity mechanisms.

The study of cortical plasticity has greatly benefited from animal research tracking changes at levels ranging from synapses and single cells to neuronal populations and networks. Recent developments in multielectrode recording and optical imaging techniques allow for recording neuronal population activity in awake and behaving animals in more relevant behavioral contexts. In addition, brain imaging studies in humans have opened up new avenues for studying functional reorganizations within widely distributed cortical networks. The main purpose of this chapter is to present current ideas and recent advances in animal and human research on somatosensory and motor representational plasticity, conceived as a neural substrate of skill acquisition and postlesion behavioral recovery, of particular relevance to clinical rehabilitation practices.

Reshaping of somatosensory and motor maps by intensive practice

Cortical areas have the capacity for being constantly reshaped by novel experience and behavioral context. In a pioneer study, Jenkins et al [46] demonstrated with monkeys conditioned to repeatedly contact a rotating disk with two finger tips in daily sessions spanning several months that the resulting cutaneous stimulation specifically increased the area of representation of the stimulated skin surfaces and decreased the sizes of constituent neurons' receptive fields within layer IV of area 3b in the primary somatosensory cortex (S1). Similar changes occurring within specific neural representations related to the animal's behavior were recorded in more naturalistic contexts, for example, from rats engaged in nursing [106] or exposed to enriched environments promoting tactile experience [19]. In addition, we showed that S1 map changes induced by episodic nursing were reversible [79]. These studies showed a refinement of the somatotopic organization, as repeated sensory stimulation tended to further fractionate the cortical areas into discrete functional modules. In all those studies, the selective expansion of cutaneous representational territories occurred at the expense of adjacent cutaneous or proprioceptive representations, thus revealing the competitive nature of the underlying neural networks.

Experience-driven remodeling in the somatosensory cortex originally revealed by electrophysiological mapping in animal studies has been confirmed by brain imaging in humans. For example, a magnetoencephalography (MEG) study with expert violinists showed a significant enlargement in the S1 representation for the digits in contact with the violin strings [29]. This expansion was correlated with the age at which the person had started to play. Fine motor skill practice in humans has also been shown to be accompanied by a recruitment of additional M1 cortical region serving the practiced movements [39, 49]. In contrast, functional magnetic resonance imaging (fMRI) studies with professional piano players suggest that long-term motor practice results in a smaller number of voxels activated by overtrained complex finger movements in M1, supplementary motor area, premotor cortex, and superior parietal lobule. This study raises the possibility that smaller neuronal populations of cortex motor fields are recruited to control movement sequences in overtrained subjects, who can potentially perform the movements more easily and efficiently [43, 54].

Amazingly, representational changes within somatosensory maps can be induced by short-term practice or experience, as demonstrated after a few days of nursing behavior [106] or even after a few hours of skin stimulation [80]. A mapping study by transcranial magnetic stimulation (TMS) showed that musically naive subjects trained on a five-finger exercise on a piano keyboard in daily sessions of two hours over five days exhibited an increased excitability and an enlargement of the cortical representation of the finger flexor and extensor muscles, as their performance improved [73]. These early changes were transient, as a return to baseline was recorded after a weekend of rest. We have also shown that salient, short-duration, daily experience related to a tactile discrimination task can induce representational alterations in the S1 that prevail over those resulting from long-term experience promoted by housing in enriched or impoverished environments [110].

The rapid onset and reversibility of changes in motor cortex outputs to motoneuron pools is compatible either with a process of unmasking [44, 81] and remasking of existing connections or with the rapid turnover of both dentritic spines [92] and synapses [51] (for a review, see [2]). Longer-term structural changes like the increase in dendritic branching patterns of motor and sensory cortical neurons associated with motor training, initially revealed by Greenough et al [37], may facilitate more stable morphological changes in intracortical and subcortical networks, as skills consolidate and become automatic.

Cortical map remodeling as a neural substrate for acquisition of perceptual and motor abilities

The heuristic concept that neural representations embedded in primary somatosensory and motor cortices in the adult brain are dynamically maintained through use-dependent mechanisms and remodeled by intensive practice raised the question of the perceptual and behavioral relevance of experience-dependent changes in cortical organization. How do alterations in cortical representations translate into improvement or degradation in perceptual and motor abilities?

With monkeys trained on a digit dexterity task, we showed that acquisition of a new manual skill during a few hours of daily practice over a 2-month period was accompanied by discrete changes in the hand representation of the S1 [107, 109]. These alterations consisted of a representational expansion as well as refinement related to smaller, less overlapping neuronal receptive-fields, both of which were restricted to the cutaneous representations of the tips of the fingers most intensively engaged in the manual task that required fine somatosensory feedback (Fig. 1). The same digital dexterity task that required learning of precise temporal coordination of muscles and joints to produce fine digit movements led to an expansion of the M1 motor representation of these skilled movements, whereas nonchallenging motor practice did not result in significant map changes [71]. In this mapping study, representational changes that paralleled manual skill learning tended to form enlarged functional units serving synergistic movements rather than more fractionated representations dedicated to differentiated muscle finger control. However, this may reflect a limitation of this intracortical stimulation technique.

An MEG study revealed that the topographically disorganized finger S1 representation in patients with congenital syndactyly was associated with a misperception of the webbed digits [67]. Similarly, the hand representation in blind subjects using several fingers in daily Braille reading practice was reported to be topographically disrupted [87]. These multiple-digit blind readers misperceived the fingers touched by a light tactile stimulus, consistent with the idea that a representational disruption correlates with degraded perceptual abilities. Conversely, this representation was normally organized in both sighted controls and blind subjects using only one finger for Braille reading, for whom no mislocalization was observed.

An earlier study had shown an inverse relationship between the extent of cortical representation of the digit skin regions engaged in a frequency discrimination of tactile stimuli and the perceptual discrimination threshold [76], although, in that study, improvement in performance was primarily related to changes in the temporal response properties of cortical neurons. This finding is consistent with the results obtained from humans subjected to a hebbian coactivation protocol of simultaneous pairing of tactile stimuli [74]. Improvement in spatial tactile discrimination was correlated with a lateral shift of the index finger N20-dipole on the postcentral gyrus, suggesting a representational expansion. The same protocol, in fMRI studies, revealed a linear relationship between learning-induced improvement in individual discrimination thresholds and functional enlargement of the finger representation in the S1 [40, 75]. An fMRI study showed that expansion of the active area in the S1 was correlated with the subjects sensorimotor ability [39], consistent with our electrophysiological mapping study cited above [109]. Similarly, practicing a complex tapping task over several months was found to induce an increase by about 25% in the fMRI activation area in the M1, in nonmusicians [49]. When nonmusicians and highly skilled pianists were subjected to a novel tapping task, the latter exhibited a rapid increase in M1 activation area, which was not observed in nonmusicians [43], suggesting an effect of prepractice experience.

In their fMRI study, Karni et al [49] showed that the area of response induced in the M1 by a finger sequence practiced over several weeks was significantly larger than that elicited by a nonpracticed finger sequence. The area of evoked response for the trained sequence was confined to the hand representation, whose overall extent remained constant. These findings raised the question of whether the learning-related changes were sustained by a neuronal population competing for cortical output or alternatively sharing cortical space with other

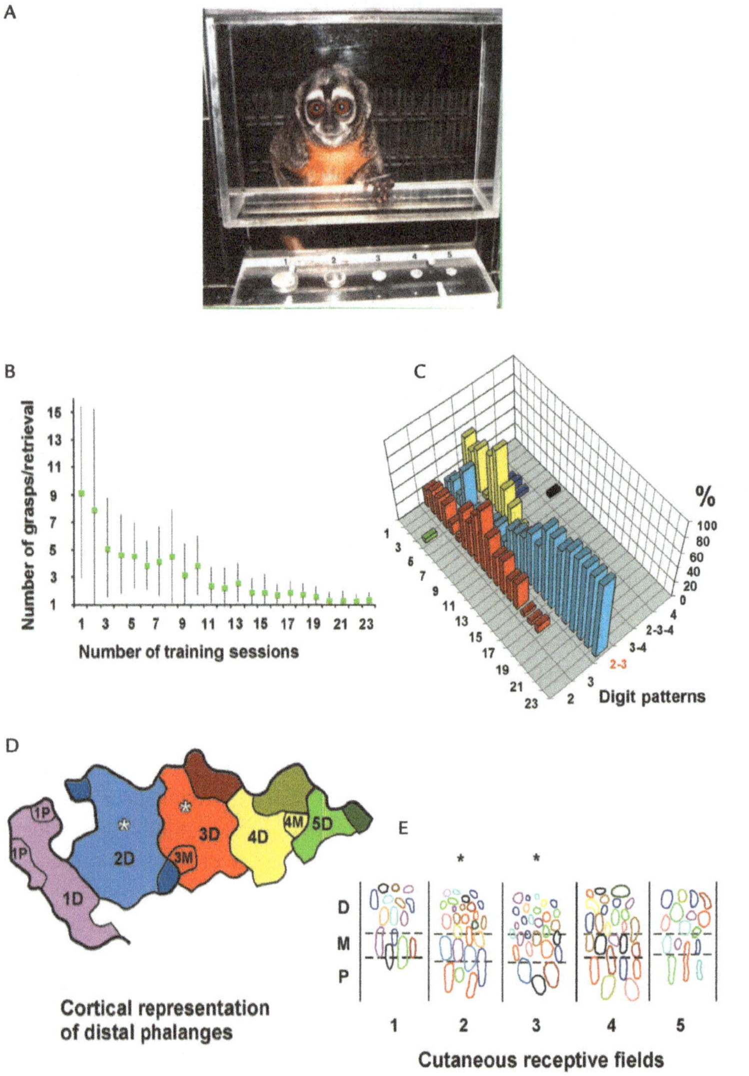

Fig. 1. Representational changes induced by a digital dexterity task in area 3b of the primary somatosensory cortex. **(A)** Experimental device (Klüver board) used to assess digital dexterity in an owl monkey. **(B)** Succes of pellet retrieval from the smallest food well (mean number of grasp attempts per successful retrieval; vertical bars are standard deviations) as a function of the number of training sessions. **(C)** Frequencies of the different combinations of fingers used, for each daily session, for 20 trials of pellet retrievals from the smallest food well. Note that the combination of digits 2 and 3 was gradually selected as the most efficient. **(D)** Cartoon maps restricted to the representations of the 5 digit tips. Map elaboration is based on the properties ("submodality", location, and size) of the receptive fields of neurons recorded within layer IV. Fingertips that were preferentially used for pellet retrievals from the small wells in the final days of practice are indicated by asterisks. The areal extent of representation of these digits was enlarged. **(E)** Cutaneous receptive fields defined on the proximal (*P*), medial (*M*), and distal (*D*) phalanges of the digits. Receptive fields are artificially separated (they normally overlap with one another) to facilitate their visualization. Note that receptive fields on the tips of the trained digits (2 and 3) are smaller than those on the other digits less frequently used in the retrieval task (Modified from [109])

neuronal ensembles assigned to different movements. Hlustik et al [39] investigated the changes of movement representations induced in the M1 and S1 areas by practiced or unpracticed tasks, over a 3-week period. They reported a gradual improvement of motor performance both on the practiced finger sequence and on unpracticed single-finger and wrist movements. In agreement with the findings reported in two separate monkey mapping studies investigating the effects of the same manual dexterity task on the S1 and M1 hand representations [71, 109], Hlustik et al [39] reported a gradual expansion of both the M1 and the S1 areas activated during finger movements. However, this expansion was observed both with the practiced finger sequence and with unpracticed hand movements. Interestingly, the degree of overlap among representations of individual movements in the M1 and S1 areas increased, but more prominently in the fine-skill group, suggesting that a substantial sharing of primary motor and somatosensory representations contributed to the fine-skill learning. The sharing of cortical territories reported by Hlustik et al [39] is reminiscent of an earlier 2-deoxyglucose study of the mouse somatosensory cortex [52]. That study showed a learning-dependent and selective enlargement of the cortical representation of one row of vibrissae. This enlarged representation overlapped with an unchanged cortical representation of adjacent rows of vibrissae not stimulated during the training. On the basis of this finding, one can propose the heuristic hypothesis that several representations may coexist as dynamic networks embedded in the same cortical field and should not necessarily be considered as competing for cortical space in an exclusive way.

The expansion of M1 representations of the unpracticed simple movements observed by Hlustik et al [39] may not be at variance with the cortical expansion restricted to trained hand movements that was first described for monkeys [71]. Indeed, the latter study was based on intracortical stimulation aimed at mapping the M1 output to muscles and, thus, was not specifically related to the finger movements naturally produced in the learned manual dexterity task. It is therefore plausible that motor map expansion might contribute to improved control of the trained muscles when used in nonpracticed elementary finger movements. According to Hlustik et al [39], both behavioral and brain imaging data in their study supported the conclusion that neural changes associated with practicing and learning skilled movements can be generalized to other motor tasks. Under this assumption, movement segments previously formed by combining neuronal units through learning may be more easily retrieved for new motor skill acquisitions. Along the same lines, in adult owl monkeys, trained to detect differences in the frequency of a tactile flutter-vibration stimulus delivered to a constant skin locus, increased cutaneous receptive-field sizes were observed that were not restricted to the representation of the trained digit. Indeed, receptive fields were also enlarged for the representations of other nearby hand surfaces, suggesting that improvements in behavioral performance might also transfer to adjacent, untrained skin surfaces [76].

Attention, expectation, and motivation contribute to reshaping of cortical representations

Importantly, the investigation of Recanzone et al [76] revealed that cortical changes related to tactile learning did not take place when attention was directed to auditory stimuli, simultaneously delivered during the tactile discrimination task. This finding prompted the question of the respective contribution of the sensory pattern elicited by intensive practice, versus attention and perceptual context, in the use-dependent alteration of sensory representations. For rats trained on a discrimination task, we found that the greater the ability to differentiate tactile textures during locomotion on floorboards, the larger the cortical zone serving the glabrous skin surfaces of the forepaw involved in this behavior and the smaller the receptive fields of the constituent neurons [110]. These relationships were only observed in animals actively engaged in the most challenging perceptual context, i.e., in the discrimination of gradually diminishing roughness contrasts. In the animals that performed better in a less demanding perceptual context, the extent of the forepaw map was correlated with the duration of training, but not with the discrimination performance. In addition, sensory stimulation per se, with a randomly distributed reward, was found to contribute only partially to receptive-field sharpening. Sensory experience associated with rewarded roughness discrimination, which demanded the animals' attention and motivation, contributed to a supplementary reduction in receptive-field sizes. Therefore, relationships between use-dependent cortical map reorganization and changes in perceptual abilities appear to be more complex than previously postulated. Indeed, for subjects trained to discriminate the orientation of synchronous tactile stimuli on digits 1 and 5, for 1 hour per day over a 4-week period, high-resolution EEG data indicated that the representations of D1 and D5 in the S1 were further apart after training, thus suggesting a representational expansion for the stimulated digits [3]. In contrast, when the stimulation was not given in a discrimination context the distance between the cortical representations of the stimulated fingers was decreased and the subjects exhibited a tendency to mislocalize near-threshold tactile stimuli of one finger to the distant finger co-stimulated during training. Thus, in different tasks, repetitive stimulation within short daily periods over 4 weeks produced opposite effects on the spatial relationship of digit representations within the activation map. The direction of the effect depended on whether the subjects were passive during functional assessment or engaged in a discrimination task. In a follow-up magnetic source imaging study [4], subjects were required to detect the direction of motion of a tactile stimulus either across a single digit (finger condition) or across the tips of adjacent fingers (hand condition) while identical stimulus patterns were delivered in both conditions. The cortical representation of digit 2 was segregated from that of digits 3–5 in the finger relative to the hand condition. This segregation occurred at the onset of practice, thus excluding a training effect. Therefore, changing the focus of attention had a significant effect on the functional tuning of receptive fields.

I reported above that naive subjects trained on a piano keyboard displayed reorganized cortical representations of the finger flexor and extensor muscles [73]. A similar, but less prominent increase was recorded in subjects who mentally repeated the motor tasks over five days [72]. Mental rehearsal of practice that is known to activate some of the same central neural structures required for the performance of the actual movements was thus hypothesized to be sufficient to promote the modulation of neural circuits involved in the early stages of motor-skill learning, in conformity with the beneficial effect of mental practice commonly reported by musicians or athletes. One should therefore consider that real practice and mental rehearsal if combined could potentially improve rehabilitative programs.

There is, however, evidence that massive practice which repeatedly induces localized activation patterns can lead to representational reorganization independently of behavioral context and attention. Repetitive exposure to prominent changes in the temporal patterning of sensory inputs can induce rapid and reversible cortical remodeling leading to improved performances, with no attention or reinforcement or even input salience required (for a review, see [104]). This form of cortical plasticity suggests a bottom-up process which could operate through a nonassociative learning like sensitization. In contrast, the cortical changes related to improvement in goal-directed discrimination performances, e.g., to meet the requirements of a task of increasing complexity, would be context specific and require top-down processes. Both attention and motivation leading to contextual modulation of input and proper consolidation of plastic changes would be required for further improvement of perceptual skills.

Collectively, these findings corroborate the notion that cortical malleability resulting from a spatial redistribution and retuning of receptive fields reflects a self-reorganization process involving selective changes in the local recruitment of neuronal populations, through segregation and desegregation mechanisms. This ongoing reallocation of neuronal resources that reflects cortical integration of behaviorally relevant sensory inputs is a requirement for adaptive behavioral changes. However, some significant improvement in perceptual abilities may occur even when no changes in cortical topographic representations and only minor alterations in neuronal response characteristics are detected. Therefore, improvement of complex perceptual abilities would result from a contextual modification of various neuronal response properties within dynamic sensory representations embedded in distributed and tightly interconnected neural networks.

Adverse effects of abnormal temporal input patterning on cortical map organization

It is now well established that experience-driven reorganization of cortical sensory representations is strongly influenced by the degree of local synchrony among sensory inputs delivered to the cortex. The synaptic mechanisms involved in this timing-based neuroplasticity have been thoroughly investigated (for a review, see [104]). The so-called "dark side" or maladaptive cortical plasticity manifests itself when an unusually intense stimulation or forceful use of digits disrupts the normal temporal pattern of inputs. Consequently, deleterious changes occur in cortical maps, such as a fusion of normally segregated representations or a degradation of the topographic arrangement within somatosensory and motor maps, which may underlie involuntary impairments of coordination and individual control of fingers. For instance, experimentally induced focal hand dystonia resulting from massively repeated synchronous stimulation of adjacent digits was found to produce such representational and sensorimotor changes [6]. Indeed, representational disruption of the hand representation in the S1 was shown for musicians affected by focal hand dystonia [1, 30]. Focal dystonia of musicians [3] or writer's cramp [82, 83] also results in impaired temporal and spatial discrimination at the finger tips. However, these neurological dysfunctions may be reversed by a training-based remediation. Indeed, insights into the influence of temporal patterning of inputs on cortical map remodeling have inspired a therapy for focal hand dystonia of professional musicians with long-standing symptoms. These treated patients displayed a marked and significant improvement in repertoire performance [12] accompanied by a normalization of the somatotopic finger representations in the S1, as assessed by an MEG study [13].

Deleterious effects of disuse on cortical map features

Sensory deprivation has also been used as a way to investigate the maladaptive effects of experience-dependent cortical reorganization. In an electrophysiological cortical mapping study, we reported that the topographic organization of somatosensory maps in the S1 cortex were severely disrupted in rats housed in an impoverished environment for about three months or subjected to forelimb immobilization for one week. The overall area serving the forepaw cutaneous representation was strongly reduced and interspersed with cortical sectors driven by proprioceptive inputs [20]. Cutaneous map shrinkage and receptive-field enlargement were reported for rats submitted to hindlimb suspension for 2 weeks [27, 57]. This reduction in representational areal extent was later corroborated by cytochrome oxydase staining [14]. These findings are consistent with studies of adult rats showing that whisker ablation or repeated plucking decreased cytochrome oxydase staining in layer IV of the corresponding S1 barrel fields [56, 103] and that acute whisker removal reduces 2-deoxyglucose uptake in the functionally deafferented barrels [28]. Representational disruption occurs in S1 maps in senescent rats [86]. This disruption, which also results from a natural decrease in tactile exploration and mobility with aging, particularly in an impoverished environment, can be substantially offset by enriched environments [21]. Deprivation-induced cortical remodeling is not restricted to the somatosensory cortex. Indeed, in a mapping study using transcranial magnetic stimulation, ankle immobilization in humans was found to decrease the area devoted to the anterior tibialis muscle in M1 [59]. This motor map reduction was correlated with the duration of immobilization and was not accompanied by an alteration in spinal excitability or motor threshold. Interestingly, the motor map shrinkage was quickly reversed by voluntary muscle contraction.

Cortical reorganization following stroke: from local remodeling to widespread recruitment of new areas

In the early 1950s, using surface stimulation electrodes in macaques, Glees and Cole provided the first demonstration that after a focal lesion targeting the motor representation of the thumb in the M1, a new representation emerged within a cortical zone adjacent to the lesion site. Later, an intracortical microstimulation study [70] showed that after a subtotal ischemic lesion targeting the M1 hand representation in squirrel monkeys, the size of the remaining sectors of the hand map substantially decreased, giving way to expanding proximal motor representations. In contrast, using microelectrode recordings from owl monkeys, Jenkins et al [47] reported that, after a focal ischemic lesion that permanently destroyed the cutaneous representation of two digit tips in area 3b of S1, a new representation serving the deprived skin surfaces emerged in the peri-infarct cortical sectors bordering the lesion. The discrepancies between these two studies may be accounted for by differences in the extent of the ischemic damage (smaller in the somatosensory cortex) and the effect of disuse of the affected hand after injury on the motor cortex. Clinical investigations have confirmed that the spared, peri-infarct cortex may be involved in neurological recovery [22, 45, 91]. But in those animal and clinical studies, the recovery developed spontaneously, while no attempt was made to monitor sensorimotor activity over the postoperative time. A TMS study in stroke patients revealed a reduced excitability in the motor cortex near the site of the injury, with a decreased cortical representation of the impaired movements [5, 93]. Such an effect may have resulted from the injury-induced disuse of the affected limb.

After a focal cortical infarct damaging a substantial part of the forepaw representation in the S1 cortex in rats, we found that the cutaneous representations were surprinsingly well preserved in the perilesional zones in the animals exposed to an enriched environment for three weeks, in contrast to the cutaneous representations in rats exposed to an impoverished environment, in which they were further deteriorated, relative to the initial postinjury map [105] (Fig. 2). Interestingly, after a single session of physiotherapy of stroke patients, the cortical motor output to the paretic muscles was significantly expanded and manual dexterity was improved [61]. The authors reported that both motor output map and motor function changes were partially reversed one day later. A number of studies have tackled the issue of the influence of training on cortical map plasticity following brain damage. Nudo and Milliken [70] confirmed that, in contrast to untrained monkeys, in animals that benefited from a rehabilitative training involving the impaired forelimb, the undamaged sectors of the distal forelimb motor map were preserved or even increased. This finding suggests that the efficacy of local network neurons projecting to hand motoneurons was maintained in these trained animals. The S1 areas that cooperate to mediate complex tactile functions are extensively interconnected: area 3b projects mainly to areas 1 and 2, whereas area 1 is also reciprocally connected with areas 2 and 3b. Therefore, this network provides a substrate for distributed changes after a focal lesion affecting one of its constituent areas. We used the same rehabilitative manual dexterity training as in Nudo's study, but after a focal ischemic lesion that targeted the cutaneous representation of two digits in area 3b and hence impaired digital dexterity primarily engaging these digits. We found that new representations emerged several millimeters away from the lesion, in the neighboring cutaneous map of area 1. Emergent cutaneous representations were even recorded within adjacent sectors of the noncutaneous area 3a that formerly received only proprioceptive inputs [107, 109]. The pattern of cortical changes across monkeys was found to be idiosyncratic, depending on individual digit use and retrieval strategies. Indeed, the skin surfaces that regained a representation in ectopic cortical regions were located on the fingers crucially engaged in the retrieval task during the rehabilitative training (Figs. 3 and 4). Importantly, as the manual dexterity recovered, the animals grad-

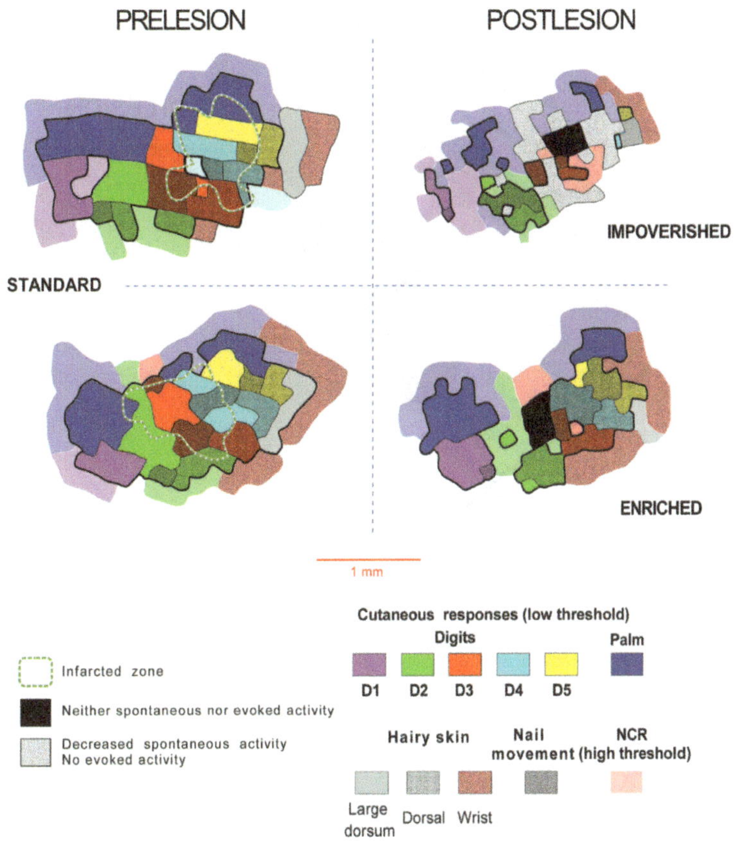

Fig. 2. Effects of housing conditions on the reorganization of the somatotopic maps after focal ischemic injury in the S1 area. Representative forepaw maps obtained prior to (*left*) and 3 weeks after the lesion induction (*right*) in two rats housed in standard environments before the lesion and in impoverished or enriched environments after the lesion. The injured areas defined on the basis of neuronal recording are outlined on the prelesion maps (dashed lines). The cortical sectors electrophysiologically silent or displaying a decreased spontaneous activity with no evoked response are illustrated on the postlesion maps. Note that the forepaw representation was best preserved of in the rat housed in an enriched environment, whereas a subsequent degradation was observed in the perilesion area in the rat housed in an impoverished environment (Modified from [105])

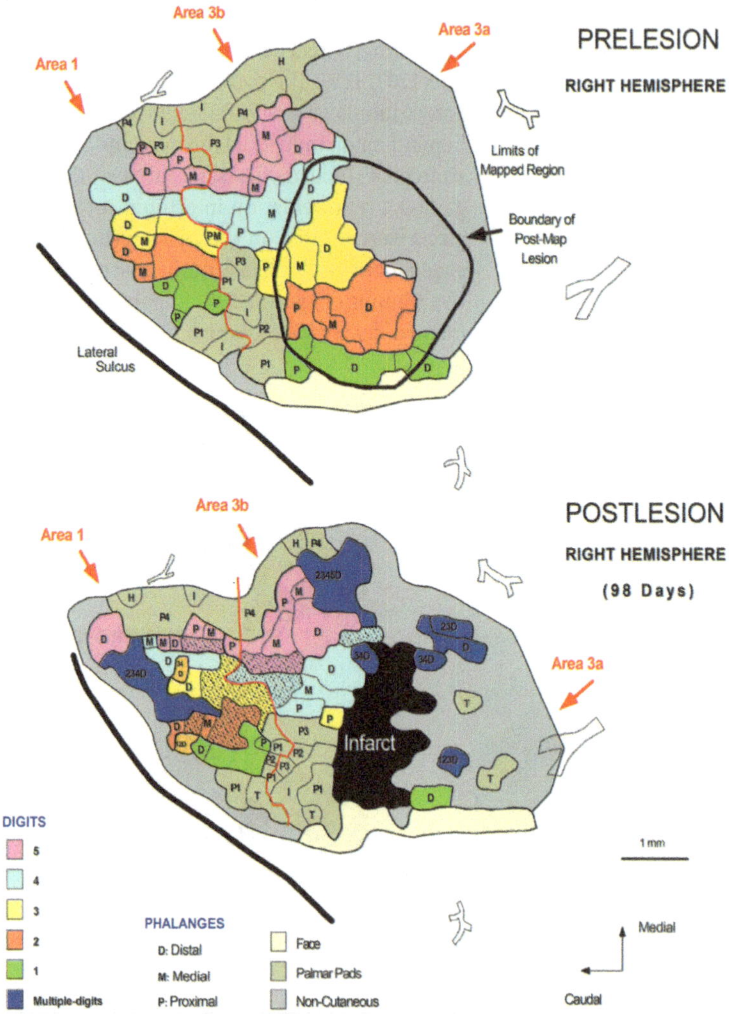

Fig. 3. Recovery of manual dexterity accompanied by extensive reorganization of the hand representation in the S1 area after focal cortical ischemia. Topographic representations of the hand skin surfaces before (*top*) and 98 days (*bottom*) after induction of a cortical lesion in the digit representation of area 3b, in an owl monkey. The infarcted zone determined on the basis of electrophysiological recordings over the first two hours after the lesion induction is outlined on the prelesion map. The tips of digits 2 and 3 were most frequently employed in the final phases of recovered dexterity of the contralateral hand on the food pellet retrieval. *1–5* Digits (*1* thumb); *D*, *M*, and *P* distal, middle, and proximal phalanges; multiple-digit representations are shown (e.g., *23* indicates cortical zones in which neurons displayed receptive fields located on the tips of digits 2 and 3); *P1–P4* palmar pads at the bases of the digits; *H* hypothenar eminence; *I* insular zone in the center of the palm; *T* thenar eminence; the zones over which no cutaneous responses could be evoked are indicated in light gray. Dorsal skin representational zones are shown as dotted areas. The red line marks the border between areas 3b and 1 border. The black infarct area delimits the zone over which neither cortical spontaneous nor driven neuronal discharges could be recorded in the postlesion mapping experiment. Constant vascular landmarks are shown in each map in the experimental hemisphere, to facilitate comparisons. Note the emergence of new cutaneous representational zones in both area 3a and area 1 (Modified from [108])

ually reused the digits that had regained a tactile sensitivity through this experience-dependent somatosensory map remodeling. This study provided the first evidence for an experience-driven substitution between discrete somatosensory areas that probably mediated restoration of fine sensorimotor regulation following a focal ischemic injury. Along the same lines, studies in stroke patients that benefited from rehabilitation for several weeks revealed a substantial recruitment of motor representations in the damaged hemisphere [15, 60, 93].

Frost and colleagues [32] have documented how an ischemic infarct resulting in a partial or complete destruction of the M1 hand area affected the ventral premotor area (PMv), which sends intracortical projections to the M1. The authors reported that lesions damaging more than 50% of the hand area in the M1 induced an expansion of the PMv hand representation. The area of expansion could reach 50% in the case where the lesion affected the whole hand area. This compensatory reorganization in a distant premotor area after injury to the M1 is presumed to provide a neural substrate for the postlesion recovery of fine manual skills. Five months after ischemic injury to the M1 hand area, cortical map reorganization in PMv was found to be accompanied by a proliferation of PMv terminal fields and the occurrence of retrogradely labeled cell bodies within area 1/2 of the S1, presumably in the hand representation zone, suggesting the formation of new connections within these remote areas [23]. Furthermore, alterations were found in the trajectory of axons originating from the PMv near the site of the lesion. The rewiring of corticocortical connections involved in transmission of new cutaneous and proprioceptive inputs between the PMv and area 1/2 of the S1 may have played a compensatory role in the behavioral recovery documented by Frost et al [32].

Electrophysiological cortical mapping has the advantage of probing cortical changes with a high spatial resolution. However, as the number of mapping sessions per animal is limited and the area that can be mapped is spatially restricted, this procedure does not allow as-

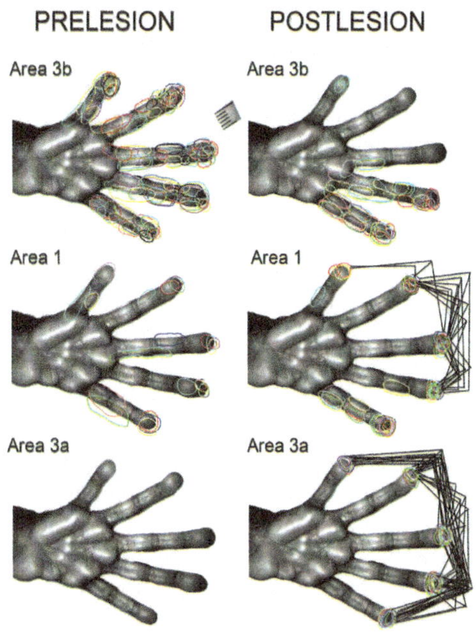

Fig. 4. Emergence of new cutaneous receptive fields (RFs) paralleling recovery of manual dexterity in areas 1 and 3a, after focal cortical ischemia to the hand representation in S1. Cutaneous receptive fields before the lesion and after the functional recovery was complete, defined on digital surfaces of the hand used for pellet retrieval. Note that densely located on the tips of multiple digits numerous cutaneous receptive fields emerged in both area 1 and area 3a during this postlesion period. However, stronger responses were commonly recorded for stimulation applied on digits 2 and 3. Solid lines interconnect multiple receptive fields identified for individual recording sites. Receptive fields located on the palm are not shown (Modified from [108])

sessment of the time course of changes over spatially distributed cortical areas. By contrast, brain imaging techniques including TMS, fMRI, EEG, MEG, and PET permit such an extensive exploration, but with a more limited spatial resolution. Those brain imaging techniques have provided evidence that subcortical or cortical strokes induce widespread recruitment of anatomical and functionally interconnected regions in the ipsilateral or contralateral hemispheres assumed to sustain functional recovery in patients [18, 97]. For instance, in chronic stroke patients with focal subcortical damage, a temporary disruption of activity by transcranial magnetic stimulation in the premotor cortex of the ipsilesional hemisphere induced a deterioration in the use of the impaired limb [5, 31, 98], thus suggesting that the spared regions may have participated in behavioral recovery.

Typically, after a subcortical stroke, brain imaging studies have documented prominent activations in the unaffected hemisphere in the early stages of recovery, while cortical activations were found to shift towards the affected hemisphere in the late stages [15, 22, 65]. Interestingly, in a PET study Calautti et al [9] reported that, 2 months after a unilateral striatocapsular infarction, overactivations take place bilaterally in the hand area of the primary sensorimotor cortices and in the premotor and supplementary motor areas of the unaffected hemisphere during a thumb-to-index tapping task with the paretic hand. Overactivations remained, but were less prominent in the primary sensorimotor cortices and premotor of the affected hemisphere at 8 months, indicating a decrease in the recruitment of the bilateral motor networks and a significant change in the interhemispheric balance as the functional recovery progressed. However, new overactivations were found at 8 months in some patients, in the left prefrontal areas, the putamen, and the premotor cortex suggesting late-appearing compensatory reorganization. Despite relative homogeneity in the infarct size and location, as well as in clinical symptoms, this idiosyncratic pattern of compensatory overactivation after injury to the corticospinal pathway revealed interindividual differences in sensorimotor strategies and cognitive processes to reinstate the motor performance.

Despite a relative consensus that functional recruitment of compensatory networks evolves over time as recovery progresses, there is no general agreement as to whether the contralesional activations sustain a better or faster recovery [63]. For example, an fMRI study in stroke patients with hemiparesis caused by infarctions in the corticospinal tract showed that recovery of the paretic hand was associated with a relative increase in activity in the contralateral (ipsilesional) sensorimotor cortex compared with the ipsilateral (contralesional), and a relative decrease in the prefrontal and the ipsilateral posterior parietal regions [65]. In a PET analysis, Calautti et al [10] corroborated these findings by assessing changes in motor abilities (thumb-to-index tapping) and laterality index in patients with subcortical, striatocapsular infarction. They found that S1-M1 activation tended to shift toward the unaffected hemisphere over time, but that this shift was associated with lesser functional recovery. A longitudinal fMRI exploration in a rat model of stroke that induced acute dysfunction of the contralateral forelimb indicated that functional recovery was related to a decreased involvement of the contralesional hemisphere and a gradual reinstatement of responses in the peri-infarction zones [25]. Consistent with this finding, other studies have shown that the best recovery for paresis was achieved if the sensorimotor network normally subserving the impaired functions regained functional activity and was reintegrated in the active neural network [8, 11, 33, 62, 63, 96].

Collectively, these studies have shown that within the first few days to weeks of the recovery process, an increased activity takes place in motor areas bilaterally, but primarily in the contralesional hemisphere. Then in a later stage of the recovery process, spanning 3 to 6 months, contralesional activation is often reduced, while a more focused activation occurs in peri-infarct and other spared motor regions in the injured hemisphere [16, 65, 69, 95]. All of these remote changes may be accounted for by sustained alterations of the neuronal excitability of distant areas following cortical damage, through upregulation of *N*-methyl-D-aspartate receptors and downregulation of gamma-aminobutyric acid A receptors, which has been documented in both the ipsilesional and contralesional hemispheres [77, 99, 100]. Nevertheless, seeking a correlation between the extent of contralesional activation and the speed, degree, or quality of recovery should not obscure the functions mediated by a given area and its potential implication in postlesion recovery. For instance, fMRI data showed that an early recruitment of the supplementary motor area and ipsilesional inferior Brodmann area 40 was positively correlated with motor recovery, whereas activation of the prefrontal cortex and parietal cortex in the contralesional hemisphere predicted a slower and less complete recovery [62].

Rehabilitation-induced improvements were found to be associated with contralesional hemisphere activations [24, 50, 85]. Furthermore, Loubinoux and colleagues [63] investigated prognosis factors indicative of the quality of the outcome in the case of capsular lesions. Indeed, the higher the early activation in the ipsilesional Brodmann area 4, S1, and insula, the better the recovery one year later. Their findings also suggested that there is a benefit associated with increasing ipsilesional M1 activity shortly after stroke, as a rehabilitative approach in mildly impaired patients. Interestingly, a recent fMRI study indicates that the pattern of brain activation present in the first few days after stroke correlates with subsequent motor recovery, thus suggesting that rehabilitative intervention may facilitate the recovery process by targeting the structures engaged early on after stroke [66].

Protective effects of postlesion training, harmfulness of overuse during critical period

The elucidation of use-dependent cortical reorganization and recovery-associated plastic changes has inspired poststroke motor rehabilitation procedures. It has been argued that behavioral recovery may be enhanced if compensatory strategies leading to learned nonuse and presumably disadvantageous cortical reorganization following brain injury are counteracted [58, 90]. Therefore, constraint-induced movement therapy (CIMT) involving a forced use of the impaired limb by animals and stroke patients has received particular attention. Indeed, relatively rapid beneficial effects of CIMT on recovery have been reported [68, 89, 94, 101, 102].

Early implementation of CIMT was thought to be beneficial, as early use of the affected limb may minimize or prevent learned nonuse. In addition, early training of the impaired limb is likely to promote and optimize postlesion neuroplasticity. However, initiation of CIMT [53] or early training without immobilization of the intact forelimb [78] in the very early stage of focal ischemia was aslo shown to increase lesion volume and to impede motor recovery of the affected limb of rats. In contrast, constraining the intact limb [41] or training the affected limb seven days after the onset of brain damage resulted in the best performances and did not induce enlarged cortical infarct volumes. These studies raised the concern that an early postlesion vulnerable period may exist. The involvement of *N*-methyl-D-aspartate receptors in this use-dependent exaggeration of brain injury has been suggested [42]. After a focal ischemic lesion to the forepaw area in the S1, cortical recordings from rats exposed to an impoverished environment for three weeks showed an expansion of the ischemic zone and a compression and a fragmentation of the remaining cutaneous forepaw representations within the spared cortical sectors surrounding the lesion. In contrast, in animals housed in enriched conditions the ischemic zone did not grow and only a limited compression of the forepaw map was found, with a preservation of most representational sectors [105]. We thus postulated that whereas intensive training within a critical time window after focal cortical ischemia may be detrimental for the peri-infarct tissue, and consequently for behavioral recovery, moderate stimulation started early after the lesion would have protective effects on peri-infarct cortical representations. In fact, we showed that early cutaneous stimulations delivered to the forepaw for short daily periods during the first postlesion week were sufficient to induce the preservation effects reported for the rats in an enriched environment. In addition, we observed that these effects were less pronounced when a similar stimulation regimen was delivered during the second postlesion week, whereas a lack of stimulation resulted in the outward expansion of the ischemic zone along with the severe loss of cortical representations reported in previous studies. Furthermore, the preservation effect within the forepaw map appeared to be specifically related to the stimulated skin surfaces (Fig. 5).

A direct causal relationship between structural or functional changes in the brain and improvement of sensorimotor abilities has yet to be demonstrated. Nevertheless, using TMS, Liepert et al [60, 61] were the first to show that CIMT enlarged the initially smaller-than-normal ipsilesional motor map of the paretic hand in stroke patients, whereas opposite changes were recorded in the contralesional motor map, thus rebalancing the hand motor representations between the two hemispheres. Shifts of the center of the output map in the affected hemisphere were compatible with the recruitment of adjacent brain areas. In addition, the amount of map expansion was found to correlate with motor ability improvement [60, 84], thus suggesting a causal link, consistent with the findings from nonhuman primate

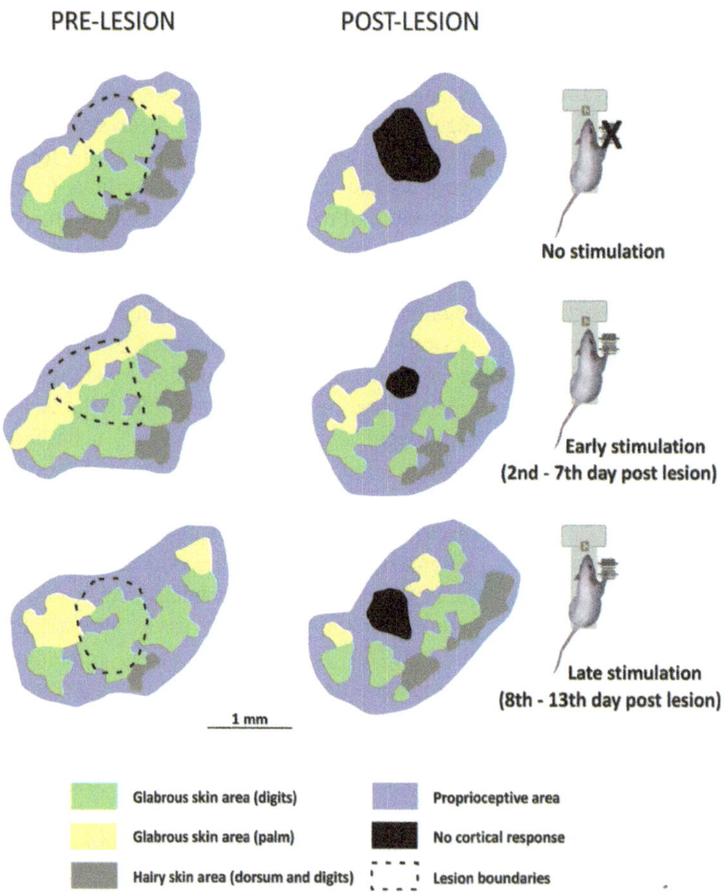

Fig. 5. Selective effects of tactile stimulation on the remodeling of the somatotopic maps in the S1 cortex after focal ischemic injury. Individual pre- and postlesion maps recorded from rats whose contralateral forepaw was either not stimulated or stimulated during the first or second postlesion week on a rotating textured cylinder. Stimulations were delivered on the glabrous skin of digits over two daily sessions of 30 min interrupted by a 15 min resting period. Representation of the stimulated skin surfaces was best preserved when tactile stimulations were delivered over the first postlesion week

studies described in this review. Six months after a CIMT, motor performance remained at a high level, whereas the cortical area sizes in the two hemispheres became almost identical, indicating a return toward a normal balance of interhemisphere excitability [61]. After combining restraint of the unaffected limb with gradual exercises for the affected limb in stroke patients, the extent of improvement in hand function was found to be correlated with increases in fMRI activity in the premotor and secondary somatosensory cortices contralateral to the affected hand and in superior posterior regions of the cerebellar hemispheres bilaterally [48]. This finding suggests that the therapy-induced recruitment in sensorimotor regions was associated with successful motor rehabilitation.

It is noteworthy that the degree of change in contralesional M1 activation in stroke patients during the early period of CIMT has predictive value for the motor recovery achieved by the end of therapy [26]. However, changes in brain activation related to the motor improvement induced by CIMT vary over time and among individual stroke patients. Differences in the brain reorganization patterns which may underlie the motor improvement induced by CIMT are very likely accounted for by differences in the infarct and its size and whether white matter is damaged or not [38]. There is, however, a controversy relating to whether the extent of injury to the corticospinal tract affects the magnitude of motor gain in response to CIMT [34, 55]. This question is of importance, as a recent study pointed out that in rats subjected to forced use of the impaired forelimb after lesion of the corticospinal tract, recovery of motor function was accompanied by greater density of corticospinal tract axon collaterals terminating in the denervated spinal cord [64].

Conclusions

The findings reviewed in this chapter clearly show that, while local experience-dependent changes in cortical circuitry underlie new skill acquisition, restricted cortical or subcortical damage affects the functional contribution of more distributed cortical areas. Such damage triggers complex time-evolving changes that are relevant to behavioral recovery. Taking advantage of use-dependent plasticity of neural structures, early initiated rehabilitative procedures seem to improve the potential for recovery. To translate animal research findings into controlled clinical trials and to gain further insight into the neurobiology of rehabilitation, future studies should combine various techniques to probe the causal link between experience-induced structural and functional brain changes and the recovery of sensory, motor, and cognitive abilities. A greater understanding of how idiosyncratic changes are related to the recovery process would facilitate the development of new therapeutic interventions that seek to optimize individual treatments or rehabilitative procedures.

References

[1] Bara-Jimenez W, Catalan MJ, Hallett M, Gerloff C (1998) Abnormal somatosensory homunculus in dystonia of the hand. Ann Neurol 44: 828–831

[2] Barnes SJ, Finnerty GT (2010) Sensory experience and cortical rewiring. Neuroscientist 16: 186–198

[3] Braun C, Schweizer R, Elbert T, Birbaumer N, Taub E (2000) Differential activation in somatosensory cortex for different discrimination tasks. J Neurosci 20: 446–450

[4] Braun C, Haug M, Wiech K, Birbaumer N, Elbert T, Roberts LE (2002) Functional organization of primary somatosensory cortex depends on the focus of attention. Neuroimage 17: 1451–1458

[5] Butefisch CM, Kleiser R, Seitz RJ (2006) Post-lesional cerebral reorganisation: evidence from functional neuroimaging and transcranial magnetic stimulation. J Physiol 99: 437–454

[6] Byl NN, Merzenich MM, Jenkins WM (1996) A primate genesis model of focal dystonia and repetitive strain injury: I. Learning-induced dedifferentiation of the representation of the hand in the primary somatosensory cortex in adult monkeys. Neurology 47: 508–520

[7] Byl NN, Merzenich MM, Cheung S, Bedenbaugh P, Nagarajan SS, Jenkins WM (1997) A primate model for studying focal dystonia and repetitive strain injury: effects on the primary somatosensory cortex. Phys Ther 77: 269–284

[8] Calautti C, Baron JC (2003) Functional neuroimaging studies of motor recovery after stroke in adults: a review. Stroke 34: 1553–1566

[9] Calautti C, Leroy F, Guincestre JY, Baron JC (2001) Dynamics of motor network overactivation after striatocapsular stroke: a longitudinal PET study using a fixed-performance paradigm. Stroke 32: 2534–2542

[10] Calautti C, Leroy F, Guincestre JY, Marie RM, Baron JC (2001) Sequential activation brain mapping after subcortical stroke: changes in hemispheric balance and recovery. Neuroreport 12: 3883–3886

[11] Calautti C, Naccarato M, Jones PS, Sharma N, Day DD, Carpenter AT, Bullmore ET, Warburton EA, Baron JC (2007) The relationship between motor deficit and hemisphere activation balance after stroke: a 3T fMRI study. Neuroimage 34: 322–331

[12] Candia V, Schafer T, Taub E, Rau H, Altenmuller E, Rockstroh B, Elbert T (2002) Sensory motor retuning: a behavioral treatment for focal hand dystonia of pianists and guitarists. Arch Phys Med Rehabil 83: 1342–1348

[13] Candia V, Wienbruch C, Elbert T, Rockstroh B, Ray W (2003) Effective behavioral treatment of focal hand dystonia in musicians alters somatosensory cortical organization. Proc Natl Acad Sci USA 100: 7942–7946

[14] Canu MH, Stevens L, Falempin M (2007) Effect of hindlimb suspension on activation and MHC content of triceps brachii and on the representation of forepaw on the sensorimotor cortex. Exp Neurol 203: 521–530

[15] Carey JR, Kimberley TJ, Lewis SM, Auerbach EJ, Dorsey L, Rundquist P, Ugurbil K (2002) Analysis of fMRI and finger tracking training in subjects with chronic stroke. Brain 125: 773–788

[16] Carey LM, Abbott DF, Egan GF, O'Keefe GJ, Jackson GD, Bernhardt J, Donnan GA (2006) Evolution of brain activation with good and poor motor recovery after stroke. Neurorehabil Neural Repair 20: 24–41

[17] Carmichael ST (2003) Plasticity of cortical projections after stroke. Neuroscientist 9: 64–75

[18] Chollet F, DiPiero V, Wise RJ, Brooks DJ, Dolan RJ, Frackowiak RS (1991) The functional anatomy of motor recovery after stroke in humans: a study with positron emission tomography. Ann Neurol 29: 63–71

[19] Coq JO, Xerri C (1998) Environmental enrichment alters organizational features of the forepaw representation in the primary somatosensory cortex of adult rats. Exp Brain Res 121: 191–204

[20] Coq JO, Xerri C (1999) Impoverishment and sensorimotor restriction deteriorate the forepaw cutaneous map in the primary somatosensory cortex of adult rats. Exp Brain Res 129: 518–531

[21] Coq JO, Xerri C (2001) Sensorimotor experience modulates age-dependent alterations of the forepaw representation in the rat primary somatosensory cortex. Neuroscience 104: 705–715

[22] Cramer SC, Nelles G, Benson RR, Kaplan JD, Parker RA, Kwong KK, Kennedy DN, Finklestein SP, Rosen BR (1997) A functional MRI study of subjects recovered from hemiparetic stroke. Stroke 28: 2518–2527

[23] Dancause N, Barbay S, Frost SB, Plautz EJ, Chen D, Zoubina EV, Stowe AM, Nudo RJ (2005) Extensive cortical rewiring after brain injury. J Neurosci 25: 10167–10179

[24] Dechaumont S, Marque P, Carel C, Albucher JF, Cassol E, Chollet F, Loubinoux I (2004) Neural correlates of sensory and proprioceptive integration in the controlesional hemisphere of poor recovered patients with subcortical stroke? An fMRI study. Neurology 62: A541, S566.005

[25] Dijkhuizen RM, Ren J, Mandeville JB, Wu O, Ozdag FM, Moskowitz MA, Rosen BR, Finklestein SP (2001) Functional magnetic resonance imaging of reorganization in rat brain after stroke. Proc Natl Acad Sci USA 98: 12766–12771

[26] Dong Y, Dobkin BH, Cen SY, Wu AD, Winstein CJ (2006) Motor cortex activation during treatment may predict therapeutic gains in paretic hand function after stroke. Stroke 37: 1552–1555

[27] Dupont E, Canu MH, Langlet C, Falempin M (2001) Time course of recovery of the somatosensory map following hindpaw sensory deprivation in the rat. Neurosci Lett 309: 121–124

[28] Durham D, Woolsey TA (1978) Acute whisker removal reduces neuronal activity in barrels of mouse SmL cortex. J Comp Neurol 178: 629–644
[29] Elbert T, Pantev C, Wienbruch C, Rockstroh B, Taub E (1995) Increased cortical representation of the fingers of the left hand in string players. Science 270: 305–307
[30] Elbert T, Candia V, Altenmuller E, Rau H, Sterr A, Rockstroh B (1998) Alteration of digital representations in somatosensory cortex in focal hand dystonia. Neuroreport 9: 3571–3575
[31] Fridman EA, Hanakawa T, Chung M, Hummel F, Leiguarda RC, Cohen LG (2004) Reorganization of the human ipsilesional premotor cortex after stroke. Brain 127: 747–758
[32] Frost SB, Barbay S, Friel KM, Plautz EJ, Nudo RJ (2003) Reorganization of remote cortical regions after ischemic brain injury: a potential substrate for stroke recovery. J Neurophysiol 89: 3205–3214
[33] Fujii Y, Nakada T (2003) Cortical reorganization in patients with subcortical hemiparesis: neural mechanisms of functional recovery and prognostic implication. J Neurosurg 98: 64–73
[34] Gauthier LV, Taub E, Mark VW, Perkins C, Uswatte G (2009) Improvement after constraint-induced movement therapy is independent of infarct location in chronic stroke patients. Stroke 40: 2468–2472
[35] Gauthier LV, Taub E, Perkins C, Ortmann M, Mark VW, Uswatte G (2008) Remodeling the brain: plastic structural brain changes produced by different motor therapies after stroke. Stroke 39: 1520–1525
[36] Glees P, Cole J (1950) Recovery of skilled motor functions after small repeated lesions in motor cortex in macaque. J Neurophysiol 13: 137–148
[37] Greenough WT, Larson JR, Withers GS (1985) Effects of unilateral and bilateral training in a reaching task on dendritic branching of neurons in the rat motor-sensory forelimb cortex. Behav Neural Biol 44: 301–314
[38] Hamzei F, Liepert J, Dettmers C, Weiller C, Rijntjes M (2006) Two different reorganization patterns after rehabilitative therapy: an exploratory study with fMRI and TMS. Neuroimage 31: 710–720
[39] Hlustik P, Solodkin A, Noll DC, Small SL (2004) Cortical plasticity during three-week motor skill learning. J Clin Neurophysiol 21: 180–191
[40] Hodzic A, Veit R, Karim AA, Erb M, Godde B (2004) Improvement and decline in tactile discrimination behavior after cortical plasticity induced by passive tactile coactivation. J Neurosci 24: 442–446
[41] Humm JL, Kozlowski DA, James DC, Gotts JE, Schallert T (1998) Use-dependent exacerbation of brain damage occurs during an early post-lesion vulnerable period. Brain Res 783: 286–292
[42] Humm JL, Kozlowski DA, Bland ST, James DC, Schallert T (1999) Use-dependent exaggeration of brain injury: is glutamate involved? Exp Neurol 157: 349–358
[43] Hund-Georgiadis M, von Cramon DY (1999) Motor-learningrelated changes in piano players and non-musicians revealed by functional magnetic-resonance signals. Exp Brain Res 125: 417–425
[44] Jacobs KM, Donoghue JP (1991) Reshaping the cortical motor map by unmasking latent intracortical connections. Science 251: 944–947
[45] Jaillard A, Martin CD, Garambois K, Lebas JF, Hommel M (2005) Vicarious function within the human primary motor cortex? A longitudinal fMRI stroke study. Brain 128: 1122–1138
[46] Jenkins WM, Merzenich MM, Ochs MT, Allard T, Guic-Robles E (1990) Functional reorganization of primary somatosensory cortex in adult owl monkeys after behaviorally controlled tactile stimulation. J Neurophysiol 63: 82–104
[47] Jenkins WM, Merzenich MM, Recanzone G (1990) Neocortical representational dynamics in adult primates: implications for neuropsychology. Neuropsychologia 28: 573–584
[48] Johansen-Berg H, Dawes H, Guy C, Smith SM, Wade DT, Matthews PM (2002) Correlation between motor improvements and altered fMRI activity after rehabilitative therapy. Brain 125: 2731–2742
[49] Karni, A, Meyer G, Jezzard P, Adams MM, Turner R, Ungerleider LG (1995) Functional MRI evidence for adult motor cortex plasticity during motor skill learning. Nature 377: 155–158
[50] Kimberley TJ, Lewis SM, Auerbach EJ, Dorsey LL, Lojovich JM, Carey JR (2004) Electrical stimulation driving functional improvements and cortical changes in subjects with stroke. Exp Brain Res 154: 450–460
[51] Knott GW, Quairiaux C, Genoud C, Welker E (2002) Formation of dendritic spines with GABAergic synapses induced by whisker stimulation in adult mice. Neuron 34: 265–273
[52] Kossut M, Siucinska E (1998) Learning-induced expansion of cortical maps – what happens to adjacent cortical representations? Neuroreport 9: 4025–4028
[53] Kozlowski DA, James DC, Schallert T (1996) Use-dependent exaggeration of neuronal injury after unilateral sensorimotor cortex lesions. J Neurosci 16: 4776–4786
[54] Krings T, Topper R, Foltys H, Erberich S, Sparing R, Willmes K, Thron A (2000) Cortical activation patterns during complex motor tasks in piano players and control subjects. A functional magnetic resonance imaging study. Neurosci Lett 278: 189–193
[55] Kuhnke N, Juenger H, Walther M, Berweck S, Mall V, Staudt M (2008) Do patients with congenital hemiparesis and ipsilateral corticospinal projections respond differently to constraint-induced movement therapy? Dev Med Child Neurol 50: 898–903
[56] Land PW, Simons DJ (1985) Metabolic activity in SmI cortical barrels of adult rats is dependent on patterned sensory stimulation of the mystacial vibrissae. Brain Res 341: 189–194
[57] Langlet C, Canu MH, Falempin M (1999) Short-term reorganization of the rat somatosensory cortex following hypodynamia-hypokinesia. Neurosci Lett 266: 145–148

[58] LeVere T (1980) Recovery of function after brain damage. A theory of the behavioral deficit. Physiol Psychol 8: 297–308
[59] Liepert J, Tegenthoff M, Malin JP (1995) Changes of cortical motor area size during immobilization. Electroencephalogr Clin Neurophysiol 97: 382–386
[60] Liepert J, Miltner WH, Bauder H, Sommer M, Dettmers C, Taub E, Weiller C (1998) Motor cortex plasticity during constraint-induced movement therapy in stroke patients. Neurosci Lett 250: 5–8
[61] Liepert J, Graef S, Uhde I, Leidner O, Weiller C (2000) Training-induced changes of motor cortex representations in stroke patients. Acta Neurol Scand 101: 321–326
[62] Loubinoux I, Carel C, Pariente J, Dechaumont S, Albucher JF, Marque P, Manelfe C, Chollet F (2003) Correlation between cerebral reorganization and motor recovery after subcortical infarcts. Neuroimage 20: 2166–2180
[63] Loubinoux I, Dechaumont-Palacin S, Castel-Lacanal E, De Boissezon X, Marque P, Pariente J, Albucher JF, Berry I, Chollet F (2007) Prognostic value of FMRI in recovery of hand function in subcortical stroke patients. Cereb Cortex 17: 2980–2987
[64] Maier IC, Baumann K, Thallmair M, Weinmann O, Scholl J, Schwab ME (2008) Constraint-induced movement therapy in the adult rat after unilateral corticospinal tract injury. J Neurosci 28: 9386–9403
[65] Marshall RS, Perera GM, Lazar RM, Krakauer JW, Constantine RC, DeLaPaz RL (2000) Evolution of cortical activation during recovery from corticospinal tract infarction. Stroke 31: 656–661
[66] Marshall RS, Zarahn E, Alon L, Minzer B, Lazar RM, Krakauer JW (2009) Early imaging correlates of subsequent motor recovery after stroke. Ann Neurol 65: 596–602
[67] Mogilner A, Grossman JA, Ribary U, Joliot M, Volkmann J, Rapaport D, Beasley RW, Llinás RR (1993) Somatosensory cortical plasticity in adult humans revealed by magnetoencephalography. Proc Natl Acad Sci USA 90: 3593–3597
[68] Myint JM, Yuen GF, Yu TK, Kng CP, Wong AM, Chow KK, Li HC, Chun Por W (2008) A study of constraint-induced movement therapy in subacute stroke patients in Hong Kong. Clin Rehabil 22: 112–124
[69] Nhan H, Barquist K, Bell K, Esselman P, Odderson IR, Cramer SC (2004) Brain function early after stroke in relation to subsequent recovery. J Cereb Blood Flow Metab 24: 756–763
[70] Nudo RJ, Milliken GW (1996b) Reorganization of movement representations in primary motor cortex following focal ischemic infarcts in adult squirrel monkeys. J Neurophysiol 75: 2144–2149
[71] Nudo RJ, Milliken GW, Jenkins WM, Merzenich MM (1996) Use-dependent alterations of movement representations in primary motor cortex of adult squirrel monkeys. J Neurosci 16: 785–807
[72] Pascual-Leone A (2001) The brain that plays music and is changed by it. Ann N Y Acad Sci 930: 315–329
[73] Pascual-Leone A, Nguyet D, Cohen LG, Brasil-Neto JP, Cammarota A, Hallett M (1995) Modulation of muscle responses evoked by transcranial magnetic stimulation during the acquisition of new fine motor skills. J Neurophysiol 74: 1037–1045
[74] Pleger B, Dinse HR, Ragert P, Schwenkreis P, Malin JP, Tegenthoff M (2001) Shifts in cortical representations predict human discrimination improvement. Proc Natl Acad Sci USA 98: 12255–12260
[75] Pleger B, Foerster AF, Ragert P, Dinse HR, Schwenkreis P, Malin JP, Nicolas V, Tegenthoff M (2003) Functional imaging of perceptual learning in human primary and secondary somatosensory cortex. Neuron 40: 643–653
[76] Recanzone GH, Merzenich MM, Jenkins WM (1992) Frequency discrimination training engaging a restricted skin surface results in an emergence of a cutaneous response zone in cortical area 3a. J Neurophysiol 67: 1057–1070
[77] Redecker C, Wang W, Fritschy JM, Witte OW (2002) Widespread and long-lasting alterations in GABA(A)-receptor subtypes after focal cortical infarcts in rats: mediation by NMDA-dependent processes. J Cereb Blood Flow Metab 22: 1463–1475
[78] Risedal A, Zeng J, Johansson BB (1999) Early training may exacerbate brain damage after focal brain ischemia in the rat. J Cereb Blood Flow Metab 19: 997–1003
[79] Rosselet C, Zennou-Azogui Y, Xerri C (2006) Nursing-induced somatosensory cortex plasticity: temporally decoupled changes in neuronal receptive field properties are accompanied by modifications in activity-dependent protein expression. J Neurosci 26: 10667–10676
[80] Rosselet C, Zennou-Azogui Y, Escoffier G, Kirmaci F, Xerri C (2008) Nursing-induced new input timing remodels cortical maps after skin flap rotation. Eur J Neurosci 27: 1245–1260
[81] Sanes JN, Wang J, Donoghue JP (1992) Immediate and delayed changes of rat motor cortical output representation with new forelimb configurations. Cereb Cortex 2: 141–152
[82] Sanger TD, Tarsy D, Pascual-Leone A (2001) Abnormalities of spatial and temporal sensory discrimination in writer's cramp. Mov Disord 16: 94–99
[83] Sanger TD, Pascual-Leone A, Tarsy D, Schlaug G (2002) Nonlinear sensory cortex response to simultaneous tactile stimuli in writer's cramp. Mov Disord 17: 105–111
[84] Sawaki L, Butler AJ, Leng X, Wassenaar PA, Mohammad YM, Blanton S, Sathian K, Nichols-Larsen DS, Wolf SL, Good DC, Wittenberg GF (2008) Constraint-induced movement therapy results in increased motor map area in subjects 3 to 9 months after stroke. Neurorehabil Neural Repair 22: 505–513
[85] Schaechter JD, Kraft E, Hilliard TS, Dijkhuizen RM, Benner T, Finklestein SP, Rosen BR, Cramer SC (2002)

Motor recovery and cortical reorganization after constraint-induced movement therapy in stroke patients: a preliminary study. Neurorehabil Neural Repair 16: 326–338

[86] Spengler F, Godde B, Dinse HR (1995) Effects of ageing on topographic organization of somatosensory cortex. Neuroreport 6: 469–473

[87] Sterr A, Muller MM, Elbert T, Rockstroh B, Pantev C, Taub E (1998) Perceptual correlates of changes in cortical representation of fingers in blind multifinger Braille readers. J Neurosci 18: 4417–4423

[88] Stowe AM, Plautz EJ, Eisner-Janowicz I, Frost SB, Barbay S, Zoubina EV, Dancause N, Taylor MD, Nudo RJ (2007) VEGF protein associates to neurons in remote regions following cortical infarct. J Cereb Blood Flow Metab 27: 76–85

[89] Taub E, Miller NE, Novack TA, Cook EW 3rd, Fleming WC, Nepomuceno CS, Connell JS, Crago JE (1993) Technique to improve chronic motor deficit after stroke. Arch Phys Med Rehabil 74: 347–354

[90] Taub E, Crago JE, Burgio LD, Groomes TE, Cook EW, DeLuca SC, Miller NE (1994) An operant approach to rehabilitation medicine: overcoming learned nonuse by shaping. J Exp Anal Behav 61: 281–293

[91] Teasell RW, Kalra L (2005) What's new in stroke rehabilitation: back to basics. Stroke 36: 215–217

[92] Trachtenberg JT, Chen BE, Knott GW, Feng G, Sanes JR, Welker E, Svoboda K (2002) Long-term in vivo imaging of experience-dependent synaptic plasticity in adult cortex. Nature 420: 788–794

[93] Traversa R, Cicinelli P, Bassi A, Rossini PM, Bernardi G (1997) Mapping of motor cortical reorganization after stroke. A brain stimulation study with focal magnetic pulses. Stroke 28: 110–117

[94] van der Lee JH, Wagenaar RC, Lankhorst GJ, Vogelaar TW, Deville WL, Bouter LM (1999) Forced use of the upper extremity in chronic stroke patients: results from a single-blind randomized clinical trial. Stroke 30: 2369–2375

[95] Ward NS, Brown MM, Thompson AJ, Frackowiak RS (2004) The influence of time after stroke on brain activations during a motor task. Ann Neurol 55: 829–834

[96] Ward NS, Newton JM, Swayne OB, Lee L, Thompson AJ, Greenwood RJ, Rothwell JC, Frackowiak RS (2006) Motor system activation after subcortical stroke depends on corticospinal system integrity. Brain. 129: 809–819

[97] Weder B, Seitz RJ (1994) Deficient cerebral activation pattern in stroke recovery. Neuroreport 5: 457–460

[98] Werhahn KJ, Conforto AB, Kadom N, Hallett M, Cohen LG (2003) Contribution of the ipsilateral motor cortex to recovery after chronic stroke. Ann Neurol 54: 464–472

[99] Witte OW (1998) Lesion-induced plasticity as a potential mechanism for recovery and rehabilitative training. Curr Opin Neurol 11: 655–662

[100] Witte OW, Stoll G (1997) Delayed and remote effects of focal cortical infarctions: secondary damage and reactive plasticity. Adv Neurol 73: 207–227

[101] Wolf SL, Lecraw DE, Barton LA, Jann BB (1989) Forced use of hemiplegic upper extremities to reverse the effect of learned nonuse among chronic stroke and head-injured patients. Exp Neurol 104: 125–132

[102] Wolf SL, Winstein CJ, Miller JP, Taub E, Uswatte G, Morris D, Giuliani C, Light KE, Nichols-Larsen D (2006) Effect of constraint-induced movement therapy on upper extremity function 3 to 9 months after stroke: the EXCITE randomized clinical trial. JAMA 296: 2095–2104

[103] Wong-Riley MT, Welt C (1980) Histochemical changes in cytochrome oxidase of cortical barrels after vibrissal removal in neonatal and adult mice. Proc Natl Acad Sci USA 77: 2333–2337

[104] Xerri C (2008) Imprinting of idiosyncratic experience in cortical sensory maps: neural substrates of representational remodeling and correlative perceptual changes. Behav Brain Res 192: 26–41

[105] Xerri C, Zennou-Azogui Y (2003) Influence of the postlesion environment and chronic piracetam treatment on the organization of the somatotopic map in the rat primary somatosensory cortex after focal cortical injury. Neuroscience 118: 161–177

[106] Xerri C, Stern JM, Merzenich MM (1994) Alterations of the cortical representation of the rat ventrum induced by nursing behavior. J Neurosci 14: 1710–1721

[107] Xerri C, Coq JO, Merzenich MM, Jenkins WM (1996) Experience-induced plasticity of cutaneous maps in the primary somatosensory cortex of adult monkeys and rats. J Physiol (Paris) 90: 277–287

[108] Xerri C, Merzenich MM, Peterson BE, Jenkins W (1998) Plasticity of primary somatosensory cortex paralleling sensorimotor skill recovery from stroke in adult monkeys. J Neurophysiol 79: 2119–2148

[109] Xerri C, Merzenich MM, Jenkins W, Santucci S (1999) Representational plasticity in cortical area 3b paralleling tactual-motor skill acquisition in adult monkeys. Cereb Cortex 9: 264–276

[110] Xerri C, Bourgeon S, Coq JO (2005) Perceptual context-dependent remodeling of the forepaw map in the SI cortex of rats trained on tactile discrimination. Behav Brain Res 162: 207–221

New insights into the neurobiology of language from functional brain imaging

Pascale Tremblay, Anthony St. Dick and Steven L. Small

Introduction

Undoubtedly, language is among the most celebrated hallmarks of human cognition. Even though we perceive, produce, and comprehend language, and do so seemingly effortlessly every day of our lives, the underlying neural mechanisms for language remain far from understood. With the cognitive revolution of the last century, it became a common viewpoint that language is a modular system segregated from other functional systems in the nervous system. This notion combined with the findings from neuropsychological "lesion analysis" studies from earlier research led to the notion that these modules are instantiated in localized brain regions of the left inferior frontal, temporal and inferior parietal regions of the human brain. The most commonly cited characterization of this system, sometimes referred to as the Broca-Wernicke-Geschwind model, is represented by the iconic diagram of a white matter pathway (the arcuate fasciclus; AF) connecting the posterior superior temporal region ("Wernicke's area") involved in receptive language to the posterior part of the inferior frontal gyrus ("Broca's area") involved in expressive language (Fig. 1A). The 19th century studies of patients presenting language difficulties following brain injury have had a tremendous influence in the field of language neurobiology. For instance, the classic description of two severely dysfluent patients (Leborgne and Lelong) by the French neurologist Paul Broca in 1861 led to the longstanding belief that the posterior two thirds of the inferior frontal gyrus ("Broca's Area") is the motor center for language. The lesion method, however, is not without faults. For example, recent MRIs of Leborgne's and Lelong's brain revealed extensive lesions also involving the insula and associated perisylvian white matter [16, 25]. This indicates that the patients' syndrome did not necessarily result from injury to Broca's area. While studies of brain lesions and concomitant behavioral syndromes have had, and continue to have, a tremendous influence in the field of language neurobiology, interpretation of lesion data are complicated by the size of the lesions (which are often extensive) and compensatory mechanisms of plasticity that occur in the brain following injury. In the past fifteen years, the use of state-of-the-art neuroimaging, neurophysiological, and brain stimulation methods has enhanced the precision with which we can investigate language in the brain, and has contributed to rapid progress in the understanding of the neural basis of language. In fact, the consensus of how the brain processes language has shifted in three fundamental ways: (1) there is an increasing consensus that the brain is not organized into dissociable regions for production and comprehension, but rather that language functions are distributed

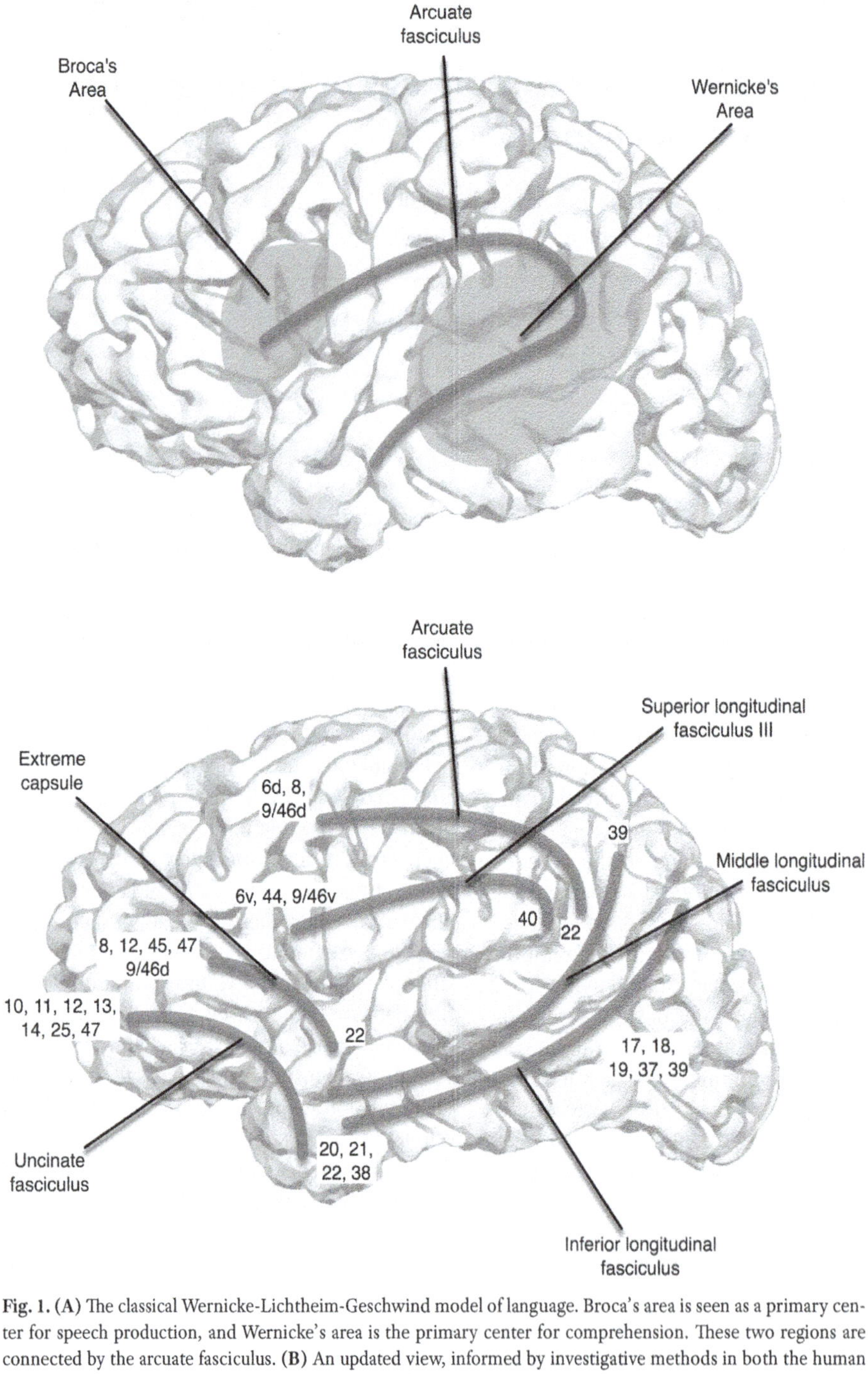

Fig. 1. (A) The classical Wernicke-Lichtheim-Geschwind model of language. Broca's area is seen as a primary center for speech production, and Wernicke's area is the primary center for comprehension. These two regions are connected by the arcuate fasciculus. **(B)** An updated view, informed by investigative methods in both the human and monkey, suggests that as many as six dissociable fiber pathways may contribute to language processing in the brain. Numbers indicate Brodmann Areas and potential regions of origin and termination

into large-scale cortical and subcortical networks; (2) there is an increasing focus on hodology [34], or how brain regions within this distributed network are connected by particular fiber pathways; (3) there an increasing acceptance that cortical and subcortical regions involved in processing language are also involved in other cognitive and sensory-motor domains traditionally considered to be non-linguistic. In this chapter we review the organization and anatomy of language, focusing on the crosstalk among language components (speech production, speech perception, speech comprehension) and among functional systems.

Language networks: above and beyond compartmentalization

The identification of the brain regions and fiber pathways involved in the production, perception, and comprehension of language is a fundamental problem in neuroscience. As discussed, historically the focus has been on two functional "centers" for language in the brain: Broca's area for production and Wernicke's area for comprehension, with the arcuate fasciculus fiber pathway connecting the two regions. The compartmentalization of language into independent "modules" is unfortunate, as it gives the inaccurate impression that the different processes leading to language occur serially, or at least without much interaction. In this section, we review the neural organization of speech production, and of speech perception culminating with language understanding. Special emphasis is placed on the overlap of the brain mechanisms that implement these various levels of language processing.

The production of speech

Speech production is an exquisitely complex and multistage process. It begins with a (prelexical) intention to communicate, continues to the translation of this message into lexical units (words) which in turn need to be temporally ordered (sequenced) and encoded phonologically [56a]. These processes together conclude with the production of words. This final output stage is inherently quite complex in its own right, as it demands the close coordination of multiple sensory and motor components, including the respiratory system (which generates the power source necessary to produce speech), the laryngeal system (which converts the airflow into a sound by setting the vocal folds in vibration (i.e., *phonation*), and the articulatory system (which changes the configuration of the vocal tract to convert the laryngeal output into sequences of vowels and consonants). Hence, the neural architecture for speech production is extraordinarily complex, including multiple cortical and subcortical control centers, six cranial nerve fibers and their associated nuclei (facial, hypoglossal, trigeminal, glossopharyngeal, vagus, and accessory), a substantial number of muscles covering the abdomen, neck, face, mouth and larynx, and an even larger set of sensory receptors in joints, tendons and muscles. And yet, despite this complexity, the chain of events that leads to the production of speech occurs within several hundreds of milliseconds. Indeed, mature speakers may produce as many as 14 phonemes per second, i.e., between six and nine syllables per second [53].

Some of the neural mechanisms for speech production have been elucidated, but a number remain poorly understood. One of the reasons for this is that imaging the brain during speech production comes with a number of challenges that are not present for studies of receptive language. This is particularly true for electroencephalography (EEG) and functional magnetic resonance imaging (fMRI), which are susceptible to movement correlated signal, or (in the case of fMRI) magnetic field variations resulting from motion [10]. Structural MRI looking at brain lesions and their impact on behavior does not present these problems, and data collected using this methodology have led to reevaluation of long accepted beliefs about how the brain coordinates speech production. For example, in 1996, Dronkers and colleagues [24] showed that, in a group of 25 patients with articulatory planning deficits (i.e., apraxia of speech), all patients had a lesion in the insular cortex, but not all had a lesion in Broca's area. This finding demonstrated that Broca's area is not the only cortical center important for speech production.

Recent solutions to the problem of movement for fMRI, including sparse-sampling imaging protocols (e.g. [42]), have provided data that are consistent with Dronkers and colleagues. These studies also suggest a prominent role for the insula in speech production [66, 72, 11, 71]; for a review, see [2]). Additional brain regions are also implicated for single word repetition and more complex word generation tasks. These include the primary sensory and motor areas of the precentral and postcentral gyri and sulci, the inferior frontal gyrus, the ventral premotor cortex, the medial motor areas (cingulate motor area, supplementary and pre-supplementary motor areas), the insula, basal ganglia and the cerebellum (e.g. [66, 97, 37, 1, 88, 89, 3]). The idea that the machinery for speech production is "special", i.e., that it is specialized for the specific task of producing syllables and words, has been advocated for at least a generation [57, 99]. However, recent evidence supports the opposite view, namely that a general sensory-motor system is involved in multiple tasks, including speaking, swallowing and other oro-facial movements [75, 12, 89].

With new data come new theoretical models, and a growing number of models of language production are now taking into account the neural complexity inherent to speech production. For example, Riecker et al [72] have proposed a dual system for speech production, with a *preparatory* loop including the supplementary motor area, insula, superior cerebellum, and dorsolateral frontal cortex, and an *executive* loop including the primary motor cortex, thalamus, basal ganglia and inferior cerebellum. Guenther and colleagues [43, 44] have proposed a detailed model of speech production (DIVA) that focuses on the role of sensory feedback (auditory, somatosensory) in speech acquisition and production. The model links superior temporal areas to inferior parietal and inferior frontal areas, and includes a contribution of the cerebellum to feed-forward modeling for speech production. As the authors acknowledge, at the neurobiological level, DIVA is incomplete and requires additional data. Despite it shortcomings, however, DIVA and other models are trying to account for an astounding complexity of neural processes that implement the production of speech, and the full model will need to take into account all aspects of language, including detailed speech planning and production mechanisms as well as comprehension mechanisms.

In sum, it is clear that the neural architecture of speech production reflects the inherent complexity of this process, and goes far beyond the so-called Broca's area, involving multiple cortical and sub-cortical domain-general control centers.

The perception and comprehension of speech: from acoustic waves to meaning

Although speech perception is often described as the processing of the sub-lexical units that form the speech stream (i.e., syllables, phonemes), the ultimate goal of speech perception is to comprehend language and to communicate. In this section, we review what is known about the neurobiology of both perception and comprehension.

The first neural signals relevant for speech originate in the inner ear and proceed via the vestibulo-cochlear nerve (auditory nerve; eighth cranial nerve) to the brainstem, then to the inferior colliculus of the midbrain, to the medial geniculate nucleus of the thalamus, and, finally, to the primary auditory cortex located within the transverse temporal gyrus of Heschl. Despite numerous sub-cortical relays, electrophysiological recordings demonstrate that the first cortical manifestations of sounds are promptly observed, approximately 50–200ms after stimulus onset. The primary auditory cortex (A1) receives tonotopically-organized projections from the thalamus (see, for example Hackett, [45], for a review) and is, in turn, also organized tonotopically, with higher frequencies located more medially. The tonotopy of A1, however, is more diffuse than that of the inner ear, suggesting that frequency analysis is completed in the lower levels of the auditory pathway. Moreover, brain imaging studies have shown that the primary auditory cortex responds to the presentation of speech sounds, but no more so than to the presentation noise bursts with similar acoustic properties, suggesting that A1 is not specialized for the processing of speech [98].

From the primary auditory area, the auditory signal is sent to auditory association areas. There are currently two main accounts of how the sounds travel from A1 to be further processed. Rauschecker and colleagues have proposed that there exist a ventral route and a dorsal routes for sound processing. According to this view, a ventral auditory route is involved in auditory object identification and in speech perception [69] leading to the term auditory "what"-stream [70]. The ventral route includes A1 as well as anterior superior temporal gyrus (aSTG), a region that has been shown to be sensitive to "voice" [8, 9] and to vowels [64], and a portion of the inferior frontal gyrus (pars opercularis and triangularis) [69]. This account also postulates a dorsal stream, involved in spatial processing of sounds, in both monkeys and humans, which has been referred to as an auditory "where"-stream [70]. This route includes A1, the planum temporale (PT), which lies immediately posterior to the transverse temporal sulcus on the superior temporal plane, the posterior parietal areas, and ends in the premotor and prefrontal cortex (corresponding to Brodmann's areas 6 and 8) [68]. Within this route, the inferior parietal lobule and superior frontal sulcus are most responsive to spatial information, while PT responds to spatial information just as much as it responds to non-spatial information, providing partial support for the existence of a auditory "where"-stream [4].

Hickok and Poeppel [48, 50, 51] have also proposed a dual route system for language. According to this view, all sounds first undergo spectral and phonological analyses in the dorsal STG and posterior STS. From there, the processing of sounds diverges into a dorsal route for auditory-motor transformation, and a ventral route for auditory language comprehension. The dorsal route is involved in auditory-motor transformations, articulation, and speech perception. This route includes connections from PT to the PMv, pIFG and insula. According to Hickok and Poeppel, PT, rather than being involved in spatial processing as suggested by Rauschecker et al [70], is a key component of the auditory-motor transformation process. A body of evidence originating mainly from brain imaging experiments supports this hypothesis. For instance, brain imaging studies have shown that PT is active during overt speech production [95, 97, 87, 11, 71], but also during silent speech production or speech rehearsal, which does not involve self-generated auditory feedback [14, 97, 49, 17]. Furthermore, it has been shown that the caudal part of PT is more strongly active for sub-vocal rehearsal of auditory stimuli than for the perception of auditory stimuli (e.g., [14, 49]), suggestive of a role in auditory-motor transformation. Thus, whereas the ventral-dorsal model of Rauschecker and colleagues focuses on localizing and identifying sounds in general, the ventral-dorsal model of Hickok and Poeppel proposes a dorsal route for speech perception, and a ventral route for semantic processing and speech comprehension.

Investigations of processing routes for language have revealed an extensive overlap in the neural basis of speech production and speech comprehension. For example, it has been shown that passive listening to syllables and phonemes activates frontal motor regions within and around the ventral precentral sulcus, in the region controlling mouth movements [94, 68, 93]. Passive watching of videos of a speaker telling a story also activates the ventral premotor cortex (PMv), more so than listening to the same stories without seeing the talker, suggesting a role for the premotor cortex in recognizing the talker's articulatory gestures [81]. TMS studies have revealed that stimulation of the left primary motor cortex in the region controlling the face during both passive speech listening and viewing results in the enhancement of motor-evoked potentials (MEP) recorded from the lips or tongues [33, 82, 90, 91]. Furthermore, when applied to the PMv, TMS interferes with the discrimination of speech sounds presented in noise [61]. In sum, brain imaging and TMS findings indicate a role for the PMv and adjacent pIFG—typically implicated in speech production—in speech perception. The current debate centers around whether these motor speech mechanisms are essential for, or simply supportive of, speech perception. More recent data suggests recruitment of motor cortex during perception when speech is difficult to understand, but not when it is easy to understand. For example, repetitive TMS to PMv has no effect on participants' ability to perceive/categorize speech sounds in the absence of ambient noise [82, 76], suggesting that involvement of PMv may not be critical for speech perception under many natural circumstances, but that it may have a contribution under difficult situations, for instance, in a noisy environment, or while performing a difficult phonological task.

The sounds of speech form words, which in turn form sentences, which in turn lead to comprehension of the message of a speaker. Although this process seems straightforward, in reality it is difficult to clearly determine where speech perception ends and language comprehension begins. Nevertheless, a general organizational principle appears to be that more posterior temporal regions are involved in sound processing while anterior and inferior temporal regions are involved in semantics and language comprehension processes. For instance, according to the dual route model of language proposed by Hickok and Poeppel [48, 50, 51], all sounds undergo spectral and phonological analyses in the dorsal and mid posterior superior temporal sulcus (STS). These analyses are followed by access to semantic representations through a lexical interface involving the posterior part of the middle temporal gyrus (pMTG) and inferior temporal sulcus (pITS). Higher aspects of syntactic and compositional semantics (for instance, sentence level semantics) involve the anterior MTG and aSTG. In addition to temporal sites, there is also evidence from brain imaging and brain stimulation studies that the anterior inferior frontal gyrus (aIFG) may also be involved in semantic processing. Petersen and colleagues [65] were among the first to show, using positron emission tomography (PET), that word generation activates the aIFG more strongly than less semantically taxing tasks such as word repetition, suggesting a role in semantic processing for this region. A number of fMRI studies show that when response selection during language tasks relies on semantic processing, the aIFG is activated [52, 3, 88, 89]. Moreover, Devlin et al [22] demonstrated using transcranial magnetic stimulation (TMS) that stimulation of the aIFG results in delayed performance on a semantic decision task, but not on a perceptual decision task. Similarly, Gough et al [41] showed that TMS to the aIFG leads to delayed behavioral performance during a synonym judgment task, but not during a homophone judgment task. Taken together, these studies suggest that the aIFG is involved semantic analysis.

Adding to the complexity of semantic processing in the brain, according to advocates of "embodied semantics," is the observation that understanding the meaning of action words and sentences also recruits motor circuits required to produce that action. By analogy with the macaque, this process is thought to involve mechanisms analogous to those involving mirror neurons. Mirror neurons are individual neurons that respond to both action execution and action observation [23, 38, 73]. In the macaque, neurons with this dual property have been found in the ventral premotor cortex (area F5) and in the inferior parietal lobe. Several brain-imaging studies have shown activation in primary motor and premotor cortex during passive language tasks (e.g., [46, 85, 6]). Brain stimulation experiments have also shown somatotopic modulation of the motor cortex during the processing of sentences [15], and words [67].

Taken together, these results suggest that the motor system may contribute to language comprehension, although it is possible that activation in motor areas during language tasks is not critical for semantic analysis of linguistic stimuli, but instead represents an associational discharge that is not causal to comprehension [58].

As we have seen in this section, the neural basis of speech perception and auditory language comprehension involve a number of components spanning most of the neocortex. We have shown that while speech production, perception and comprehension are associated with some distinct regions, they also exhibit a large degree of overlap.

Fiber pathways important for language

The previous section demonstrates that a large number of brain regions are active during speech production, speech perception and language comprehension, thereby revealing the remarkable complexity of the neural architecture of language. But with the growing consensus that language is distributed into large-scale cortical and subcortical networks [62], there has also been an increasing focus on hodology [34], or how brain regions within this distributed network are connected by particular fiber pathways. That is, there is a renewed acknowledgement that connectivity provides critical insights into function. This focus has been driven by investigations using diffusion tensor imaging (DTI) and intraoperative electrical stimulation in humans, and anterograde tract tracing in the rhesus monkey. These three methods provide both conflicting and complementary information about the fiber pathways important for language. Studies using these methodologies have suggested six fiber pathways that are potentially important for language (Fig. 1B). These are (1) the third subcomponent of the superior longitudinal fasciculus (SLFIII); (2) the arcuate fasciculus (AF); (3) the middle longitudinal fasciculus (MdLF); (4) the inferior longitudinal fasciculus (ILF); (5) the uncinate fasciculus (UF); and (6) extreme capsule (EF).

Fiber tract identification in the human: diffusion tensor imaging (DTI) and intraoperative electrical stimulation

Historically, the examination of fiber pathways in the human has been accomplished with gross dissection methods in the postmortem brain [21, 19]. More recent methods allow examination of the living brain, and considerable effort has been expended to map the cerebral fiber pathways *in vivo* using diffusion tensor imaging (DTI; [63]). The DTI method takes advantage of the anisotropic (directional) nature of diffusion of water molecules in neural fibers, which can be measured with a specific MRI pulse sequence. Because water molecules flow along the direction of the fiber paths, the measure of fractional anisotropy (a measure of diffusion anisotropy) is higher in white matter than in gray matter. Further, the direction of fractional anisotropy can be traced across voxels to map fiber pathways, a procedure known as tractography [7]. This procedure has been used to map long association fiber pathways thought to be involved in language.

Several pathways for language that have been identified in the historical literature (e.g., SLF, AF) have been investigated using DTI [18, 40], and additional pathways have been defined. For example, Makris and colleagues [59] recognized three components of the SLF, which they also distinguished from the AF. They argued that the third SLF subcomponent, SLFIII, which may connect the posterior inferior frontal gyrus with the supramarginal gyrus of the inferior parietal lobe, is involved in the articulatory component of language. This function is typically associated with the AF, classically thought to connect Broca's and Wernicke's areas [29]. Instead, Makris et al argued that the AF connects the posterior temporal cortex with more dorsal frontal cortex, and might be involved in localizing the source of auditory information in space. Thus, it is not, strictly speaking, a language pathway. This dissociation of the AF and SLF fibers is also supported by more recent DTI studies [35, 74, 77].

Additional language pathways identified by DTI include the extreme capsule, middle longitudinal fasciculus (MdLF), and inferior longitudinal fasciculus (ILF; [35, 77]). For example, in their fMRI/DTI study, Saur and colleagues [77] provided evidence that repetition of pseudowords was associated with SLFIII and AF pathways, in conjunction with the MdLF (coursing the length of the superior and middle temporal lobe). This comprised a "dorsal route" involved in auditory-motor representation of speech sounds. Mapping sounds to meaning, indexed by sentence comprehension, was associated with the MdLF, the extreme capsule (connecting the anterior inferior frontal cortex with the anterior superior temporal cortex) and the ILF (coursing the length of the inferior temporal lobe). This comprised a "ventral route" of semantic processing. These findings provide evidence for the relevant fiber pathways that connect the brain regions comprising the "dorsal" and "ventral" language routes discussed earlier (cf. [48, 50, 51]).

Despite these advances in understanding fiber pathways in the human brain, DTI tractography has some serious shortcomings that are often minimized or ignored in the literature [5, 86]. First, in order to perform tractography, a seed region of interest (ROI) must be selected. Thus, DTI begins with anatomical knowledge derived from postmortem studies, and requires *a priori* hypotheses about the course of the fiber tract. Such hypotheses are based on potentially erroneous conclusions of earlier histological and dissection preparations, leading to the perpetuation of such errors into the DTI literature [78]. DTI also assumes that a single diffusion tensor defines each voxel, but this assumption is invalid where grey/white matter or white matter/cerebrospinal fluid overlap (i.e., partial volume averaging) and in cases where there are crossing fiber tracts. Pixels with partial volumes or crossing fibers will appear hypointense, and such errors accumulate along the length of the trajectory path [55, 84]. These issues can lead to several problems, including the premature termination of a fiber, the identification of nonexistent fiber tracts, or the misidentification of two or more fiber tracts as one tract [5]. Methods are being developed to deal with some of these issues (e.g., diffusion spectrum imaging to deal with crossing fibers; [80, 92]), but it remains important to consider the conclusions of DTI studies in light of those reached using other methodologies.

In addition to DTI, extensive studies of the fiber pathways connecting brain regions involved in language have been conducted using intraoperative electrical stimulation [28, 29, 30, 31, 32, 47, 60]. This technique, used during surgery of awake patients, involves stimulating, with an electrode, certain areas of exposed white matter during performance of a task ([83] for a review of the method). If the electrical stimulation results in disruption of a particular task (e.g., picture naming, counting), the pathway is determined to be involved in that task. Thus, electrical stimulation provides information about the function of the pathway, which is information that DTI by itself cannot provide.

These studies have generally supported the DTI findings, suggesting that the more dorsal pathways (i.e., SLF and AF) are involved in phonological and articulatory processes, and the more ventral pathways are involved in semantics (i.e., fiber pathways of the inferior temporal lobe). For example, electrical stimulation of the white matter under the inferior frontal, inferior parietal, and posterior superior temporal cortex results in phonemic paraphasias (i.e., mispronunciation; [28]) and also speech arrest [28, 29]. With respect to semantic processing, electrical stimulation of white matter coursing the inferior temporal cortex induces semantic paraphasias in response to a picture naming task (i.e., instead of labeling the target picture, the patient responds to the picture with words that are either in the same category as the target picture, that are antonyms of the target, or that have associative or functional proximity to the target). The induction of semantic paraphasia occurs across the extent of the pathway (beneath occipito-temporal, insular, and frontal cortex; [31, 60]). Stimulation of a third pathway connecting the anterior temporal lobe with the orbitofrontal cortex, the UF, did not result in any deficit in semantic processing [32]. Interpretation of this null finding, though, should proceed with caution. Due to its connectivity with the anterior temporal lobe, it is possible that the UF is involved in other linguistic functions, and that the semantic task used in this study (i.e., picture naming) has insufficient sensitivity to detect the function of this pathway. Therefore, the status of the UF as a language pathway remains an open empirical question.

Notably, although Duffau and colleagues [30] proposed a distinction between the ILF and a putative inferior occipital-frontal fasciculus (IOFF) connecting the occipital cortex to the frontal cortex, the existence of such an uninterrupted occipital-frontal pathway is disputed by recent investigations using anterograde tract tracing in the rhesus monkey [78], which we discuss in the next section.

Intraoperative electrical stimulation has both advantages and disadvantages for identifying fiber pathways involved in language. A major advantage is the precision of the method, in both spatial and functional terms—i.e., the method can identify, *in vivo*, areas of white matter that are necessary to accomplish certain linguistic tasks to a degree that surpasses what can be learned from more gross lesions of the same pathways. But a major limitation is that there is no way to determine, with certainty, the origin and termination sites of the fiber pathways. Tract tracing methodologies using radioactive tracers are the only available methods for reliably identifying the origin and termination sites of fiber pathways, but their use is limited to animal studies.

Fiber tract identification in the rhesus monkey: anterograde tract tracing

The anterograde tract tracing method takes advantage of the orthograde transport (transport away from the cell body) of radioactively labeled substances along the axon. Injection of radioactive compound in an animal brain is followed by histological analysis of the tissue, revealing both the labeled fibers and their terminations [54]. Recently, Schmahmann and Pandya [78] conducted a comprehensive study of fiber pathways of the rhesus monkey brain. Based on the proposed correspondence of homologous regions of the monkey and human brain, these authors identified the SLF III, the extreme capsule, and the MdLF as important fiber pathways for language. Notably, two prominent fiber pathways that have been identified using other methods are missing from this list. The first is the IOFF, and its very existence as a distinct fiber path is disputed by the rhesus data. Schmahmann and Pandya instead argue that the rostral extension of the ILF is in reality the UF and the extreme capsule fiber pathways, and that these three fiber pathways comprise what is thought to be the IOFF. More pertinent to both historical [21, 39] and contemporary [26, 27, 32, 36, 40] investigations of language, though, is the proposal that the AF is not a language pathway. The rhesus data instead suggest that the AF links caudal temporal cortex with dorsal and lateral prefrontal regions that are more distal to the classic perisylvian language cortex of the frontal lobe. These findings complement some of the human DTI work (e.g., [35, 59, 74, 77]), but are in conflict with findings from the postmortem dissection and electrical stimulation studies. Given the prominence of the AF for language, these discrepant findings suggest interesting avenues for future research.

Although anterograde tract tracing has the advantages over the other methods in that (1) it can reliably identify origin and termination sites for distant fiber pathways, and (2) it allows a precision that is not approached by other methods, the major disadvantage is its restriction to use in animals, which limits its utility for studying language. That said, much can be gained from considering the three methods in complement.

Summary of fiber pathways for language

In summary, the available evidence seems to be converging on the involvement of several fiber pathways for language (Fig. 1B). The pathways we have discussed exclude those additional cortico-striatal, cortico-thalamic, and cortico-pontocerebellar pathways likely involved in the execution of speech (reviewed elsewhere in [30, 79]). However, for speech perception and comprehension the core pathways connect the inferior frontal, inferior parietal, and superior and middle temporal cortices. These include the SLF, MdLF, extreme capsule, and the ILF. Whether the SLF pathway can be identified as two pathways—an SLFIII and AF pathway—and whether both the SLFIII and AF participate in language, is a matter of debate. The status of the UF as a language pathway also awaits additional investigation. Further debate centers on the status of an uninterrupted pathway through the ventral temporal cortex connecting occipital, temporal, and frontal regions, potentially involved in semantic processing (i.e., the IOFF). The relevant monkey data suggest that this pathway may in fact be comprised of the dissociable ILF, UF, and extreme capsule fiber paths.

Conclusion

To summarize, in this brief survey of current thinking on the neurobiological basis of language, we suggest that language, which is one of the most celebrated higher functions of the human brain, is deeply rooted in distributed networks that work in synchrony to perceive, produce and comprehend language. We summarize literature that demonstrates that many of the regions involved in any of these levels of organization (i.e. perception, production and comprehension) are typically also involved in the other levels, and that with modern tools and analytical methods, it is no longer necessary to investigate individual levels in isolation from the others.

References

[1] Abrahams S, Goldstein LH, Simmons A, Brammers MJ, Williams SCR, Giampietro VP et al (2003) Functional magnetic resonance imaging of verbal fluency and confrontation naming using compressed image acquisition to permit overt responses. Hum Brain Mapp 20: 29–40

[2] Ackermann H, Riecker A (2004) The contribution of the insula to motor aspects of speech production: a review and a hypothesis. Brain Lang 89: 320–328

[3] Alario FX, Chainay H, Lehericy S, Cohen L (2006) The role of the supplementary motor area (SMA) in word production. Brain Research 1076: 129–143

[4] Arnott SR, Binns MA, Grady CL, Alain C (2004) Assessing the auditory dual-pathway model in humans. Neuroimage 22: 401–408

[5] Assaf Y, Pasternak O (2008) Diffusion tensor imaging (DTI)-based white matter mapping in brain research: A review. J Mol Neurosci 34: 51–61

[6] Aziz-Zadeh L, Koski L, Zaidel E, Mazziotta J, Iacoboni M (2006) Lateralization of the human mirror neuron system. J Neurosci 26: 2964–2970

[7] Basser PJ, Pajevic S, Pierpaoli C, Duda J, Aldroubi A (2000) In vivo fiber tractography using DT-MRI data. Magn Reson Med 44: 625–632

[8] Belin P, Zatorre RJ (2003) Adaptation to speaker's voice in right anterior temporal lobe. Neuroreport 14, 2105–2109

[9] Belin P, Zatorre RJ, Ahad P (2002) Human temporal-lobe response to vocal sounds. Brain Res Cogn Brain Res 13(1): 17–26

[10] Birn RM., Bandettini PA, Cox RW, Jesmanowicz A, Shaker R (1998) Magnetic field changes in the human brain due to swallowing or speaking. Magn Reson Med 40: 55–60

[11] Bohland JW, Guenther FH (2006) An fMRI investigation of syllable sequence production. Neuroimage 32(2): 821–841

[12] Bookheimer SY, Zeffiro TA, Blaxton TA, Gaillard PW, Theodore WH (2000) Activation of language cortex with automatic speech tasks. Neurology 55(8): 1151–1157

[13] Brauer J, Anwander A, Friederici AD (2010) Neuroanatomical prerequisites for language functions in the maturing brain. Cereb Cortex DOI. 10.1093/cercor/bhq108

[14] Buchsbaum B, Hickok G, Humphries C (2001) Role of left posterior superior temporal gyrus in phonological processing for perception and production. Cogn Sci 25: 663–678

[15] Buccino G, Riggio L, Melli G, Binkofski F, Gallese V, Rizzolatti G (2005) Listening to action-related sentences modulates the activity of the motor system: a combined TMS and behavioral study. Brain Res Cogn Brain Res 24(3): 355–363

[16] Cabanis E, Iba-Zizen M, Abelanet R, Monod-Broca P, Signoret J (1994) "Tan-Tan" the first Paul Broca's patient with "Aphemia" (1861): CT (1979), and MRI (1994) of the brain. In: Picard SGL (ed) The refresher course of the ESNR: language and the aphasias. European Society of Neuroradiology, Nancy, pp 9–22
[17] Callan DE, Tsytsarev V, Hanakawa T, Callan AM, Katsuhara M, Fukuyama H et al (2006) Song and speech: brain regions involved with perception and covert production. Neuroimage 31(3): 1327–1342
[18] Catani M, Howard RJ, Pajevic S, Jones DK (2002) Virtual in vivo interactive dissection of white matter fasciculi in the human brain. NeuroImage 17: 77–94
[19] Clarke E, O'Malley CD (1996) The human brain and spinal cord: A historical study illustrated by writings from antiquity to the twentieth century, 2nd edn. Norman, San Francisco
[20] Corfield DR, Murphy K, Josephs O, Fink GR, Frackowiak RS, Guz A et al (1999) Cortical and subcortical control of tongue movement in humans: a functional neuroimaging study using fMRI. J Appl Physiol 86(5): 1468–1477
[21] Déjerine JJ (1895) Anatomie des centres nerveux. Rueff et Cie, Paris
[22] Devlin JT, Matthews PM, Rushworth MFS (2003) Semantic processing in the left inferior prefrontal cortex: A combined functional magnetic resonance imaging and transcranial magnetic stimulation study. J Cogn Neurosci 15(1): 71–84
[23] di Pellegrino G, Fadiga L, Fogassi L, Gallese V, Rizzolatti G (1992) Understanding motor events: a neurophysiological study. Exp Brain Res 91(1): 176–180
[24] Dronkers NF (1996) A new brain region for coordinating speech articulation. Nature 384: 159–161
[25] Dronkers NF, Plaisant O, Iba-Zizen MT, Cabanis EA (2007) Paul Broca's historic cases: high resolution MR imaging of the brains of Leborgne and Lelong. Brain 130(Pt 5): 1432–1441
[26] Dubois J, Hertz-Pannier L, Cachia A, Mangin JF, Le Bihan D, Dehaene-Lambertz G (2009) Structural asymmetries in the infant language and sensori-motor networks. Cereb Cort 19: 414–423
[27] Duffau H (2008) The anatomo-functional connectivity of language revisited: New insights provided by electrostimulation and tractography. Neuropsychologia 46: 927–934
[28] Duffau H, Capelle L, Sichez N, Denvil D, Lopes M, Sichez J-P et al (2002) Intraoperative mapping of the subcortical language pathways using direct stimulations: An anatomo-functional study. Brain 125: 199–214
[29] Duffau H, Gatignol P, Denvil D, Lopes M, Capelle L (2003) The articulatory loop: Study of subcortical connectivity by electrostimulation. NeuroReport 14: 2005–2008
[30] Duffau H, Gatignol P, Mandonnet E, Capelle L, Taillandier L (2008) Intraoperative subcortical stimulation mapping of language pathways in a consecutive series of 115 patients with grade II glioma in the left dominant hemisphere. J Neurosurg 109: 461–471
[31] Duffau H, Gatignol P, Mandonnet E, Peruzzi P, Tzourio-Mazoyer N, Capelle L (2005) New insights into the anatomo-functional connectivity of the semantic system: A study using cortico-subcortical electrostimulations. Brain 128: 797–810
[32] Duffau H, Gatignol P, Moritz-Gasser S, Mandonnet E (2009) Is the left uncinate fasciculus essential for language? J Neurol 256: 382–389
[33] Fadiga L, Craighero L, Buccino G, Rizzolatti G (2002) Speech listening specifically modulates the excitability of tongue muscles: a TMS study. Eur J Neurosci 15(2): 399–402
[34] Ffyche DH, Catani M (2005) Beyond localization: From hodology to function. Philos Trans Roy Soc: Biol Sci 360: 767–779
[35] Frey S, Campbell JSW, Pike GB, Petrides M (2008) Dissociating the human language pathways with high angular resolution diffusion fiber tractography. J Neurosci 28(45): 11435–11444
[36] Friederici AD (2009) Pathways to language: Fiber tracts in the human brain. Trends Cogn Sci 13: 175–181
[37] Fu CH, Morgan KJS, Williams SC, Andrew C, Vythelingum GN et al (2002) A functional magnetic resonance imaging study of overt letter verbal fluency using a clustered acquisition sequence: greater anterior cingulate activation with increased task demand. Neuroimage 17(2): 871–879
[38] Gallese V, Fadiga L, Fogassi L, Rizzolatti G (1996) Action recognition in the premotor cortex. Brain 119(2): 593–609
[39] Geschwind N (1970) The organization of language and the brain. Science 170: 940–944
[40] Glasser MF, Rilling JK (2008) DTI tractography of the human brain's language pathways. Cereb Cort 18: 2471–2482
[41] Gough PM, Nobre AC, Devlin JT (2005) Dissociating linguistic processes in the left inferior frontal cortex with Transcranial Magnetic Stimulation. J Neurosci 25(35): 8010–8016
[42] Gracco VL, Tremblay P, Pike B (2005) Imaging speech production using fMRI. Neuroimage 26(1): 294–301
[43] Guenther FH (2006) Cortical interactions underlying the production of speech sounds. J Commun Disord 39: 350–365
[44] Guenther FH, Ghosh SS, Tourville JA (2006) Neural modeling and imaging of the cortical interactions underlying syllable production. Brain Lang 96: 280–301
[45] Hackett TA (2011) Information flow in the auditory cortical network. Hear Res 271(1–2): 133–146
[46] Hauk O, Johnsrude I, Pulvermuller F (2004) Somatotopic representation of action words in human motor and premotor cortex. Neuron 41(2): 301–307

[47] Henry RG, Berman JI, Nagarajan SS, Mukherjee P, Berger MS (2004) Subcortical pathways serving cortical language sites: Initial experience with diffusion tensor imaging fiber tracking combined with intraoperative language mapping. NeuroImage 21: 616–622
[48] Hickok G, Poeppel D (2000) Towards a functional neuroanatomy of speech perception. Trends Cogn Sci 4(4): 131–138
[49] Hickok G, Buchsbaum B, Humphries C, Muftuler T (2003) Auditory-motor interaction revealed by fMRI: speech, music, and working memory in area Spt. J Cogn Neurosci 15(5): 673–682
[50] Hickok G, Poeppel D (2004) Dorsal and ventral streams: A framework for understanding aspects of the functional anatomy of language. Cognition 92(1–2): 67–99
[51] Hickok G, Poeppel D (2007) The cortical organization of speech processing. Nat Rev Neurosc 8(5): 393–402
[52] Hirshorn EA, Thompson-Schill SL (2006) Role of the left inferior frontal gyrus in covert word retrieval: neural correlates of switching during verbal fluency. Neuropsychologia 44(12): 2547–2557
[53] Kent RD (2000) Research on speech motor control and its disorders: a review and prospective. J Commun Disord 33(5): 391–427; quiz 428
[54] Köbbert C, Apps R, Bechmann I, Lanciego JL, Mey J, Thanos S (2000) Current conecpts in neuroanatomical tracing. Prog Neurobiol 62: 327–351
[55] Lazar M, Alexander AL (2003) An error analysis of white matter tractography methods: Synthetic diffusion tensor field simulations. NeuroImage 20(2): 1140–1153
[56] Levelt WJ (1999) Models of word production. Trends Cogn Sci 3(6): 223–232
[56a] Levelt WJ, Roelofs A, Meyer AS (1999). A theory of lexical access in speech production. Behav Brain Sci 22(1): 1–38; discussion 38–75
[57] Liberman AM, Mattingly IG (1985) The motor theory of speech perception revised. Cognition 21(1): 1–36
[58] Mahon BZ, Caramazza A (2008) A critical look at the embodied cognition hypothesis and a new proposal for grounding conceptual content. J Physiol Paris 102: 59–70
[59] Makris N, Kennedy DN, McInerney S, Sorensen AG, Wang RP, Caviness VS et al (2005) Segmentation of subcomponents within the superior longitudinal fascicle in humans: A quantitative, *in vivo*, DT-MRI study. Cereb Cort 15: 854–869
[60] Mandonnet E, Nouet A, Gatignol P, Capelle L, Duffau H (2007) Does the left inferior longitudinal fasciculus play a role in language? A brain simulation study. Brain 130: 623–629
[61] Meister IG, Wilson SM, Deblieck C, Wu AD, Iacoboni M (2007) The essential role of premotor cortex in speech perception. Curr Biol 17: 1692–1696
[62] Mesulum MM (1998) From sensation to cognition. Brain 121(6): 1013–1052
[63] Mori S, van Zijl PCM, Oishi K, Faria AV (2010) MRI atlas of human white matter, 2nd edn. Elsevier Science, New York
[64] Obleser J, Boecker H, Drzezga A, Haslinger B, Hennenlotter A, Roettinger M et al (2006) Vowel sound extraction in anterior superior temporal cortex. Hum Brain Mapp 27(7): 562–571
[65] Petersen SE, Fox PT, Posner MI, Mintun M, Raichle ME (1988) Positron emission tomographic studies of the cortical anatomy of single-word processing. Nature 331(6157): 585–589
[66] Price CJ, Moore CJ, Humphreys GW, Frackowiak RS, Friston KJ (1996) The neural regions sustaining object recognition and naming. Proc Biol Sci 263(1376): 1501–1507
[67] Pulvermuller F, Hauk O, Nikulin VV, Ilmoniemi RJ (2005) Functional links between motor and language systems. Eur J Neurosci 21(3): 793–797
[68] Pulvermuller F, Huss M, Kherif F, Moscoso del Prado Martin F, Hauk O, Shtyrov Y (2006) Motor cortex maps articulatory features of speech sounds. Proc Natl Acad Sci USA 103: 7865–7870
[69] Rauschecker JP, Scott SK (2009) Maps and streams in the auditory cortex: nonhuman primates illuminate human speech processing. Nat Neurosci 12(6): 718–724
[70] Rauschecker JP, Tian B (2000) Mechanisms and streams for processing of "what" and "where" in auditory cortex. Proc Natl Acad Sci USA 97(22): 11800–11806
[71] Riecker A, Brendel B, Ziegler W, Erb M, Ackerman H (2008) The influence of syllable onset complexity and syllable frequency on speech motor control. Brain Lang 107(2): 102–113
[72] Riecker A, Mathiak K, Wildgruber D, Erb M, Hertrich I, Grodd W et al (2005) fMRI reveals two distinct cerebral networks subserving speech motor control. Neurology 64: 700–706
[73] Rizzolatti G, Fadiga L, Gallese V, Fogassi L (1996) Premotor cortex and the recognition of motor actions. Brain Research. Cogn Brain Res 3(2): 131–141
[74] Rushworth MFS, Behrens TEJ, Johansen-Berg H (2006) Connection patterns distinguish 3 regions of human parietal cortex. Cereb Cort 16: 1418–1430
[75] Saarinen T, Laaksonen H, Parviainen T, Salmelin R (2006) Motor cortex dynamics in visuomotor production of speech and non-speech mouth movements. Cereb Cort 16(2): 212–222
[76] Sato M, Tremblay P, Gracco VL (2009) A mediating role of the premotor cortex in phoneme segmentation. Brain Lang 111(1): 1–7
[77] Saur D, Kreher BW, Schnell S, Kümmerer D, Kellmeyer, P, Vry M-S et al (2008) Ventral and dorsal pathways for language. Proc Natl Acad Sci USA 105(46): 18035–18040
[78] Schmahmann JD, Pandya DN (2006) Fiber pathways of the brain. Oxford University Press, New York
[79] Schmahmann JD, Pandya DN (2008) Disconnection

syndromes of basal ganglia, thalamus, and cerebrocerebellar systems. Cortex 44: 1037–1066

[80] Schmahmann JD, Pandya DN, Wang RP, Dai G, D'Arceuil H, de Crespigny AJ et al (2007) Association fibre pathways of the brain: Parallel observations from diffusion spectrum imaging and autoradiography. Brain 130: 630–653

[81] Skipper JI, Nusbaum HC, Small SL (2005) Listening to talking faces: motor cortical activation during speech perception. Neuroimage 25(1): 76–89

[82] Sundara M, Namasivayam AK, Chen R (2001) Observation-execution matching system for speech: a magnetic stimulation study. Neuroreport 12(7): 1341–1344

[83] Szelényi A, Bello L, Duffau H, Fava E, Feigl GC, Galanda M et al (2010) Intraoperative electrical stimulation in awake craniotomy: Methodological aspects of current practice. Neurosurg Focus 28(2): E7

[84] Tench CR, Morgan PS, Wilson M, Blumhardt LD (2002) White matter mapping using diffusion tensor MRI. Magn Reson Med 47: 967–972

[85] Tettamanti M, Buccino G, Saccuman MC, Gallese V, Danna M, Scifo P et al (2005) Listening to action-related sentences activates fronto-parietal motor circuits. J Cogn Neurosci 17(2): 273–281

[86] Tournier J-D, Calamante F, King MD, Gadian DG, Connelly A (2002) Limitations and requirements of diffusion tensor fiber tracking: An assessment using simulations. Magn Reson Med 47: 701–708

[87] Tourville JA, Reilly KJ, Guenther FH (2008) Neural mechanisms underlying auditory feedback control of speech. Neuroimage 39(3): 1429–1443

[88] Tremblay P, Gracco VL (2006) Contribution of the frontal lobe to externally and internally specified verbal responses: fMRI evidence. Neuroimage 33(3): 947–957

[89] Tremblay P, Gracco VL (2009) On the selection of words and oral motor responses: evidence of a response-independent fronto-parietal network. Cortex 46(1): 15–28

[90] Watkins K, Paus T (2004) Modulation of motor excitability during speech perception: the role of Broca's area. J Cogn Neurosci 16(6): 978–987

[91] Watkins KE, Strafella AP, Paus T (2003) Seeing and hearing speech excites the motor system involved in speech production. Neuropsychologia 41(8): 989–994

[92] Wedeen VJ, Wang RP, Schmahmann JD, Benner T, Tseng WYI, Dai G et al (2008) Diffusion spectrum magnetic resonance imaging (DSI) tractography of crossing fibers. NeuroImage 41: 1267–1277

[93] Wilson SM, Iacoboni M (2006) Neural responses to non-native phonemes varying in producibility: Evidence for the sensorimotor nature of speech perception. Neuroimage 33(1): 316–325

[94] Wilson SM, Saygin AP, Sereno MI, Iacoboni M (2004) Listening to speech activates motor areas involved in speech production. Nat Neurosci 7(7): 701–702

[95] Wise R, Chollet F, Hadar U, Friston K, Hoffner E, Frackowiak R (1991) Distribution of cortical neural networks involved in word comprehension and word retrieval. Brain 114: 1803–1817

[96] Wise RJ, Greene J, Büchel C, Scott SK (1999) Brain regions involved in articulation. Lancet 353: 1057–1061

[97] Wise RJ, Scott SK, Blank SC, Mummery CJ, Murphy K, Warburton EA (2001) Separate neural subsystems within "Wernicke's area". Brain 124: 83–95

[98] Zatorre RJ, Evans AC, Meyer E, Gjedde A (1992) Lateralization of phonetic and pitch discrimination in speech processing. Science 256: 846–849

[99] Ziegler W (2003) Speech motor control is task-specific. Evidence from dysarthria and apraxia of speech. Aphasiology 17: 3–36

Neural basis of memory

Matthew A. Lambon Ralph

"You see," he [Sherlock Holmes] explained, "I consider that a man's brain originally is like a little empty attic, and you have to stock it with furniture as you choose. A fool takes in all the lumber of every sort that he comes across, so that the knowledge which might be useful to him gets crowded out, or at best is jumbled up with a lot of other things so that he has a difficulty in laying his hands upon it. Now the skilful workman is very careful indeed as to what he takes into his brain-attic. ... It is a mistake to think that that little room has elastic walls and can distend to any extent. Depend upon it, there comes a time when for every addition of knowledge you forget something that you knew before. It is of the highest importance, therefore, not to have useless facts elbowing out the useful ones."

From *A Study in Scarlet*, Sir Arthur Conan Doyle (1887)

Although not intended as a neuroscience theory of memory, Sherlock Holmes's working hypothesis contains elements within it that still resonate with contemporary theories of memory and its neural basis. For example, most theories and computational models of memory agree that our long-term representations are formed through a gradual learning process based upon our experiences. And, in that sense, Holmes is correct to note that our knowledge base (which we now call *semantic memory*) is a distillation or reflection of verbal and nonverbal experience. The second observation, which also finds a parallel in contemporary computational neuroscience, is that there is a danger that new learning might interfere with established long-term, representations – something which is now known as "catastrophic interference" [27]. As will be described below, the neural architecture used for the formation and consolidation of memories seems to have been formulated in such a way so as to minimise the danger that new facts might "elbow" out the old ones. Finally, there is an important assumption underlying Holmes's idiosyncratic approach to learning, namely that memory is not simply a passive register which records previous events and facts, but instead it is a productive database which we use when we communicate, when we use objects or conduct other nonverbal activities, and it influences our perceptions of the world (famously, and as he constantly reminds Watson, Holmes's heightened powers of perception of crime scenes is nothing to do with better eyesight but rather his knowledge base alters and improves his perception of the scene). These facts have clear implications when considering neurological patients or the potential impact of neurosurgery; when this productive database becomes degraded or dysfunction then it not only impacts on patients' ability to remember past events or facts (a key ingredient in their own personality and identity) but also on their ability to communicate and function in an effective and efficient manner.

A key factor, missed in this quote, is that memory is not a singular process. Indeed, Tulving (1972) is credited in the modern literature to have been the first to propose that memory should be divided into separable types and since then, neuropsychological studies have shown that these different aspects of memory can dissociate in patients with different locations of damage [14]. Although various divisions of memory have been proposed over the years, it is still common to follow Tulving's suggestion and to distinguish two broad types of long-term memory: episodic memory and semantic memory. Each of these is considered in turn below.

Episodic memory

Anterograde amnesia and the medial temporal lobes

Episodic memory refers to a sub-type of long-term representation, which allows us to code information in relation to specific events. As such, a critical characteristic of episodic memory is that it is time-related and allows us to distinguish between different events (e.g., last year's family holiday vs. a recent business trip) or even similar events occurring at different times (e.g., different annual family holidays to the same location). There is a now a wealth of evidence that the medial temporal lobes and especially the hippocampus, are crucial in the initial registration of new episodes and information. Although this can be precisely demonstrated in animal ablation studies, it was most famously described in the seminal case-series studies by Scoville, Penfield and Milner [40], which included patient HM. In an attempt to treat patients with extreme and disabling temporal lobe epilepsy, Scoville explored the results of bilateral hippocampectomy. Whilst this did improve the patients' level of epilepsy, it left them with a profound anterograde amnesia such that their ability to remember new events or to learn new facts was devastated. The detailed assessment of HM and other patients with bilateral medial temporal resection highlighted the fact that not all aspects of memory are compromised such that previously-learnt information about events, geographical routes and semantic memory are left at least relatively intact.

Two further key neuroscience and clinical insights arose from these seminal studies. The first is that unilateral hippocampectomy does not lead to the same profound and chronic degree of amnesia [40]. Such patient can initially present with mild amnesia and formal testing can demonstrate mild anterograde amnesia in the chronic phase, but it is never on the same scale as that found in the bilaterally-resected patients. Whilst this has had clear implications for neurosurgical practice, the comparison suggests that the neural basis of episodic memory is based upon a bilateral system in which memories are represented in at least a semi-redundant fashion across the two hemispheres. Presumably, this kind of neural architecture has the benefit that the resultant memory system is relatively robust to the effects of localised brain damage. In comparison, it takes bilateral resection or a bilateral disease with a focus upon the medial temporal lobes (as found, for example, in early prodromal Alzheimer's disease) to generate considerable anterograde amnesia.

The third observation, that we can glean from these and related studies of medial temporal lobe damage, is that the episodic vs. semantic distinction does not seem to hold and instead a better division might be the age of the memory. So, for example, HM's memory deficit was not limited to episodic memory alone – he was able to recall events (episodic) from his childhood and, in addition, he was not able to learn new facts or vocabulary (semantics) following his surgery. As a consequence, it seems more sensible to conclude that (a) the medial temporal lobe structures are critical in the initial registration and coding of new episodes, facts and conjunctions of other types of multi-modal information; and (b) over time the medial temporal system becomes less critical in the recall of information such that, eventually, there is no need for hippocampal involvement at all. Indeed, contrastive neuropsychological studies of early Alzheimer's disease (AD) vs. semantic dementia (SD) are instructive. In the early phase, these two diseases form a neurological contrast between bilateral medial temporal lobe dysfunction (in AD) vs. bilateral inferior-lateral, anterior neocortical atrophy (in SD: [14]). Neuropsychologically, this leads to a clear double dissociation which is a crucial feature for differential diagnosis [28] – the mild AD patients have a significant anterograde amnesia for new events and facts yet relatively preserved semantic (long-term retrograde) memory, whilst the opposite is found in semantic dementia. There are also some indications that these two disorders generate a different profile of episodic recall with respect to the age of the memories: mild AD patients demonstrate the Ribot effect in which recall of old memories (from childhood and young adulthood) is better than very poor recall of recent events. The reverse is found in SD patients, who demonstrate poor detailed recall of events from their childhood but excellent knowledge for very recent events (within the past two years: [18]). As per the nature of the graded breakdown of semantic representations in this condition (see below), it would appear that some vestiges of information remain for the older events which can be elicited by using picture-based stimuli and verbal prompts [47].

Consolidation from new to old memories: the complementary learning systems theory

Assuming that the age of a memory is a crucial factor in the neural basis of memory, then it is important to consider how memories are first encoded and then how they are consolidated. There are at least four critical challenges for the brain to overcome in this regard: (a) we have an ability to bring together and code multimodal and verbal information from one brief event for later accurate recall – thus the system has to integrate across multiple sources of sensory and verbal experience, rapidly and accurately; (b) information about each new event must be kept insulated from previous experiences, even if such events are similar, otherwise we would confuse elements of different events/time-points with each other (e.g., we are able to recall specific information about individual trips from the commute to work, even though the other aspects of the regular journey – e.g., train, time, routine – are identical); (c) in the longer term, facts and other semantic information needs to be coded on a long-term basis (e.g., we are able to recall facts that were first learnt when we were children); and (d) certain types of knowledge (e.g., semantics) are built-up in a gradual manner through repeated exposures and experiences. As such, it is crucial to integrate across separate learning episodes and across different types of sensory and verbal experience in order to generate "coherent" concepts. For example, our knowledge of the concept *bread* will have been accrued over many (probably daily) experiences throughout life and in different contexts (making, eating, preparing, buying, giving, breaking, etc.). Despite the multifaceted-nature of our experience with this object, we do not have separate memories for each occasion and each type of sensory experience, but instead these are integrated into a single coherent, sophisticated concept – to the extent that it becomes impossible to remember when and how the different elements of the meaning were originally learnt.

Investigations from computational neuroscience have generated a neuroanatomically-constrained, computational account which is able to meet these challenges [27]. McClelland et al's computational model was built upon the neuroanatomical distinction described above – namely, a functional system generated from the interaction of medial temporal regions and neocortex. Each of these systems was implemented to have complementary characteristics – the medial temporal component utilised a sparse-encoding which licenses key aspects for initial coding of new events or information. Because they are sparse, it is possible to use each representation: (i) to link together disparate pieces of information arising in multiple sensory and verbal domains; (ii) to do so rapidly without distortion of the experiences; and (iii) to keep each experience isolated and distinct from all other events, irrespective of how superficially similar they are. The cortical learning system was based upon "distributed" representations in which information is coded across a series of processing units rather than assigned to a single unit. This type of representation is not capable of effective and accurate rapid learning because, by itself, rapid learning makes long-term information vulnerable to interference from new learning (something known as "catastrophic interference"). Distributed representations, however, solve the challenges associated with long-term storage and the formation of integrated semantic memory. If learning is slow and the different events are interleaved, then distributed representations are capable of coding a vast quantity of information and maintaining this information in the long-term [38]. Distributed representations are an effective method for coding knowledge because they are robust to the effects of local damage or inefficiencies (something known as "graceful degradation" [11]). This is because no one unit is critical in the reactivation of the representation. In addition, distributed representations are able to code the statistical structure across all learning experiences. As such, this licenses the formation of coherent concepts through the amalgamation of separate experiences into a single representation and permits accurate generalisation between similar concepts, irrespective of their superficial similarity [23].

When these two complementary systems work together, they meet the four challenges noted above. Specifically, in the first phase, the sparse medial temporal lobe representations are able to rapidly code and isolate individual events and experiences. These sources of information can then be coded more gradually by the distributed representations in the neocortex thereby generating long-term, robust and generalisable memories.

Semantic memory

Having discussed episodic memory and the consolidation of information over time, it is important to consider Tulving's second main sub-type of long-term memory – semantic memory. Semantic memory encompasses the meaning of all types of verbal and nonverbal stimuli including words, pictures, objects and faces. As well as underpinning comprehension it also allows us to express knowledge in a wide variety of domains, both verbal (e.g., naming and verbal definitions) and non-verbal (e.g., drawing and object use). As such it is integral to our everyday lives and impairments of semantic memory are extremely debilitating. A key question for neuroscience research, therefore, is which parts of the brain support semantic memory and how do they function?

Long before the term "semantic memory" was introduced, Wernicke and Meynert (see [9]) were interested in how the brain formed and reactivated concepts – a process they referred to as "conceptualisation". Meynert and Wernicke's model of conceptualisation made the following assumptions: (a) that the building blocks of concepts were modality-specific engrams (stores of information) localised to the cortical areas responsible for the corresponding sensory, motor or verbal domain; (b) that these modality-specific engrams, in widespread brain regions, were fully interconnected; and (c) that this web of connections was the basis of conceptualisation – a specific concept being represented by the co-activation of all its associated engrams (see Fig. 1, Panel A for a pictorial representation of this idea). For example, if you taste an apple (even with your eyes closed), the taste-specific engram will automatically activate all of the other associated modality-linked engrams, enabling your brain to retrieve other knowledge concerning the object: its visual form, likely colour, name, presence of seeds, how you would peel it, and so on. In this proposal, Wernicke and Meynert argued that – unlike forms of agnosia and aphasia – central disorders of conceptualisation only occurred as a consequence of global brain damage (dementia) because only such widespread cortical damage would disrupt the distributed engram re-activation process. This "distributed" theory of semantic memory is, perhaps, still the most prevalent in contemporary neuroscience [26]. Indeed, it has many appealing characteristics including the fact that there is no mystery about the source and formation of concepts – they are the distillation of our multimodal experience.

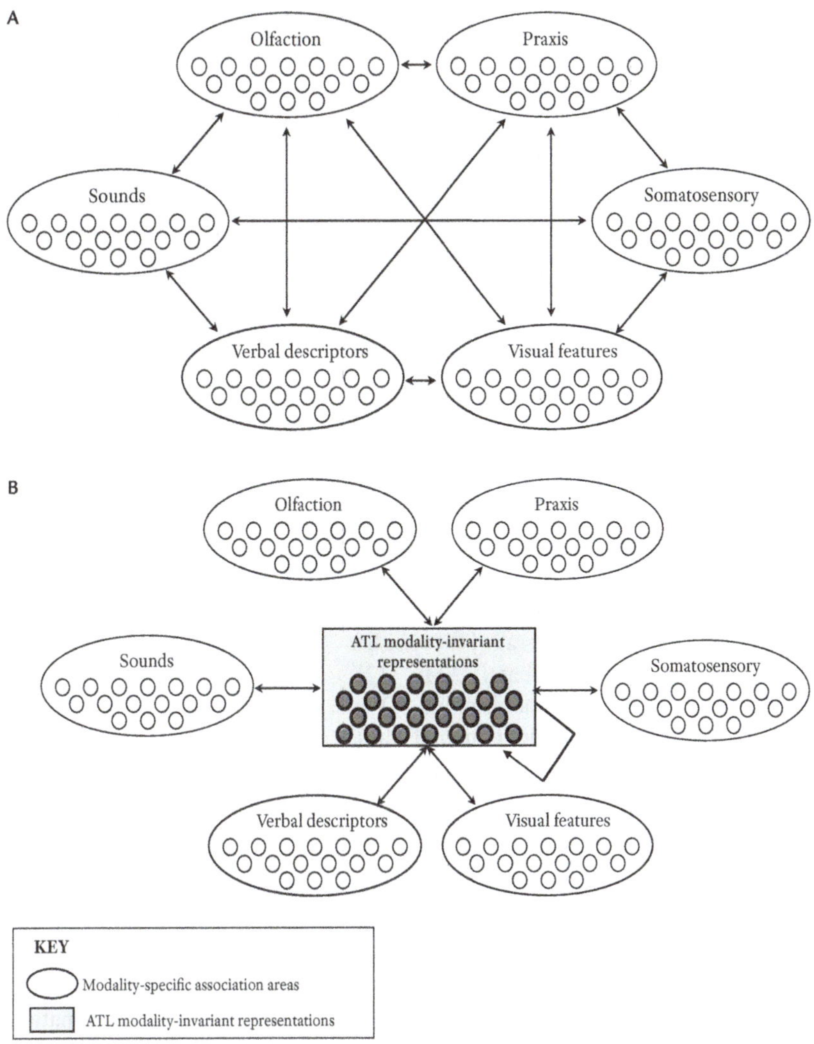

Fig. 1. **(A)** Wernicke-Meynert model of conceptualisation **(B)** Hub-and-spoke semantic system [37]

A key prediction of the distributed models of semantic memory is that generalised conceptual impairment should not be associated with localised brain damage – and only follow from brain-wide degeneration. Instead, localised brain damage to specific engrams/modality-specific information sources should lead to selective impairments including category-specific deficits (e.g., deterioration of praxis engrams should produce poor knowledge and abilities with tools) or agnosias. Whilst there is considerable empirical support for this part of the story, there is now mounting evidence that, in addition to these modality-specific sources of information, key regions in the anterior temporal lobe play a vital role in the summation and formation of concepts. The first modern evidence for this hypothesis came in the form of a seminal study by Warrington [46]. She described three patients with progressive brain disease resulting in a range of semantic deficits that cut across different modalities but were nonetheless restricted to the domain of semantics. Snowden et al [46] subsequently identified the condition as part of the spectrum of fronto-temporal dementia and gave it the label, "semantic dementia" (SD). There was little or no neuroanatomical information about the patients in Warrington's original study but modern structural neuroimaging has demonstrated that semantic dementia is consistently associated with selective atrophy of the anterior temporal lobes bilaterally [19] – with a particular focus on the basal ATL [12]. Furthermore, the hypometabolism in SD is largely restricted to the anterior temporal lobes, thus indicating that the associated cognitive disorder is attributable to damage in this region rather than to widespread functional abnormality [29, 39].

The data from SD provide a strong counterpoint to the distributed theories of semantic memory in two further ways. First, unlike patients with agnosia or category-specific semantic impairments, SD patients present with selective yet multimodal semantic deficits [4]. Various detailed neuropsychological studies have demonstrated that, although basic perception/sensation is preserved in each modality, the SD patients have increasing difficulties in activating the meaning associated with each stimulus [25, 32]. As a result the patients have poor comprehension and production of information across all verbal and nonverbal domains, including words, objects, pictures, sounds, smells, and touch, etc. in receptive abilities, and speech, object use, drawing, etc. in expression. Returning to our apple example, from above – SD patients are able to process its visual form (e.g., distinguishing it from another visually similar item) or its taste (e.g., differentiating it from the taste of a pear) but they have reduced ability to access its meaning, making them unable to demonstrate how it is prepared, how and where it might grow, and to name it. The second key characteristic of the semantic impairment in SD is that the deficit is a graded phenomenon in which concepts and the boundaries between concepts gradually "dissolve" or "dim", rather than dropping out abruptly [19, 37]. Thus, the neural system for semantic memory is not like an encyclopaedia with separate entries for each concept but rather it is built on more distributed representations such that concepts degrade in a gradual fashion in the face of brain damage.

The "hub-and-spoke" model of conceptualisation

These characteristics of SD can be understood in terms of the anterior temporal lobe (ATL) supporting the formation of modality-invariant representations as the neural hub of the semantic system. Specifically, this theory retains the notion of a brain-wide network of modality-specific sources of information but links them all through a set of ATL modality-invariant representations. In the domain of computational neuroscience, Rogers et al [37] demonstrated that the central component of a hub-and-spoke framework, which draws together modality-specific information, will generate modality-invariant semantic representations. As can be seen from Fig. 1 (Panel B), the Rogers et al model is an extension of the Meynert-Wernicke framework. Information arising in each specific modality (e.g., the apple's shape, colour, smell, praxis, name, verbal descriptors, etc.) is coded in the corresponding specific cortical sensory or motor or language region. Information from the modality-specific regions is fused together through the intermediate ATL modality-invariant representations. Rogers et al were able to demonstrate that simulated damage to these hub units reproduced the core features of semantic dementia. That is, increasing degrees of damage produced gradual decline in performance of any task requiring conceptual knowledge and, due to its central processing role, damage to this hub component of semantic memory resulted in verbal and nonverbal receptive and expressive deficits.

Convergent evidence for the ATL hub-and-spoke semantic framework.

Whilst the data arising from semantic dementia clearly implicate the bilateral ATL in semantic representation, this area is often overlooked or even disputed in other research on semantic memory [17, 26]. Several factors probably account for this situation. First, classical aphasiological models have never associated anterior temporal lobe regions with comprehension disorders – patients with Wernicke's aphasia typically have damage to the left posterior middle temporal and superior temporal gyri, whilst patients with transcortical sensory aphasia have damage to the left temporoparietal or prefrontal cortices [1, 2]. Second, following unilateral resection of the temporal pole, epilepsy patients do not have semantic impairment or at least not to the same degree as SD patients [16]. Third, fMRI studies of semantic tasks rarely activate anterior temporal lobe regions but, in line with the aphasiological models, find activation in left temporoparietal and prefrontal regions [7].

Recent studies indicate that these observations are not contradictory with the results from semantic dementia. First, direct comparisons of SD and aphasia-related comprehension impairments show that whilst both conditions can lead to impairment of multimodal semantic cognition, there is a qualitative difference between the patient groups; SD results from a gradual dissolution or "dimming" of the semantic representations themselves whilst aphasic patients with multimodal comprehension disorders have impairment to the mechanisms that control or shape the activation of task-relevant information rather than damage to semantic knowledge per se [6, 20]. This indicates that semantically-driven behaviour (which we have come to refer to as "semantic cognition") is comprised of two key, interacting components: (a) the core semantic representations and (b) "semantic control" - executive-control mechanisms that interact with the underlying semantic representations to produce task- and time-appropriate activation of key knowledge for the specific task in hand. This is consistent with functional neuroimaging which shows that left temporoparietal and inferior prefrontal regions are involved in the control or selection mechanisms that underpin a variety of cognitive processes including semantic cognition [13, 31]. Second, results from the outcome of epilepsy-related resections are complicated by two factors: (a) long-standing epilepsy might lead to changes in neural organisation and, indeed, recent imaging studies have shown that white matter connectivity and neurotransmitter function are significantly altered in this condition [15, 36]; and (b) this procedure is unilateral whilst SD patients have bilateral temporal lobe atrophy. Other neurological disorders, such as herpes simplex virus encephalitis, do produce semantic impairment when damage affects the same bilateral temporal lobe regions as semantic dementia [21] and, like amnesia in the context of unilateral vs. bilateral hippocampectomy (see above), it would appear that significant semantic degradation only arises in the context of bilateral ATL damage [23].

Given these debates on the neural basis of semantic memory, we recently embarked on testing the validity of the hub-and-spoke framework through the application of other neuroscience techniques, including repetitive transcranial magnetic stimulation (rTMS) and distortion-corrected fMRI. In a series of studies, we have now been able to demonstrate that rTMS of the lateral ATL in neurologically-intact participants generates results that directly parallel the SD data. Specifically, ATL rTMS slows both receptive (e.g., synonym judgement) and expressive (e.g., naming) semantic tasks. Like the SD patients, this effect appears to be selective in that the same stimulation does not interfere with non-semantic tasks of equal difficulty [35]. In addition, both verbal and nonverbal semantic abilities are affected by lateral ATL stimulation [33]. Finally, it is possible to use rTMS to probe the function of different regions within the same participants (an impossibility in neuropsychology where the lesion is under the control of nature rather than the experimenter). We used this characteristic of rTMS in order to probe the roles of hub vs. spoke representation [34]. As predicted, we found that stimulation of the lateral ATL generated a category-general slowing of naming times (i.e., slowed naming of both living and nonliving entities), whereas stimulation of the inferior parietal lobule generated a category-specific impairment (selective slowing of manipulable objects and tools) related to the praxis information that is coded in this region.

Which areas of the ATL are critical in semantic function?

To finish this chapter, it is important to consider the neuroanatomical definition of the anterior temporal lobe, more carefully. Many authors have used the term "ATL" to refer simply to those regions primarily affected in SD (e.g., [30]). This lacks specificity given that SD implicates a rather broad region incorporating a large proportion of the rostral half of the temporal lobe. Neuroanatomically speaking, the prospect of this ATL region existing as a single unified functional entity seems unlikely. In considering the area typically affected in SD, it is possible to identify at least eight cytoarchitecturally distinguishable albeit graded sub-regions [5]. In fact, it has been shown more recently that, in the temporopolar cortex alone, there is a minimum of 7 distinct sub-areas [8].

The use of fMRI is the obvious choice in generating greater anatomical precision, not only due to its superior spatial resolution (in comparison to PET and MEG) but also because there are limits in the extent to which SD studies and TMS can be used to probe differential function within the ATL. SD is a neurodegenerative disease and, as such, there is a graded distribution of tissue loss across the affected region rather than the absolute boundary found in acute stroke, for example. The local effects of rTMS are much more spatially-specific than brain disease, however, it is anatomically impossible to map out the inferior surface or polar cap of the temporal lobes using TMS. It is important, however, to be able to probe the anterior-inferior surface given that (i) intracranial recordings suggest that these regions show relatively early and semantically-selective neural activity [24], (ii) a number of PET functional imaging studies find language-related activations in this region [41] and (iii) SD atrophy is particularly pronounced in this area [12, 39].

It is important to note, however, that until recently fMRI and other neuroimaging studies have been somewhat silent over the involvement of the ATL in semantic memory. A formal meta-analysis showed that at least some of this absence of evidence reflects technical or methodological issues [45]. These include the fact that the sensitivity of fMRI is not constant across the brain. The ATL and adjacent orbitofrontal cortex reside near air-bone interfaces that cause inhomogeneities in the magnetic field leading to geometric image distortion and signal loss when using conventional gradient-echo EPI [7]. By utilising recent improvements of engineering (e.g., parallel receiver coils) and image acquisition-processing techniques (post-acquisition, k-space spatial-correction of spin echo EPI data:[10, 44]), we have been able to probe the ATL of neurologically-intact participants whilst undertaking the same semantic tasks as those used in the neuropsychological assessment of SD patients and in the parallel rTMS investigations. These have demonstrated that there are at least two key areas within the ATL that are critical for semantic processing: (a) the basal anterorolateral temporal region (including the anterior third of the fusiform and inferior temporal gyrus) and (b) the anterior STS/STG [3, 44]. Intriguingly, the basal ATL area (which is one of the most affected regions in uncorrected gradient EPI) has been implicated previously in other neuroscience literatures not only as the "basal temporal language region" (from PET neuroimaging and intracortical electrode studies: [41]) but also as the site for semantic, visually-invariant activation [24] – clearly fitting with the modality-invariant representations encoded within the semantic hub of the Rogers et al computational framework.

References

[1] Alexander MP, Hiltbrunner B, Fischer RS (1989) Distributed anatomy of transcortical sensory aphasia. Arch Neurol 46: 885–892

[2] Berthier ML (2001) Unexpected brain-language relationships in aphasia: Evidence from transcortical sensory aphasia associated with frontal lobe lesions. Aphasiology 15: 99–130

[3] Binney RJ, Embleton KV, Jefferies E, Parker GJM, Lambon Ralph MA (2010). The ventral and inferolateral aspects of the anterior temporal lobe are crucial in semantic memory: Evidence from a novel direct comparison of distortion-corrected fMRI, rTMS, and semantic dementia. Cereb Cortex 20: 2728–2738

[4] Bozeat S, Lambon Ralph MA, Patterson K, Garrard P, Hodges JR (2000) Non-verbal semantic impairment in semantic dementia. Neuropsychologia 38: 1207–1215

[5] Brodmann K (1909) Vergleichende Lokalisationslehre der Grosshirnrinde. Barth, Leipzig

[6] Corbett F, Jefferies E, Ehsan S, Lambon Ralph MA (2009) Different impairments of semantic cognition in semantic dementia and semantic aphasia: evidence from the non-verbal domain. Brain 132: 2593–2608

[7] Devlin JT, Russell RP, Davis MH, Price CJ, Wilson J, Moss HE et al (2000) Susceptibility-induced loss of signal: Comparing PET and fMRI on a semantic task. Neuroimage 11: 589–600

[8] Ding S, Van Hoesen GW, Cassell MD, Poremba A (2009) Parcellation of human temporal polar cortex: A combined analysis of multiple cytoarchitectonic, chemoarchitectonic, and pathological markers. J Comp Neurol 514: 595–623

[9] Eggert GH (1977) Wernicke's works on aphasia: A source-book and review. Mouton, The Hague

[10] Embleton KV, Haroon HA, Morris DM, Lambon Ralph MA, Parker GJM (2010) Distortion correction for diffusion weighted MRI and tractography in the temporal lobes. Hum Brain Mapp (in press)

[11] Farah MJ, McClelland JL (1991) A computational model of semantic memory impairment: Modality specificity and emergent category specificity. J Exp Psychol General 120: 339–357

[12] Galton CJ, Patterson K, Graham KS, Lambon Ralph MA, Williams G, Antoun N et al (2001) Differing patterns of temporal atrophy in Alzheimer's disease and semantic dementia. Neurology 57: 216–225

[13] Garavan H, Ross TJ, Li SJ, Stein EA (2000) A parametric manipulation of central executive functioning. Cereb Cortex 10: 585–592

[14] Graham KS, Hodges JR (1997) Differentiating the roles of the hippocampal complex and the neocortex in long-term memory storage: Evidence from the study of semantic dementia and Alzheimer's disease. Neuropsychology 11: 77–89

[15] Hammers A, Koepp MJ, Richardson MP, Hurlemann R, Brooks DJ, Duncan JS (2003) Grey and white matter flumazenil binding in neocortical epilepsy with normal MRI. A PET study of 44 patients. Brain 126: 1300–1318

[16] Hermann B, Davies K, Foley K, Bell B (1999) Visual confrontation naming outcome after standard left anterior temporal lobectomy with sparing versus resection of the superior temporal gyrus: A randomized prospective clinical trial. Epilepsia 40: 1070–1076

[17] Hickok G, Poeppel D (2007) The cortical organization of speech processing. Nat Rev Neurosci 8: 393–402

[18] Hodges JR, Graham KS (1998) A reversal of the temporal gradient for famous person knowledge in semantic dementia: implications for the neural organisation of long-term memory. Neuropsychologia 36: 803–825

[19] Hodges JR, Patterson K (2007) Semantic dementia: a unique clinicopathological syndrome. Lancet Neurology 6: 1004–1014

[20] Jefferies E, Lambon Ralph MA (2006) Semantic impairment in stroke aphasia vs. semantic dementia: A case-series comparison. Brain 129: 2132–2147

[21] Lambon Ralph MA, Lowe C, Rogers TT (2007) Neural basis of category-specific semantic deficits for living things: evidence from semantic dementia, HSVE and a neural network model. Brain 130: 1127–1137

[22] Lambon Ralph MA, Cipolotti L, Manes F, Patterson K (2010a) Taking both sides: Do unilateral, anterior temporal-lobe lesions disrupt semantic memory? Brain 133: 3243–3255
[23] Lambon Ralph MA, Sage K, Jones RW, Mayberry EJ (2010b) Coherent concepts are computed in the anterior temporal lobes. Proc Natl Acad Sci 107: 2717–2722
[24] Liu H, Agam Y, Madsen JR, Kreiman G (2009) Timing, timing, timing: Fast decoding of object information from intracranial field potentials in human visual cortex. Neuron 62: 281–290
[25] Luzzi S, Snowden JS, Neary D, Coccia M, Provinciali L, Lambon Ralph MA (2007) Distinct patterns of olfactory impairment in Alzheimer's disease, semantic dementia, frontotemporal dementia, and corticobasal degeneration. Neuropsychologia 45: 1823–1831
[26] Martin A (2007) The representation of object concepts in the brain. Ann Rev Psychol 58: 25–45
[27] McClelland JL, McNaughton BL, O'Reilly RC (1995) Why there are complementary learning-systems in the hippocampus and neocortex: Insights from the successes and failures of connectionist models of learning and memory. Psychol Rev 102: 419–457
[28] Mioshi E, Dawson K, Mitchell J, Arnold R, Hodges JR (2006) The Addenbrooke's Cognitive Examination Revised (ACE-R): a brief cognitive test battery for dementia screening. Int J Geriatr Psychiatry 21: 1078–1085
[29] Nestor PJ, Fryer TD, Hodges JR (2006) Declarative memory impairments in Alzheimer's disease and semantic dementia. Neuroimage 30: 1010–1020
[30] Patterson K, Nestor PJ, Rogers TT (2007) Where do you know what you know? The representation of semantic knowledge in the human brain. Nat Rev Neurosci 8: 976–987
[31] Peers PV, Ludwig CJH, Rorden C, Cusack R, Bonfiglioli C, Bundesen C et al (2005) Attentional functions of parietal and frontal cortex. Cereb Cortex 15: 1469–1484
[32] Piwnica-Worms KE, Omar R, Hailstone JC, Warren JD (2010) Flavour processing in semantic dementia. Cortex 46: 761–768
[33] Pobric G, Jefferies E, Lambon Ralph MA (2010a) Amodal semantic representations depend on both anterior temporal lobes: Evidence from repetitive transcranial magnetic stimulation. Neuropsychologia 48: 1336–1342
[34] Pobric G, Jefferies E, Lambon Ralph MA (2010b) Category-specific versus category-general semantic impairment induced by transcranial magnetic stimulation. Curr Biol 20: 964–968
[35] Pobric GG, Jefferies E, Lambon Ralph MA (2007) Anterior temporal lobes mediate semantic representation: Mimicking semantic dementia by using rTMS in normal participants. Proc Natl Acad Sci 104: 20137–20141
[36] Powell HWR, Parker GJM, Alexander DC, Symms MR, Boulby PA, Wheeler-Kingshott CAM et al (2007) Abnormalities of language networks in temporal lobe epilepsy. Neuroimage 36: 209–221
[37] Rogers TT, Lambon Ralph MA, Garrard P, Bozeat S, McClelland JL, Hodges JR et al (2004) The structure and deterioration of semantic memory: A neuropsychological and computational investigation. Psycholl Rev 111: 205–235
[38] Rogers TT, McClelland JL (2004) Semantic cognition: A parallel distributed processing approach. Cambridge, The MIT Press, MA
[39] Rohrer JD, Warren JD, Modat M, Ridgway GR, Douiri A, Rossor MN et al (2009) Patterns of cortical thinning in the language variants of frontotemporal lobar degeneration. Neurology 72: 1562–1569
[40] Scoville WB, Milner B (1957) Loss of recent memory after bilateral hippocampal lesions. J Neurol Neurosurg Psychiatry 20: 11–21
[41] Sharp DJ, Scott SK, Wise RJS (2004) Retrieving meaning after temporal lobe infarction: The role of the basal language area. Ann Neurol 56: 836–846
[42] Snowden JS, Goulding PJ, Neary D (1989) Semantic dementia: A form of circumscribed cerebral atrophy. Behav Neurol 2: 167–182
[43] Tulving E (1972) Episodic and semantic memory. In: Tulving E and Donaldson W (eds) Organisation of memory. Academic Press, London
[44] Visser M, Embleton KV, Jefferies E, Parker GJ, Lambon Ralph MA (2010a) The inferior, anterior temporal lobes and semantic memory clarified: Novel evidence from distortion-corrected fMRI. Neuropsychologia 48: 1689–1696
[45] Visser M, Jefferies E, Lambon Ralph MA (2010b) Semantic processing in the anterior temporal lobes: A meta-analysis of the functional neuroimaging literature. J Cogn Neurosci 22: 1083–1094
[46] Warrington EK (1975) The selective impairment of semantic memory. Quart J Exp Psychol 27: 635–657
[47] Westmacott R, Leach L, Freedman M, Moscovitch M (2001) Different patterns of autobiographical memory loss in semantic dementia and medial temporal lobe amnesia: a challenge to consolidation theory. Neurocase 7: 37–55

New insights into neurocognition provided by brain mapping: visuospatial cognition

Michel Thiebaut de Schotten and Paolo Bartolomeo

The last century has shown a tremendous amount of progress in neurosurgery going from blind and frequently lethal brain surgery to intraoperative electrical stimulations (IES) for a functional living (in vivo) brain mapping. IES combined with good neuropsychological assessments allows a much better mapping of brain functions, resulting in a clear definition of the borders of a brain resection. Thus, the risks of definitive postoperative neurological deficits can be significantly decreased. Nevertheless, the difficulty to assess nonlanguage cognitive functions during the operations has led to an underestimation of the functional importance of the right hemisphere.

An overview of the historical context of brain stimulations, with a special focus on recent advancements in visuospatial mapping, constitutes the subject matter of this chapter. We first describe the invention of brain stimulation and its application to the neurosurgical practice. We then survey the importance of the right hemisphere for spatial processing. Finally, we review the new insights into visuospatial cognition provided by IES and lesion-based brain mapping. Hopefully, the ideas expressed in this chapter will encourage the practice of awake brain surgery of the right hemisphere and emphasize the importance to assess visuospatial functions with IES.

Intraoperative stimulations of the "dominant" hemisphere

In 1922, Wilder Penfield concluded one of his letters to his mother, "Brain surgery is a terrible profession. If I did not feel it will become very different in my lifetime I should hate it." [47, p. 93]. At that time no method was able to discriminate a functional from a nonfunctional area during the operation, and neurosurgery on brain tumors was frequently unsuccessful. Thus, when the brain excision was too large, patients came out from the operation permanently disabled. Conscientious neurosurgeons chose then to leave a significant portion of the tumor in place, which improved the patient outcome, but that tumor grew and killed patients months or years later [63].

Penfield's prediction had come true, and neurosurgery made giant steps ahead during his lifetime, beginning with the improvement of surgical and antiseptic techniques that dropped the rates of mortality significantly. But the tipping point that changed the neurosurgical practice was the clinical introduction of intraoperative electrical stimulation (IES). IES consists in the temporary perturbation of restricted regions (5 mm) around the tumor by electrical stimuli applied directly on the brain surface. At the moment of the IES, if the patient reports a peripheral sensation or an involuntary movement, the brain area is labeled as functional and will be spared during the operation. Thus, this approach is useful to delimit the functional areas of the brain so that neurosurgeons may be accurately informed about the functional borders of brain tumor excisions [61]. However, Penfield was not the first to apply IES on the human brain. As he acknowledges in one of his papers [64], the credit has to go to Roberts Bartholow. Following previous reports of faradizations (i.e., applications of faradic current to stimulate nerves) of the brains of living animals, such as those by Fritsch and Hitzig in 1870 [32], Bartholow, an American surgeon, described the first application of electrical stimulations to the living human brain: "Mary's health has always been good until thirteen months ago, when a small ulcer appeared on the scalp (...) when she was an infant, she had fallen into the fire, her scalp was badly burned, and the hair was never reproduced (...) The skull is eroded and has disappeared over a space of two inches in diameter, where the pulsations of the brain are plainly seen (...) as the brain has been deeply penetrated by incisions made for the escape of pus, it was supposed that fine needles could be introduced without material injury to the cerebral matter (...). Observation 1. (...) when the needle points were engaged in the dura mater, Mary declared, in answer to repeated questions, that she felt no pain (...) Mechanical irritation of the cerebral matter produced no results on motility or sensibility of the extremities. Observation 2. *To test faradic reaction of the surface of the dura mater.* Two needles insulated were introduced into left side until their points were well engaged in the dura mater. When the circuit was closed, distinct muscular contractions occurred in the right arm and leg. The arm was thrown out, the finger extended, and the leg was projected forward (...) Observation 3. *To test faradic reaction of the posterior lobes.* (...) Mary complained of a very strong and unpleasant feeling of tingling in both right extremities, especially in the right arm, which she seized with the opposite hand and rubbed vigorously." [3, p. 310-311 (Fig. 1)

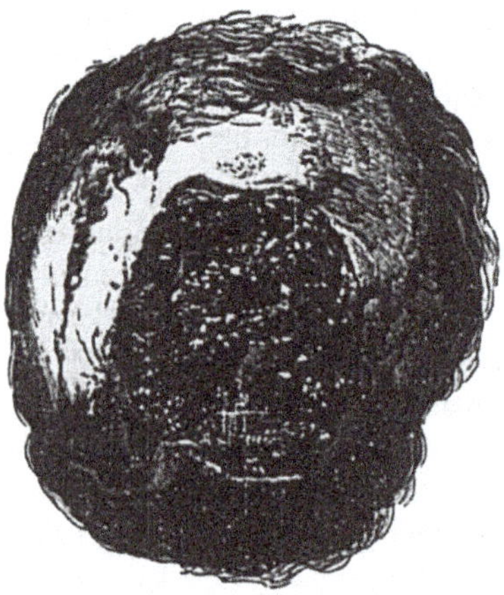

Fig. 1. Original drawing of the head of Mary (viewed from the top), patient of Roberts Bartholow, who described for the first time electrical stimulations on the human living brain [3]

Fifty years later in Breslau, Foerster and Penfield applied electrical stimulation to the clinical practice of neurosurgery, in order to map functions on the surface of the living brain [31]. This approach, when applied to brain surgery, provided two benefits. It significantly reduced the amount of critical functional brain area removed, minimizing definitive postoperative neurological deficits. And it was the first direct scientific approach to localize the functions of regions in the human living brain.

A few years later, at his newly built Montreal Neurological Institute, Penfield used IES on the cerebral cortex to pinpoint the localization of motor and somatic representation (Fig. 2) [64]. He also induced speech arrest [66] and, most impressively, elicited vivid memories in stimulated patients [65].

Arthur Ward, after his training in neurosurgery with Wilder Penfield at the Montreal Neurological Institute, introduced the practice of IES on awake epilepsy patients to the study of neurophysiology at the University of Washington [79].

In the United States, the next generation of neurosurgeons followed this approach, combining the knowledge from early neuroanatomists – [11], [18], [49] –, neurologists [41], and later from neuropsychologists [57] to map out the different functions of the living brain.

By means of IES on awake patients, the following phenomena were elicited: anomia (impaired object naming) [60], alteration of verbal working memory [56], and different brain localizations for language in bilingual patients [59].

Most importantly, the functional localization of the different aspects of language were shown to be distributed among a large network of cortical areas [58]. As a result an assumption was made that these areas were supported by long-range white-matter pathways connecting the frontal and temporal lobes (arcuate fasciculus) and the frontal and parietal lobes (superior longitudinal fasciculus) [13, 74]. The anatomo-functional correlations between data obtained by IES of the white-matter pathways and post-

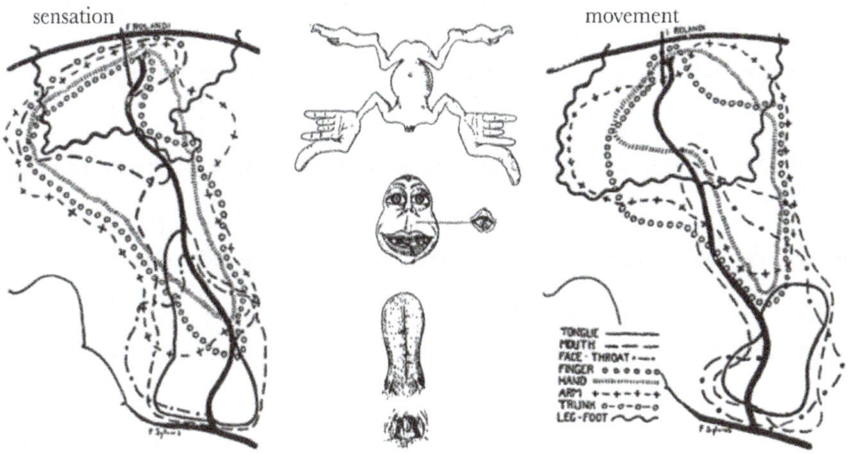

Fig. 2. Wilder Penfield's representation of the areas of stimulations, associated with precise sensory response on the left and motor response on the right hemisphere in 126 patients. Note that motor and sensory areas are not strictly separated by the central (or rolandic) sulcus. The diagram in the middle, which Penfield labeled sensory and motor homunculus [64], is meant as a representation of the comparative size of brain areas associated with different body parts in term of IES results (i.e., a large representation of a body part corresponds to larger area of brain associated with it) (Reproduced with permission from Oxford University Press)

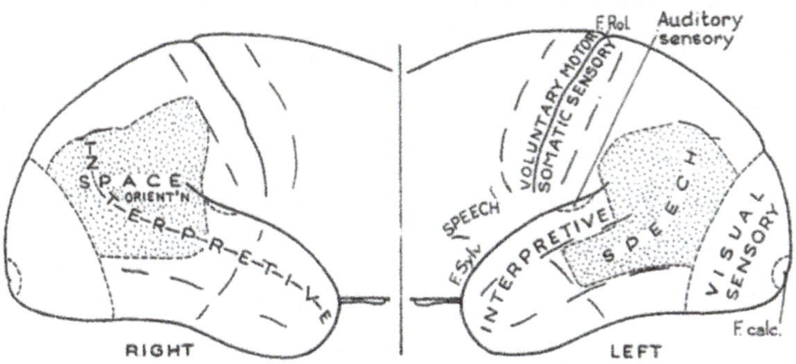

Fig. 3. Wilder Penfield drawing of the areas dedicated to space and speech. "Lateral surfaces of the posterior parts of both hemispheres of a human adult. On the dominant side, local interference-aphasia is produced by a stimulating electrode in the areas marked "speech". Active responses, produced by an electrode on other parts of this interpretive cortex, are of two types – experiential or interpretive. The area marked "space orientation" on the nondominant side (*right*) was outlined by study of the results of cortical excision. Complete removal produces visuospatial orientation impairment (i.e., neglect) in contradistinction to the aphasia produced by destruction of the homologous area on the dominant hemisphere" [62, p 308] (Reproduced with permission from Springer)

operative MRI confirmed in all patients the existence of common pathways that seem essential to language [25]. Even though there was cooperation between neurosurgeons and neuropsychologists, the majority of the studies focused mainly on the mapping of speech functions in the "dominant" left hemisphere.

This focus on the left hemisphere resulted from early fascination and intensive investigation. The attractive simplicity of the Wernicke-Lichteim model [48], combining Broca's hypothesis on the role of the left inferior frontal gyrus in speech [10], Wernicke's hypothesis on the role of the posterior left superior temporal gyrus in the language comprehension, and the white-matter link between these two regions for verbal repetition [80], stimulated the research on language function in the brain. Hence, visuospatial assessment by IES was neglected, overshadowed by the level of interest in the left hemisphere in terms of language

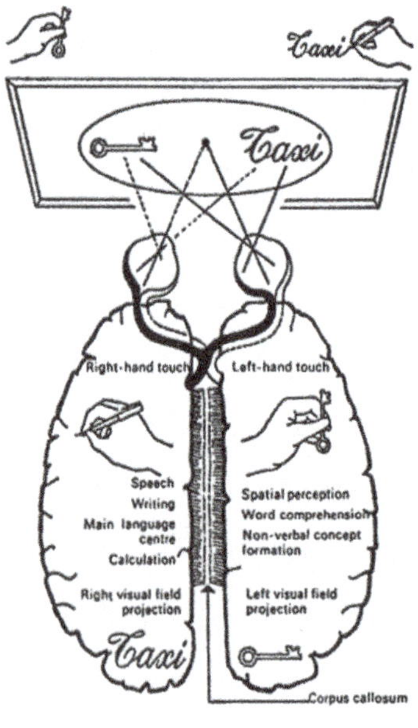

Fig. 4. Roger W. Sperry's illustration of the functional specialization of both cerebral hemispheres. The right hemisphere is shown as dominant for spatial perception, word comprehension, and non-verbal concept formation [17] (Illustration from the Nobel Committee for Physiology or Medicine, based on "Impact of Science on Society" published by UNESCO. © the Nobel Committee)

functions. Thus, researchers failed to realize the importance of the right hemisphere.

Interestingly, at the end of his career, Penfield outlined the results of cortical excision of a crucial area in the right hemisphere that, when damaged, led to visuospatial orientation impairment (Fig. 3) (i.e., neglect) [62].

The neglected hemisphere

Whilst it is clear that the left hemisphere has an essential role for language, studies on split-brain patients indicated the right hemisphere networks are crucial for visuospatial processes [35] (Fig. 4).

Visuospatial processing is a broad label referring to a heterogeneous family of processes concerned with the visual interactions with the environment. As with language, several anatomo-functional dissociations have been observed in the visuospatial domain.

For example, response time paradigms and functional magnetic resonance imaging (fMRI) helped dissect distinct forms of visuospatial attention [30, 68, 69].

When attention is dragged by surprise automatically to an unexpected location, fMRI shows increased BOLD (blood oxygen level-dependent; depending on increased oxygen extraction, which should reflect increased neural functioning) response in a ventral attentional network, including the inferior parietal cortex and the inferior and middle frontal gyri, especially in the right hemisphere [15].

When, on the other hand, attention is strategically and voluntarily oriented towards visual targets, more dorsal and bilateral fronto-parietal networks (including the intraparietal sulcus and the frontal eye field) show increased BOLD response [15, 55].

Finally, the general level of arousal or alertness is correlated with more medial cortical regions, principally centered on the cingular gyrus [52].

Damage to these networks in the right hemisphere leads to severe deficits of attentional selection and perceptual consciousness.

For example, patients with vascular strokes in the right hemisphere frequently show signs of left hemispatial neglect [1, 9, 39, 42, 67], "a fairly clear-cut syndrome of inattention directed toward the controlesional visual field" [36]. Neglect patients seem to live in a halved world: they do not eat from the left part of their dish or bump their body into obstacles situated on their left. When copying a linear drawing, they fail to copy the left part of the whole scene or of objects therein. The patients' gaze tends to be captured by right-sided, ipsilesional objects, as if they exerted a sort of "magnetic" attraction due to an imbalance of the attentional distribution between the left and the right hemifield [34].

A quick way to assess the severity of left hemispatial neglect is to ask patients to mark with a pencil the center of a horizontal line [2, 70]. Typically the bisection mark made by neglect patients deviates toward the right extremity of the line as if the left side of the line was underestimated in comparison to the right side. In contrast, patients with visual field defects (left homonymous hemianopia) but without neglect tend to produce bisection errors in the opposite direction (i.e., toward the left) [22], as if they were trying to compensate their sensory deficit by directing their gaze towards the left extremity.

Thus, left hemispatial neglect can be clearly dissociated from hemianopia with a simple line bisection test. Finally, patients with an association of left hemianopia and left hemispatial neglect display rightward deviations even greater than those by patients with "pure" left hemispatial neglect, probably because neglect prevents them from establishing a compensatory strategy for hemianopia [21] (Fig. 5).

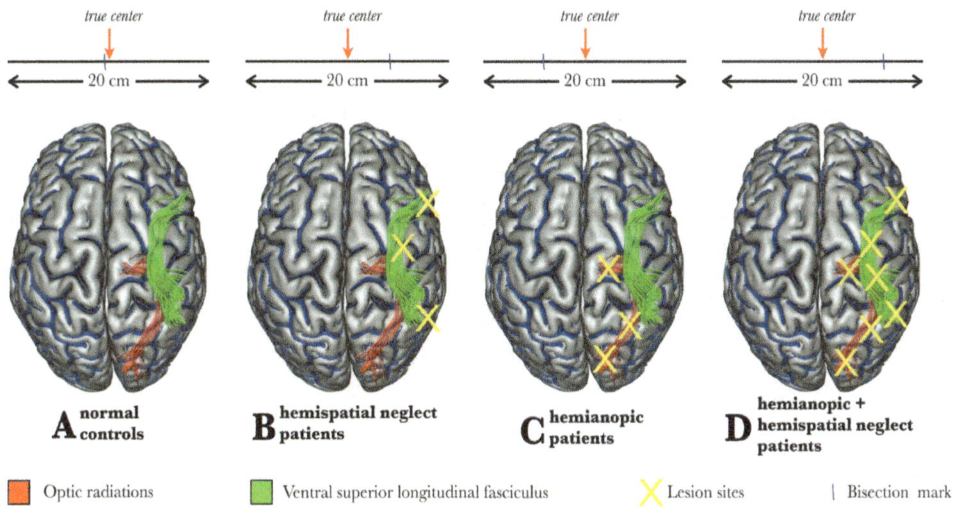

Fig. 5. Pattern of performance in line bisection. **(A)** Normal controls slightly deviate toward the left of the true center of the line (2 mm), a phenomenon reported as pseudoneglect effect and considered as the consequence of the right hemisphere dominance for spatial processing [43]. **(B)** Hemispatial neglect patients show a strong deviation toward the right of the true center of the line [2, 70]. **(C)** Hemianopic patients deviate toward the left side of the line [22]. **(D)** Patients combining neglect with hemianopia demonstrate the greatest deviations toward the right of the true center of the line [21]

Intraoperative assessment of visuospatial functions

IES of visuospatial functions began when a French neurosurgeon trained to use IES on awake patients and a research team dedicated to the study of visuospatial functions met in the historical hospital of the Pitié-Salpêtrière in Paris. Together, they set up a procedure to reduce the probability of a removal of brain areas crucial for visuospatial processing and consequently to preserve patients from developing signs of postoperative hemispatial neglect [72]. Additionally, the use of this protocol allowed them to gather direct evidence on the localization and the functional organization of visuospatial processing in the living human brain.

Two patients were assessed during brain surgery, by asking them to bisect 20 cm long horizontal lines. Patients deviated rightward, upon electrical perturbations of the right supramarginal gyrus (the rostral subdivision of the inferior parietal lobule) and of the caudal part of the right superior temporal gyrus, but performed accurately when more rostral portions of the right superior temporal gyrus or the right frontal eye field were perturbed. More importantly, the strongest deviations occurred in one patient upon perturbations of a white-matter region in the depth of the right inferior parietal lobule, after most of the tumor had been removed. Diffusion tensor tractography on postoperative MRI showed that the tract whose perturbations had brought about the maximal rightward deviation likely corresponds to a branch of the right superior longitudinal fasciculus, the most important frontoparietal pathway. Thus, in this study, functional frontoparietal perturbations dramatically disrupted the distribution of visuospatial attention in the right hemisphere, consistent with previous findings obtained with nonhuman primates [33] and with human patients who had experienced a stroke [4, 20, 46]. Clinically, the neurosurgeon was careful not to remove the regions in which perturbations had pro-

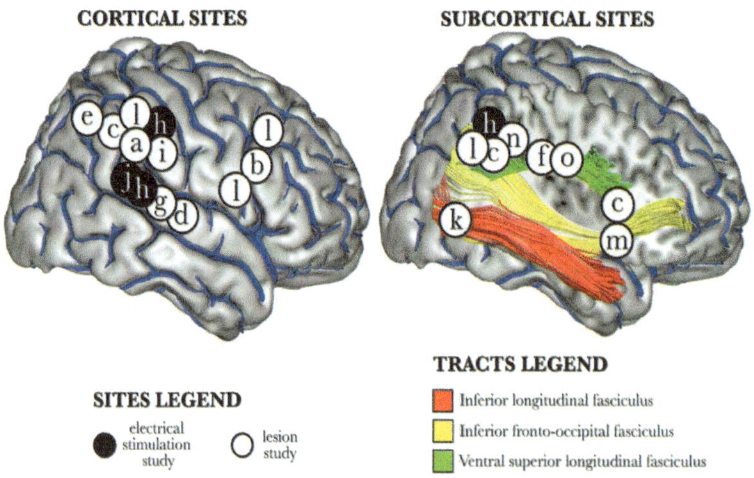

Fig. 6. The mapping of visuospatial functions in the human brain has been driven by group studies of the anatomy of the lesion or IES causing signs of neglect. Meta-analysis of the lesion sites reported in several published papers. Most of the cortical regions reported are parietal (label, reference: *a*, 77; *c*, 46; *e*, 53; *i*, 16; *h*, 72; *l*, 14) or frontal (*b*, 40; *c*, 46; *l*, 14). Other critical areas are also described in the posterior part of the superior temporal gyrus (*d*, 44; *g*, 45; *h*, 72; *j*, 37). Most of the white-matter areas reported belong to the frontoparietal white matter (*c*, 46; *f*, 20; *h*, 72; *l*, 14; *n*, 71; *o*, 78). The frontotemporal inferior longitudinal fasciculus (*k*, 8) and the inferior fronto-occipital fasciculus (*m*, 76) have also been reported as critical white matter pathways leading to hemispatial neglect when disconnected

voked rightward shifts of the subjective line center. As a consequence, a few days after surgery patients showed no signs of hemispatial neglect.

More recently, a study reported the cases of two patients with a right-hemisphere low-grade glioma who underwent neurosurgery [71]. Unfortunately, IES for visuospatial processing was not performed. Both patients showed severe signs of postoperative left hemispatial neglect. In both these cases, postoperative diffusion tensor imaging tractography revealed a disconnection of the frontoparietal pathway.

Those case reports suggest that the combination of preoperative tractography with perioperative mapping of visuospatial function can significantly improve the functional outcome of the patients.

As with all techniques of brain–behavior analysis, direct brain stimulation has limitations. The sites and the number of stimulations are dictated by clinical needs and are often dismayingly limited for the researcher. Phenomena of cortical plasticity, frequent with low-grade gliomas [19, 24, 54, 75], can complicate the interpretation of the mapping data. However, these limitations are not the same as with other methods, such as the lesion studies with nonhuman primates and humans. In the case of visuospatial functions, evidence from all of these approaches [12, 20, 33] converges in underlining an important role of the frontoparietal pathway. However, the study of distinct, parallel networks should also be considered in the future. In particular, lesions to the inferior longitudinal fasciculus [8] and the inferior fronto-occipital fasciculus [76] have been reported in stroke patients with left hemispatial neglect (Fig. 6).

Patients with damage to the left hemisphere may also show signs of contralateral neglect, albeit rarely [6]. According to some theories [38, 50], each hemisphere processes information coming from the contralateral space, but the right hemisphere can also deal with ipsilateral information, albeit slightly less efficiently [51]. Hence, the right hemisphere can compensate to a certain extent unilateral lesions of the left hemisphere, thus giving left-brain-damaged patients some ability to explore the right hemispace. Such compensation may be more difficult for intraoperative testing, where the patient performs the test during the occurrence of the virtual temporary damage induced by IES.

In order to explore the brain areas dedicated to space in the left hemisphere, we collected preliminary data on two right-handed patients with low-grade gliomas in the left temporoparietal region [73]. Patients marked with a pencil the center of a horizontal line with their right, dominant hand during direct cortical and subcortical electrical stimulation. The stimulation of the caudal left superior temporal gyrus and its subcortical white matter, but not the left supramarginal gyrus, determined leftwards deviations on line bisection (Fig. 7). Further work is needed to confirm these preliminary results, but they seem to suggest that left-hemisphere networks for spatial processing are similar, but not identical, to right-hemisphere ones.

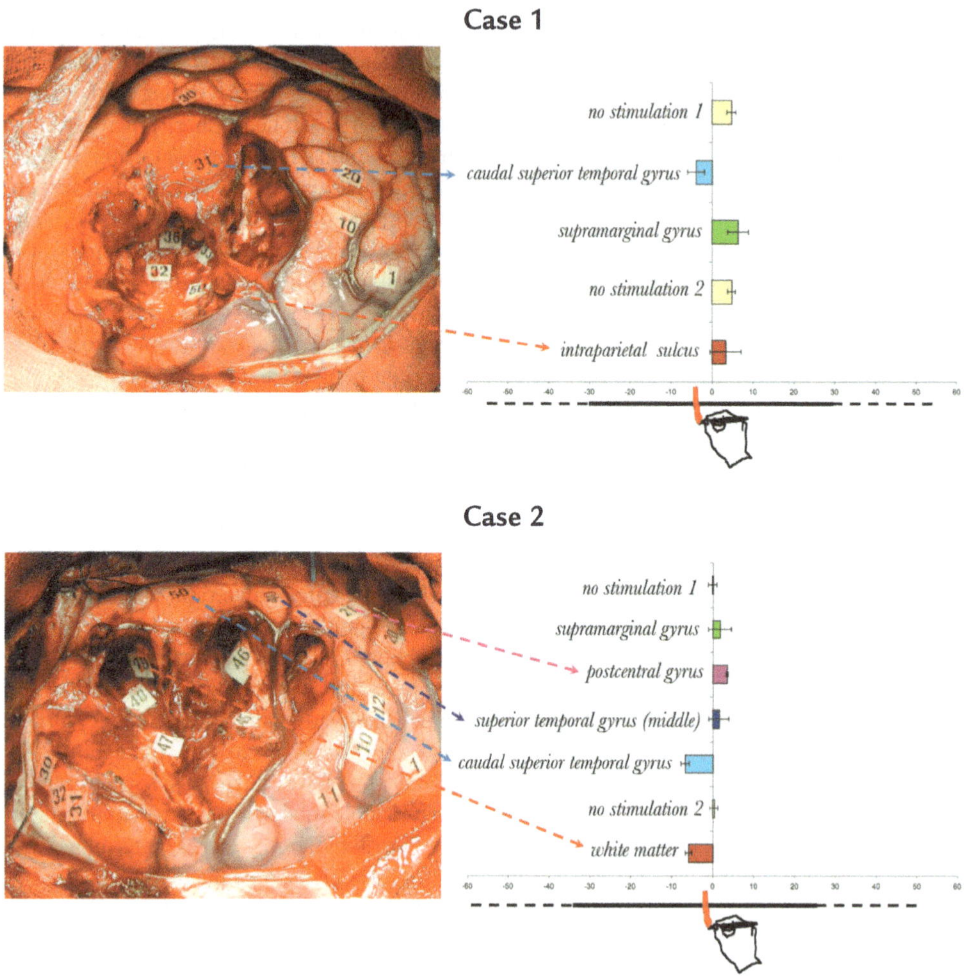

Fig. 7. Performance of two patients in line bisection during IES of the left hemisphere. A picture of the surgical field is shown on the left; mean deviations (in millimeters) with 95% confidence intervals during stimulations are reported on the right

Discussion

IES has rendered possible the study of the living brain functions while preserving functional areas in patients [28]. It also pinpointed directly the importance of the white-matter networks for specific cognitive functions [25–27, 29, 72]. Despite its potential, IES has long remained limited to few functions such as sensory motor capacities and speech abilities. Furthermore, awake IES is too often performed solely during surgery of the left hemisphere. Such a situation does not render justice to the complexity and importance of the functions which belong to the right hemisphere, such as spatial processing, complex and nonlinguistic perceptual tasks, emotion, affect, and paralinguistic aspects of communication [7, 51]. Other right-hemisphere functions are likely to be discovered in the future. Penfield and Perot [65], in their review of 1.288 cases of focal electrical stimulation of the human cerebral cortex, found that highly organized visual or auditory events, which they labeled as "experiential responses", such as seeing people in the room or hearing a song, were exclusively evoked by stimulations applied to the cortex of the temporal lobe. As Brenda Milner once remarked to one of the present authors, the majority of temporal sites whose stimulation evoked experiential responses seemed localized in the right hemisphere. This intriguing possibility, which suggests a deep involvement of the right hemisphere in conscious experience, has never been formally tested. Thus, whereas IES of language functions has advanced significantly; IES of other "high-level" cognitive functions still needs much investigation to reach a similar level of understanding.

Recent mapping results demonstrated that visuospatial functions are distributed on a large frontoparietal network of cortical areas interconnected by long-range association pathways [4, 20, 23, 46, 72]. Surgery-induced visuospatial deficits can be prevented by preserving the cortical areas and subcortical connections dedicated to these functions [5, 71].

The next steps in visuospatial mapping and cognition will be to understand the behavioral dissociations in neglect according to the different sites of lesion and to explore which anatomical nodes between the sensory input and the motor output are crucial to make the visuospatial cognition possible. IES of the right hemisphere is likely to become a key tool to precisely and directly map these complex cognitive abilities on brain structures and networks.

Ackowledgements

We thank the NatBrainLab (*http://www.natbrainlab.com*) and Paul McGillivray for useful comments. This work was supported by the Marie Curie Intra-European Fellowships for Career Development from the European Seventh Framework Program (FP7), the Bettencourt Schueller Fondation and the Agence Nationale de Recherche [project CAFORPFC, no. ANR-09-RPDOC-004-01 and project HM-TC, no. ANR-09-EMER-006].

References

[1] Anton G (1889) Ueber die Selbstwahrnehmung der Herderkrankungen des Gehirns durch den Kranken bei Rindenblindheit und Rindentaubheit. Arch Psychiatr 32: 86–127

[2] Axenfeld T (1915) Hemianopische Gesichtsfeldstörungen nach Schädelschüssen. Klin Monatsbl Augenheilkd 55: 126–143

[3] Bartholow R (1874) Art. I. – Experimental investigations into the functions of the human brain. Am J Med Sci 66: 305–313

[4] Bartolomeo P, Thiebaut de Schotten M, Doricchi F (2007) Left unilateral neglect as a disconnection syndrome. Cereb Cortex 17: 2479–2490

[5] Bartolomeo P, Thiebaut de Schotten M, Duffau H (2007) Mapping of visuospatial functions during brain surgery: a new tool to prevent unilateral spatial neglect. Neurosurgery 61: E1340. DOI 10.1227/01.neu. 0000306126. 46657.79

[6] Beis JM, Keller C, Morin N, Bartolomeo P, Bernati T, Chokron S, Leclercq M, Louis-Dreyfus A, Marchal F,

Martin Y, Perennou D, Pradat-Diehl P, Prairial C, Rode G, Rousseaux M, Samuel C, Sieroff E, Wiart L, Azouvi P; French Collaborative Study Group on Assessment of Unilateral Neglect (GEREN/GRECO) (2004) Right spatial neglect after left hemisphere stroke: qualitative and quantitative study. Neurology 63: 1600–1605

[7] Benowitz LI, Bear DM, Rosenthal R, Mesulam MM, Zaidel E, Sperry RW (1983) Hemispheric specialization in nonverbal communication. Cortex 19: 5–11

[8] Bird C, Malhotra P, Parton A, Coulthard EJ, Rushworth MFS, Husain M (2006) Visual neglect after right posterior cerebral artery infarction. J Neurol Neurosurg Psychiatry 77: 1008–1012

[9] Brain RW (1941) Visual disorientation with special reference to lesion of the right brain hemisphere. Brain 64: 244–272

[10] Broca P (1863) Localisation des fonctions cérébrales. Siège du langage articulé. Bull Soc Anthropol Paris I Ser 4: 200–208

[11] Burdach KF (1819–1826) Vom Baue und Leben des Gehirns und Rückenmarks. Dyk, Leipzig

[12] Buschman TJ, Miller EK (2007) Top-down versus bottom-up control of attention in the prefrontal and posterior parietal cortices. Science 315: 1860–1862

[13] Catani M, Jones DK, ffytche DH (2005) Perisylvian language networks of the human brain. Ann Neurol 57: 8–16

[14] Committeri G, Pitzalis S, Galati G, Patria F, Pelle G, Sabatini U, Castriota-Scanderbeg A, Piccardi L, Guariglia C, Pizzamiglio L (2007) Neural bases of personal and extrapersonal neglect in humans. Brain 130: 431–441

[15] Corbetta M, Shulman GL (2002) Control of goal-directed and stimulus-driven attention in the brain. Nat Rev Neurosci 3: 201–215

[16] Corbetta M, Kincade M, Lewis C, Snyder A, Sapir A (2005) Neural basis and recovery of spatial attention deficits in spatial neglect. Nat Neurosci 8: 1603–1610

[17] Deglin VL (1976) Our split brain. UNESCO Courier 29(1): 4–14, 16, 31–32

[18] Déjerine J (1895) Anatomie des centres nerveux. Rueff, Paris

[19] Desmurget M, Bonnetblanc F, Duffau H (2007) Contrasting acute and slow-growing lesions: a new door to brain plasticity. Brain 130: 898–914

[20] Doricchi F, Tomaiuolo F (2003) The anatomy of neglect without hemianopia: a key role for parietal-frontal disconnection? Neuroreport 14: 2239–2243; Erratum in Neuroreport 15: 217 (2004)

[21] Doricchi F, Galati G, DeLuca L, Nico D, D'Olimpio F (2002) Horizontal space misrepresentation in unilateral brain damage. I. Visual and proprioceptive-motor influences in left unilateral neglect. Neuropsychologia 40: 1107–1117

[22] Doricchi F, Onida A, Guariglia P (2002) Horizontal space misrepresentation in unilateral brain damage. II. Eye-head centered modulation of visual misrepresentation in hemianopia without neglect. Neuropsychologia 40: 1118–1128

[23] Doricchi F, Thiebaut de Schotten M, Tomaiuolo F, Bartolomeo P (2008) White matter (dis)connections and gray matter (dys)functions in visual neglect: gaining insights into the brain networks of spatial awareness. Cortex 44: 983–995

[24] Duffau H (2008) Brain plasticity and tumors. Adv Tech Stand Neurosurg 33: 3–33

[25] Duffau H, Capelle L, Sichez N, Denvil D, Lopes M, Sichez JP, Bitar A, Fohanno D (2002) Intraoperative mapping of the subcortical language pathways using direct stimulations. An anatomo-functional study. Brain 125: 199–214

[26] Duffau H, Gatignol P, Denvil D, Lopes M, Capelle L (2003) The articulatory loop: study of the subcortical connectivity by electrostimulation. Neuroreport 14: 2005–2008

[27] Duffau H, Gatignol P, Mandonnet E, Peruzzi P, Tzourio-Mazoyer N, Capelle L (2005) New insights into the anatomo-functional connectivity of the semantic system: a study using cortico-subcortical electrostimulations. Brain 128: 797–810

[28] Duffau H, Lopes M, Arthuis F, Bitar A, Sichez JP, Van Effenterre R, Capelle L (2005) Contribution of intraoperative electrical stimulations in surgery of low grade gliomas: a comparative study between two series without (1985–96) and with (1996–2003) functional mapping in the same institution. J Neurol Neurosurg Psychiatry 76: 845-851

[29] Duffau H, Thiebaut de Schotten M, Mandonnet E (2008) White matter functional connectivity as an additional landmark for dominant temporal lobectomy. J Neurol Neurosurg Psychiatry 79: 492–495

[30] Fan J, Wu Y, Fossella JA, Posner MI (2001) Assessing the heritability of attentional networks. BMC Neurosci 2: 14

[31] Foerster O, Penfield W (1930) The structural basis of traumatic epilepsy and result of radical operation. Brain 53: 99–119

[32] Fritsch G, Hitzig E (2009) Electric excitability of the cerebrum (Über die elektrische Erregbarkeit des Grosshirns). Epilepsy Behav 15: 123-130

[33] Gaffan D, Hornak J (1997) Visual neglect in the monkey. Representation and disconnection. Brain 120: 1647–1657

[34] Gainotti G, D'Erme P, Bartolomeo P (1991) Early orientation of attention toward the half space ipsilateral to the lesion in patients with unilateral brain damage. J Neurol Neurosurg Psychiatry 54: 1082–1089

[35] Gazzaniga MS, Bogen JE, Sperry RW (1965) Observations on visual perception after disconnexion of the cerebral hemispheres in man. Brain 88: 221–236

[36] Geschwind N (1965) Disconnexion syndromes in animals and man. II. Brain 88: 585–644

[37] Gharabaghi A, Fruhmann Berger M, Tatagiba M, Karnath HO (2006) The role of the right superior temporal gyrus in visual search – insights from intraoperative electrical stimulation. Neuropsychologia 44: 2578–2581; Erratum in Neuropsychologia 45: 465 (2007)

[38] Heilman KM, Van Den Abell T (1980) Right hemisphere dominance for attention: the mechanism underlying hemispheric asymmetries of inattention (neglect). Neurology 30: 327–330

[39] Holmes G (1918) Disturbances of visual orientation. Br J Ophthalmol 2: 449–468, 516–520

[40] Husain M, Kennard C (1996) Visual neglect associated with frontal lobe infarction. J Neurol 243: 652–657

[41] Jackson H (1915) On the physiology of language. Brain 38: 59–64

[42] Jackson JH (1876) Case of large cerebral tumour without optic neuritis and left hemiplegia and imperception. R Lond Ophthalmic Hosp Rep 8: 434–444

[43] Jewell G, McCourt ME (2000) Pseudoneglect: a review and meta-analysis of performance factors in line bisection tasks. Neuropsychologia 38: 93–110

[44] Karnath HO, Ferber S, Himmelbach M (2001) Spatial awareness is a function of the temporal not the posterior parietal lobe. Nature 411: 950–953

[45] Karnath HO, Fruhmann Berger M, Küker W, Rorden C (2004) The anatomy of spatial neglect based on voxelwise statistical analysis: a study of 140 patients. Cereb Cortex 14: 1164–1172

[46] Leibovitch FS, Black SE, Caldwell CB, Ebert PL, Ehrlich LE, Szalai JP (1998) Brain-behavior correlations in hemispatial neglect using CT and SPECT: the Sunnybrook Stroke Study. Neurology 50: 901–908

[47] Lewis J (1981) Something hidden. Doubleday Canada, Toronto

[48] Lichtheim L (1885) On aphasia. Brain 7: 433–484

[49] Mayo H (1827) Brain and spinal chord in man. Burgess and Hill, London

[50] Mesulam MM (1981) A cortical network for directed attention and unilateral neglect. Ann Neurol 10: 309–325

[51] Mesulam MM (1999) Spatial attention and neglect: parietal, frontal and cingulate contributions to the mental representation and attentional targeting of salient extrapersonal events. Philos Trans R Soc Lond B Biol Sci 354: 1325–1346; Erratum in Philos Trans R Soc Lond B Biol Sci 354: 2083 (1999)

[52] Mesulam MM, Nobre AC, Kim YH, Parrish TB, Gitelman DR (2001) Heterogeneity of cingulate contributions to spatial attention. Neuroimage 13: 1065–1072

[53] Mort DJ, Malhotra P, Mannan SK, Rorden C, Pambakian A, Kennard C, Husain M (2003) The anatomy of visual neglect. Brain 126: 1986–1997

[54] Noble M, Dietrich J (2004) The complex identity of brain tumors: emerging concerns regarding origin, diversity and plasticity. Trends Neurosci 27: 148–154

[55] Nobre AC, Sebestyen GN, Gitelman DR, Mesulam MM, Frackowiak RS, Frith CD (1997) Functional localization of the system for visuospatial attention using positron emission tomography. Brain 120: 515–533

[56] Ojemann GA (1971) Alteration in nonverbal short term memory with stimulation in the region of the mammillothalamic tract in man. Neuropsychologia 9: 195–201

[57] Ojemann GA (1986) Mapping of neuropsychological language parameters at surgery. Int Anesthesiol Clin 24: 115–131

[58] Ojemann GA, Mateer C (1979) Human language cortex: localization of memory, syntax, and sequential motor-phoneme identification systems. Science 205: 1401–1403

[59] Ojemann GA, Whitaker HA (1978) The bilingual brain. Arch Neurol 35: 409–412

[60] Ojemann GA, Fedio P, Van Buren JM (1968) Anomia from pulvinar and subcortical parietal stimulation. Brain 91: 99–116

[61] Penfield W (1936) Epilepsy and surgical therapy. Arch Neurol Psychiatry 36: 449–484

[62] Penfield W (1972) The electrode, the brain and the mind. Z Neurol 201: 297–309

[63] Penfield W (1977) No man alone: a neurosurgeon's life. Little, Brown, Boston

[64] Penfield W, Boldrey E (1937) Somatic motor and sensory representation in the cerebral cortex of man as studied by electrical stimulation. Brain 60: 389–440

[65] Penfield W, Perot P (1963) The Brain record of auditory and visual experience. A final summary and discusssion. Brain 86: 595–696

[66] Penfield W, Rasmussen T (1949) Vocalization and arrest of speech. Arch Neurol Psychiatry 61: 21–27

[67] Pick A (1898) Über allgemeine Gedächtnisschwäche als unmittelbare Folge cerebraler Herderkrankung. Beiträge zur Pathologie und pathologischen Anatomie des Centralnervensystems mit Bemerkungen zur normalen Anatomie desselben. Karger, Berlin

[68] Posner MI, Cohen Y, Rafal RD (1982) Neural systems control of spatial orienting. Philos Trans R Soc Lond B Biol Sci 298: 187–198

[69] Posner MI, Walker JA, Friedrich FJ, Rafal RD (1984) Effects of parietal injury on covert orienting of attention. J Neurosci 4: 1863–1874

[70] Schenkenberg T, Bradford DC, Ajax ET (1980) Line bisection and unilateral visual neglect in patients with neurologic impairment. Neurology 30: 509–517

[71] Shinoura N, Suzuki Y, Yamada R, Tabei Y, Saito K, Yagi K (2009) Damage to the right superior longitudinal fasciculus in the inferior parietal lobe plays a role in spatial neglect. Neuropsychologia 47: 2600–2603

[72] Thiebaut de Schotten M, Urbanski M, Duffau H, Volle E, Levy R, Dubois B, Bartolomeo P (2005) Direct evidence for a parietal-frontal pathway subserving spatial awareness in humans. Science 309: 2226–2228; Erratum in Science 317: 597 (2007)

[73] Thiebaut de Schotten M, Urbanski M, Duffau H, Volle E, Dubois B, Levy R, Bartolomeo P (2007) Networks for spatial attention in the human left hemisphere: evidence from intraoperative stimulation. In: Cognitive Neuroscience Society Annual Meeting Program 2007. Cognitive Neuroscience Society, Davis, CA poster D22

[74] Thiebaut de Schotten M, Kinkingnehun S, Delmaire C, Lehéricy S, Duffau H, Thivard L, Volle E, Levy R, Dubois B, Bartolomeo P (2008) Visualization of disconnection syndromes in humans. Cortex 44: 1097–1103
[75] Thiel A, Herholz K, Koyuncu A, Ghaemi M, Kracht LW, Habedank B, Heiss WD (2001) Plasticity of language networks in patients with brain tumors: a positron emission tomography activation study. Ann Neurol 50: 620–629
[76] Urbanski M, Thiebaut de Schotten M, Rodrigo S, Catani M, Oppenheim C, Touzé E, Chokron S, Méder JF, Levy R, Dubois B, Bartolomeo P (2008) Brain networks of spatial awareness: evidence from diffusion tensor imaging tractography. J Neurol Neurosurg Psychiatry 79: 598–601
[77] Vallar G, Perani D (1986) The anatomy of unilateral neglect after right-hemisphere stroke lesions. A clinical/CT-scan correlation study in man. Neuropsychologia 24: 609–622
[78] Verdon V, Schwartz S, Lovblad KO, Hauert CA, Vuilleumier P (2010) Neuroanatomy of hemispatial neglect and its functional components: a study using voxel-based lesion-symptom mapping. Brain 133: 880–894
[79] Ward AJ (1961) The epileptic neuron. Epilepsia 2: 70–80
[80] Wernicke C (1874) Der aphasische Symptomencomplex. Eine psychologische Studie auf anatomischer Basis. Cohn und Weigert, Breslau

Preoperative and postoperative functional magnetic resonance imaging and intraoperative assessment of mental spatial transformations in patients undergoing surgery for brain tumors

Anne S. Champod, Emily Ferreira, Céline Amiez, Penelope Kostopoulos, D. Louis Collins, Rolando Del Maestro, and Michael Petrides

Introduction

Improvements in neurosurgical techniques have contributed to an increase in life expectancy among patients with brain tumors [12], thereby increasing the need for the development of management strategies focused on maximizing independence and quality of life postoperatively [15]. The main goal of neurosurgery is to maximize the extent of the resection, which has been associated with improved survival rates [5, 6, 27, 29], while minimizing potential sensory, motor, and cognitive deficits [17]. Cognitive impairments following neurosurgery can have a far-reaching impact on the patients' daily functioning and quality of life postoperatively. The presence of cognitive impairments in neurological populations, and more specifically the presence of spatial dysfunctions, deficits in working memory and executive processes, has been linked to increased functional disability [9, 31]. As such, cognitive functioning is increasingly recognized as an important outcome measure in patients with brain tumors [46].

The use of functional neuroimaging techniques as a preoperative tool to preserve function in patients with brain tumors is still fairly limited to the mapping of the location of motor, sensory, and language primary areas [4, 21, 22, 24, 25, 28, 32, 33, 47]. These sensorimotor and language functions cover a limited range of the human brain processes that are of crucial importance for normal daily functioning and that can be impaired following surgery.

Ongoing improvements in our understanding of the role of specific brain regions in human cognition are opening the door to the development of new preoperative neuroimaging and intraoperative tools that have the potential to become important for improving the patients' quality of life following surgery. More specifically, the combination of findings from studies of lesions in human and monkey brains with the functional data obtained from neuroimaging of human brains currently allows the development of sophisticated tools that can be used for preoperative neurosurgical mapping of a large range of cognitive functions. Accordingly, we recently developed new pre- and intraoperative tools to assess the function of the premotor cortex that is centrally involved in the selection between competing motor responses on the basis of learned conditional rules [3]. This cognitive process has been shown to depend on the integrity of the rostral part of the dorsal premotor cortex (PMdr) in monkey brains [36, 37] and in human brains with frontal damage that includes the premotor cortex [23, 38]. We first developed a functional magnetic resonance imaging (fMRI) protocol with a visuo-motor conditional task to assess function in the PMdr in healthy control participants [2]. We subsequently used this fMRI protocol preoperatively in 4 patients with brain tumors that were located close to this region and showed specific activity increases in the vicinity of the tumors, that is, in the PMdr. These preoperative fMRI data were transferred to a neuronavigation system together with the patients' structural MR images, allowing the neurosurgeon to maximize the extent of resection while preserving the functional region within the PMdr that was identified preoperatively. Behavioral performances on the visuo-motor conditional task remained unchanged following surgery, suggesting that preoperative fMRI may be a useful procedure to assist in preserving cognitive function in patients that will undergo brain surgery for the removal of a tumor.

Another domain of cognitive processing that should be assessed, particularly in patients that will undergo surgery close to or within the posterior parietal region, is visuospatial cognition. For example, mental rotation, which is the ability to manipulate mentally objects in space, is a critical aspect of spatial cognition that has been widely investigated in healthy individuals, but not in patients undergoing neurosurgery [50]. Findings from lesion studies indicate a crucial role of the posterior parietal cortex in visuospatial cognition [42]. For example, lesions involving the posterior parietal cortex have been shown to impair performance on spatial tasks such as mental rotation tasks [19, 40]. More recently, it has been shown that transcranial magnetic stimulation (TMS) in the left or right posterior parietal cortex selectively altered mental rotation performance [7, 34]. Furthermore, functional neuroimaging studies have consistently highlighted the role of the intraparietal sulcus (IPS) region in mental rotation [1, 26, 30, 39, 50]. Recent fMRI studies have also demonstrated the central role of the IPS region in the manipulation (i.e., mental reordering) of information in working memory [10, 11]. Thus, lesion and TMS studies, as well as functional neuroimaging, indicate the critical role of the posterior parietal cortex in visuospatial cognition, with a specific role of the IPS region in mental transformation, which can be studied by mental rotation paradigms.

Considering the importance of visuospatial cognition in everyday life, we developed pre-, intra- and postoperative tools for the evaluation of mental rotation function in neurosurgical candidates with brain tumors close to or invading the parietal lobe. In this chapter, we will discuss the development of the functional neuroimaging protocol in healthy individuals for clinical purposes and the subsequent application of this protocol in patients with brain tumors for neurosurgical mapping of posterior parietal function.

Clinical material and methods

Cognitive paradigm

We first developed a cognitive paradigm with a group of healthy control participants. This paradigm consisted of a mental rotation condition and a visuospatial control condition (Fig. 1). In both the mental rotation and the visuospatial control tasks, the participants viewed pairs of figures composed of three-dimensional cubes and had to decide whether the two figures constituting the pair were the "same" or "different". The figures used were similar to those initially developed by Shepard and Metzler [43], who introduced the mental rotation paradigm into cognitive science.

On each trial of the mental rotation task (Fig. 1a), the participants viewed two figures composed of three-dimensional cubes that were either the same but one was rotated at varying angles (from 45 to 270 degrees) with respect to the other around its vertical axis ("same" trials), or they were mirror images of each other ("different" trials). The participants had to mentally rotate one of the figures to decide if it could be aligned with the other so that they could respond whether the two figures were the same or different. Numerous behavioral studies have demonstrated that there is a linear relationship between the angle of rotation of one of the figures with respect to the other and the participant's response time, indicating that this task is performed by mentally rotating one of the figures. The participants reported their decision by pressing the left button of a computer mouse if the two figures were the same although rotated with respect to one another or the right button if the figures were different.

In the visuospatial control task (Fig. 1b), the pairs of stimuli that the subjects viewed on each trial were identical figures composed of three-dimensional cubes that were not rotated with respect to one another. Dots were placed on the surface of different sections of the cubes on both figures and the participants were instructed to determine whether the dots were located on the same sections on both figures ("same" trials) or not ("different" trials). Thus, in this condition, there was no need to rotate

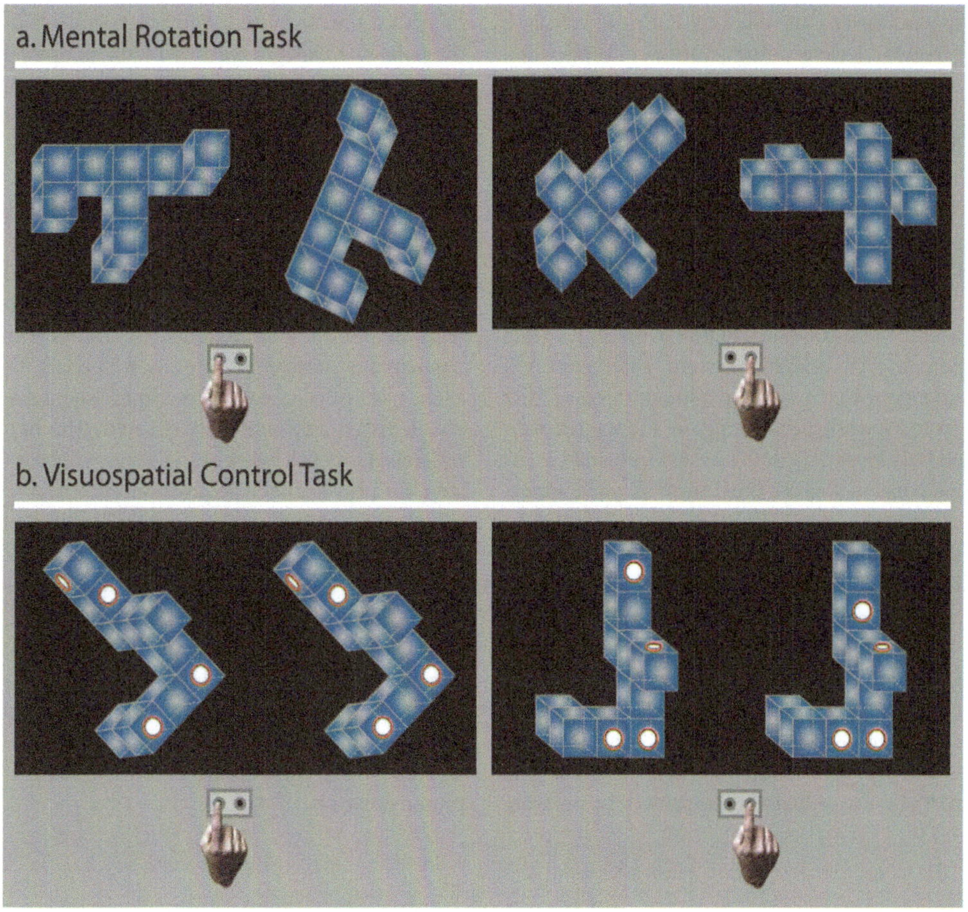

Fig. 1. Mental rotation (**a**) and visuospatial control (**a**) tasks. In both tasks, the participants viewed pairs of two-dimensional renderings of three-dimensional cube figures and had to determine whether the two figures were the "same" (left panels) or "different" (right panels). (**a**) In the mental rotation task, the pairs of figures were either the same but rotated with respect to one another ("same" trials), or they were mirror images of each other ("different" trials). Stimuli were rotated at varying angles (from 45 to 270 degrees) with respect to one another around their vertical axis. The participants were instructed to mentally rotate one of the figures into alignment with the other in order to decide if they were the same or different. (**a**) In the visuospatial control task, the pairs of stimuli were identical, not rotated, and had dots placed on their surface. The participants had to determine whether the dots were located on the same parts of both figures ("same" trials) or not ("different" trials). In this condition, there was no need to rotate figures into alignment

the figures into alignment, but rather reach the decision that the figures were the "same" or "different" on the basis of the location of the dots on the cubes. This task therefore provided control for various aspects of the mental rotation task, including the visual properties of the stimuli, eye movements and evaluation process required to reach a "same or different" decision, and the motor act of pressing a button to indicate the decision. The presentation of the stimuli and the recordings of the motor responses were controlled by E-Prime software version 1.1 (Psychology Software Tools, Inc., Sharpsburg, PA, USA).

Neuroimaging procedure

The neuroimaging procedure used for healthy control participants will first be described, followed by the minimal adjustment that was required for the patients. We first scanned 10 healthy control participants with a 1.5-tesla Sonata MR imaging system (Siemens AG, Erlangen, Germany) and a block-design fMRI protocol while performing the mental rotation and visuospatial control tasks. Informed written consent was obtained from all participants according to the institutional guidelines established by the ethics committee of the Montreal Neurological Hospital and Institute. One hour prior to the scanning session, the participants were familiarized with and practiced the mental rotation and visuospatial control tasks on a computer. Each trial was composed of the presentation of a pair of stimuli for 3 seconds followed by a 1-second intertrial interval. The participants provided their answer by pressing the appropriate mouse button during the presentation of each pair of stimuli. After a high-resolution T1-weighted anatomical scan (whole head, 1 mm^3 isotropic resolution), 5 sequences of 125 images each (18 oblique T2* gradient echo planar images; voxel size, 3.4 by 3.4 by 3.4 mm; repetition time, 3.5 s; echo time, 45 ms; flip angle, 90°) sensitive to the blood oxygenation level-dependent (BOLD) signal were acquired. Visual stimuli were presented through a liquid-crystal display projector with a mirror system and the responses of the participants were recorded with an optical MR compatible computer mouse. During each functional sequence, 6 blocks of 4 trials for each task were presented in the same order (i.e., 4 mental rotation trials followed by 4 visuospatial control trials). We did not alternate the order of presentation of the blocks of trials during scanning in order to make it as simple as possible for the patients, given that the two tasks shared the same sequence of events. The participants performed a total of 24 trials for each task during each functional sequence which lasted approximately 6.5 min. The first trial onset for each block of trials was synchronized with the scanner acquisition via a trigger signal generated by the scanner. Behavioral and imaging data were acquired for all trials.

The only adjustment from the above neuroimaging procedure that was required for the patients was to increase the time allowed to provide their response on each trial by up to 7 seconds. The pair of stimuli remained on the screen until the patients responded. The reason for this increase was to compensate for any cognitive or motor slowness.

Table 1. Patient characteristics and tumor locations[a]

Case nr.	Age (yr)	Sex[b]	Tumor type (grade)	Location of tumor[c]
1	31	F	astrocytoma (II)	SPL
2	46	F	oligodendroglioma (II)	SPL
3	48	F	oligodendroglioma (II)	SPL
4	42	M	glioblastoma multiforme (IV)	SPL
5	51	M	ganglioglioma (II)	P frontal/SPL
6	41	M	oligodendroglioma (II)	P temporo-parieto-occipital

[a] All tumors were in the left hemisphere
[b] F, female; M, male
[c] P, posterior; SPL, superior parietal lobule

Patient testing

We examined preoperatively 6 patients whose brain tumors were located close to the IPS region. Table 1 summarizes the location of the tumor in each patient. Given that this region plays an important role in the mental transformation of visual images, we used fMRI and the mental rotation task to establish preoperatively the region essential for this function in each patient. During tumor resection, we evaluated the performance of these patients on the mental rotation task. Four patients underwent postoperative fMRI 2 months after surgery to confirm that the functional region shown to be involved in the mental rotation task during the preoperative fMRI had not been removed during resection and that increased activity was still observed within this region, postoperatively. The other 2 patients lived outside of the province and could not come back to Montreal for their postoperative scans. The period of 2 months was chosen to avoid any effect of swelling or other surgical factors that can be observed in the early weeks after the surgery.

Neurosurgical procedure

Before the tumor resection in each patient, the preoperative fMRI data were transferred to a neuronavigation system (Surgical Navigation Network; ISG Technologies, Salina, Kans., U.S.A.) together with the patient's structural MR images. These data were registered to the patient in a 2-step process. First, corresponding anatomical landmarks (such as the canthus, meatus, and bridge of the nose) were identified in the MRI image data on the computer, and on the patient with a three-dimensional (3-D) pointer, to calculate the patient-to-image transformation. Afterwards, points on the surface of the skin of the patient were marked and used to refine the patient-to-image transformation by comparing them to the skin surface extracted from the MRI data. Once the dura mater was opened, the neurosurgeon was able to localize the functional peak of activity in relation to the tumor location and to adjust the surgical approach throughout the surgery (Fig. 2).

The brain shift induced during surgery was measured with the neuronavigation system in the operating room. At the beginning of surgery, the patient's anatomy and functional images were aligned with the procedure described above. During surgery, any brain shift could be estimated by identifying one or more landmarks (such as vessel bifurcations) on the patient's brain with the tracked 3-D pointer. The spatial location of the 3-D pointer was then compared with the coordinates of the corresponding landmark in the 2-D images and 3-D renderings of the patient's brain. The difference between these positions enabled precise estimation of any brain shift. If the shift was deemed significant, the neurosurgeon could re-register the patient's images to the shifted brain with these landmarks. In this manner, the borders of the functional region observed in the IPS region could always be compared to the anatomical landmarks (sulci and blood vessels) represented on the 3-D rendered images on the computer.

Intraoperative behavioral assessment

We developed an intraoperative procedure to evaluate the patients' performance on the mental rotation task and, thus, the function of the posterior parietal cortex during tumor resection. The surgery was performed after induction of anaesthesia using neuroleptics and while the patient was awake and alert. On the day of the surgery, we assessed the performance of the patients on the mental rotation task on 3 different occasions: (1) 30 minutes before the surgery, in the patient's room, to obtain a baseline performance; (2) in the operating room, after the induction of anaesthesia but before the surgical intervention, to test the effect of anaesthesia on performance; and (3) throughout the surgery at regular intervals to establish whether the cognitive processing underlying the performance of the task was preserved during the tumor resection.

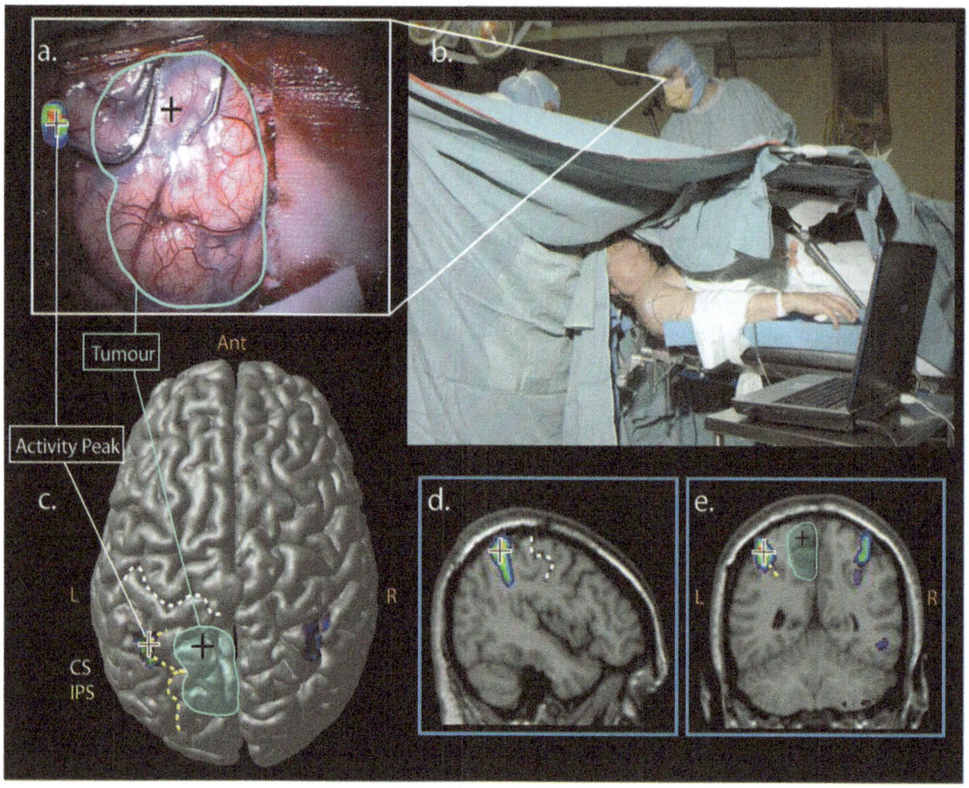

Fig. 2. Neuronavigation system used during neurosurgery. Overview of the neuronavigation system and setting used during neurosurgery. (**a**) Intraoperative photograph representing the view through the microscope as seen by the neurosurgeon in case 5. (**b**) Photograph of the setting used for testing patients during surgery. (**c–e**) The location of the tumor and the focus of functional activity (activity peak) within the IPS region are illustrated on the cortical surface (**c**), on the sagittal view (**d**), and on the coronal view (**e**) of the left hemisphere of the patient in case 5. The tumor location is shown in turquoise. The black crosses indicate the same point in panels (**a**, **c**, and **e**). The white crosses indicate the same point in panels (**a**, **c**, **d**, and **e**). In panels (**c**, **d**, and **e**), the sulci are color-coded as follows: white dashed line, central sulcus (*CS*); yellow dashed line, intraparietal sulcus (*IPS*). *L*, left; *R*, right; *Ant*, anterior

Data analysis for healthy control participants and patients

Statistical analysis for healthy participants and patients. The BOLD signal obtained during the mental rotation task was compared with the BOLD signal in the visuospatial control task, allowing us to identify the regions that showed greater response in the mental rotation task. The first 3 volumes of each functional sequence were discarded due to the T1 saturation effects. Images from all imaging sequences were first realigned with the Analysis of Functional NeuroImages (AFNI) image registrations software using the third frame of the first sequence as a reference [16], and then smoothed with a MINC blurring software (mincblur) using a 6 mm full-width half-maximum isotropic Gaussian kernel. Subsequently, all images were transformed into the Montreal Neurological Institute standard proportional stereotaxic space [45] by in-house dedicated software [13]. Functional and anatomical data were then merged for each healthy participant and patient separately to localize regions of significant activation.

The statistical analysis of the fMRI data was based on a linear model with correlated errors

[49] (available at http://www.math.mcgill.ca/keith/fmristat). The resulting t-statistic images were thresholded for significance by the minimum value given by a Bonferroni correction and random field theory to account for multiple comparisons [48]. Significance was assessed on the basis of the spatial extent of consecutive voxels. A predicted cluster of voxels with a volume extent of >175 mm^3 with a t-value of >3 was significant ($p < 0.05$) corrected for multiple comparisons by the method of Friston et al [20].

Comparison between pre- and postoperative MRI data for patients. To compare the pre- and postoperative MRI data, it was necessary to account for any change in brain shape occurring after surgery that might confound the direct comparison of the functional regions. Nonlinear registration of the pre- and postoperative anatomical images was used to estimate deformations of brain tissue due to the tumor. The ANIMAL (Anatomical Nonlinear Image Matching and Automated Labelling) software was used to compute a dense 3-D warping field to best match the pre- and postoperative anatomical MRI data [14]. The procedure estimates deformation vectors on a 3-D grid that best match local neighborhoods by cross-correlation. Once estimated, the warping field was used to resample the preoperative image data (anatomical and functional) in the space of the postoperative data to enable direct comparisons.

Results

fMRI data for healthy control participants

We first tested the fMRI protocol with 10 healthy control participants to confirm the prediction from earlier work that there will be increased activity within the IPS region in relation to the performance of the mental rotation task. We compared the BOLD signal obtained in the mental rotation block of trials (which required the visuospatial manipulation of images) with that of the visuospatial control block of trials (which required the same visual comparison and motor action but no manipulation). For this comparison, significantly increased activity was observed in the predicted region of the posterior parietal lobe, that is, within the horizontal segment of the IPS region, bilaterally (x = –38, y = –42, z = 42, t = 10.399; x = 36, y = –50, z = 62, t = 10.252).

We subsequently looked at this comparison for individual participants to ensure that these peaks of increased activity were observed in each one of them given that the protocol was developed to localize functional areas in individual patients. For each participant, significant increase in activity was located within the horizontal segment of the IPS region, posterior to the postcentral sulcus and anterior to the stereotaxic coordinate y = –60 (see Fig. 3 for an illustration of the peak of increased activity in a representative participant). The functional imaging protocol therefore proved to be a sensitive and robust means of localizing functional activity within the posterior parietal region involved in mental spatial transformations.

Preoperative behavioral and fMRI imaging data for patients

All patients were able to perform the mental rotation task and the visuospatial control task. The mean success rates varied from 65 to 98%

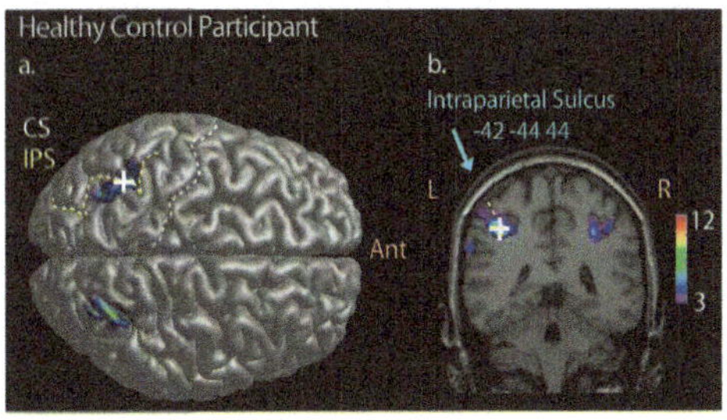

Fig. 3. Increased activity in the posterior parietal region during mental rotation in a representative healthy control participant. The activity is based on the contrast mental rotation minus visuospatial control condition. The focus of the functional activity (activity peak) is in the horizontal segment of the IPS illustrated on the cortical surface (**a**) and on the coronal view (**b**) of a healthy participant's brain. The white crosses indicate the same point in panels (**a**) and (**b**). For color codes of sulci and definitions of abbreviations, see caption of Fig. 2

for the mental rotation task and from 75 to 99% for the visuospatial control task.

We compared the BOLD signal obtained in the mental rotation block of trials with that of the visuospatial block of trials preoperatively in all 6 patients. The location of activity peaks obtained by fMRI was assessed on the basis of their stereotaxic coordinates, as well as the sulcal and gyral patterns of the posterior parietal region. We did not solely rely on the stereotaxic coordinates of the peaks because brain tumors exert pressure on the neighboring tissue and thus shift the location of sulci and gyri. In all patients, the functional region involved in mental spatial transformations was localized within the horizontal segment of the IPS region (see Table 2 for the stereotaxic coordinates of the preoperative peaks in all 6 patients and Fig. 4 for an illustration of the regions of increased activity in the 4 patients who took part in the preoperative fMRI, the intraoperative testing procedure, and the postoperative fMRI). In all 6 patients the functional activity was located in the vicinity of the tumor and never within the tumor. Furthermore, in all patients, there were also various peaks of activity that were far from the tumor site and constitute part of the network of areas that have been demonstrated to be involved in visuospatial transformations in healthy participants. Evidently, not all these peaks of activity represent areas that are essential to the task at hand. We focused here on the functional region located within the posterior parietal cortex that has been shown to be centrally involved in the performance of such a task (i.e., the IPS region).

Neurosurgical procedure and intraoperative behavioral data

The fMRI data were available to the neurosurgeon for each patient 2–3 days before the surgery. The operative procedure was planned so that the part of the tumor that was located closest to the functional peak of activity within the IPS region was to be resected last (Fig. 4). The success rate of the 6 patients varied from 93.75% (15 of 16) to 100% (16 of 16 trials) on the mental rotation task 30 min before the surgery. In the surgery room, the patients performed 1 to 2 blocks of 8 mental rotation trials after the induction of anaesthesia, and a number of blocks of 8–16 trials during the tumor resection in intervals that averaged 35 min. With the exception of a transient drop in performance after the induction of anaesthesia in patient 1 ($F_{(9,142)} = 3.445$; $p < 0.001$), the success rate on the mental rotation task did not vary significantly at any time of the surgery in any of the 6 patients (patient 2: $F_{(7,56)} = 0.684$; $p > 0.685$, not significant [ns]; patient 3: $F_{(11,84)} = 0.636$; $p > 0.793$, ns; patient 4: $F_{(19,140)} = 1.590$; $p > 0.066$, ns; patient 5: $F_{(12,91)} = 0.833$; $p > 0.616$, ns; patient 6: $F_{(8,63)} = 0.875$; $p > 0.542$, ns; ANOVA). For patient 1, it was noted that the anaesthesia induced significant sleepiness for a period of approximately 20 min. A state of optimal alertness and recovery of function was observed before the begin-

Table 2. Pre- and postoperative MRI data of BOLD signal maxima

Case nr.	Preoperative maximum				Postoperative maximum			
	Stereotaxic coordinate (mm)			t-value[a]	Stereotaxic coordinate (mm)			t-value[a]
	x	y	z		x	y	z	
1	-43	-50	57	13.75*	-37	-51	56	6.75*
2	-43	-53	57	8.50*	-38	-54	56	13.06*
3	-43	-44	67	13.05*	-43	-41	57	10.70*
4	-46	-46	48	5.20*	-34	-52	60	2.40
5	-47	-46	48	3.82*				
6	-29	-46	40	3.064*				

[a] Asterisk indicates significance at p < 0.05 corrected for multiple comparisons

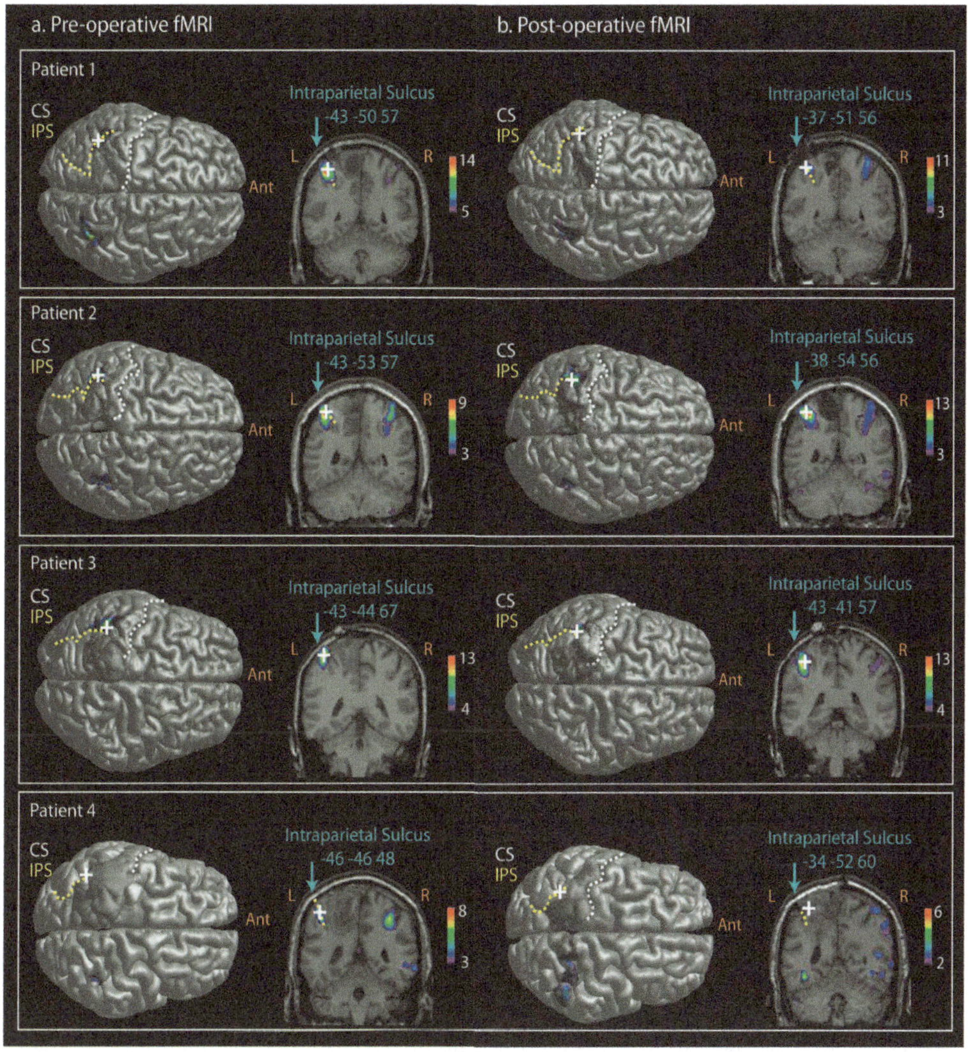

Fig. 4. Preoperative and postoperative activity in the posterior parietal region during mental rotation in 4 patients. Preoperative (**a**) and postoperative fMRI (**b**) data showing the focus of activity within the horizontal segment of the IPS region in each patient. The gray shaded area on the cortical surface shows the tumor location. The foci of activity correspond to the exact location of the peaks described in Table 2. White crosses indicate the location of the activity increase in the IPS region. The stereotaxic coordinates provide the location of the activity peak in the Montreal Neurological Institute standard stereotaxic space. The color scale indicates the t value range. (**b**) Deformation of the cortical surface shows the extent of the tumor resection. The data demonstrate that, in each patient, the region involved in mental rotation was not removed during the tumor resection. The functional activity was observed in the same region of the IPS preoperatively (**a**) and postoperatively (**b**). For color codes of sulci and definitions of abbreviations, see caption of Fig. 2

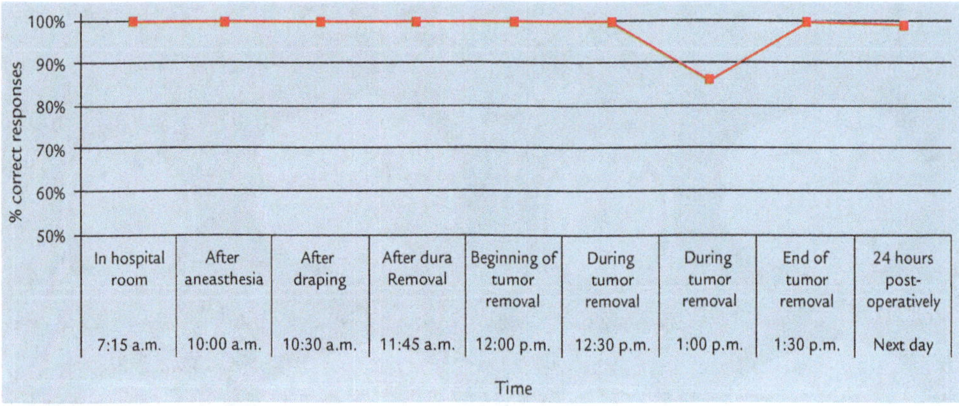

Fig. 5. Intraoperative behavioral performance of a representative patient on the mental rotation task. The performance of this patient (case 5) on the mental rotation task was assessed on 3 separate occasions: 30 minutes before surgery; in the operating room, after the induction of anaesthesia but before the surgical intervention; and during the tumor resection. In each testing session, the patient performed 16 mental rotation trials. The performance was at 100% before the surgery and remained stable throughout the surgical procedure with the exception of a minor fluctuation during the resection

ning of the tumor resection. This transient fluctuation of alertness following the induction of anaesthesia was also observed with the use of a different intraoperative protocol [3] and seems to require special attention to ensure a return to baseline performance before the tumor resection. Figure 5 displays the behavioral results obtained throughout this procedure in a representative patient (patient 5).

Postoperative behavioral, MRI and fMRI data

The mean success rates on the mental rotation task for each patient postoperatively indicate that there was no decline in performance on this task following surgery. Performance remained stable in 3 patients (patient 1: $F_{(1,438)} = 1.94$; $p > 0.165$, ns; patient 2: $F_{(1,478)} = 0.588$; $p > 0.444$, ns; patient 3: $F_{(1,482)} = 1.90$; $p > 0.169$, ns; ANOVA) and even improved in one patient (patient 4: $F_{(1,406)} = 14.37$; $p < 0.000$, significant; ANOVA). It is important to mention that the preoperative MRI in patient 4 demonstrated significant oedema surrounding the tumor and very close to the region of functional activity associated with the performance of the mental rotation task. The postoperative MRI demonstrated a reduction of swelling, which could explain the improvement in performance on the mental rotation task postoperatively.

These results are consistent with the fact that, as seen in the postoperative structural MRIs of the 4 patients (patients 1–4), the neurosurgeon preserved during the surgery the functional region involved in the mental rotation function as defined with the preoperative fMRI (Fig. 4b). Reports of the surgery procedure written by the neurosurgeon described that the functional region involved in mental rotation identified preoperatively was preserved in the 2 patients who could not come back to Montreal postoperatively (patients 5 and 6). Their performance on the task 24 h after surgery remained at the preoperative level (see Fig. 5 for the intraoperative performance of patient 5).

We compared the BOLD signal obtained in the mental rotation block of trials with that of the visuospatial block of trials postoperatively in the 4 patients who were scanned both pre- and postoperatively to confirm that increased activity would be observed in the same IPS region that was identified preoperatively. In all 4 patients, the functional region involved in mental spatial transformations was localized within the same region of the horizontal segment of the IPS identified preoperatively in the patients and in the 10 healthy participants (see Table 2 for the stereotaxic coordinates of the postoperative peaks and Fig. 4b for an illustration of the postoperative peaks of increased activity in the 4 patients who took part in every aspect of the procedure).

Discussion

In neurosurgical clinical practice, language and somatomotor functions are commonly assessed pre- and intraoperatively in patients undergoing tumor resection [4, 8, 25, 28, 44, 47]. Apart from language, other cognitive functions, such as executive and spatial cognitive processing are typically neglected in this neurosurgical population despite the clear relevance for everyday functioning of preserving as much cognitive processing as is compatible with a successful operation. Over the years, survival rates in primary brain tumors have increased due to the advancement in medical science and technology [12]. Surgery is one of the first-line treatments in primary brain tumors and it is increasingly being recognized that more resources should be devoted to the development of management strategies focused on the improvement of the patients' independence and quality of life postoperatively [18].

By combining findings from lesion and neuroimaging studies, we have previously shown that the pre- and intraoperative assessment of specific aspects of executive functioning may be useful for surgical planning in order to minimize potential neurological postoperative deficits in this domain [3]. Here, we discussed the development of a pre- and postoperative fMRI and intraoperative behavioral paradigm for the assessment of mental spatial transformations which is of particular relevance to the preservation of function in neurosurgical candidates with tumors invading or close to the parietal cortex. The ability to manipulate objects mentally that relies heavily on the parietal cortex in interaction with the frontal cortex [10, 11] is important to successful postoperative adjustment in multiple daily life situations. For example, efficiently navigating or rearranging objects in space, such as parking a car, require the ability to transform mentally objects so as to understand the implications of any movement of these objects in space. Mental rotation paradigms have been extensively used to study the neural correlates of mental spatial transformations in healthy participants. Lesion and TMS studies have demonstrated a crucial role of the posterior parietal cortex in the performance of such tasks [7, 19, 34, 40] and functional neuroimaging studies have identified the IPS region as a key region in the manipulation of information in different cognitive domains [10, 11, 26, 30, 39, 41, 50].

On the basis of this prior knowledge of the function of the IPS region, we designed a mental rotation task that would be suitable for use with patients with brain tumors near or within the posterior parietal cortex. We developed an fMRI protocol with this task for a group of 10 healthy normal participants that produced reliable and robust functional activity within the IPS region that could be reproduced on a subject by subject basis. We then tested preoperatively by this fMRI protocol 6 patients with brain tumors invading the parietal lobe. The fMRI results obtained for the 6 patients showed that the functional region involved in the mental rotation task was focused on the same IPS region. Importantly, in all 6 patients, this functional region was located in the proximity of their brain tumors (Fig. 4). Consequently, the preoperative fMRI data allowed the neurosurgeon to plan the surgery on the basis of information about a still functional posterior parietal region surrounding the tumor.

The transfer of preoperative fMRI results to the operating room through an integrated image-guided neuronavigation system allowed the neurosurgeon to optimize the approach for tumor resection in patients to preserve the relevant functional posterior parietal cortex. We were able to demonstrate the usefulness of this methodology by testing 4 of the patients postoperatively on the mental rotation task. These patients showed no decline in performance postoperatively and the functional activity related to the mental rotation in the IPS region was observed pre- and postoperatively.

The behavioral results obtained during the intraoperative assessment revealed that such a procedure can be helpful to the neurosurgeon during the brain tumor resection, without interfering with the surgical procedure. A major change in the performance of the patient during the tumor resection could be important in deciding whether to modify the approach used for the surgical resection or to terminate the resection. Since the resection of the part of the tumor closest to the functional activity was left at the end of the operative procedure (see Methods), we were in a position to monitor the performance of the patient throughout the surgery in an efficient manner and to provide useful information to the neurosurgeon during the operation.

Functional neuroimaging techniques are commonly used to minimize postoperative risk to language, sensory and motor function in patients with brain tumors [4, 21, 22, 24, 25, 28, 32, 33, 47]. The findings from this study and from our previous work focusing on executive functions [3] demonstrate that such procedures can also be applied to cognitive functions other than language. The present study introduces a new paradigm to localize and evaluate reliably by fMRI the ability of patients with brain tumors invading or close to the parietal lobe to manipulate mentally or transform visuospatial material. The procedure developed here may be useful in minimizing potential postoperative deficits in patients with tumors in the parietal cortex. In the future, such procedures may prove useful, just as traditional mapping of motor, somatosensory, and language functions, in the safe resection of brain tumors.

Acknowledgments

This work was funded by the National Sciences and Engineering Research Council of Canada, the Canadian Institutes of Health Research, the Brain Tumour Foundation of Canada, the Franco Di Giovanni, the B-Strong, the Alex Pavanel Family and the Raymonde and Tony Boeckh Funds for Brain Tumour Research. We thank the Montreal English School Board, the Brainstorm and Colannino Foundations for financial support. A.S.C was supported by a scholarship award from the Canadian Institutes of Health Research. R.D.M. holds the William Feindel Chair in Neuro-Oncology at McGill University, and M.P. holds a James McGill Research Chair; both are Killam Scholars at the Montreal Neurological Institute.

References

[1] Alivisatos B, Petrides M (1997) Functional activation of the human brain during mental rotation. Neuropsychologia 35: 111–118

[2] Amiez C, Kostopoulos P, Champod AS, Petrides M (2006) Local morphology predicts functional organization of the dorsal premotor region in the human brain. J Neurosci 26: 2724–2731

[3] Amiez C, Kostopoulos P, Champod AS, Collins DL, Doyon J, Del Maestro R, Petrides M (2008) Preoperative functional magnetic resonance imaging assessment of higher-order cognitive function in patients undergoing surgery for brain tumors. J Neurosurg 108: 258–268

[4] Ashtari M, Perrine K, Elbaz R, Syed U, Thaden E, McIlree C, Dolgoff-Kaspar R, Clarke T, Diamond A, Ettinger A (2005) Mapping the functional anatomy of sentence comprehension and application to presurgical evaluation of patients with brain tumor. AJNR Am J Neuroradiol 26: 1461–1468

[5] Berger MS, Deliganis AV, Dobbins J, Keles GE (1994) The effect of extent of resection on recurrence in patients with low grade cerebral hemisphere gliomas. Cancer 74: 1784–1791

[6] Berger MS, Rostomily RC (1997) Low grade gliomas: functional mapping resection strategies, extent of resection, and outcome. J Neurooncol 34: 85–101

[7] Bestmann S, Thilo KV, Sauner D, Siebner R, Rothwell JC (2002) Parietal magnetic stimulation delays visuomotor mental rotation at increased processing demands. Neuroimage 17: 1512–1520

[8] Bogomolny DL, Petrovich NM, Hou BL, Peck KK, Kim MJJ, Holodny AI (2004) Functional MRI in the brain tumor patient. Top Magn Reson Imaging 15: 325–335

[9] Buxbaum LJ, Ferraro MK, Veramonti T, Farne A, Whyte J, Ladavas E, Frassinetti F, Coslett HB (2004) Hemispatial neglect: subtypes, neuroanatomy, and disability. Neurology 62: 749–756

[10] Champod AS, Petrides M (2007) Dissociable roles of the posterior parietal and the prefrontal cortex in manipulation and monitoring processes. Proc Natl Acad Sci USA 104: 14837–14842

[11] Champod AS, Petrides M (2010) Dissociation within the frontoparietal network in verbal working memory: a parametric functional magnetic resonance imaging study. J Neurosci 30: 3849–3856

[12] Claus EB, Black PM (2006) Survival rates and patterns of care for patients diagnosed with supratentorial low-grade gliomas: data from the SEER program, 1973–2001. Cancer 106: 1358–1363

[13] Collins DL, Neelin P, Peters TM, Evans AC (1994) Automatic 3D intersubject registration of MR volumetric data in standardized Talairach space. J Comput Assist Tomogr 18: 192–205

[14] Collins DL, Holmes CJ, Peters TM, Evans AC (1995) Automatic 3-D model-based neuroanatomical segmentation. Hum Brain Mapp 3: 190–208

[15] Correa DD (2006) Cognitive functions in brain tumor patients. Hematol Oncol Clin North Am 20: 1363–1376

[16] Cox RW, Jesmanowicz A (1999) Real-time 3D image registration for functional MRI. Magn Reson Med 42: 1014–1018

[17] Duffau H (2005) Lessons from brain mapping in surgery for low-grade glioma: insights into associations between tumour and brain plasticity. Lancet Neurol 4: 476–486

[18] Duffau H (2009) Surgery of low-grade gliomas: towards a functional neurooncology. Curr Opin Oncol 21: 543–549

[19] Farah MJ, Hammond KM (1988) Mental rotation and orientation-invariant object recognition: dissociable processes. Cognition 29: 29–46

[20] Friston KJ, Holmes AP, Poline JB, Grasby PJ, Williams SC, Frackowiak RS, Turner R (1995) Analysis of fMRI time-series revisited. Neuroimage 2: 45–53

[21] Gumprecht H, Ebel GK, Auer DP, Lumenta CB (2002) Neuronavigation and functional MRI for surgery in patients with lesion in eloquent brain areas. Minim Invasive Neurosurg 45: 151–153

[22] Hall WA, Kim P, Truwit CL (2009) Functional magnetic resonance imaging-guided brain tumor resection. Top Magn Reson Imaging 19: 205–212

[23] Halsband U, Freund HJ (1990) Premotor cortex and conditional motor learning in man. Brain 113: 207–222

[24] Henson JW, Gaviani P, Gonzalez RG (2005) MRI in treatment of adult gliomas. Lancet Oncol 6: 167–175

[25] Hirsch J, Ruge M, Kim K, Correa DD, Victor JD, Relkin NR, Labar DR, Krol G, Bilsky MH, Souweidane MM, DeAngelis LM, Gutin PH (2000) An integrated functional magnetic resonance imaging procedure for preoperative mapping of cortical areas associated with tactile, motor, language, and visual functions. Neurosurgery 47: 711–722

[26] Jordan K, Heinze H-J, Lutz K, Kanowski M, Jäncke L (2001) Cortical activations during the mental rotation of different visual objects. Neuroimage 13: 143–152

[27] Keles GE, Lamborn KR, Berger MS (2001) Low-grade hemispheric gliomas in adults: a critical review of extent of resection as a factor influencing outcome. J Neurosurg 95: 735–745

[28] Klein D, Olivier A, Milner B, Zatorre RJ, Johnsrude I, Meyer E, Evans AC (1997) Obligatory role of the LIFG in synonym generation: evidence from PET and cortical stimulation. Neuroreport 8: 3275–3279

[29] Laws ER, Parney IF, Huang W, Anderson F, Morris AM, Asher A, Lillehei KO, Bernstein M, Brem H, Sloan A, Berger MS, Chang S (2003) Glioma Outcomes Investigators. Survival following surgery and prognostic factors for recently diagnosed malignant glioma: data from the Glioma Outcomes Project. J Neurosurg 99: 467–473

[30] Milivojevic B, Hamm JP, Corballis MC (2009) Functional neuroanatomy of mental rotation. J Cogn Neurosci 21: 945–959

[31] Mok VCT, Wong A, Lam WWM, Fan YH, Tang WK, Kwok T, Hui AC, Wong KS (2004) Cognitive impairment and functional outcome after stroke associated with small vessel disease. J Neurol Neurosurg Psychiatry 75: 560–566

[32] Mueller WM, Yetkin FZ, Hammeke TA, Morris GL 3rd, Swanson SJ, Reichert K, Cox R, Haughton VM (1996) Functional magnetic resonance imaging mapping of the motor cortex in patients with cerebral tumors. Neurosurgery 39: 515–520; discussion 520–521

[33] Nelson L, Lapsiwala S, Haughton VM, Noyes J, Sadrzadeh AH, Moritz CH, Meyerand ME, Badie B (2002) Preoperative mapping of the supplementary motor area in patients harboring tumors in the medial frontal lobe. J Neurosurg 97: 1108–1114

[34] Pelgrims B, Andres M, Olivier E (2009) Double dissociation between motor and visual imagery in the posterior parietal cortex. Cereb Cortex 19: 2298–2307

[35] Petersen SE, Robinson DL, Currie JN (1989) Influences of lesions of parietal cortex on visual spatial attention in humans. Exp Brain Res 76: 267–280

[36] Petrides M (1982) Motor conditional associative-learning after selective prefrontal lesions in the monkey. Behav Brain Res 5: 407–413

[37] Petrides M (1986) The effect of periarcuate lesions in the monkey on the performance of symmetrically and asymmetrically reinforced visual and auditory go, no-go tasks. J Neurosci 6: 2054–2063

[38] Petrides M (1997) Visuo-motor conditional associative learning after frontal and temporal lesions in the human brain. Neuropsychologia 35: 989–997

[39] Podzebenko K, Egan GF, Watson JDG (2002) Widespread dorsal stream activation during a parametric mental rotation task, revealed with functional magnetic resonance imaging. Neuroimage 15: 547–558

[40] Ratcliff G (1979) Spatial thought, mental rotation and the right cerebral hemisphere. Neuropsychologia 17: 49–54

[41] Richter W, Ugurbil K, Georgopoulos A, Kim SG (1997) Time-resolved fMRI of mental rotation. Neuroreport 8: 3697–3702

[42] Sack AT (2009) Parietal cortex and spatial cognition. Behav Brain Res 202: 153–161
[43] Shepard RN, Metzler J (1971) Mental rotation of three-dimensional objects. Science 171: 701–703
[44] Smits M, Visch-Brink E, Schraa-Tam CK, Koudstaal P, van der Lugt A (2006) Functional MR imaging of language processing: an overview of easy-to-implement paradigms for patient care and clinical research. Radiographics 26 (Suppl 1): S145–158
[45] Talairach J, Tournoux P (1988) Co-planar stereotaxic atlas of the human brain: 3-dimentional proportional system: an approach to cerebral imaging. Thieme, Stuttgart
[46] Taphoorn MJB, Klein M (2004) Cognitive deficits in adult patients with brain tumours. Lancet Neurol 3: 159–168
[47] Tie Y, Suarez RP, Whalen S, Radmanesh A, Norton IH, Golby AJ (2009) Comparison of blocked and event-related fMRI designs for pre-surgical language mapping. Neuroimage 47 (Suppl 2): T107–115
[48] Worsley K, Marrett S, Neelin P, Vandal AC, Friston KJ, Evans AC (1996) A unified statistical approach for determining significant signals in images of cerebral activation. Hum Brain Mapp 4: 58–73
[49] Worsley KJ, Liao CH, Aston J, Petre V, Duncan GH, Morales F, Evans AC (2002) A general statistical analysis for fMRI data. Neuroimage 15: 1–15
[50] Zacks JM (2008) Neuroimaging studies of mental rotation. J Cogn Neurosci 20: 1–19

New insights into neurocognition provided by brain mapping: Social cognition and theory of mind

John S. Bellerose, Miriam H. Beauchamp, and Maryse Lassonde

Introduction

Social cognition refers to the mental processes that are used to perceive and process social cues, stimuli, and environments, and has traditionally been studied by social, developmental, and neuropsychologists from conceptual and behavioral perspectives. Advances in brain imaging technology within the last two decades have also provided a means of investigating the neural underpinnings of these three interrelated concepts. Researchers from the field of social neurosciences have recently started to map abstract sociocognitive concepts in the brain, and the value and practical applications of this cortical mapping are wide ranging, diverse, and profound. In particular, greater insight into the neural substrates of social cognition is beneficial to medical practice in explaining the sociocognitive deficits that may result from neurosurgical interventions for conditions such as epilepsy and brain tumors. Furthermore, knowledge of the brain regions and circuits that underlie uniquely human social skills can deepen our understanding of neurological and psychiatric illnesses in which the presenting symptoms include socioemotional and sociocognitive deficits, such as in autism or schizophrenia. To illustrate these points, this chapter will focus on two central components of social cognition, theory of mind and moral reasoning, in the context of human brain mapping. These concepts will first be defined and distinguished from one another, before we turn to a study of their development and neural substrates. External factors that have bearing on neurocognition, such as culture and gender, will also be highlighted. Lastly, to underscore the importance of these issues for the clinical community, we will illustrate how sociocognitive abilities are affected in a range of neurological and psychiatric conditions, what implications such impairments may have for our understanding of neurocognition, and how knowledge of sociocognitive substrates may help inform clinical practice, such as neurosurgery.

Scourfield and colleagues [27] define social cognition as "those aspects of higher cognitive function which underlie smooth social interactions by understanding and processing interpersonal cues and planning appropriate responses." As such, it includes a vast range of functions. Within these, one of the processes most central to appropriate social and emotional functioning is the theory of mind (ToM), or "mentalizing", which is the ability to attribute mental states (beliefs, intents, desires, pretending, knowledge) to oneself and others, and to understand that others have mental states that are different from one's own. ToM is a subcomponent of social cognition because it refers specifically to inferences about other people's mental states and is distinct from other social cues such as eye gaze and gestures, which can also be used to make judgments in social situations [10]. ToM has been found to influence and subserve many other sociocognitive abilities, in particular, moral reasoning and cognitive empathy.

Neuroanatomical and neurofunctional substrates of sociocognitive functions

Given the centrality of ToM and related cognitive functions to appropriate social cognition, knowledge of their neural substrates is key to understanding how disruptions in brain structure and function can cause social problems throughout development. Yet, neuroimaging studies attempting to elucidate the brain regions involved in sociocognitive abilities have yielded some discrepancies across studies. In general, studies have shown that the anterior temporal lobes are activated in a wide range of social cognition tasks. In addition, electrophysiological studies of monkeys have shown that the superior temporal cortex, the orbitofrontal cortex and the amygdala are reliably involved in social cognition [9]. However, more specificity is needed to distinguish particular sociocognitive processes from each other, as functions such as ToM, empathy, and moral reasoning may be differentially affected by brain disruption. After reviewing a large number of ToM studies, Carrington and Bailey [10] posit that there may be "core" regions activated for most ToM abilities and more "peripheral" regions recruited on the basis of the specific demands of some ToM tasks. This suggestion is based on the observation that methodological variability does not account for the varied results produced by neuroimaging studies thus far [5, 10]. The following regions have been identified as "core" or critical structures sustaining ToM abilities: medial prefrontal cortex (mPFC, including anterior medial regions), superior temporal sulcus (STS), temporoparietal junction (TPJ), as well as the para- and anterior cingulate cortices [10, 30]. Additionally, Carrington and collaborators posit that the control of shared representations is involved in ToM abilities. The concept of shared representations stems from the study of "mirror neurons", through which it has been suggested that the brain systems involved in representing one's own mental states may be the same as those representing others' mental states. Mirror neurons were first discovered in monkeys whereby it was observed that a monkey's premotor cortex was activated both when the animal engaged in a motor action and when it observed another monkey performing a motor action [12, 15]. Scientists studying social cognition with humans have since extrapolated that there may also be mirror neurons for sociocognitive abilities that fire when mental representations are shared (e.g., theory of mind). Indeed, there is some evidence that the same brain circuits are activated when one is thinking about one's own mental states and when one is reflecting on another person's mental states [16, 26]. In fact, scientists have suggested that this similarity in activation is what makes us capable of ToM. If this is true, then there must be a way of differentiating between one's own mental states and that of others, and the anterior medial frontal cortex and TPJ may play a crucial role in this regard. Specifically, the anterior medial frontal cortex appears to control shared representations, whereas the TPJ seems to govern self–other differentiation [30]. Other studies corroborate this idea, demonstrating that the mPFC and the TPJ (especially the right TPJ) show the most consistent activation across a variety of ToM tasks [10, 27]. The mPFC is activated in 90% of ToM studies [9]. Of note, studies conducted by Gallagher and Frith [13, 14] found that ToM solicits the following brain areas: the mPFC, STS, TPJ, and the temporal poles. Remarkably, although the amygdala seems to play a role in ToM, it is not consistently activated across studies. Carrington and Bailey [10] speculate that the amygdala may be involved in the development of ToM but may not be as necessary for the maintenance of this ability later in life. This suggestion partly stems from the observation that early damage to the amygdala has been linked with deficits in ToM [28]. The temporal poles have also shown some activation during ToM tasks but, once again, not in a consistent way, and their specific role (whether it be critical or part of a larger circuitry that gives rise to ToM) remains ill-defined. So far, researchers have identified the role of the temporal poles within ToM as being activated when individuals are presented with complex social stimuli (e.g., a story or social vignette) in which they must evaluate another person's intentions, beliefs, and emotions [23]. In essence, knowledge gained from neuroimaging studies of ToM suggest that although there appears to be a core circuit related to this process, the brain areas recruited are also task-dependent. In addition, investigators interested in mapping the neural substrates of ToM must keep in mind that even though all ToM tasks assess some aspect of ToM function, variations in task design may result in additional regions being recruited, such as those related to linguistic abilities.

ToM is closely linked to a number of other sociocognitive abilities. In particular, it underlies moral reasoning, which allows individuals to consider the world around them and to make decisions about right and wrong. Moral reasoning is tightly linked to ToM because for individuals to engage in appropriate moral behaviors, they must have the ability to understand and represent another person's perspective (ToM). Although both ToM and moral reasoning refer to consideration of other's mental states, in moral reasoning the focus is on considering intention and justification for moral action, rather than on understanding another's beliefs. Research into the neural correlates of moral reasoning to date has shown that the mPFC, TPJ, orbitofrontal regions, cingulate cortices, STS, and the amygdala are involved when moral judgments about a social situation are made [22, 33, 37, 38]. Since these brain regions are recruited specifically for moral versus nonmoral information, it is likely that one automatically infers mental states in order to process moral facts [37]. This is an interesting conjecture given that many of the regions recruited during moral reasoning are also recruited during ToM tasks (Fig. 1).

In an attempt to reconcile the somewhat heterogeneous neuroimaging findings generated in relation to ToM and moral reasoning, some researchers have looked to other internal and external factors, such as development, gender, and culture, that may influence social cognition. Singer [29] provides a good argument for focusing on developmental aspects of social cognition: a child's mentalizing abilities develop by the age of four years, clearly much before the brain regions sustaining those abilities have fully matured. Indeed, the

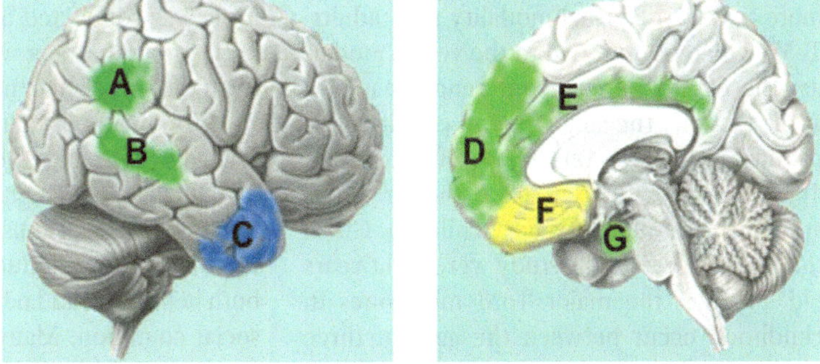

Fig. 1. Overlap (green) between areas commonly reported to be activated in studies of theory of mind (blue) and moral reasoning (yellow). *A* temporo-parietal junction; *B* superior temporal sulcus; *C* temporal pole; *D* medial prefrontal cortex; *E* cingulate cortex; *F* orbitofrontal cortex; *G* amygdala

behavioral aspects of ToM development are well established and there is ample evidence in the literature that ToM abilities change qualitatively during childhood and well into the teenage years. First-order beliefs (attributing a belief to another person) generally develop by four years of age, while second-order beliefs (attributing a belief about another person's belief) generally emerge between the ages of six and ten. However, given that scientists, especially those in the field of social neuroscience, have only begun to look at the neural underpinnings of ToM development, studies that have looked at this aspect specifically are scarce. Interestingly, Kobayashi and colleagues [17] comparing children and adults on ToM tasks found that the bilateral TPJ and the right inferior parietal lobule are areas that are important for ToM during both childhood and adulthood. However, in the child group, there was more activity in the right superior temporal gyrus, the right temporal poles, the cuneus, and the right ventromedial PFC. In contrast, in the adult group, there was more activity in the left amygdala compared to the child group. The authors suggest that the regions that were more activated in the child group may represent areas that are recruited by the cognitive precursors of ToM during development. Furthermore, the same study examined whether the performance of both groups could be affected by the modality in which the ToM task was presented. The findings indicate that children's ToM seems to be more tied to the visual modality than adults' ToM, which is more tied to the verbal modality. These developmental differences may hint at the fact that the language and cognitive capabilities adults rely on when confronted with ToM tasks are different from the strategies available to and employed by children. Given that the children in the study were eight years old and that the major ToM milestones in childhood occur between the ages of three and seven, future studies looking at children in that age range will help to further elucidate developmental aspects in the neural substrates of ToM. In a similar context, studies have been conducted with blind individuals in order to examine if being deprived of a modality (i.e., vision) affects brain regions recruited for ToM tasks. Bedny et al [3] found that the regions solicited during ToM tasks for congenitally blind adults are the same as those recruited by sighted adults. Namely, the blind individuals showed activations in the biltateral TPJ, the mPFC, the precuneus, and the anterior STS. The authors conclude that for congenitally blind and sighted adults, inferring mental states that are based on seeing is an akin process. That is, visual experience is not a requisite for being able to neurally represent another person's experiences. In blind individuals, the same brain regions are recruited for ToM tasks, regardless of the modality solicited. Those findings further suggest that at least some aspects of ToM development are based on inborn influences and on more abstractly represented experiences, independent of the modality in which the information was acquired.

Gender differences have been studied extensively in relation to discrepancies in performance on ToM tasks. Behavioral studies have provided evidence that women perform better on ToM tasks than do men, regardless of age, and these gender differences are apparent even in children [10]. As with developmental aspects of ToM, the literature on neuroimaging and gender differences for ToM is scant and constitutes a rich area for future research endeavors. One of the few imaging studies on this topic was conducted by Krach and colleagues [19], who found that women elicit more activity in the mPFC than men during ToM tasks.

Proponents of cultural neuroscience have started to delve into the question of the impact of culture (and language) on sociocognitive functions. Lack of consideration of these external factors may influence interpretations of both behavioral and neurofunctional aspects of social cognition. Many studies fail to control for these variables, and this may introduce significant methodological bias, as noted by Kobayashi and colleagues [18]. The authors studied a group of bilingual Japanese-English-speaking children and compared them to a group of monolingual English-speaking children on a ToM task. They found that only bilingual children tested with Japanese stories showed activation in one of the typical four ToM areas (mPFC). Surprisingly, even monolingual English children did not show any activation in the TPJ. Based on these results, Perner and Aichhorn [25] speculate that when presented with a ToM task, the activation seen in the TPJ could be influenced by environmental factors such as culture or language. However, how each of these factors possibly contributes to these differences in activation is still unknown. Their argument is founded on previous studies that have shown that bilingual children perform better on false-belief tasks than monolingual children; that performance on false-belief tasks is related to verbal intelligence; and that deaf children with language delays also exhibit delays in performance on false-belief tasks [25]. Of note, generalization across ToM tasks is still premature because no studies have examined other types of ToM tasks. In interpreting these results, researchers and clinicians should also bear in mind that there may be important interindividual differences in performance on ToM tasks regardless of external factors. As Carrington and Bailey [10] state: "some individuals consistently show insight into mental states and adapt their behavior accordingly, whereas others show less awareness" (p 2330). Therefore, "normal" individual differences may in part explain the lack of consistency across neuroimaging studies. Otsuka et al [24] provide evidence in favor of this argument, demonstrating a correlation between performance on a ToM task and activation in the anterior STS, and thus suggesting that the anterior STS might reflect individual differences in ToM. However, given the correlational nature of their study, future studies need to be conducted to establish the cause-and-effect relation between these two elements.

Clinical insights into the neural substrates of the theory of mind

A discussion of the neuroanatomical and functional bases of social cognition would be incomplete without consideration of clinical conditions in which ToM deficits are common occurrences. As such, developmental, neurological, and psychiatric conditions that affect sociocognitive functioning provide a rich source for understanding the link between neurological processes and socioemotional function (Table 1). One neurological condition in which important ToM deficits have been observed is acquired brain injury. In their review, Martín-Rodríguez and León-Carrión [21] found that ToM deficits in adults with frontal lobe lesions due to acquired brain injury followed a particular task-related hierarchy. Deficits were most severe on ToM tasks in which an understanding of indirect speech (i.e., sarcasm, metaphors, irony) was required. Less severe, but nonetheless important deficits were found on social faux pas tasks, followed by second-order and first-order belief ToM tasks. This descending order of severity does not fit the "hierarchy of ToM tasks hypothesis" [19]. According to the hypothesis, in some disorders those skills that develop later in life should be most impaired by neural disruption [2, 8, 31]. However, the review suggests that individuals with acquired brain injury show fewer deficits on faux pas tasks than on tasks requiring an understanding of indirect speech even though one's understanding of faux pas develops later in life. In addition, the review found that frontal lobe lesions affect the performance on faux pas tasks, while right hemisphere damage has an impact on the understanding of indirect speech and faux pas. However, recent neuroimaging studies question this link between right-hemisphere damage and ToM deficits in acquired brain injury and emphasize the importance of the left hemisphere as well as the connection between the two hemispheres [32]. Neuroimaging studies on this topic would be greatly aided by behavioral studies looking at whether ToM abilities

Table 1. Clinical manifestations of ToM deficits and brain circuits involved in specific pathologies

Pathology	Brain circuitry involved	Problem	ToM deficits
Acquired brain injury	Frontal lobes	Lesion	Indirect speech, faux pas, second-order beliefs, first-order beliefs
Parkinson'sd isease	Dorso-lateral prefrontal-striatal circuitry	Early atrophy	Cognitive ToM deficits
	Fronto-striatal-limbic circuitry	Protracted degeneration	Affective ToM deficits
Schizophrenia	Medial prefrontal cortex, amygdala, temporoparietal junction	Abnormal activation	General deficits in mental state inference; "hyper-ToM"
	Inferior parietal lobule	Abnormal activation	Abnormal source monitoring
	Dopamine circuitry	General transmission impairment	Erroneous contextual information
	Cortico-cerebellar-thalamic-cortical circuit	Inappropriate recruitment	General ToM deficits
Bipolar disorder	Left anterior cingulate, precuneus, cuneus bilaterally	High individual variability; consistent activations do not include regions typically recruited for ToM	Restricted ability to mentalize; decreased cognitive flexibility
Autism spectrum disorders	Medial prefrontal cortex, amygdala, anterior cingulate cortex, medial temporal lobe, inferior parietal lobe	Abnormal activation	General ToM deficits
	Supplementary motor area	Inappropriate recruitment of lower-order processing brain areas	Difficulty with perception of specific social stimuli
	Right temporoparietal junction	Grey matter abnormalities	General ToM deficits

are independent of other general cognitive capacities (language, executive functions). The specific impact of severe traumatic brain injury on ToM skills has also been explored in children. Findings indicate that children between the ages of 5 and 7 years with severe traumatic brain injury show developmental lag on tasks of first- and second-order mental states comparable to their peers, suggesting stagnation or even regression of existing ToM skills [35]. Additionally, Walz and colleagues [36] found that even when strong predictors of ToM abilities, such as age, task-specific cognitive demands, and verbal skills were taken into account, children who sustained severe traumatic brain injury still performed more poorly than typically developing peers and children suffering from orthopedic injuries. Another study conducted by the same group corroborates the aforementioned findings in that they found the same results in a younger group (3 years old) of children having sustained severe traumatic brain injury [25].

Parkinson's disease is also the focus of scientific scrutiny in relation to ToM abilities. In order to understand the ToM deficits in Parkinson's disease, Bodden et al [6] propose that a distinction be made between "cognitive ToM", in which one intellectually understands mental states, and "affective ToM", which provides an appreciation of others' emotions. Of note, ToM deficits are primarily observed in individuals with advanced Parkinson's disease, with cognitive ToM deficits appearing first, followed by deficits in affective ToM once the disease has progressed further. Neuroanatomically, early emerging deficits in cognitive ToM appear to be attributed to early atrophy of the dorsolateral prefrontal-striatal circuitry, whereas later deficits in affective ToM seem to be related to the protracted degeneration of the frontostriatal-limbic circuitry. In this regard different etiologies of Parkinson's disease may result in different ToM deficits.

Psychiatric illnesses have also received attention in relation to ToM deficits. In particular, such deficits are well documented for individuals with schizophrenia and have been investigated via neuroimaging. Brunet-Gouet and Decety [9] found abnormal prefrontal and amygdala activation in individuals with schizophrenia, which translated behaviorally into general deficits in mental state inference. Furthermore, abnormal activity in the inferior parietal lobule was identified and this resulted in abnormal source monitoring. That is, individuals with schizophrenia are impaired in monitoring their actions and when making self-other distinctions. Additionally, erroneous processing of contextual information by individuals with schizophrenia performing ToM tasks has been attributed to a general impairment of dopamine transmission in the brain. Brunet-Gouet and Decety [9] conclude that "the inability to maintain a stable and relevant social context for the processing of one's own or other people's mental representations and the impairment of the ability to modify this type of context may arise from acute neurotransmitter disturbances" (p. 86). A positron emission tomography study conducted by Andreasen and colleagues [1] supports previous studies which have found that individuals with schizophrenia have a deficit in soliciting the cortico-cerebellar-thalamic-cortical circuit. Similarly, the areas of increased brain activity (as observed through blood flow) found in their study suggest that these individuals may need to utilize resources in the right hemisphere to compensate for impairments in the left hemisphere. These finding beg the question: Is this a problem in brain lateralization or use of different strategies? Individuals with schizophrenia might be using a more "logical" process rather than a "social" process for ToM tasks. Walter and colleagues' fMRI study examining the relation between ToM and paranoid schizophrenia yielded similar results; there was a general dysfunction of the ToM circuitry in the brain of individuals with paranoid schizophrenia [34]. The authors argue that as soon as a social element is present, the difference in neural activation in individuals suffering from schizophrenia versus their healthy counterparts becomes obvious, especially in the mPFC and the TPJ, the structures that have most consistently been associated with ToM. Given that the mPFC and the TPJ have been respectively attributed the roles of distinguishing between one's own mental states and those of others, as well as communicating intentions, Walter and colleagues [34] posit that individuals with schizophrenia are not able to differentiate between their own mental states and those of others, leading to "hyper-ToM". As stated by the authors: "schizophrenic patients perceive agency where others see none [...] the ToM module in paranoid schizophrenia might malfunction because it is overactive from the start and thus isn't well suited to distinguish properly between mental and physical states" (p. 175). Benedetti and colleagues' [4] results corroborate the aforementioned statements in that individuals with schizophrenia present abnormal neural activations in mPFC and temporal structures while performing ToM tasks. Furthermore, these differences in activation in the mPFC are present early on in the evolution of the disease, leading the authors to suggest that they might have identified a biological basis for the sociocognitive problems affecting this psychiatric population.

Bipolar disorder also constitutes a psychiatric condition of interest in terms of the neural substrates of ToM. Malhi and colleagues [20] found that individuals suffering from euthymic bipolar disorder do not show the same consistent, widespread activation that is seen in healthy individuals performing ToM tasks. In fact, even within the bipolar group, only the left anterior cingulate, the precuneus, and the cuneus bilaterally showed some consistency in activation. Of note, the activations found in the patient group do not include structures that are typically activated during ToM tasks in healthy individuals, such as the mPFC and the TPJ. Although individuals suffering from bipolar disorder seem capable of solving ToM tasks, they may not have the level of understanding and appreciation that is seen in healthy peers; they seem to lack cognitive flexibility in that they do not seem capable of changing their cognitive perspective to suit particular social situations.

Given that sociocognitive deficits are the hallmark of autism spectrum disorder, researchers have focused on the neural correlates of ToM deficits in this population. In particular, Di Martino and colleagues [11] published a meta-analysis of the neuroimaging literature on autism spectrum disorder in relation to sociocognitive tasks (not restricted to ToM tasks). These researchers found that there is a consistent pattern of hypoactivation and abnormal activity in regions typically involved in ToM, such as the mPFC, the amygdale, and the anterior cingulate cortex. Likewise, there seems to be inappropriate soliciting of lower-order processing regions such as the supplementary motor area in lieu of higher-order processing regions such as the anterior cingulate and pre-supplementary motor area. Additionally, Brieber and colleagues [7] identified gray-matter abnormalities close to the right TPJ, which may explain the ToM deficits observed for individuals with autism spectrum disorder. Further abnormalities were found in the medial temporal and inferior parietal lobe.

Conclusions

The investigation of the neural correlates of sociocognitive abilities is still in its infancy given that the technology permitting this endeavor is relatively recent. However, substantive literature in relation to the neuroanatomical and functional bases of social cognition is now emerging, with a specific focus on ToM. The brain regions involved in ToM abilities have received particular attention and studies have yielded relatively consistent results; the mPFC and the TPJ seem to be the key players in ToM abilities, with several other brain regions assigned "peripheral" roles. To further knowledge of the neural underpinnings of ToM, researchers have also examined neurological and psychiatric populations in which sociocognitive deficits were common clinical symptoms of the illness. Although this investigation has been fruitful, much research is still needed to further elaborate the brain abnormalities observed in relation to the presenting behavioral deficits. Clinically, research into the neural substrates of social cognition is beneficial to medical practice, such as neurosurgery, given that the goal of a surgery is to maximize the removal of unhealthy tissue while minimizing the cognitive deficits and impact on patients' quality of life. Although ToM is an ability that is more abstract and more difficult to evaluate during surgery than other abilities (such as language) that neurosurgeons aim to preserve, it is of critical importance because it underlies social competence. Disruption of ToM and its related sociocognitive skills may lead to social isolation, psychological distress, and socially maladaptive behaviors. Neuroanatomical and functional knowledge of underlying circuitry may guide neurosurgeons so that they may preserve fundamental social abilities.

References

[1] Andreasen NC, Calarge CA, O'Leary DS (2008) Theory of mind and schizophrenia: a positron emission tomography study of medication-free patients. Schizophr Bull 34: 708–719

[2] Baron-Cohen S, O'Riordan M, Stone V, Jones R, Plaisted K (1999) Recognition of faux pas by normally developing children and children with Asperger syndrome or high-functioning autism. J Autism Dev Disord 29: 407–418

[3] Bedny M, Pascual-Leone A, Saxe RR (2009) Growing up blind does not change the neural bases of "theory of mind". Proc Natl Acad Sci USA 106: 11312–11317

[4] Benedetti F, Bernasconi A, Bosia M, Cavallaro R, Dallaspezia S, Falini A, Poletti S, Radaelli D, Riccaboni R, Scotti G, Smeraldi E (2009) Functional and structural brain correlates of theory of mind and empathy deficits in schizophrenia. Schizophr Res 114: 154–160

[5] Blakemore SJ (2008) The social brain in adolescence. Nat Rev Neurosci 9: 267–277

[6] Bodden ME, Dodel R, Kalbe E (2010) Theory of mind in Parkinson's disease and related basal ganglia disorders: a systematic review. Mov Disord 25: 13-27

[7] Brieber S, Neufang S, Bruning N, Kamp-Becker I, Remschmidt H, Herpertz-Dahlmann B, Fink GR, Konrad K (2007) Structural brain abnormalities in adolescents with autism spectrum disorder and patients with attention deficit/hyperactivity disorder. J Child Psychol Psychiatry 48: 1251–1258

[8] Brüne M (2003) Theory of mind and the role of IQ in chronic disorganized schizophrenia. Schizophr Res 60: 57–64

[9] Brunet-Gouet E, Decety J (2006) Social brain dysfunctions in schizophrenia: a review of neuroimaging studies. Psychiatry Res 148: 75–92

[10] Carrington SJ, Bailey AJ (2009) Are there theory of mind regions in the brain? A review of the neuroimaging literature. Hum Brain Mapp 30: 2313–2335

[11] Di Martino A, Ross K, Uddin LQ, Sklar AB, Castellanos FX, Milham MP (2009) Functional brain correlates of social and nonsocial processes in autism spectrum disorders: an activation likelihood estimation meta-analysis. Biol Psychiatry 65: 63–74

[12] di Pellegrino G, Fadiga L, Fogassi L, Gallese V, Rizzolatti G (1992) Understanding motor events: a neurophysiological study. Exp Brain Res 91: 176–180

[13] Frith CD, Frith U (2006) The neural basis of mentalizing. Neuron 50: 531–534

[14] Gallagher HL, Frith CD (2003) Functional imaging of "theory of mind". Trends Cogn Sci 7: 77–83

[15] Gallese V, Fadiga L, Fogassi L, Rizzolatti G (1996) Action recognition in the premotor cortex. Brain 119: 593–609

[16] Kaplan JT, Iacoboni M (2006) Getting a grip on other minds: mirror neurons, intention understanding, and cognitive empathy. Soc Neurosci 1: 175–183

[17] Kobayashi C, Glover GH, Temple E (2007) Children's and adults' neural bases of verbal and nonverbal "theory of mind". Neuropsychologia 45: 1522–1532

[18] Kobayashi C, Glover GH, Temple E (2007) Cultural and linguistic effects on neural bases of "theory of mind" in American and Japanese children. Brain Res 1164: 95–107

[19] Krach S, Blumel I, Marjoram D, Lataster T, Krabbendam L, Weber J, van Os J, Kircher T (2009) Are women better mindreaders? Sex differences in neural correlates of mentalizing detected with functional MRI. BMC Neurosci 10: 9

[20] Malhi GS, Lagopoulos J, Das P, Moss K, Berk M, Coulston CM (2008) A functional MRI study of Theory of Mind in euthymic bipolar disorder patients. Bipolar Disord 10: 943–956

[21] Martin-Rodriguez JF, Leon-Carrion J (2010) Theory of mind deficits in patients with acquired brain injury: a quantitative review. Neuropsychologia 48: 1181–1191

[22] Moll J, de Oliveira-Souza R, Bramati IE, Grafman J (2002) Functional networks in emotional moral and nonmoral social judgments. Neuroimage 16: 696–703

[23] Olson IR, Plotzker A, Ezzyat Y (2007) The enigmatic temporal pole: a review of findings on social and emotional processing. Brain 130: 1718–1731

[24] Otsuka Y, Osaka N, Ikeda T, Osaka M (2009) Individual differences in the theory of mind and superior temporal sulcus. Neurosci Lett 463: 150–153

[25] Perner J, Aichhorn M (2008) Theory of mind, language and the temporoparietal junction mystery. Trends Cogn Sci 12: 123–126

[26] Rizzolatti G, Fabbri-Destro M (2008) The mirror system and its role in social cognition. Curr Opin Neurobiol 18: 179–184

[27] Scourfield J, Martin N, Lewis G, McGuffin P (1999) Heritability of social cognitive skills in children and adolescents. Br J Psychiatry 175: 559–564

[28] Shaw P, Lawrence EJ, Radbourne C, Bramham J, Polkey CE, David AS (2004) The impact of early and late damage to the human amygdala on "theory of mind" reasoning. Brain 127: 1535–1548

[29] Singer T (2006) The neuronal basis and ontogeny of empathy and mind reading: review of literature and implications for future research. Neurosci Biobehav Rev 30: 855–863

[30] Spengler S, von Cramon DY, Brass M (2009) Control of shared representations relies on key processes involved in mental state attribution. Hum Brain Mapp 30: 3704–3718

[31] Stone VE, Baron-Cohen S, Knight RT (1998) Frontal lobe contributions to theory of mind. J Cogn Neurosci 10: 640–656

[32] Tesink CM, Buitelaar JK, Petersson KM, van der Gaag RJ, Kan CC, Tendolkar I, Hagoort P (2009) Neural correlates of pragmatic language comprehension in autism spectrum disorders. Brain 132: 1941–1952

[33] Van Overwalle F (2008) Social cognition and the brain: a meta-analysis. Hum Brain Mapp 30: 829–858

[34] Walter H, Ciaramidaro A, Adenzato M, Vasic N, Ardito RB, Erk S, Bara BG (2009) Dysfunction of the social brain in schizophrenia is modulated by intention type: an fMRI study. Soc Cogn Affect Neurosci 4: 166–176

[35] Walz NC, Yeates KO, Taylor HG, Stancin T, Wade SL (2009) First-order theory of mind skills shortly after traumatic brain injury in 3- to 5-year-old children. Dev Neuropsychol 34: 507–519

[36] Walz NC, Yeates KO, Taylor HG, Stancin T, Wade SL (2010) Theory of mind skills 1 year after traumatic brain injury in 6- to 8-year-old children. J Neuropsychol 4: 181–195

[37] Young L, Saxe R (2009) An fMRI investigation of spontaneous mental state inference for moral judgment. J Cogn Neurosci 21: 1396–1405

[38] Young L, Cushman F, Hauser M, Saxe R (2007) The neural basis of the interaction between theory of mind and moral judgment. Proc Natl Acad Sci USA 104: 8235–8240

Surgical applications

Neurocognitive outcome and resective brain tumor surgery in adults

Martin Klein and Philip C. De Witt Hamer

Introduction

Patient performance is of particular importance to evaluate treatment outcome in the circumstances of incurable neurological disease. This is the case for patients with gliomas for whom palliation of symptoms and sustained or improved quality of life are equally important goals of treatment as prolonged survival and postponed tumor progression. Evaluation of treatment in brain tumor patients should therefore focus beyond oncological endpoints, and should also aim at avoiding adverse treatment effects on the normal brain to ensure optimal social and professional functioning. Functional outcome can be considered as a construct with several dimensions. These dimensions include neurological, cognitive, professional, and social performance of an individual, which can be represented by the health related quality of life (HRQOL). Cognitive functioning is one of the critical outcome measures because subclinical cognitive impairment can have a large impact on the daily life of patients [88, 127]. Even mild cognitive difficulties can have functional and psychiatric consequences -especially when persistent and left untreated. Deficits in specific cognitive domains such as inattention, dysexecutive function, and impaired processing speed may affect HRQOL. For example, cognitive impairment negatively affects professional reintegration, interpersonal relationships, and leisure activities. In addition, fear of future cognitive decline may also negatively affect HRQOL. Compared to the classical oncological endpoints, evaluation of HRQOL and cognitive functioning is more time-consuming for the care provider and more burdensome for the patient. Besides, given the relatively low incidence of glial brain tumors and the often ultimately fatal outcome of the disease, the interest in HRQOL and cognitive functions emerged relatively late in these patients [32]. Cognitive functioning and HRQOL, however, are not only useful as outcome measures in clinical trials for brain tumor patients. They may also serve as an early indicator of disease progression and have prognostic significance, thereby providing additional arguments in clinical decision making.

Resective brain surgery is one of several treatment modalities for patients with brain tumors. As such, resective brain surgery requires balancing of functional outcome with oncological goals. For adequate preoperative counseling ultimately detailed quantitative knowledge of postoperative functioning in time for an individual patient would be required. This knowledge is however unavailable and is customary translated in general terms from a subjective overall risk assessment. Obviously, the weighing of this balance will differ substantially with the natural history of diseases for which resective brain surgery is considered. For instance, what is considered an acceptable risk for a loss of function after a hemorrhage from a potentially lethal arteriovenous malformation may be considered unacceptable in resective brain surgery for obtaining seizure freedom. A vast body of literature exists describing the impact of resective brain surgery on neurological outcome, such as motor strength and language, in patients with brain tumors, epilepsy and other parenchymal brain lesions. The impact of resective brain surgery on cognitive outcome, such as learning, memory and executive functions, and HRQOL has not been systematically determined, however. While this chapter discusses the impact of the tumor, epilepsy, radiotherapy, antiepileptic drugs, and steroids on cognitive outcome, the focus will primarily be on the impact of resective brain tumor surgery and on the application of cognitive tasks in intraoperative brain mapping procedures.

Factors affecting neurocognitive functioning

Many factors potentially influence neurocognitive functioning of patients with brain lesions. In attempting to determine the isolated effect of resective surgery on cognition, the multifactorial processes involved should be recognized. These factors include premorbid level of cognitive functioning, distant mechanical effects on the normal brain by the lesion, epilepsy, medication, and other oncological treatments.

Brain tumor effects

Cognitive deficits can be induced by several mechanisms. Firstly, a brain tumor can induce compression of normal brain either directly or indirectly by reactive edema. The influence of distant mechanical effects on cognition is examplified by cognitive improvement after removal of non-invasive lesions such as meningiomas [131] or arachnoidal cysts [139] and cognitive improvement even after cranioplasty [2]. However, long-term cognitive outcome in WHO grade I meningiomas patients is affected by antiepileptic drug use and tumor location [23], but not by the use of radiotherapy [133]. Secondly, the invasion of parenchymal glial tumors directly into functional brain regions or indirectly by disconnection of structures can further contribute to cognitive deficits [10, 106, 127].

Epilepsy effects

Often, epileptic seizures are the first symptom of a brain tumor and may result in morbidity and decreased quality of life, even if the tumor is not progressing [61]. Thus, treatment with antiepileptic drugs (AEDs) is clearly indicated for brain tumor patients with preoperative tumor-related seizures. The mechanism and pattern of seizures is determined by the tumor type, the tumor location and peritumoral and genetic changes in brain tumor patients [132]. Apart from the effects of the tumor itself, cognitive function can be impaired by the seizures [25]. An increased epilepsy burden has been found to adversely affect a broad range of cognitive functions [61] even to a larger extent than radiation therapy [62]. Attention, processing speed, and executive deficits are notable sequelae of seizures and AEDs in patients with brain tumors [17, 61, 62]. However, the literature is inconsistent on this point. Other investigators, however, did not detect any apparent supra-ordinate effect of seizures on cognition across multiple cognitive domains assessed even in a postsurgical sample [131].

Surgery effects

To determine the effect of resective surgery on neurocognitive outcome, ideally a homogeneous population of patients with a similar brain lesion in a similar brain region is examined at an individual level by standardized neuropsychological assessment at various time intervals after a similar surgical procedure and compared to preoperative baseline measurements. Several issues contribute to a deviation from this ideal situation in practice mainly due to heterogeneity in the multiple factors that contribute to cognitive functioning. Firstly, premorbid and baseline preoperative level of cognitive functioning is variable in patients. Secondly, the signs and symptoms of disease, such as seizure pattern, are variable, despite similar lesions in similar locations. Thirdly, progression of disease after surgery can substantially affect cognitive functioning and the timing and slope of progression usually varies between patients. For instance, the rate of volume progression and anaplastic transformation of low-grade gliomas is highly variable despite similar histopathology and location. Fourthly, a surgical procedure is often combined with other treatments that potentially influence cognitive outcome. Fifthly, repeated neuropsychological examination is subject to learning effects that are unlikely distributed homogeneously in the patient population.

Although between-group comparisons of neurosurgery patients with well-matched controls permit useful estimates of the incidence of neuropsychological impairment that are associated with neurosurgery, they do not allow for characterization of cognitive outcome at the individual level. Based on group outcomes, a widely-used criterion for defining significant change from baseline in postoperative test scores (i.e., one that requires 1.5 standard deviation change) can be applied to individual patients indicating a clinically relevant rather than a statistically significant change.

To review cognitive outcome after resective brain surgery, the discussion that follows is structured by disease entity. As such, publications will be discussed describing resective surgery for temporal lobe epilepsy and brain tumors, respectively.

Cognitive outcome has been reported systematically after temporal lobe surgery in patients with intractable epilepsy that is not tumor-related. In these studies, cognitive function such as memory or verbal fluency can improve with adequate seizure control after temporal resection [7, 77, 91, 115, 116, 124, 137]. Less extensive selective resection of the mesiotemporal structures seems to correlate with better memory outcome compared with more extensive temporal lobectomy according to some groups [50, 92, 114], whereas others have reported conflicting observations. For instance, an underpowered randomized study failed to detect differential cognitive outcome after selective amygdalohippocampectomy and transtemporal approach [74] in concordance with observational data [45, 140]. Furthermore, dominant temporal lobe resections have been correlated with verbal memory decline in a subset of patients [22, 42, 57, 58, 76, 105, 120], whereas non-dominant temporal lobe resections were correlated with visuospatial memory decline [31, 33, 57, 66, 93, 101, 118].

After extra-temporal resective surgery for intactable epilepsy variable cognitive outcome has been reported. After unilateral removal of frontal cortex cognition was either unchanged [63, 73], or specific cognitive domains were impaired, such as reaction time [49, 64], impulsivity [86], advance information utilization [3], conditional learning [99], or search and retrieval strategies [53]. Furthermore, identification of faces and categorization of emotional facial expression was impaired after either frontal or temporal cortex resection [11]. Olfactory identification was impaired following unilateral excision of the temporal lobe or the orbitofrontal cortex on either side [56].

Cognitive outcome has not been systematically assessed for resective brain surgery in patients with brain tumors, although several interesting observations have been done in smaller observational cohort studies.

Firstly, cognitive improvement has been observed in several studies after brain tumor resection. Long term improvement of verbal memory compared to preoperative assessment has been reported after low-grade glioma resections in frontal premotor and anterior temporal areas [12, 40, 128], usually after a transient immediate postoperative worsening. Accordingly, long term improvement in attentional functions resulting in faster and more accurate performance, has been observed after surgical removal of frontal meningiomas, [37, 69, 131]. This attentional improvement was not related to level of edema, brain compression, or lesion size, but rather to localization, such that patients with meningiomas of the falx cerebri or frontobasal region demonstrated most favorable improvement. Global cognitive improvement has also been observed after surgical resection of high-grade glioma [15, 84].

Secondly, in some studies stable cognitive performance was observed after brain tumor resection. For instance, patients with tumors of the third ventricle demonstrated cognitive impairment in memory, executive functioning and fine manual speed prior to surgery, without worsening of cognition after surgical removal [36, 100]. Out of several executive tasks, only letter fluency performance was impaired in patients after glioma surgery in left frontal locations compared with right frontal and posterior lesions [136]. Visuospatial processing in patients after resective glioma surgery in left and right, frontal and parietal locations was comparable to that of normal subjects according to one study [55] and impaired spatial and positional memory processing was demonstrated in patients with tumors in the right posterior parietal cortex or in the frontal cortex according to others [60, 96].

Thirdly, a number of studies have demonstrated cognitive deficits in specific domains after brain tumor removal. For instance, some patients demonstrated minor deterioration in attention after resection of parenchymal frontal or precentral tumors [12, 43] and resection of the right prefrontal cortex rather than the left was associated with a selective attentional impairment in Stroop test performance [134]. After resection of the supplementary motor area, patients exhibited impaired procedural learning and agraphia [1, 113]. Subsets of patients with resections involving the frontal lobe demonstrated a variety of deficits. For instance, impaired sequence ordering of novel material was observed particularly in right-sided lesions, while recognition memory was unaffected [123], and planning and executive impairment, irrespective of side, site, and size [95, 135]. Furthermore, severe executive deficits in a reward learning task were observed in patients after bilateral fronto-orbital resections for various tumor types [52] and impaired virtual planning of real life activities after resections in the left and right prefrontal cortex, which could not be explained by memory deficits [44, 87].

It comes as no surprise that brain tumor patients have feelings of anxiety, depression, and future uncertainty as psychological reactions to the disease [19, 119, 126]. These mood disturbances may lead to deficits in attention, vigilance, and motivation that subsequently affect several cognitive domains [4]. Mood changes are more common in brain tumor patients than in patients with other neurological diseases [5] and might be related to tumor location [71]. Unilateral surgical removal of prefrontal cortex, including the fronto-orbital or anterior cingulate cortex, has resulted in emotional dysregulation with impaired voice and face expression identification in patients with various brain lesions including brain tumors [51]. Furthermore, deficits in recognizing emotional facial expression were observed after surgical removal of brain tumors that involved both heteromodal and limbic/paralimbic cortices [138]. Concordantly, impairment of arousal and emotional valence was demonstrated after resective surgery in various brain regions, but particularly in the right temporoparietal region [97]. This emotional impairment can have an impact on social and professional performance. Negative mood changes were observed after brain tumor resection involving heteromodal cortices located either prefrontal or temporoparietal, whereas positive mood changes were observed after lateral frontal resections [54]. Mood states did not correlate with laterality of the resection, tumor grading or lesion size. Social interactions also depend on the ability of theory of mind, i.e. to attribute mental states such as beliefs, intents, desires, pretending and knowledge to others and to understand that these beliefs, desires and intentions can be different from one's own. The theory of mind ability was significantly impaired in patients with either right or left frontal lobe resections for various reasons, including brain tumors, which could not be related to executive or memory functioning [111]. Furthermore, amnesia correlated with bilateral damage of the fornices after removal of third ventricle tumors or of the mammilary bodies after craniopharyngeoma removal [80, 125]. Transient amusia has been observed after resection of Heschl gyrus of the right hemisphere in glioma surgery [112].

Other treatments and medication as a cause of cognitive deficits

Radiotherapy

Late-delayed encephalopathy is an irreversible and serious complication that follows radiotherapy by several months to many years and may take the form of local radionecrosis, diffuse leukoencephalopathy, and cerebral atrophy. Neurocognitive disturbances are the hallmark of the diffuse encephalopathy [8]. The severity of neurocognitive deficits ranges from mild or moderate neurocognitive deficits to neurocognitive deterioration leading to dementia. Patients with mild to moderate neurocognitive deficits have attention or short-term memory disturbances as main features. Both the clinical picture and the incidence of this complication are hard to define exactly as studies on this subject vary greatly in the neuropsychological test procedures, the populations studied, and the duration of follow-up [8]. There is a relation between neurocognitive status and cerebral atrophy and leukoencephalopathy [26, 103]. According to a review of 18 clinical studies [18], severe neurocognitive deterioration, leading to dementia with subcortical features as expressed by progressive mental slowing and deficits in attention and memory, occurred in at least 92 of 748 patients treated with radiotherapy. In these cases, MRI shows diffuse atrophy with ventricular enlargement as well as severe confluent white-matter abnormalities [89]. A more recent study [62] that showed that the use of radiotherapy was associated with poor neurocognitive function on only a few tests and not restricted to one specific neurocognitive domain, however, suggests that neurocognitive deficits in low-grade glioma survivors should not be attributed to radiotherapy, but rather to the tumor itself or other treatment factors. Serious memory deficits, however, are still to be expected with fraction doses exceeding 2 Gy [62].

While short-term follow-up studies show limited or transient effects of radiotherapy [127], a number of studies in patients with long survival of low-grade glioma of more than 5 years following radiotherapy concluded that radiotherapy in low-grade glioma patients poses a significant risk of long-term leukoencephalopathy and neurocognitive impairment. Surma-Aho and co-workers [122] reported on patients with long survival after low-grade glioma (mean follow-up 7 years) who had more neurocognitive deficits after early radiotherapy than controls without radiotherapy. Moreover, leukoencephalopathy on MRI was more severe in the group with postoperative irradiation. A recent follow-up of the Klein et al (2002) study [62] demonstrated that all tumor progression-free low-grade glioma patients that had irradiation had neurocognitive deterioration 13 years after radiotherapy while all patients without irradiation remained stable [26]. Moreover, an increase in radiological abnormalities was found only in the irradiated group.

Antiepileptic drugs

Risks of cognitive side-effects of antiepileptic drugs can add to previous cognitive decline by surgery or radiotherapy, and therefore appropriate choice and dose of antiepileptic drug is crucial. The classical antiepileptic drugs (phenytoin, carbamazepine, and valproic acid) are known to decrease cognitive functioning [27, 82]. Importantly, these drugs may also have pharmacological interactions with the therapies used in brain tumor patients [79, 94] and thus potentially affect survival. These drugs may result in impairment of attention and cognitive slowing, which can subsequently have effects on memory by reducing the efficiency of encoding and retrieval [82]. The importance of the classical antiepileptic drugs as a risk factor for cognitive deficits has been reported in a study on low-grade glioma; [61] in a group of 156 long-term survivors without signs of tumor recurrence, deficits in information processing speed, psychomotor functioning, executive function, and working memory capacity were significantly related to the use of antiepileptic drugs. As patients in this study who took antiepileptic drugs had cognitive disturbances even in the absence of seizures, the use of drugs primarily affects cognitive function. Moreover, AED use in low-grade glioma patients may be associated with highly elevated levels of fatigue [121], which in itself is also associated with poorer cognitive outcome. Several new generation AEDs, like oxcarbazepine [78] and levetiracetam as add-on therapy [24] appear to have fewer adverse cognitive effects than the classical agents. Of the newer agents, topiramate is associated with the greatest risk of cognitive impairment, although this risk is decreased with slow titration and low target doses [81, 83]. It appears to be safe to switch patients from phenytoin to levetiracetam monotherapy following craniotomy for supratentorial glioma [70].

Steroids

Mounting evidence suggests pleiotropic adverse effects of corticosteroids including central nervous system compromise, which are often misdiagnosed or underestimated [35]. Corticosteroids – of which dexamethasone is most commonly used to treat brain tumors – may cause mood disturbances, psychosis, and cognitive deficits particularly in declarative memory performance. Steroid dementia is a reversible cause of cognitive deficits even in the absence of psychosis. Recent data suggest that the cognitive deficits are due to neurotoxic effects on both the hippocampal and the prefrontal areas [141]. Both short-term and long-term use of steroids has been associated with cognitive deficits [59]. More likely, cognitive deficits in brain tumor patients will be alleviated by steroids owing to the resolution of brain edema. Antipsychotics, AEDs, and antidepressants can be used to normalize mood changes associated with corticosteroids. Moreover, corticosteroid-induced mood and cognitive deficits are reversible with dose reduction or discontinuation of treatment [13].

Neurocognitive mapping

Contrary to the widely-used sensomotor and language mapping, interference with cognitive and sensory functions has only been demonstrated rarely in surgical brain mapping using electrostimulation. Cognitive tasks have not been systematically validated for routine clinical use. Albeit, many interesting observations, usually in small numbers of patients, have been reported. A selection of these, not necessarily restricted to cognitive tasks, will be reviewed here. Electrostimulation has induced experiential responses such as complex somatosensory and vestibular responses creating an out-of-body sensory illusion elicited from the right angular gyrus and superior temporal gyrus [9, 130] and evoked memories elicited from the temporal gyri [90]. Primary sensory responses were also induced by electrostimulation. For instance, in order to preserve central vision and visual fields, visual evoked potentials and awake mapping of the visual cortex inducing photic phenomena have been used to determine the margins of occipital corticectomy [20, 21, 30]. Also, interference with visual search has been observed during electrostimulation of the right superior temporal gyrus [39]. Furthermore, electrical stimulation of the same region has also induced unilateral and bilateral hearing suppression and deficit in the auditory discrimination of motion [28, 34, 117]. Crossmodal integration inference sites were localized in the dorsolateral prefrontal cortex by electrostimulation using a visual-auditory congruency task [102].

Cortical stimulation of the right inferior parietal lobule and the caudal part of the superior temporal gyrus and subcortical stimulation at the level of the superior longitudinal fascicle interfered with spatial awareness during a line bisection task [6, 129].

Electrostimulation using depth electrodes that were situated in the hippocampus has induced specific memory deficits [16], such that stimulation of the hippocampus on the dominant side induced word recognition interference, whereas stimulation on the non-dominant side induced face recognition interference. Intraoperative mapping of the dominant frontal premotor area and anterior temporal lobe has identified specific areas involved in famous face recognition [41]. Short term memory errors were observed by intra- and extraoperative stimulation of the left temporal neocortex [98]. The hippocampus has also been cooled by rinsing with cold saline intraoperatively to evaluate memory and learning performance and to determine the risk of postoperative memory disorder [67, 68].

Variations on the picture naming task for language assessment have been used for other category specific naming evaluation such as for color naming that identified sites in the dominant frontal operculum, the inferior parietal lobule and the posterior parts of the temporal gyri [109] and for naming of living objects with specific sites in the dominant posterior inferior temporal gyrus [104]. After resection of this region naming of living objects was impaired. Alternatively, auditory naming sites were identified in the dominant temporal gyri, sometimes equivalent to visual picture naming sites, but often inequivalent [46–48]. Postoperative naming decline correlated with removal of auditory naming sites in these studies. Furthermore, areas involved in reading have been identified in various cortical regions, including the lateral pre- and postcentral gyri, the inferior parietal lobule, the frontal operculum and the posterior part of the middle temporal gyrus [75, 110], as well as areas involved in writing in the dominant frontal gyri and angular gyrus [72, 107, 113]. Use of a calculation task during cortical electrostimulation has localized interference within the dominant inferior parietal lobe independent from language interference [29, 65, 108].

Proposal for standardized examination of neurocognitive outcome

Cognitive deficits in patients with brain tumors can be caused by the tumor, by tumor-related epilepsy and its treatment (surgery, radiotherapy, antiepileptics, chemotherapy, or corticosteroids), and by psychological distress. More likely, a combination of these factors will contribute to cognitive dysfunction [127].

Broadly speaking, neurocognitive examination in brain tumor patients can be carried out with a number of purposes in mind:

(1) Diagnosis for classification of the neurocognitive deficits.
(2) To direct a specific rehabilitation program or to provide driver's license legislation or professional reintegration.
(3) Treatment outcome evaluation, such as resective surgery or cognitive rehabilitation.

The selection of tests will vary with the purpose of neuropsychological examination. For example, the sensitivity to detect small changes in the level of neurocognitive functioning is a more important for determining treatment outcome of a cognitive rehabilitation, than for diagnostic classification purposes.

As cognitive function is recognized as an important outcome measure in clinical trials in glioma patients, this provides an opportunity to gather information on cognitive status in a standardized manner. These series cover the different cognitive domains - such as memory, attention, orientation, language abilities, and executive function, representing functions of both the dominant and the non-dominant hemisphere. However, a complete assessment is time consuming and may fatigue patients with brain tumors, thereby biasing results. Moreover, the reduced compliance of both patients and investigators as a consequence of these time-consuming procedures renders the test results not representative for the study population. Less time consuming alternatives such as IQ measurement or Mini-mental State Examination (MMSE) are less sensitive and less valid for adults with brain tumors. Therefore, the MMSE may underestimate the proportion of patients with actual cognitive decline and important though small changes in cognition can be missed. On the other hand, the MMSE appears to be sensitive enough to detect cognitive deficits associated with tumor progression [14].

Depending on the purpose of testing, background and baseline information is required prior to cognitive assessment to enable the neuropsychologist to maximize the opportunity for collecting useful data. These are respectively:

- The patient's demographic variables (e.g., age, handedness, education/qualifications, current/previous profession, cultural background), in order to set the context for the interpretation of current test performance.
- The patient's previous medical and psychiatric history as well as the current treatment and medication.
- The results of previous diagnostic investigations (e.g., neurological examination, EEG, CT/ MRI or functional imaging)
- The results of previous neuropsychological examinations - these can guide the selection of tests to allow for evaluation of change.
- Hetero-anamnestic perspectives on the patient, apparent current and previous deficits - often patients with brain tumors have little insight into the purposes of neuropsychological examination, and into the nature and/or extent of their cognitive deficits.

Several alternatives to formal neurocognitive examination have been attempted. These include self-reports of cognitive function, which can be reliable, but are not necessarily valid because of the lack of introspection. Furthermore, outcome scores of self-reports seem to be related to mood state rather than to neurocognitive performance [19]. For outpatients, reports of cognitive changes made by the partner or a proxy offer a potential alternative to formal cognitive examination.

Because of the multifactorial processes involved with usually a combination of cortical and subcortical lesions, epilepsy, surgery, radiotherapy, AEDs, corticosteroids, and psychological distress in an individual patient, it would be worth selecting a standardized neuropsychological examination covering a wide range of neurocognitive functions. Such a test battery has to meet the following criteria: (i) coverage of several domains with sufficient sensitivity to detect tumor and treatment effects; (ii) standardized multilingual materials and administration procedures; (iii) based on published normative data; (iv) moderate to high test-retest reliability and insensitivity to practice effects to be able to monitor changes in neurocognitive function over time; (v) availability of alternative forms ; and (vi) an administration time not exceeding 30-40 minutes [85]. The neurocognitive domains deemed essential to be evaluated include attention, executive functions, verbal memory, and motor speed.

One standardized neuropsychological examination that meets these criteria is currently in use in a number of EORTC, NCCTG, NCI-C, RTOG, and MRC multisite clinical trials (Table 1). This battery [85] has been shown to be quick and easy to administer by nonphysicians with very good compliance and motivation of patients. Evidently, local modifications of this battery can be made by adding tests depending on the goal of the neuropsychological assessment. Data that can thus be gathered both for clinical and research purposes enabling comparisons over patient groups and/or treatment regimens.

Cognitive rehabilitation

Cognitive rehabilitation refers to a set of interventions that aim to improve a person's ability to perform cognitive tasks by retraining previously learned skills and/or by teaching compensatory strategies. Common interventions for improvements in attention, memory, and executive function, as well as comprehensive programs, which combine neuropsychological and pharmacological treatment modalities suggest to be effective in patients with brain tumors [38]. Further research, however, is still needed to identify the patient and treatment factors that contribute to successful outcome, to explicate the theoretical models underlying the interventions, and to identify the extent of the clinical significance of these interventions. So far, cognitive rehabilitation interventions are a promising treatment that may contribute to improve cognitive outcome and quality of life of patients with resective surgery of brain tumors.

Conclusion

Next to neurological functioning, cognitive functioning of brain tumor patients is an important outcome measure, because cognitive impairments can have a large impact on everyday-life functioning, social functioning, and professional functioning of these patients, and thus on their HRQOL. Many factors contribute to cognitive outcome, such as direct and indirect tumor effects, seizures, medication, and oncological treatment. Both cognitive improvement and decline have been observed after resective brain surgery, depending on pathology, lesion size, localization and laterality. However, neurocognitive outcome prior and following brain tumor resection has not been systematically determined, although a feasible, quantitative assessment procedure is available and suggested in the present chapter. Intrasurgical neurocognitive mapping procedures to improve cognitive outcome also have not been systematically applied in these patients. Concerted action into studying the costs and benefits of presurgical, intrasurgical, and postsurgical cognitive assessments related to outcome of these patients is thus warranted.

Table 1: Core neurocognitive testing battery

Test	Domain measured	Outcome
Trail Making Test A	Visual scanning speed	Number of seconds to complete (0–300)
Trail Making Test B	Divided attention	Number of seconds to complete (0–300)
Controlled Oral Word Association	Verbal fluency	Age and sex-adjusted raw score (range 0–no upper limit)
Hopkins Verbal Learning Test	Verbal memory	Immediate memory of word list rehearsed three times (maximum score = 36). After 20–30 min delay, number of words correctly recalled (maximum score = 12). Recognition is number of words recognized from a longer list (maximum score = 12).
Digit Symbol Subtest of the WAIS-III	Psychomotor speed	Age-corrected subtest score (0–20).
Grooved Pegboard Test	Fine motor control for dominant and non dominant hands	Number of seconds to complete (0–300)

References

[1] Ackermann H, Daum I, Schugens MM et al (1996) Impaired procedural learning after damage to the left supplementary motor area (SMA). J Neurol Neurosurg Psychiatry 60: 94–97.

[2] Agner C, Dujovny M, Gaviria M (2002) Neurocognitive assessment before and after cranioplasty. Acta Neurochir 144: 1033–1040; discussion 1040

[3] Alivisatos B (1992) The role of the frontal cortex in the use of advance information in a mental rotation paradigm. Neuropsychol 30: 145–159

[4] Anderson SI, Taylor R, Whittle IR (1999) Mood disorders in patients after treatment for primary intracranial tumours. Br J Neurosurg 13: 480–485

[5] Andrewes DG, Kaye A, Murphy M et al (2003) Emotional and social dysfunction in patients following surgical treatment for brain tumour. J Clin Neurosci 10: 428–433

[6] Bartolomeo P, Thiebaut de Schotten M, Duffau H (2007) Mapping of visuospatial functions during brain surgery: a new tool to prevent unilateral spatial neglect. Neurosurgery 61: E1340

[7] Baxendale S, Thompson PJ, Duncan JS (2008) Improvements in memory function following anterior temporal lobe resection for epilepsy. Neurology 71: 1319–1325

[8] Béhin A, Delattre JY (2003) Neurologic sequelae of radiotherapy of the nervous system, in Schiff D, Wen PY (eds): Cancer neurology in clinical practice. Humana Press, Totowa, pp 173–192

[9] Blanke O, Ortigue S, Landis T et al (2002) Stimulating illusory own-body perceptions. Nature 419: 269–270

[10] Bosma I, Vos MJ, Heimans JJ et al (2007) The course of neurocognitive functioning in high-grade glioma patients. Neurooncol 9: 53–62

[11] Braun CM, Denault C, Cohen H et al (1994) Discrimination of facial identity and facial affect by temporal and frontal lobectomy patients. Brain Cogn 24: 198–212

[12] Braun V, Albrecht A, Kretschmer T et al (2006) Brain tumour surgery in the vicinity of short-term memory representation - results of neuronavigation using fMRI images. Acta Neurochir 148: 733–739

[13] Brown ES (2009) Effects of glucocorticoids on mood, memory, and the hippocampus. Treatment and preventive therapy. Ann NY Acad Sci 1179: 41–55

[14] Brown PD, Ballman KV, Rummans TA et al (2006) Prospective study of quality of life in adults with newly diagnosed high-grade gliomas. J Neurooncol 76: 283–291

[15] Brown PD, Maurer MJ, Rummans TA et al (2005) A prospective study of quality of life in adults with newly diagnosed high-grade gliomas: the impact of the extent of resection on quality of life and survival. Neurosurgery 57: 495–504; discussion 495–504

[16] Coleshill SG, Binnie CD, Morris RG et al (2004) Material-specific recognition memory deficits elicited by unilateral hippocampal electrical stimulation. J Neurosci 24: 1612–1616

[17] Correa DD, DeAngelis LM, Shi W et al (2007) Cognitive functions in low-grade gliomas: disease and treatment effects. J Neurooncol 81: 175–184

[18] Crossen JR, Garwood D, Glatstein E et al (1994) Neurobehavioral sequelae of cranial irradiation in adults: a review of radiation-induced encephalopathy. J Clin Oncol 12: 627–642

[19] Cull A, Hay C, Love SB et al (1996) What do cancer patients mean when they complain of concentration and memory problems? Br J Cancer 74: 1674–1679

[20] Curatolo JM, Macdonell RA, Berkovic SF et al (2000) Intraoperative monitoring to preserve central visual fields during occipital corticectomy for epilepsy. J Clin Neurosci 7: 234–237

[21] Danks RA, Aglio LS, Gugino LD et al (2000) Craniotomy under local anesthesia and monitored conscious sedation for the resection of tumors involving eloquent cortex. J Neurooncol 49: 131–139

[22] Davies KG, Bell BD, Bush AJ et al (1998) Prediction of verbal memory loss in individuals after anterior temporal lobectomy. Epilepsia 39: 820–828

[23] Dijkstra M, van Nieuwenhuizen D, Stalpers LJ et al (2008) Late neurocognitive sequelae in WHO grade I meningioma patients. J Neurol Neurosurg Psychiatry 80: 910–915

[24] Dinapoli L, Maschio M, Jandolo B et al (2009) Quality of life and seizure control in patients with brain tumor-related epilepsy treated with levetiracetam monotherapy: preliminary data of an open-label study. Neurol Sci [Epup ahead of print]

[25] Dodrill CB (2002) Progressive cognitive decline in adolescents and adults with epilepsy. Progr Brain Res 135: 399–407

[26] Douw L, Klein M, Fagel SS et al (2009) Cognitive and radiological effects of radiotherapy in patients with low-grade glioma: long-term follow-up. Lancet Neurol 8: 810–818

[27] Drane LD, Meador KJ (2002) Cognitive and behavioral effects of antiepileptic drugs. Epilepsy Behav 3: 49–53

[28] Ducommun CY, Michel CM, Clarke S et al (2004) Cortical motion deafness. Neuron 43: 765–777

[29] Duffau H, Denvil D, Lopes M et al (2002) Intraoperative mapping of the cortical areas involved in multiplication and subtraction: an electrostimulation study in a patient with a left parietal glioma. J Neurol Neurosurg Psychiatry 73: 733–738

[30] Duffau H, Velut S, Mitchell MC et al (2004) Intra-operative mapping of the subcortical visual pathways using direct electrical stimulations. Acta Neurochir 146: 265–269; discussion 269–270

[31] Dulay MF, Levin HS, York MK et al (2009) Predictors of individual visual memory decline after unilateral anterior temporal lobe resection. Neurol 72: 1837–1842

[32] Efficace F, Bottomley A (2002) Health related quality of life assessment methodology and reported outcomes in randomised controlled trials of primary brain cancer patients. Eur J Cancer 38: 1824–1831
[33] Feigenbaum JD, Polkey CE, Morris RG (1996) Deficits in spatial working memory after unilateral temporal lobectomy in man. Neuropsychologia 34: 163–176
[34] Fenoy AJ, Severson MA, Volkov IO et al (2006) Hearing suppression induced by electrical stimulation of human auditory cortex. Brain Res 1118: 75–83
[35] Fietta P, Fietta P, Delsante G (2009) Central nervous system effects of natural and synthetic glucocorticoids. Psychiatry Clin Neurosci 63: 613–622
[36] Friedman MA, Meyers CA, Sawaya R (2003) Neuropsychological effects of third ventricle tumor surgery. Neurosurg 52: 791–798; discussion 798
[37] Gazzeri R, Galarza M, Gazzeri G (2008) Giant olfactory groove meningioma: ophthalmological and cognitive outcome after bifrontal microsurgical approach. Acta Neurochir 150: 1117–1125; discussion 1126
[38] Gehring K, Sitskoorn MM, Aaronson NK et al (2008) Interventions for cognitive deficits in adults with brain tumours. Lancet Neurol 7: 548–560
[39] Gharabaghi A, Fruhmann Berger M, Tatagiba M et al (2006) The role of the right superior temporal gyrus in visual search-insights from intraoperative electrical stimulation. Neuropsychologia 44: 2578–2581
[40] Giovagnoli AR, Casazza M, Ciceri E et al (2007) Preserved memory in temporal lobe epilepsy patients after surgery for low-grade tumour. A pilot study. Neurol Sci 28: 251–258
[41] Giussani C, Roux FE, Bello L et al (2009) Who is who: areas of the brain associated with recognizing and naming famous faces. J Neurosurg 110: 289–299
[42] Gleissner U, Helmstaedter C, Schramm J et al (2004) Memory outcome after selective amygdalohippocampectomy in patients with temporal lobe epilepsy: one-year follow-up. Epilepsia 45: 960–962
[43] Goldstein B, Armstrong CL, John C et al (2003) Attention in adult intracranial tumors patients. J Clin Exp Neuropsychol 25: 66–78
[44] Goldstein LH, Bernard S, Fenwick PB et al (1993) Unilateral frontal lobectomy can produce strategy application disorder. J Neurol Neurosurg Psychiatry 56: 274–276
[45] Goldstein LH, Polkey CE (1993) Short-term cognitive changes after unilateral temporal lobectomy or unilateral amygdalo-hippocampectomy for the relief of temporal lobe epilepsy. J Neurol Neurosurg Psychiatry 56: 135–140
[46] Hamberger MJ, Goodman RR, Perrine K et al (2001) Anatomic dissociation of auditory and visual naming in the lateral temporal cortex. Neurology 56: 56–61
[47] Hamberger MJ, McClelland S 3rd, McKhann GM 2nd et al (2007) Distribution of auditory and visual naming sites in nonlesional temporal lobe epilepsy patients and patients with space-occupying temporal lobe lesions. Epilepsia 48: 531–538
[48] Hamberger MJ, Seidel WT, McKhann GM 2nd et al (2005) Brain stimulation reveals critical auditory naming cortex. Brain 128: 2742–2749
[49] Helmstaedter C, Gleibner U, Zentner J et al (1998) Neuropsychological consequences of epilepsy surgery in frontal lobe epilepsy. Neuropsychologia 36: 333–341
[50] Helmstaedter C, Richter S, Roske S et al (2008) Differential effects of temporal pole resection with amygdalohippocampectomy versus selective amygdalohippocampectomy on material-specific memory in patients with mesial temporal lobe epilepsy. Epilepsia 49: 88–97
[51] Hornak J, Bramham J, Rolls ET et al (2003) Changes in emotion after circumscribed surgical lesions of the orbitofrontal and cingulate cortices. Brain 126: 1691–1712
[52] Hornak J, O'Doherty J, Bramham J et al (2004) Reward-related reversal learning after surgical excisions in orbito-frontal or dorsolateral prefrontal cortex in humans. J Cogn Neurosci 16: 463–478
[53] Incisa della Rocchetta A, Gadian DG, Connelly A et al (1995) Verbal memory impairment after right temporal lobe surgery: role of contralateral damage as revealed by 1H magnetic resonance spectroscopy and T2 relaxometry. Neurology 45: 797–802
[54] Irle E, Peper M, Wowra B et al (1994) Mood changes after surgery for tumors of the cerebral cortex. Arch Neurol 51: 164–174
[55] Jagaroo V, Rogers MP, Black PM (2000) Allocentric visuospatial processing in patients with cerebral gliomas: a neurocognitive assessment. J Neurooncol 49: 235–248
[56] Jones-Gotman M, Zatorre RJ (1988) Olfactory identification deficits in patients with focal cerebral excision. Neuropsychologia 26: 387–400
[57] Jones-Gotman M, Zatorre RJ, Olivier A et al (1997) Learning and retention of words and designs following excision from medial or lateral temporal-lobe structures. Neuropsychologia 35: 963–973
[58] Joo EY, Han HJ, Lee EK et al (2005) Resection extent versus postoperative outcomes of seizure and memory in mesial temporal lobe epilepsy. Seizure 14: 541–551
[59] Keenan PA, Jacobson MW, Soleymani RM et al (1996) The effect on memory of chronic prednisone treatment in patients with systemic disease. Neurology 47: 1396–1402
[60] Kessels RP, Postma A, Kappelle LJ et al (2000) Spatial memory impairment in patients after tumour resection: evidence for a double dissociation. J Neurol Neurosurg Psychiatry 69: 389–391
[61] Klein M, Engelberts NHJ, Van der Ploeg HM et al (2003) Epilepsy in low-grade gliomas: the impact on cognitive functioning and quality of life. Ann Neurol 54: 514–520
[62] Klein M, Heimans JJ, Aaronson NK et al (2002) Effect of radiotherapy and other treatment-related factors on

mid-term to long-term cognitive sequelae in low-grade gliomas: a comparative study. Lancet 360: 1361–1368

[63] Koski L, Petrides M (2002) Distractibility after unilateral resections from the frontal and anterior cingulate cortex in humans. Neuropsychologia 40: 1059–1072

[64] Koski LM, Paus T, Petrides M (1998) Directed attention after unilateral frontal excisions in humans. Neuropsychologia 36: 1363–1371

[65] Kurimoto M, Asahi T, Shibata T et al (2006) Safe removal of glioblastoma near the angular gyrus by awake surgery preserving calculation ability-case report. Neurol Med-Chir 46: 46–50

[66] Lambrey S, Amorim MA, Samson S et al (2008) Distinct visual perspective-taking strategies involve the left and right medial temporal lobe structures differently. Brain 131: 523–534

[67] Lee GP, Loring DW, Smith JR et al (1995) Intraoperative hippocampal cooling and Wada memory testing in the evaluation of amnesia risk following anterior temporal lobectomy. Arch Neurol 52: 857–861

[68] Lee GP, Smith JR, Loring DW et al (1995) Intraoperative thermal inactivation of the hippocampus in an effort to prevent global amnesia after temporal lobectomy. Epilepsia 36: 892–898

[69] Leimkuhler ME, Mesulam MM (1985) Reversible go-no go deficits in a case of frontal lobe tumor. Ann Neurol 18: 617–619

[70] Lim DA, Tarapore P, Chang E et al (2009) Safety and feasibility of switching from phenytoin to levetiracetam monotherapy for glioma-related seizure control following craniotomy: a randomized phase II pilot study. J Neurooncol 93: 349–354

[71] Litofsky NS, Resnick AG (2009) The relationships between depression and brain tumors. J Neurooncol 94: 153–161

[72] Lubrano V, Roux FE, Demonet JF (2004) Writing-specific sites in frontal areas: a cortical stimulation study. J Neurosurg 101: 787–798

[73] Luerding R, Boesebeck F, Ebner A (2004) Cognitive changes after epilepsy surgery in the posterior cortex. J Neurol, Neurosurg Psychiatry 75: 583–587

[74] Lutz MT, Clusmann H, Elger CE et al (2004) Neuropsychological outcome after selective amygdalohippocampectomy with transsylvian versus transcortical approach: a randomized prospective clinical trial of surgery for temporal lobe epilepsy. Epilepsia 45: 809–816

[75] Mani J, Diehl B, Piao Z et al (2008) Evidence for a basal temporal visual language center: cortical stimulation producing pure alexia. Neurol 71: 1621–1627

[76] Martin RC, Kretzmer T, Palmer C et al (2002) Risk to verbal memory following anterior temporal lobectomy in patients with severe left-sided hippocampal sclerosis. Arch Neurol 59: 1895–1901

[77] Martin RC, Sawrie SM, Edwards R et al (2000) Investigation of executive function change following anterior temporal lobectomy: selective normalization of verbal fluency. Neuropsychology 14: 501–508

[78] Maschio M, Dinapoli L, Vidiri A et al (2009) The role side effects play in the choice of antiepileptic therapy in brain tumor-related epilepsy: a comparative study on traditional antiepileptic drugs versus oxcarbazepine. J Exp Clin Cancer Res 28: 60

[79] Maschio M, Dinapoli L, Zarabia A et al (2006) Issues related to the pharmacological management of patients with brain tumours and epilepsy. Funct Neurol 21: 15–19

[80] McMackin D, Cockburn J, Anslow P et al (1995) Correlation of fornix damage with memory impairment in six cases of colloid cyst removal. Acta Neurochir 135: 12–18

[81] Meador KJ (2006) Cognitive and memory effects of the new antiepileptic drugs. Epilepsy Res 68: 63–67

[82] Meador KJ (2002) Cognitive outcomes and predictive factors in epilepsy. Neurology 58: 21–26

[83] Meador KJ, Gevins A, Loring DW et al (2007) Neuropsychological and neurophysiologic effects of carbamazepine and levetiracetam. Neurology 69: 2076–2084

[84] Meyers CA, Berman SA, Hayman A et al (1992) Pathological left-handedness and preserved function associated with a slowly evolving brain tumor. Dev Med Child Neurol 34: 1110–1116

[85] Meyers CA, Smith JA, Bezjak A et al (2004) Neurocognitive function and progression in patients with brain metastases treated with whole-brain radiation and motexafin gadolinium: results of a randomized phase III trial. J Clin Oncol 22: 157–165

[86] Miller LA (1992) Impulsivity, risk-taking, and the ability to synthesize fragmented information after frontal lobectomy. Neuropsychologia 30: 69–79

[87] Miotto EC, Morris RG (1998) Virtual planning in patients with frontal lobe lesions. Cortex 34: 639–657

[88] Mitchell AJ, Kemp S, Benito-Leon J et al (2010) The influence of cognitive impairment on health-related quality of life in neurological disease. Acta Neuropsychiatr 22: 2–13

[89] Monje ML, Palmer T (2003) Radiation injury and neurogenesis. Curr Opin Neurol 16: 129–134

[90] Moriarity JL, Boatman D, Krauss GL et al (2001) Human «memories» can be evoked by stimulation of the lateral temporal cortex after ipsilateral medial temporal lobe resection. J Neurol, Neurosurg Psychiatry 71: 549–551

[91] Morino M, Ichinose T, Uda T et al (2009) Memory outcome following transsylvian selective amygdalohippocampectomy in 62 patients with hippocampal sclerosis. J Neurosurg 110: 1164–1169

[92] Morino M, Uda T, Naito K et al (2006) Comparison of neuropsychological outcomes after selective amygdalohippocampectomy versus anterior temporal lobectomy. Epilepsy Behav 9: 95–100

[93] Nunn JA, Graydon FJ, Polkey CE et al (1999) Differential spatial memory impairment after right temporal lobectomy demonstrated using temporal titration. Brain 122: 47–59

[94] Oberndorfer S, Piribauer M, Marosi C et al (2005) P450 enzyme inducing and non-enzyme inducing antiepilep-

tics in glioblastoma patients treated with standard chemotherapy. J Neurooncol 72: 255–260
[95] Owen AM, Downes JJ, Sahakian BJ et al (1990) Planning and spatial working memory following frontal lobe lesions in man. Neuropsychologia 28: 1021–1034
[96] Owen AM, Morris RG, Sahakian BJ et al (1996) Double dissociations of memory and executive functions in working memory tasks following frontal lobe excisions, temporal lobe excisions or amygdalo-hippocampectomy in man. Brain 119: 1597–1615
[97] Peper M, Irle E (1997) Categorical and dimensional decoding of emotional intonations in patients with focal brain lesions. Brain Lang 58: 233–264
[98] Perrine K, Devinsky O, Uysal S et al (1994) Left temporal neocortex mediation of verbal memory: evidence from functional mapping with cortical stimulation. Neurology 44: 1845–1850
[99] Petrides M (1997) Visuo-motor conditional associative learning after frontal and temporal lesions in the human brain. Neuropsychologia 35: 989–997
[100] Petrucci RJ, Buchheit WA, Woodruff GC et al (1987) Transcallosal parafornicial approach for third ventricle tumors: neuropsychological consequences. Neurosurgery 20: 457–464
[101] Pigott S, Milner B (1993) Memory for different aspects of complex visual scenes after unilateral temporal- or frontal-lobe resection. Neuropsychologia 31: 1–15
[102] Plaza M, Gatignol P, Cohen H et al (2008) A discrete area within the left dorsolateral prefrontal cortex involved in visual-verbal incongruence judgment. Cereb Cortex 18: 1253–1259
[103] Postma TJ, Klein M, Verstappen CC et al (2002) Radiotherapy-induced cerebral abnormalities in patients with low-grade glioma. Neurology 59: 121–123
[104] Pouratian N, Bookheimer SY, Rubino G et al (2003) Category-specific naming deficit identified by intraoperative stimulation mapping and postoperative neuropsychological testing. Case report. J Neurosurg 99: 170–176
[105] Rausch R, Kraemer S, Pietras CJ et al (2003) Early and late cognitive changes following temporal lobe surgery for epilepsy. Neurology 60: 951–959
[106] Reijneveld JC, Sitskoorn MM, Klein M et al (2001) Cognitive status and quality of life in patients with suspected versus proven low-grade gliomas. Neurology 56: 618–623
[107] Roux FE, Boetto S, Sacko O et al (2003) Writing, calculating, and finger recognition in the region of the angular gyrus: a cortical stimulation study of Gerstmann syndrome. J Neurosurg 99: 716–727
[108] Roux FE, Boukhatem L, Draper L et al (2009) Cortical calculation localization using electrostimulation. J Neurosurg 110: 1291–1299
[109] Roux FE, Lubrano V, Lauwers-Cances V et al (2006) Category-specific cortical mapping: color-naming areas. J Neurosurg 104: 27–37
[110] Roux FE, Lubrano V, Lauwers-Cances V et al (2004) Intra-operative mapping of cortical areas involved in reading in mono- and bilingual patients. Brain 127: 1796–1810
[111] Rowe AD, Bullock PR, Polkey CE et al (2001) "Theory of mind" impairments and their relationship to executive functioning following frontal lobe excisions. Brain 124: 600–616
[112] Russell SM, Golfinos JG (2003) Amusia following resection of a Heschl gyrus glioma. Case report. J Neurosurg 98: 1109–1112
[113] Scarone P, Gatignol P, Guillaume S et al (2009) Agraphia after awake surgery for brain tumor: new insights into the anatomo-functional network of writing. Surg Neurol 72: 223–241; discussion 241
[114] Schramm J (2008) Temporal lobe epilepsy surgery and the quest for optimal extent of resection: a review. Epilepsia 49: 1296–1307
[115] Selwa LM, Berent S, Giordani B et al (1994) Serial cognitive testing in temporal lobe epilepsy: longitudinal changes with medical and surgical therapies. Epilepsia 35: 743–749
[116] Shin MS, Lee S, Seol SH et al (2009) Changes in neuropsychological functioning following temporal lobectomy in patients with temporal lobe epilepsy. Neurol Res 31: 692–701
[117] Sinha SR, Crone NE, Fotta R et al (2005) Transient unilateral hearing loss induced by electrocortical stimulation. Neurology 64: 383–385
[118] Spiers HJ, Burgess N, Maguire EA et al (2001) Unilateral temporal lobectomy patients show lateralized topographical and episodic memory deficits in a virtual town. Brain 124: 2476–2489
[119] Stewart AL, Ware JE (eds) (1992) Measuring functioning and well-being: The medical outcomes study approach. Duke University Press, Durham, NC
[120] Stroup E, Langfitt J, Berg M et al (2003) Predicting verbal memory decline following anterior temporal lobectomy (ATL). Neurology 60: 1266–1273
[121] Struik K, Klein M, Heimans JJ et al (2009) Fatigue in low-grade glioma. J Neurooncol 92: 73–78
[122] Surma-aho O, Niemela M, Vilkki J et al (2001) Adverse long-term effects of brain radiotherapy in adult low-grade glioma patients. Neurology 56: 1285–1290
[123] Swain SA, Polkey CE, Bullock P et al (1998) Recognition memory and memory for order in script-based stories following frontal lobe excisions. Cortex 34: 25–45
[124] Takaya S, Mikuni N, Mitsueda T et al (2009) Improved cerebral function in mesial temporal lobe epilepsy after subtemporal amygdalohippocampectomy. Brain 132: 185–194
[125] Tanaka Y, Miyazawa Y, Akaoka F et al (1997) Amnesia following damage to the mammillary bodies. Neurology 48: 160–165

[126] Taphoorn MJB, Heimans JJ, Snoek FJ et al (1992) Assessment of quality of life in patients treated for low-grade glioma: a preliminary report. J Neurol Neurosurg Psychiatry 55: 372–376
[127] Taphoorn MJB, Klein M (2004) Cognitive deficits in adult patients with brain tumours. Lancet Neurol 3: 159–168
[128] Teixidor P, Gatignol P, Leroy M et al (2007) Assessment of verbal working memory before and after surgery for low-grade glioma. J Neurooncol 81: 305–313
[129] Thiebaut de Schotten M, Urbanski M, Duffau H et al (2005) Direct evidence for a parietal-frontal pathway subserving spatial awareness in humans. Science 309: 2226–2228
[130] Tong F (2003) Out-of-body experiences: from Penfield to present. Trends Cogn Sci 7: 104–106
[131] Tucha O, Smely C, Preier M et al (2003) Preoperative and postoperative cognitive functioning in patients with frontal meningiomas. J Neurosurg 98: 21–31
[132] van Breemen MS, Wilms EB, Vecht CJ (2007) Epilepsy in patients with brain tumours: epidemiology, mechanisms, and management. Lancet Neurol 6: 421–430
[133] van Nieuwenhuizen D, Klein M, Stalpers LJ et al (2007) Differential effect of surgery and radiotherapy on neurocognitive functioning and health-related quality of life in WHO grade I meningioma patients. J Neurooncol 84: 271–278
[134] Vendrell P, Junque C, Pujol J et al (1995) The role of prefrontal regions in the Stroop task. Neuropsychologia 33: 341–352
[135] Vilkki J (1992) Cognitive flexibility and mental programming after closed head injuries and anterior or posterior cerebral excisions. Neuropsychologia 30: 807–814
[136] Vilkki J, Levanen S, Servo A (2002) Interference in dual-fluency tasks after anterior and posterior cerebral lesions. Neuropsychologia 40: 340–348
[137] Wachi M, Tomikawa M, Fukuda M et al (2001) Neuropsychological changes after surgical treatment for temporal lobe epilepsy. Epilepsia 42 Suppl 6: 4–8
[138] Weniger G, Irle E (2002) Impaired facial affect recognition and emotional changes in subjects with transmodal cortical lesions. Cereb Cortex 12: 258–268
[139] Wester K, Hugdahl K (1995) Arachnoid cysts of the left temporal fossa: impaired preoperative cognition and postoperative improvement. J Neurol Neurosurg Psychiatry 59: 293–298
[140] Wolf RL, Ivnik RJ, Hirschorn KA et al (1993) Neurocognitive efficiency following left temporal lobectomy: standard versus limited resection. J Neurosurg 79: 76–83
[141] Wolkowitz OM, Lupien SJ, Bigler E et al (2004) The "steroid dementia syndrome": an unrecognized complication of glucocorticoid treatment. Ann NY Acad Sci 1032: 191–194

Functional neuroimaging in neurosurgical practice

Geert-Jan M. Rutten and Nick. F. Ramsey

Present role of fMRI in surgical practice

Historical premises

As of today, most clinicians are still being trained in a fairly localizationalist view regarding the functional anatomy of the brain. This view originated in dissection studies of patients with brain lesions at the end of the 19th century by researchers and clinicians like Gall, Lichtheim, Broca, and Wernicke. They in fact launched the era of the "diagram makers" that claimed that brain functions could be mapped out in detail anatomically and that sensorimotor and cognitive functions could be damaged independently of each other. The resulting so-called eloquent areas, which were formulated in these models, are still generally considered no-go areas in neurosurgery because of the presumed high risk of severe and permanent neurological deficits. At that time it had already been noted (initially by Wernicke) that neurological dysfunction could result from both cortical and subcortical damage [12, 29].

It is understandable that the classical view is still dominant in neurological and neurosurgical practice because this model provided – for the first time – a theoretical framework that could explain some of the neurological syndromes that had been discovered (e.g., hemiparesis, alexia or transcortical aphasia). The model is also intuitively appealing, as it states that if an area is damaged and this leads to neurological dysfunction, then this area must be critically involved in that particular function. Although there are several alternative and more recent models, it is important to note that none of these models have sufficient predictive value in clinical practice [24]. For the location of primary sensorimotor functions, the old model is fairly accurate when it is judged on clinical outcome; for cognitive functions, it is not. This can to a large extent be explained by the fact that the primary cortex has specific anatomical characteristics and a direct relationship with large subcortical fiber bundles, restricting variability and plasticity. However, even for these primary areas substantial anatomical and functional variations have been described that makes a priori localization of function on anatomical characteristics not always reliable [2, 11, 21, 62, 78, 102]. Because of the inherent interindividual and pathology-driven variability of brain areas and their interconnections, functional mapping techniques are necessary to identify each individual's critical epicenters to optimize surgical treatment [22]. In this chapter we focus on functional magnetic resonance imaging (fMRI) in the neurosurgical context. Other techniques that are also relevant, like mapping of connections with diffusion tensor imaging (DTI) and DTI-based fiber tracking, are discussed in the chapters by Bello et al and Catani and Dell'Acqua.

Properties and features of fMRI relevant for neurosurgery

Clinical use of fMRI requires validated and standardized protocols. For routine use in the clinic, the fMRI acquisition needs to be easy to perform and analyze by radiological personnel. Ultimately, analyses have to be performed with easy-to-use software, and interpretation should not require a dedicated expert. From a technical point of view, these requirements are already feasible: most scanners have software for (real-time) automatic analysis and display of results during, or immediately after, scanning. Brain activation maps can be entered into surgical guidance systems for functional neuronavigation [79]. The fact that these automated software programs are available (either commercially or as freeware) does not imply that the resulting maps are always a reliable roadmap for surgery or that expert knowledge is no longer needed. Contrary to the suggestion that is sometimes made in the literature or in commercial advertisements, there are presently no standardized and user-independent fMRI protocols that can be easily and reliably used for surgical purpose. Importantly, there are no studies available that have tested the results of commercially available fMRI analyses in comparison to either the standards used in neuroscience or clinical standards such as the Wada test or intraoperative electrical stimulation. Nevertheless, fMRI in neurosurgery has come a long way in centers with access to specialists in the field of cognitive neuroscience. In what follows, we explain the various factors that affect application of fMRI in neurosurgical planning.

Design of the fMRI experiment

The main reason that fMRI is not yet routinely used as a presurgical tool is the fact that fMRI

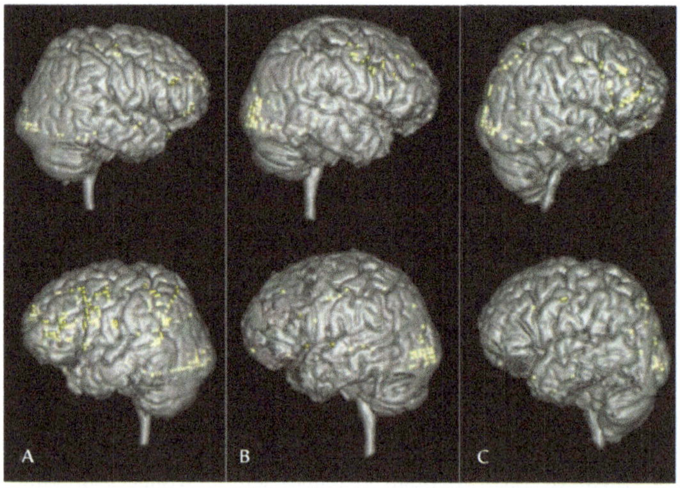

Fig. 1. (A–C) Brain activity (yellow) from the combined analysis of four different language tasks as visualized on a surface rendering of the left and right hemispheres in three patients. A lateralization index was calculated on the basis of the number of voxels in both hemispheres. This index ranges from −100 (all voxels in the right hemisphere) to 100 (all voxels in the left hemisphere) and was respectively 86 (patient A), −13 (patient B), and −64 (patient C). fMRI showed good correlation with the sodium amytal test for left (**A**), bilateral (**B**), and right (**C**) hemispheric language dominance. Such a combined analysis improves the detection power for language-related fMRI activity and yields a better correlation with sodium amytal test results than the use of individual language tasks [80]

studies typically identify more brain regions than existing clinical methods, suggesting that fMRI detects not only areas that are critical for a particular function but also areas that participate in a less critical manner in functional networks [82]. Neurosurgical use of fMRI requires very strict criteria, as both the presence and the absence of areas that harbor critical functions need to be identified with sufficient spatial resolution. Thus, the fMRI experiment has to be constructed so as to extract only the function of interest to the examiner. Most fMRI experiments follow a block design, by which two (or more) conditions are alternated over the course of the scan. Ideally, one condition contains the function of interest, while another (control) condition involves a similar set of functions except for the one of interest. Experiments that use subtraction of conditions are fairly simple to implement, are robust, and have high statistical power. For these reasons they are most often used in clinical practice [1]. However, subtraction of conditions relies on assumptions that are not always valid. One is the idea of "pure insertion", by which it is assumed that a cognitive process can be "added" to a set of existing cognitive processes without affecting them [28]. More complex task designs have been developed to target such methodological pitfalls or to analyze hemodynamic responses to individual stimuli; these designs involve multiple levels of task complexity (parametric design), measurements of single stimulus-related BOLD (blood oxygen level-dependent) responses (event-related design) or multiple task-control conditions (e.g., conjunction analyses; see also Fig. 1) [69, 71]. Although more elaborate experimental designs do indeed improve the correlation of fMRI results with clinical gold standards, the match is still far from perfect [30, 80].

What exactly is measured with fMRI?

Several techniques are available for imaging brain activity, but one in particular is generally used in clinical and cognitive neuroscience.

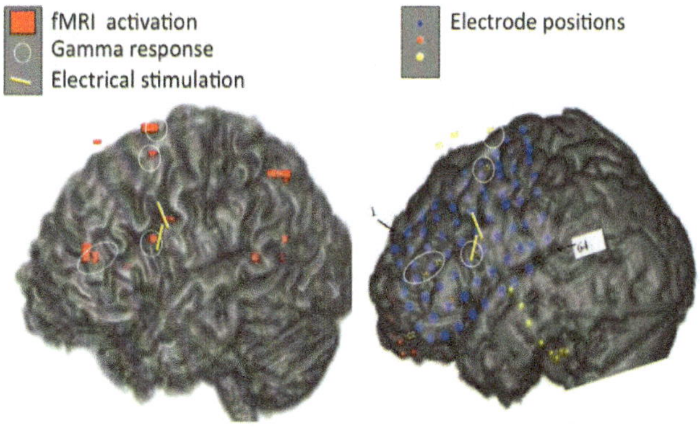

Fig. 2. Convergence of different techniques for mapping brain function. In one epilepsy patient, fMRI scans were acquired before implant of an electrode grid (for seizure localization), during performance of a working memory task (a Sternberg item recognition task [42]). Electrocorticography was conducted with the implant and the same task, to assess which brain areas exhibited a high-frequency (gamma, 65–95 Hz) response to the task. Finally, electrocortical stimulation mapping was performed for working memory. Positive sites were those where stimulation disrupted reverse production of three letters (e.g., hearing "s-k-j", replying "j-k-s") but not repetition of three letters (e.g., hearing "l-p-m", replying "l-p-m"). The procedure is detailed in ref. 48. fMRI activation (red squares) is displayed on the left, superimposed on an anatomical scan of the patient. White circles indicate electrodes where a significant gamma response was found. On the right, a rendering of the same anatomical scan is displayed with locations of the electrode grids (obtained with three-dimensional computed tomography) (data from N. Ramsey, F. Leyten, P. Van Rijen, et al, UMC Utrecht, unpubl.)

This technique, called BOLD fMRI, measures hemodynamic changes in the level of oxygenated hemoglobin, blood flow, and blood volume, and this is thought to reflect changes in neural activity. The exact relationship between vascular and neural changes remains unknown, but microelectrode recordings for both animals and humans strongly suggest that the BOLD signal correlates to local field potentials (LFPs). LFPs reflect the input and intracortical processing of a population of neurons rather than the spiking output [56]. In a recent study with microelectrode recordings of patients during epilepsy surgery, a significant correlation was found between an increase in the fMRI signal and an increase in LFPs in the 50–250 Hz range [65]. Several studies have also reported a correlation between fMRI signals and increases in the power spectrum as measured with electrocorticography (ECoG). Correlations are found for different tasks, varying from motor or auditory tasks to cognitive tasks such as working memory and language [14, 89]. An example is shown in Fig. 2 for one patient, comparing presurgical fMRI of a working memory task, with LFP responses to the same task with an implanted electrode grid and electrocortical stimulation during a similar working memory task. Figure 2 shows convergence of these measures (N. Ramsey, F. Leyten, P. Van Rijen, UMC Utrecht, unpubl. data). Increased activity in these higher (gamma) frequencies for cognitive processes is also observed by electro- and magnetoencephalography [37, 43]. The relevance for surgical planning is not yet known; a handful of exploratory studies have been published [103]. As of yet it is not clear whether LFPs match electrical stimulation (virtual lesions): if LFPs indeed correlate closely with fMRI, one can expect LFPs to also detect regions that are involved in a task but are not critical.

An important limitation of the ability to accurately localize brain functions with BOLD fMRI is caused by hemodynamic mechanisms. The signal changes associated with brain activation are dominated by medium to larger sized venous blood vessels (for a discussion, see ref. 100). This has two consequences. First, fMRI activation extends to a larger volume (several voxels or more), downstream along the draining venules and veins, than the parenchymal source of the neurovascular response. This causes fMRI activations to extend beyond the patch of neuronal tissue that is activated. Second, the focus of maximum signal changes is drawn towards the draining veins, causing an error in localization of functional events in the order of at least several millimeters (or centimeters in the case of more extensive regions of activation drained by the same veins). Special adjustments can be made to the BOLD fMRI scan technique to reduce the contribution of blood vessels (e.g., PRESTO [64]), but complete elimination is essentially impossible. Other techniques, such as spin-echo fMRI, yield better accuracy, but do so at the expense of sensitivity.

Spatial accuracy

Precise definition of the activation boundaries of fMRI areas is necessary in order to safely maximize the surgical resection. Many parameters determine the BOLD contrast and the spatial resolution of fMRI images: magnetic field strength, duration of the fMRI session, type of pulse sequence or slice thickness [101]. The eventual choice of parameters always depends on the question that needs to be answered by the fMRI experiment and constitutes a trade-off between these values. High spatial resolution is not necessarily advantageous for studies where a language lateralization index is calculated or where data are normalized and averaged across subjects for groupwise analyses. In these cases, fMRI images are sometimes smoothed to facilitate detection of brain activity at the cost of spatial precision. By electrical stimulation mapping (ESM) it has been shown that language areas can be as small as one voxel (e.g., 4 mm^3) [84]. Smoothing reduces the ability to distinguish between separate but closely positioned active brain areas and might therefore compromise detection of functional areas in individual patients. On the other hand, an increase of the spatial resolution reduces signal-to-noise contrast and this will decrease detection power for brain activity. A spatial resolution of 3 or 4 mm^3 seems adequate (and is feasible) for neurosurgical application where precise gyrus localization is the minimum requirement [82].

Absence of activation

Failure to detect activity can be caused by several factors, of which some are difficult or impossible to control. A tumor or vascular malformation can distort the brain or cause blood flow abnormalities that may alter or diminish the BOLD signal [38, 54, 87]. Under these circumstances, the absence of fMRI activation does not necessarily imply the absence of relevant neural activity. On the other hand, fMRI activity within tumor borders is not necessarily false-positive and can be functionally relevant, as has been shown by ESM [55, 76]. Other factors that can potentially influence BOLD responses are the age of the subject [36], sensorimotor or cognitive deficits [3], medication or drugs [53], or a poor task performance [88]. Task performance is a particularly relevant factor in the population of neurosurgical patients, which can strongly affect brain activation maps [52]. Patients with a paresis or cognitive impairments may suffer from a limited attention span or early fatigue and may exhibit either under- or overactivation due to disengagement or excessive effort, respectively. Optimal task performance may require prior to the scan session a practice session in which the patient is acquainted with the setting and the stimulus presentation and the experimenter can determine the feasibility of an fMRI experiment. If task performance is not monitored, the investigator is left with uncertainty about the cause of poor results: Is some brain function impaired or did the patient fail to perform the task as required? The effects of impaired performance due to brain damage on brain activation maps are a known caveat that is very difficult to solve with task-driven fMRI. A recently developed fMRI technique, resting-state functional connectivity mapping (see below), bypasses the problem of impaired task performance on activation maps, but presently the resulting functional maps are not yet reliable within individual subjects [16].

fMRI in surgical planning: review of the literature

Brain mapping in neurosurgery is predominantly performed for surgical planning of motor and language areas. The goal is to obtain a map of areas that are indispensable for normal neurological functioning. This map is usually considered a predictor for immediate and significant functional deficits when these areas are damaged. The clinical questions mainly concern the location of primary sensorimotor areas (sometimes in adjunct with the location of the motor part of the supplementary motor area [SMA]), assessment of the language-dominant hemisphere, and location of language areas.

Other (cognitive) functions are seldom asked for and are only occasionally mapped by neurosurgeons who have a special interest in functional mapping. Examples are calculation, writing, spatial attention or working memory [65, 75, 92]. This is probably due to two reasons. First, it is common neurosurgical opinion that these other cognitive functions are not easily damaged after surgery and that they are therefore not as localized and vulnerable as motor and language functions. More recent neuropsychological studies, however, clearly show that cognitive deficits are far more common than previously assumed on the basis of clinical impression and observation, both before and after surgery [31, 91]. Second, in the classical lesion studies a firm anatomical basis for most cognitive functions was never established.

Motor areas

In the absence of anatomical variations or functional reorganization it is probably safe to assume that the primary motor cortex (M1) is located on the precentral gyrus. Various anatomical landmarks have been described that help to identify the central sulcus and the precentral gyrus. On MRI scans, there are at least six of these landmarks, the "handknob" being the most robust one (in fact, this landmark was discovered because of MRI activation within this area) [102]. Under pathological conditions where a lesion can distort or destroy anatomical and functional topography, these landmarks are not useful, and functional imaging is called for.

Various rather simple motor tasks (e.g., finger tapping or hand clenching) have shown reliable activation of the M1 with fMRI. What makes clinical interpretation difficult is that there are usually several other activated areas, often in neighboring gyri. The challenge is to disentangle the M1 activation from activation in secondary motor or nonmotor areas. There are currently no fMRI tasks that can selectively activate only the primary motor cortex, so additional information is needed from other modalities to increase reliability. What is often done in practice, as a first step, is to compare the location of fMRI activity to the expected location of M1 according to anatomical landmarks. Note that this stems from the classical view of functional topography, which may not be adequate [68]. For instance, it has been shown that at least part of the primary motor cortex seems to code for specific movements rather than for a specific muscle or body part, with several sites for each functional representation instead of one [86]. In addition to that, the M1 has been postulated to participate not only in the executive but also in the preparative motor phase [15, 46]. Pathological lesions can lead to functional reorganization of motor areas, even at the level of the M1 [11, 23, 87]. This all implies that unexpected activation in fMRI maps needs to be cautiously interpreted, keeping in mind that our anatomically guided expectations may be outdated. Abnormal fMRI activation can of course reflect false-positive activations because of movement artifacts or a low statistical threshold, but it can also represent a less usual variation in normal anatomy (e.g., double precentral sulcus) and physiology (multiple representations) or it may reflect brain plasticity.

Still, there is general consensus in the literature that fMRI is a valuable tool for localization of the primary motor cortex and assessment of presurgical risks. Lehéricy et al [55] found that in 8 of 60 patients with a centrally located brain tumor it was not possible to reliably identify the precentral gyrus with anatomical landmarks only. With the help of fMRI or ESM, identification was 100%. According to their study there was a good agreement between fMRI findings and intraoperative mapping with 92% of ESM areas located within the margins of the fMRI area; the remaining ESM sites were within 15 mm of fMRI areas. Bizzi et al [9] reported a sensitivity and specificity of 88% and 87%, respectively, when fMRI hand motor function was compared to ESM (allowing for a radius of 1 cm around foci). With similar criteria, Roessler et al [74] found 100% concordance for 17 patients with low- or high-grade gliomas, which they attribute to the high field strength of their scanner (3 tesla). In conclusion, although several methodological and practical questions remain to be answered, motor fMRI can be of surgical use.

Language lateralization

The clinical gold standard for assessment of language dominance remains the amobarbital test, although this technique has serious methodological and practical flaws [78]. Several fMRI and positron emission tomography studies have tried to match outcome of the amobarbital test. To do this, most studies have calculated a lateralization index (LI) to quantify the proportion of activation in both hemispheres; this LI varies from −100 (all activation in the right hemisphere) to 100 (all activation in the left hemisphere). A cutoff value of the LI is then chosen to determine whether patients have typical or atypical language dominance. Unfortunately the variability in the reported LIs across fMRI studies is so large that every study or research institute has defined its own criteria for assessment of language dominance; there is no consensus about an optimal fMRI protocol or cutoff values for the LI. In general a good correlation is reported in the literature between both fMRI and the amobarbital test but no protocol is able to obtain complete agreement between both methods. Combining multiple fMRI language tasks is currently the best strategy and yields reproducible and reliable results. If these results indicate clear-cut left hemisphere dominance, some authors advocate that a sodium amytal test is not necessary [30, 80]. Use of only a single task is less reliable in particular for identification of the one atypical patient among the majority of typical patients [30, 80]. When atypical language dominance is suspected, activation maps require close inspection for possible mixed dominance, as frontal and temporoparietal areas can be located in different hemispheres [44, 80]. There are only few studies that have compared fMRI and the amobarbital test to the true gold standard: patient outcome. Sabsevitz et al [85] showed that preoperative fMRI predicted naming decline after left anterior temporal lobectomy. Paradoxically, in this study ESM was used to tailor the extent of the resection.

Language areas

Contemporary neurological textbooks still teach that language is served by two areas in the left hemisphere (Broca and Wernicke) that are connected by the arcuate fasciculus, despite abundant evidence that language processing depends on a network of many other subcortical and cortical areas. In reality, there are no clear functional or anatomical definitions of the areas of Broca and Wernicke [67, 98]. Although Broca's area is generally denoted as the posterior part of the left inferior frontal gyrus, damage to this area alone yields only a transient decrease of speech output but not Broca's aphasia [15]. Wernicke's area is often circularly defined as "the region that causes Wernicke's aphasia when damaged" [62, p. 37].The view that language areas are difficult to define topographically is strongly supported by the many functional neuroimaging studies that have identified widespread and overlapping networks for phonological, semantical, orthographic, and syntactic processing [27, 95]. Recent MRI-based analyses of brain-lesioned dysphasic patients confirm a wide area of potential language cortex in the left hemisphere with frontal and temporal epicenters different from those classically formulated [4]. ESM and fMRI studies show that these critical language epicenters are smaller than generally assumed (<1–2 cm^2) with multiple representations in frontal and temporoparietal areas [67].

Only few studies have meticulously compared fMRI and ESM for the purpose of language localization [26, 76, 82]. Tasks that were used with ESM were simple oral language tasks (picture naming and verb generation). General findings are the following. (1) fMRI is able to identify most of the language areas that are found with ESM. To achieve optimal detection power, the results from multiple fMRI tasks need to be combined (a minimum of three tasks seems necessary). In practice this means that fMRI can reliably predict the absence of positive ESM sites (i.e., fMRI has a very high negative predictive value). (2) fMRI finds (up to 50%) more areas than ESM, and consequently the positive predictive value is limited. (3) There is a significant variability of fMRI data across patients, tasks, and statistical methodology and this makes generalization of results across centers impossible. ESM currently remains the safest method for (sub)cortical language mapping. fMRI results are not sufficient for surgery to rely on completely when language areas are judged to be in close proximity to the surgical area of interest. As such, fMRI and electrocortical mapping are complementary techniques.

Factors impeding surgical application of fMRI

The argument that is usually given to explain the disagreement between fMRI results and existing clinical techniques is that fMRI is unable

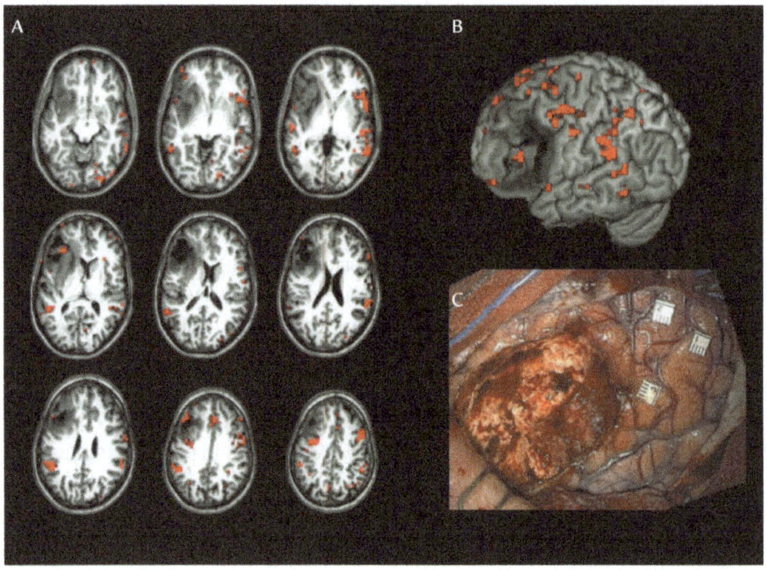

Fig. 3. 32-year-old right-handed patient with a large left-sided frontal-insular low-grade oligodendroglioma. Patient had one generalized seizure after which she complained for three days that her "memory was gone" and that she had difficulties with word finding and writing. After that period there were no deficits on clinical examination. In-depth neuropsychological investigation showed disturbances in verbal and nonverbal memory and occasionally mild problems with attention, executive functions, and language. (**A**) Axial images (left = left) with fMRI information (red voxels) that show the result of a verb generation task. The task is block-designed and consists of visual presentation of nouns (5 epochs of 9 nouns) alternated with a simple control task (looking at abstract symbols). Imaging time is 5 minutes, in which a total of 486 volume images is acquired (PRESTO, Philips Achieva 3T). There is predominant activation in the homologue area of Broca on the right side and bilaterally in premotor and temporoparietal areas. There is activation within the tumor that becomes a relative weak area of activation when the statistical threshold is set more stringently. In conclusion, fMRI suggests bilateral language representation with reorganization of left frontal language areas to perilesional areas and the contralateral hemisphere. (**B**) Rendered view of the left hemisphere. The tumor locally disturbs the rendered view. (**C**) Photograph showing the large frontal resection; the ventricle was opened. The cortical resection was radiologically complete. Subcortically the resection was stopped because of phonological paraphasias (presumably the arcuate fasciculus), and part of the tumor in the insula was left in situ. Markers indicate positive sites found with cortical mapping in the primary sensorimotor cortex and ventral premotor cortex. Subcortical sites are not shown. There were no (temporary) postoperative deficits

to differentiate between critical and noncritical areas and thus is not directly relevant for surgical planning. From this disagreement it is often concluded that fMRI cannot yet replace the existing techniques and that further research and refinement is needed in order to obtain that goal. Although this is indeed a fundamental limitation of fMRI, with stringent planning, execution, and analysis of the experiment the brain activation maps can already be a valuable adjunct in surgical planning (see Fig. 3 for an example). It is unlikely that fMRI will ever completely agree with the sodium amytal test and ESM because of fundamental differences in methodologies and outcome measures. But how reliable are the gold standards? [78]

At first glance, ESM seems very intuitive and valid: When a particular area is stimulated and the patient has difficulty performing a task, there must be a close and essential relationship between that brain area and the disturbed function. Consequently, areas where ESM is "positive" are considered to be indispensable for normal function and are not included in the resection. However, such a straightforward inference is not fully justified. For example, when the posterior part of the SMA (the SMA proper) is electrically stimulated, this will often elicit involuntary motor responses in a patient. As expected, resection results in immediate postoperative neurological deficits (hemiparesis, akinesia, mutism). However, these deficits typically resolve within several weeks or months. Thus, the fact that an area is tested positively with ESM does not necessarily imply that it is indispensable (i.e., eloquent) for that particular function (note that in this case an "eloquent area" is defined as an area that when damaged leads to permanent deficits). This questions the clinical usefulness and even the validity of ESM for its purpose, as ESM seems unable to account for functional reorganization after surgery. What probably happened in the patients with SMA resections is that contralateral secondary motor areas partially compensated for the loss of function. Indeed, such unmasking of new motor areas has been demonstrated when fMRI activation patterns were compared before and after surgery [50].

It is very likely that such a redundancy of positive ESM sites not only is present in the motor domain but also holds for other functions. There is indirect evidence for this in the language domain. For instance, several authors have claimed that a nontailored left anterior temporal lobectomy without the use of ESM does not worsen language functions [17, 35]. This conflicts with the results of ESM studies in similar groups of epilepsy patients where in approximately 20% of patients language areas were found in the dominant anterior temporal lobe [67]. Similar conflicting observations have been made for the basal temporal language area [58]. Again, functional compensation could account for this redundancy. Another explanation would be that stimulation of anterior or basal temporal areas indirectly interferes with more distant critical areas via subcortical connections.

Alternatively, electrical stimulation can also inadvertently lead to false-negative results [59, 78]. This may be caused by evaluating the wrong function for a targeted region (one can only test an a priori determined function) or not having a tasks available, giving the false impression that resection is safe [10].

For assessment of the language-dominant hemisphere, fMRI language results are compared to the sodium amytal (Wada) test as gold standard. This test temporarily disables part of a hemisphere, during which time the contralateral hemisphere is tested for language and other cognitive functions. There are several factors that can confound the interpretation of this test. Importantly, there is no agreement on outcome measures between different clinical centers. This accounts for at least some of the considerable variability in the reported incidences of typical (i.e., left-sided) and nontypical (i.e., right-sided or bilateral) language dominance. The Wada test may underestimate the incidence of bilateral language dominance, as inconsistencies have been reported with clinical outcome or the findings of ESM [41, 49, 99]. In addition to that, some authors have found evidence for a continuous distribution of language functions across hemispheres, instead of the classical dichotic model with language either left and/or right. For instance, Springer et al [90] observed a Gaussian-like distribution of fMRI-derived LI values in both healthy volunteers and epilepsy patients. This could implicate a degree of equipotentiality between hemispheres with respect to language processing that is also supported by at least some of the amobarbital studies [7, 73].

In conclusion, ESM and the sodium amytal test are currently the best techniques available to assess the immediate functional consequences of removal of a part of the brain. These techniques cannot, however, predict whether or not perilesional or distant neural networks are able to compensate for any loss of function after operation (i.e., there is a risk of false-positive results). They also have limited potential to test more complex cognitive functions or a set of different functions. To develop techniques that can achieve these goals, the functional topography of the brain needs to be better understood, and in particular the dynamic behavior of functional networks. By studying patients before and after surgery, the mechanisms of brain plasticity can be elucidated to the extent that preoperative functional neuroimaging results can be used to predict long-term postoperative sensorimotor and cognitive functions. fMRI carries the potential for this, as will be explained in the next paragraphs.

Emerging role of fMRI in clinical practice

Brain functions emerging from networks: neuroscience and clinical-practice perspectives

There is now overwhelming evidence that the classical model, even when it only serves as a metaphor, is insufficient on many grounds. At the beginning of the 20th century only few scientists were in favor of the diagram-makers, and the dominant view tended towards holism (with strong proponents such as Lashley, Goldstein, Marie, Head, and von Monakow). Marie [60] reanalyzed the classic cases of Broca and found that the lesions extended far beyond the so-called area of Broca. Moutier [63] demonstrated patients with Broca's aphasia that had lesions outside of Broca's area. Later, with the advent of computed tomography and MRI, brain damage could be localized more precisely and shown in more detail, and several other areas have since been discovered that are potentially critical for normal language functioning. Bates et al [4] used voxel-based lesion–symptom mapping to reanalyze the relationship between tissue damage and behavior on a voxel-by-voxel basis in aphasic patients. This method is comparable to that used in functional neuroimaging and largely overcomes the methodological errors that are made when patients are grouped either by lesion or by behavioral deficit, as was done in the earlier studies. Remarkably, the areas that are usually associated with language deficits (i.e., those of Broca and Wernicke) were not the epicenters that were found in that study. Fluency was most affected by lesions in the anterior part of the insula and in the parietal white matter; auditory comprehension was most affected by lesions in the middle temporal gyrus, with significant contributions seen also in the inferior parietal and dorsolateral prefrontal cortex. Other differences found in that study are that the spatial range of regions involved in language is much larger than expected, and that there is a significant variability between patients that was never accounted for in the older models. The last two aspects were also recognized by neurosurgeons early in the 20th century and were in fact the rationale for the use of intraoperative electrocortical stimulation. Penfield and later Ojemann and others published overviews of their stimulation results for language functions and found that there was no single cortical area that is consistently involved in language functions in every patient. As a consequence, they calculated a probability that reflects the chance to find a critical language area in a particular brain region. This view is fundamentally different from the classical model, where each area either is or is not associated with a particular function.

In the modern neuroscientific view, information processing for any given function is performed in a highly distributed and hierarchically organized system of multiple different brain areas [61]. In principle, such a model is not very different from the classical model, which has only a few areas and (sub)cortical connections: Still the model does not explain how exactly a function is represented in the brain but merely schematically represents the scientific findings. There is, however, a fundamental difference as compared to the older model: Due to the increased complex architecture (large number of areas, parallel design, reciprocal connections), it now becomes a practical impossibility to exactly localize a particular function [25]. The function itself seems somehow distributed over the network. This indeterminism is already present at a cellular level, where information is diffusely stored in neuronal networks (Fig. 4).

Fig. 4. (A, B) The mathematical Hopfield model for associative memory can be used as simple metaphor to illustrate some of the properties of neural networks (reproduced from ref. 13). In this particular example there are n = 400 binary neurons that are all interconnected (i.e., 400 by 399 connections) [39]. Each neuron is either in an "active" state or "in rest"; this is visualized with a white or a black voxel. The state of every neuron is continuously updated and is determined by its local input. This input is the mathematical sum of all other neurons modulated by the strenght of the (synaptic) connection. **(A)** Information can be stored in the network by use of a learning rule. Hopfield used Hebb's rule, where the strenght of the connection between neurons increases when the two neurons are in the same state and decreases when they are in different states [34]. In this example, the network has "learned" ten different patterns. **(B)** Simulation experiment where one of the patterns is mutilated by randomly changing 20% of the state of the neurons. By randomly updating the neurons the pattern is gradually recovered, illustrating the robustness of this "holistic" form of information storage

Hebb [34] was one of the first researchers to propose that the stored information can be altered via modifications in synaptic transmissions in neural networks (although quite similar ideas on this mechanism for plasticity had been formulated much earlier by Wernicke, James, and Cajal [6, 29]). Continuous modifications in neuronal networks are a sine qua non for the brain to store and update information, to acquire new skills, to optimize and automate information processing, and to adapt to structural changes (e.g., normal aging or a brain tumor). These dynamic networks, even the more simple ones, have several interesting properties that are also thought to be fundamental underlying principles for brain function. An important property is that information is stored diffusely and nonlocally, which makes the networks very robust (Fig. 4). Also, neural networks process information in a reciprocal and parallel manner; some networks are able to learn without the need for explicit rules (unsupervised learning).

Several issues are to be incorporated in the new views on functional topography based on dynamic networks: (1) a function is performed by a complex and distributed network of brain areas and therefore cannot simply be anatomically reduced to a one or a few brain areas and their connections, (2) there is a significant intersubject variation in anatomy and functional topography, (3) the structure and functional dynamics of brain networks are constantly modified under the influence of physiological or pathophysiological circumstances.

Network properties relevant for neurosurgery

Thinking in terms of large-scale networks has major implications for clinical practice. First of all, these networks connect remote areas across the brain, stressing the importance of both cortical and subcortical structures in surgical planning. Mesulam [61] postulated that cognitive functions are represented by at least five different networks: for spatial attention, language, memory and emotion, working memory and executive function, and face and object recognition (note that these functions were never represented in the classical models). Most brain functions are supported by interactions between multiple regions and this will probably increase their resistance to damage and improve the potential for reorganization.

What is also relevant for surgical planning is that a particular brain area can have functional subdivisions and can participate in different functions. This has been repeatedtly shown with ESM, for instance, for auditory and visual naming [33], reading and naming [77], writing and naming [57], or different languages sites in bilingual patients [32]. Overlap in activation between different brain functions is frequently seen in functional neuroimaging studies (Fig. 5). There is evidence that areas that participate in multiple functions are more criti-

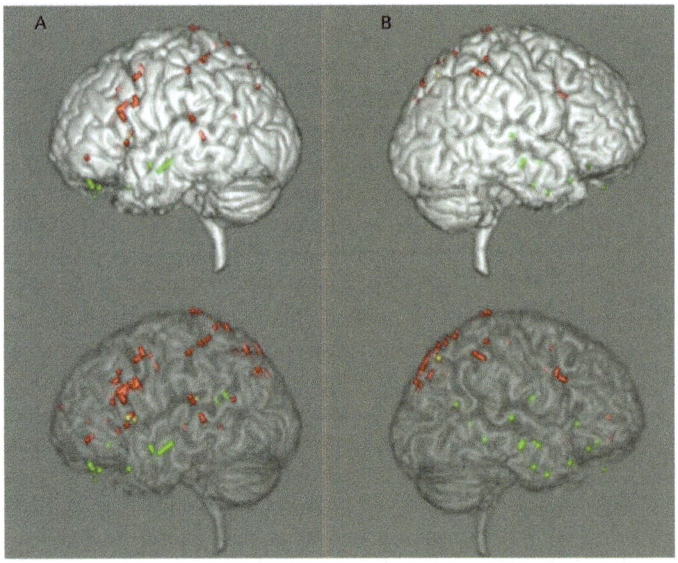

Fig. 5. Differences in regional brain activity between different fMRI language tasks. Red voxels are active in tasks that involve the processing of single words or pictures (combined results from classic language tasks: verb generation, verbal fluency, and picture naming). Green voxels represent activity that reflects processing of sentences. Yellow voxels are active in sentence processing and in one or more of the other tasks (i.e., red merged with green). The patient had left hemispheric dominance according to a sodium amytal test, in agreement with the activity that is seen from the classical language tasks that is predominantly seen in left inferior and middle frontal gyrus and left temporoparietal areas. The sentence comprehension task yielded strong bilateral activity in the temporal lobes and inferior frontal and orbitofrontal areas. **(A)** Cortical representation of fMRI activity; **(B)** activation has been made visible beneath the cortical surface (<10–15 mm)

cally involved and better correlate with the results of the clinical gold standard techniques. For instance, a conjoined analysis of different language tasks better predicts hemispheric language dominance than individual fMRI language tasks [30, 80].

The relevant clinical question is what the smallest unit is that can have clinical consequences when damaged, and whether other areas are able to compensate for any loss of function. ESM and fMRI studies suggest that this is about 1 cm^2 of cortical surface but this question is by no means resolved. To answer this question, new imaging tools need to be (further) developed that can inform us how functional networks behave and evolve under the influence of surgery or tumor growth. fMRI is the best candidate for this purpose. ESM is insufficient as it can provide us with only a static map of part of the brain and cannot be used to follow-up patients after surgery. One of the very few examples available is the landmark paper of Krainik et al [51]. They were able to show that resection of an area with fMRI activation in the posterior part of the SMA (the SMA proper) predicted an immediate but temporary functional deficit after surgery. In a follow-up paper [50] this group showed that in these low-grade glioma patients there was already preoperative reorganization in ipsilateral and contralateral premotor cortex areas (including the SMA). Although this reorganization could not prevent temporary deficits after surgery, recovery was faster and associated with increased activity in secondary motor areas in the healthy hemisphere.

fMRI methods to study network behavior

As fMRI can measure a large part or all of the brain, it is possible to assess all of the regions that are involved in the function of interest. As discussed earlier, these maps of activity do not reveal which regions are indispensable and which are involved but not indispensable. For instance, although various language fMRI studies have shown involvement of the bilateral temporal pole in prosodic aspects of language processing, removal of either pole in epilepsy patients fails to affect performance on prosodic comprehension tasks (Fig. 5) [40, 47]. One of the challenges in fMRI is to assign significance to each region within a network, but it is unlikely that this can be done with fMRI alone. Instead, systematic assessment of the effects of loss or removal of particular regions on performance and brain activity in the other regions of the network is required. Some studies have used transcranial magnetic stimulation to establish whether or not a region that was found with fMRI is critically involved in motor or language functions [81, 97]. However, transcranial magnetic stimulation may suffer from the same limitation as ESM for precise localization of function, namely, that it can disrupt distant (critical) brain areas via connections with the stimulated area. It is also unable to selectively map subcortical structures.

Extensive work has been conducted on plasticity in stroke patients with longitudinal fMRI studies. Lesions of primary motor regions have been shown to result in changes in brain activity patterns which to some extent predict recovery of motor function and which change in the course of months after the insult [96]. As of yet, it is not quite clear how reorganization contributes to motor recovery, other than a general indication that perilesional areas are important in this [96]. Damage to the M1 is associated with limited function recovery, which appears to depend on bilateral secondary motor regions.

For language function, aphasic stroke patients, or those that have recovered from their aphasia, generally show more activation in the right hemisphere than do healthy controls. There is an ongoing discussion to what extent this activation is responsible for functional improvement of patients and truly reflects critical language processing [72]. In the long run, successful regeneration from stroke seems to depend more on recovery of available (left-sided) language-related brain regions than on recruitment of new ipsi- or contralesional areas [94,

97]. An additional important factor that needs to be considered is time. Slowly growing lesions induce different and, in general, more successful patterns of reorganization than do acute lesions such as stroke [18, 93]. For example, most low-grade glioma patients have no obvious deficits (and only slight abnormalities on neuropsychological examination) despite a large lesion that is often found in eloquent areas [19]. These lesions can nevertheless often be surgically removed without new neurological deficits (Fig. 3) [5]. Different patterns of reorganization between patients have been described [19]. As of yet these patterns have not been mapped out comprehensively, but increasing use of fMRI, coupled to functional outcome, may prove particularly informative in the coming years.

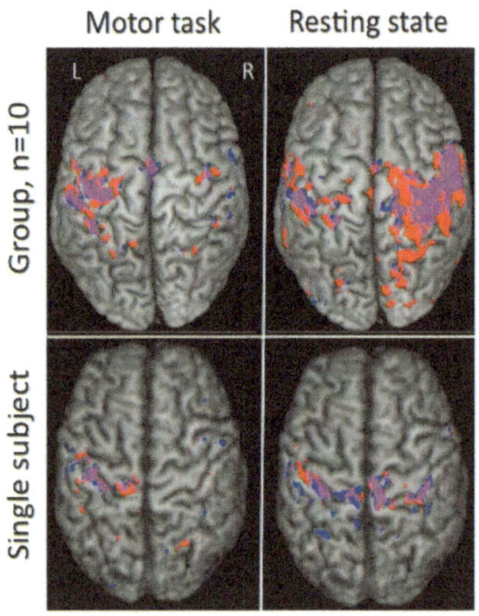

Fig. 6. Reproducibility of motor fMRI (left column) and of resting-state based motor cortex identification (right column). Data are shown for a group-average of 10 healthy subjects (top row) and for a single healthy subject (bottom row). Activity in session one is shown in red, for session 2 in blue, and overlap is shown in purple. The resting-state data were first corrected for heart rate and respiration. Next, a region of interest was defined for the right precentral gyrus in each subject. the fMRI time-series signal in that region of interest was entered as seed for subsequent correlation analysis with all voxels in the brain. Correlations within right primary sensorimotor cortex was highest, but a clear left sensorimotor region emerged also (data from G. Kristo, G.-J. Rutten, and N. Ramsey, UMC Utrecht, unpubl.)

One of the issues in detecting by fMRI any changes in function following lesions is the reproducibility of fMRI per se. Single-subject activity maps are moderately reproducible at best, with an approximate overlap of 40% of active regions [70, 83]. This is attributed to the contribution of both physiological (respiration and heart rate) and machine noise [8]. However, when comparing group-averaged maps the overlap is much better, in the order of 80% [70]. An example is shown in Fig. 6 (left column), where the match of testing a group of 10 healthy subjects twice is shown, as well as that of a single subject. The match in the group image is good and is mainly due to the fact that images of individuals are smoothed. In the individual subject, overlap is also good in the primary sensorimotor cortex, but total overlap is limited by the variable activations in nonprimary regions (which fall away in group-averaged maps). Longitudinal studies in groups can be expected to yield a more sensitive measure of change than studies in individual patients. This constitutes a problem for lesion studies, where no two patients are likely to exhibit the exact same lesion (particularly regarding white matter consisting of narrow fibers), which can only be countered by including a large number of patients in a study.

When conducting longitudinal studies with fMRI, another factor may need to be considered, namely, the effect of learning on brain function. A key feature of the brain is that it minimizes energy expenditure while improving performance. Brain activity patterns typically change dramatically after learning of a particular task such as those that involve language, motor, or memory functions. The basis for this is twofold: Learning can result in a shift from one network to another, or it can result in an increased efficiency of communication within a network [42]. For instance, remembering a wordlist will involve the cognitive control network initially, but later at retesting it will involve a network involved in retrieving learned wordlists. Practice of motor and certain cognitive tasks generally causes a significant change of activity in areas of the involved network due to an automatization of processes [42, 45]. These phenomena need to be considered in the design of tasks for a longitudinal study. One can solve this, for instance, by using different sets of stimuli for the follow-up fMRI scans (notably for language and memory tasks), or if practice effects are unavoidable, one can train patients on the task before the first scan.

Issues pertaining to tasks in fMRI can be avoided altogether. A recent approach to network dynamics makes use of fluctuations that are present in fMRI scans at rest. The origin of these low-frequency fluctuations (ca. 0.01–0.1 Hz) is under debate and may, for instance, relate to vascular instead of neuronal processes (which still makes them of clinical interest). Resting-state fMRI is conducted with the patient at rest for about 10 minutes. The fluctuations have been shown to reveal correlations between regions that belong to the same network. Such networks have been reported for motor, language, and cognitive functions, but also for networks that are more active when a subject is not performing a task. The most prominent of the last is the default mode network, and current investigations are examining whether features of this network can be used to diagnose neurological and psychiatric disorders [103]. A major advantage of resting-state fMRI is that patients do not have to perform tasks they are impaired in. It is assumed that the derived network of interest will reveal all the regions that are still contributing to the network even if actual performance on tasks is impaired. However, resting-state fMRI suffers even more the limited reproducibility seen for task performance fMRI of individual patients, precluding the use of the derived network maps for neurosurgical planning (Fig. 6, right column). Nevertheless, the absence of performance-related confounds, notably the effect of effort on brain activity maps, renders this technique a promising adjunct to functional neuroimaging techniques for surgical planning.

Future role of fMRI in clinical practice

With the advent of high-field MRI scanners in clinical settings, new techniques are becoming available for better imaging of structure and function. About 30 MRI scanners with a field strength of 7 T have been placed across the world by the three largest manufacturers. These scanners are regarded as experimental at this point, and the 7 T research centers collaborate with each other and with manufacturers

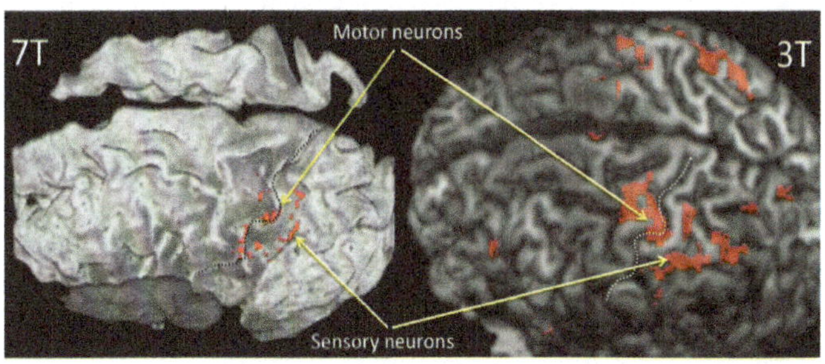

Fig. 7. Comparison of motor fMRI between 3 T and 7 T field strength. Activation is projected onto anatomy for a single patient with epilepsy. Higher signal-to-noise rate at 7 T makes smaller voxels possible. At 7 T there is virtually no signal from medium to larger blood vessels, improving accuracy of functional mapping. Motor activity is shown for a voxel size of 1.5 mm isotropic at 7 T (left) and for a voxel size of 4 mm isotropic at 3 T (right). Note the increase in spatial detail at higher field strength (data from J. Siero, N. Petridou, D. Hermes, and N. Ramsey, UMC Utrecht, unpubl.)

to expedite the development of clinically applicable scans. Already, amazingly detailed anatomical and functional scans have been developed and are now tested in patients with brain disorders. For functional MRI we expect significant improvement in accuracy of function localization for two reasons. First, the high field causes almost complete signal loss in medium to larger blood vessels, reducing the problem of mislocalizing brain activity due to signals from those vessels. Second, a signal from brain tissue is twice as intense as at 3 T, allowing for measuring brain structure and function at a higher resolution. For instance, whereas at 3 T reliable results require a voxel size of about 3 mm for fMRI, at 7 T, voxels can be as small as 1.5 mm (Fig. 7). The combination of these effects is expected to result in significantly higher accuracy of activity maps.

A wealth of brain functions has been investigated in the cognitive neuroscience community. For many of these, the studies focus on (increasingly) subtle elements of functions (e.g., processing of faces versus houses). For neurosurgery, each fMRI paradigm has to be matched with a neuropsychological test for deficits that can be applied effectively within the constraints imposed by the surgical time frame and environment. This is obviously not an easy task, given the very few functions that are tested intraoperatively. For language function, a multitude of paradigms has been published, and the various awake-surgery groups also apply several different paradigms. Surgical practice would greatly benefit from evidence-based standardized paradigms that have been demonstrated to fulfill the requirements best in terms of sensitivity, specificity, and association with well-defined deficits (and whether deficits are temporary or not). This can be achieved only in a collaborative activity, as many patients are required, involving neurosurgeons who are willing to invest in acquiring the necessary data. Currently a European network of medical specialists in the field of low-grade glioma treatment (the European Low Grade Glioma Network, www.braintumours.eu) is developing standardized protocols that allow for developing and validating paradigms. Apart from motor and language functions, paradigms are currently being tested for working memory and for spatial attention, deficits of which are manifested as impaired ability to process complex information and hemi-neglect, respectively. The former can be expected to lead to more conserva-

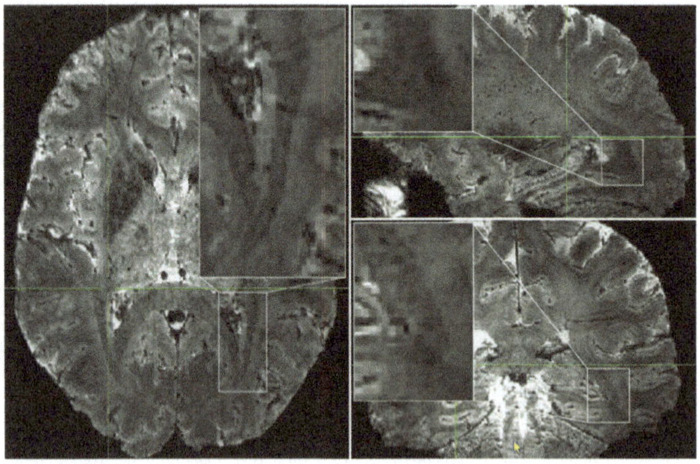

Fig. 8. Enhanced contrast within white matter at 7 T. The picture displays enhanced contrast between fibers of the optic radiation and other fibers in one healthy subject. Contrast is enhanced by a custom-made T2*-weighted ultrahigh resolution (0.5 mm isotropic) scan (data from N. Petridou, UMC Utrecht, unpubl.)

tive surgery in the prefrontal cortex given growing insight in this kind of deficit before and after resection [10, 48]. The latter may, in contrast, lead to more opportunities for resection within the parietal cortex, as it would reduce the present uncertainty about the whereabouts of regions serving spatial attention.

With the development of sophisticated techniques for mapping fiber tracts comes the promise of pre- and intraoperative localization of function-critical subcortical tissue. At present the exact course of specific tracts is not certain, due to difficulties in following them reliably in DTI datasets. This is partially due to the fact that the resolution of DTI is not even close to the diameter of nerve bundles, and partially to the difficulty of tracking fibers at sharp bends (kissing versus crossing). Resolution of DTI is, however, constantly improving due to hardware developments (notably parallel MRI headcoils) and new contrasts emerging from high-field MRI. Figure 8 shows the delineation of the optic radiation discovered in ultrahigh resolution images at 7 T. The brain shift remains a major problem for using preoperatively acquired fiber tract maps, but this problem may (at last) become resolved with the advent of sensitive three-dimensional ultrasound imaging technology.

Given the fact that patients can recover from functional deficits following a pathophysiological or surgical lesion , neurosurgical practice would benefit greatly from knowing which regions can be resected without permanent loss of function [22]. This can only be achieved if information about functional losses and their recovery potential would be available in a coherent format. Neurosurgery can play a crucial role in building a brain atlas of regions that can or cannot be resected. Such an atlas requires detailed registration of the pre- and postsurgical and long-term status of motor, sensory, and cognitive functions of patients, together with detailed structural and functional brain images before and after surgery. The European Low Grade Glioma Network is working towards a standardized multimodal protocol that enables building of a brain plasticity atlas. Once a protocol has been established, it will take several years to include and assess data from enough patients to start building, initially with a focus on motor and language functions. Ultimately, the long-term effects of surgery should be predicted with structural and functional neuroimaging techniques prior to surgery to optimize survival and quality of life for each individual patient.

References

[1] Aguirre GK, D'Esposito M (1999) Experimental design for brain fMRI. In: Moonen CTW, Bandettini PA (eds) Functional MRI. Springer, Berlin, pp 369–380

[2] Alkadhi H, Kollias SS, Crelier GR, Golay X, Hepp-Reymond M-C, Valavanis A (2000) Plasticity of the human motor cortex in patients with arteriovenous malformations: a functional MR imaging study. Am J Neuroradiol 21: 1423–1433

[3] Atlas SW, Howard RS 2nd, Maldjian J, Alsop D, Detre JA, Listerud J et al (1996) Functional magnetic resonance imaging of regional brain activity in patients with intracerebral gliomas: findings and implications for clinical management. Neurosurgery 38: 329–338

[4] Bates E, Wilson SM, Saygin AP, Dick F, Sereno MI, Knight RT, Dronkers NF (2003) Voxel-based lesion-symptom mapping. Nat Neurosci 6: 448–450, 2003

[5] Benzagmout M, Gatignol P, Duffau H (2007) Resection of World Health Organization grade II gliomas involving Broca's area: methodological and functional considerations. Neurosurgery 61: 741–752

[6] Berlucchi G, Buchtel HA (2009) Neuronal plasticity: historical roots and evolution of meaning. Exp Brain Res 192: 307–319

[7] Binder JR, Swanson SJ, Hammeke TA, Morris GL, Mueller WM, Fischer M et al (1996) Determination of language dominance using functional MRI: a comparison with the Wada test. Neurology 46: 978–984

[8] Birn RM, Smith MA, Jones TB, Bandettini PA (2008) The respiration response function: the temporal dynamics of fMRI signal fluctuations related to changes in respiration. Neuroimage. 40: 644–654

[9] Bizzi A, Blasi V, Falini A, Ferroli P, Cadioli M, Danesi U et al (2008) Presurgical functional MR imaging of language and motor functions: validation with intraoperative electrocortical mapping. Radiology 248: 579–589

[10] du Boisgueheneuc G, Levy R, Volle E, Seassau M, Duffau H, Kinkingnehun S et al (2006) Functions of the left superior frontal gyrus in humans: a lesion study. Brain 129: 3315–3328

[11] Carpentier AC, Constable RT, Schlosser MJ, de Lotbinière A., Piepmeier JM, Spencer DD, Awad IA (2001) Patterns of functional magnetic resonance imaging activation in association with structural lesions in the rolandic region: a classification system. J Neurosurg 94: 946–954

[12] Catani M, Jones DK, ffytche DH (2005) Perisylvian language networks of the human brain. Ann Neurol 57: 8216

[13] Coolen ACC, Jonker HJJ (1991) Introduction to neural networks [in Dutch]. Utrecht University, Utrecht

[14] Crone NE, Miglioretti DL, Gordon B, Lesser RP (1998) Functional mapping of human sensorimotor cortex with electrocorticographic spectral analysis. II. Event-related synchronization in the gamma band. Brain 121: 2301–2315

[15] Crutcher MD, Alexander GE (1990) Movement-related neuronal activity selectively coding either direction or muscle pattern in three motor areas of the monkey. J Neurophysiol 64: 151–163

[16] Damoiseaux JS, Rombouts SA, Barkhof F, Scheltens P, Stam CJ, Smith SM, Beckmann CF (2006) Consistent resting-state networks across healthy subjects. Proc Natl Acad Sci USA 103: 13848–13853

[17] Davies KG, Maxwell RE, Beniak TE, Destafney E, Fiol ME (1995) Language function after temporal lobectomy without stimulation mapping of cortical function. Epilepsia 36: 130–136

[18] Desmurget M, Bonnetblanc F, Duffau H (2007) Contrasting acute and slow-growing lesions: a new door to brain plasticity. Brain 130: 898–914

[19] Duffau H (2008) Brain plasticity and tumors. Adv Tech Stand Neurosurg 33: 3–33

[20] Duffau H, Lopes M, Arthuis F, Bitar A, Sichez JP, Van Effenterre R, Capelle L (2005) Contribution of intraoperative electrical stimulations in surgery of low grade gliomas: a comparative study between two series without (1985-96) and with (1996-2003) functional mapping in the same institution. J Neurol Neurosurg Psychiatry 76: 845-851

[21] Duffau H (2001) Acute functional reorganisation of the human motor cortex during resection of central lesions: a study using intraoperative brain mapping. J Neurol Neurosurg Psychiatry 70: 506–513

[22] Duffau H (2005) Lessons from brain mapping in surgery for low-grade glioma: insights into associations between tumour and brain plasticity. Lancet Neurol 4: 476–486

[23] Duffau H, Denvil D, Capelle L (2002) Long term reshaping of language, sensory, and motor maps after glioma resection: a new parameter to integrate in the surgical strategy. J Neurol Neurosurg Psychiatry 72: 511–516

[24] Ellis AW, Young AW (1996) Human cognitive neuropsychology. Psychology Press, Hove

[25] Felleman DJ, Van Essen DC (1991) Distributed hierarchical processing in the primate cerebral cortex. Cereb Cortex 1: 1–47

[26] FitzGerald DB, Cosgrove GR, Ronner S, Jiang H, Buchbinder BR, Belliveau JW et al (1997) Location of language in the cortex: a comparison between functional MR imaging and electrocortical stimulation. AJNR Am J Neuroradiol 18: 1529–1539

[27] Friederici AD, Ruschemeyer SA, Hahne A, Fiebach CJ (2003) The role of left inferior frontal and superior temporal cortex in sentence comprehension: localizing syntactic and semantic processes. Cereb Cortex. 13: 170–177

[28] Friston KJ, Price CJ, Fletcher P, Moore C, Frackowiak RS, Dolan RJ (1996) The trouble with cognitive subtraction. Neuroimage 4: 97–104

[29] Gage N, Hickok G (2005) Multiregional cell assemblies, temporal binding and the representation of conceptual knowledge in cortex: a modern theory by a "classical" neurologist, Carl Wernicke. Cortex 41: 823–832

[30] Gaillard WD, Balsamo L, Xu B, McKinney C, Papero PH, Weinstein S et al (2004) fMRI language task panel improves determination of language dominance. Neurology 63: 1403–1408

[31] Gehring K, Sitskoorn MM, Aaronson NK, Taphoorn MJ (2008) Interventions for cognitive deficits in adults with brain tumours. Lancet Neurol 7: 548–560

[32] Giussani C, Roux FE, Lubrano V, Gaini SM, Bello L (2007) Review of language organisation in bilingual patients: what can we learn from direct brain mapping? Acta Neurochir (Wien) 149: 1109–1116; discussion 1116

[33] Hamberger MJ, Goodman RR, Perrine K, Tamny T (2001) Anatomic dissociation of auditory and visual naming in the lateral temporal cortex. Neurology 56: 56–61

[34] Hebb DO (1949) The organization of behavior: a neuropsychological theory. Wiley, New York

[35] Hermann BP, Wyler AR, Somes G (1991) Language function following anterior temporal lobectomy. J Neurosurg 74: 560–566

[36] Hesselmann V, Zaro Weber O, Wedekind C, Krings T, Schulte O, Kugel H et al (2001) Age related signal decrease in functional magnetic resonance imaging during motor stimulation in humans. Neurosci Lett 308: 141–144

[37] Hirata M, Goto T, Barnes G, Umekawa Y, Yanagisawa T, Kato A et al (2010) Language dominance and mapping based on neuromagnetic oscillatory changes: comparison with invasive procedures. J Neurosurg 112: 528–538

[38] Holodny AI, Schulder M, Liu WC, Wolko J, Maldjian JA, Kalnin AJ (2000) The effect of brain tumors on BOLD functional MR imaging activation in the adjacent motor cortex: implications for image-guided neurosurgery. AJNR Am J Neuroradiol 21: 1415–1422

[39] Hopfield JJ (1982) Neural networks and physical systems with emergent collective computational abilities. Proc Natl Acad Sci USA 79: 2554–2558

[40] Humphries C, Love T, Swinney D, Hickok G (2005) Response of anterior temporal cortex to syntactic and prosodic manipulations during sentence processing. Hum Brain Mapp 26: 128–138

[41] Hunter KE, Blaxton TA, Bookheimer SY, Figlozzi C, Gaillard WD, Grandin C et al (1999) ^{15}O water positron emission tomography in language localization: a study comparing positron emission tomography visual and computerized region of interest analysis with the Wada test. Ann Neurol 45: 662–665

[42] Jansma JM, Ramsey NF, Slagter HA, Kahn RS (2001) Functional anatomical correlates of controlled and automatic processing. J Cogn Neurosci 13: 730–743

[43] Jokisch D, Jensen O (2007) Modulation of gamma and alpha activity during a working memory task engaging the dorsal or ventral stream. J Neurosci 27: 3244–3251

[44] Kamada K, Takeuchi F, Kuriki S, Todo T, Morita A, Sawamura Y (2006) Dissociated expressive and receptive language functions on magnetoencephalography, functional magnetic resonance imaging, and amobarbital studies. Case report and review of the literature. J Neurosurg 104: 598–607

[45] Karni A, Meyer G, Jezzard P, Adams MM, Turner R, Ungerleider LG (1995) Functional MRI evidence for adult motor cortex plasticity during motor skill learning. Nature 377: 155–158

[46] Kawashima R, Roland PE, O'sullivan BT (1994) Fields in human motor areas involved in preparation for reaching, actual reaching, and visuomotor learning: a positron emission tomography study. J Neurosci 14: 3462–3474

[47] Kho KH, Indefrey P, Hagoort P, van Veelen CW, van Rijen PC, Ramsey NF (2008) Unimpaired sentence comprehension after anterior temporal cortex resection. Neuropsychologia 46: 1170–1178

[48] Kho KH, Rutten GJ, Leijten FS, van der Schaaf A, van Rijen PC, Ramsey NF (2007) Working memory deficits after resection of the dorsolateral prefrontal cortex predicted by functional magnetic resonance imaging and electrocortical stimulation mapping. Case report. J Neurosurg 106: 501–505

[49] Kho KH, Leijten FS, Rutten GJ, Vermeulen J, van Rijen PC, Ramsey NF (2005) Discrepant findings for Wada test and functional magnetic resonance imaging with regard to language function: use of electrocortical stimulation mapping to confirm results. Case report. J Neurosurg 102: 169–173

[50] Krainik A, Duffau H, Capelle L, Cornu P, Boch AL, Mangin JF et al (2004) Role of the healthy hemisphere in recovery after resection of the supplementary motor area. Neurology 62: 1323–1332

[51] Krainik A, Lehéricy S, Duffau H, Capelle L, Chainay H, Cornu P et al (2003) Postoperative speech disorder after medial frontal surgery: role of the supplementary motor area. Neurology 60: 587–594

[52] Krings T, Topper R, Willmes K, Reinges MH, Gilsbach JM, Thron A (2002) Activation in primary and secondary motor areas in patients with CNS neoplasms and weakness. Neurology 58: 381–390

[53] Laurienti PJ, Field AS, Burdette JH, Maldjian JA, Yen YF, Moody DM (2002) Dietary caffeine consumption modulates fMRI measures. Neuroimage 17: 751–757

[54] Lehéricy S, Biondi A, Sourour N, Vlaicu M, du Montcel ST, Cohen L et al (2002) Arteriovenous brain malformations: is functional MR imaging reliable for studying language reorganization in patients? Initial observations. Radiology 223: 672–682

[55] Lehéricy S, Duffau H, Cornu P, Capelle L, Pidoux B, Carpentier A et al (2000) Correspondence between functional magnetic resonance imaging somatotopy and individual brain anatomy of the central region: comparison with intraoperative stimulation in patients with brain tumors. J Neurosurg 92: 589–598

[56] Logothetis NK, Pauls J, Augath M, Trinath T, Oeltermann A (2001) Neurophysiological investigations of the basis of the fMRI signal. Nature 412: 150–157

[57] Lubrano V, Roux FE, Demonet JF (2004) Writing-specific sites in frontal areas: a cortical stimulation study. J Neurosurg 101: 787–798

[58] Lüders H, Lesser RP, Hahn J, Dinner DS, Morris HH, Wyllie E, Godoy J (1991) Basal temporal language area. Brain 114: 743–754

[59] Mandonnet E, Winkler PA, Duffau H (2010) Direct electrical stimulation as an input gate into brain functional networks: principles, advantages and limitations. Acta Neurochir (Wien) 152: 185–193

[60] Marie P (1906) Revision de la question de l'aphasie. Sem Med 26: 241–247, 493–500, 565–571

[61] Mesulam MM (1990) Large-scale neurocognitive networks and distributed processing for attention, language, and memory. Ann Neurol 28: 597–613

[62] Mesulam MM (2000) Principles of behavioral and cognitive neurology. Oxford University Press; Oxford

[63] Moutier F (1908) L'aphasie de Broca [thesis]. G Steinheil, Paris

[64] Neggers SF, Hermans EJ, Ramsey NF (2008) Enhanced sensitivity with fast three-dimensional blood-oxygen-level-dependent functional MRI: comparison of SENSE-PRESTO and 2D-EPI at 3 T. NMR Biomed 21: 663–676

[65] Ojemann GA (1988) Effect of cortical and subcortical stimulation on human language and verbal memory. Res Publ Assoc Res Nerv Ment Dis 66: 101–115

[66] Ojemann GA, Corina DP, Corrigan N, Schoenfield-McNeill J, Poliakov A, Zamora L, Zanos S (2010) Neuronal correlates of functional magnetic resonance imaging in human temporal cortex. Brain 133: 46–59

[67] Ojemann GA, Ojemann JG, Lettich E, Berger MS (1989) Cortical language localization in left, dominant hemisphere: an electrical stimulation mapping investigation in 117 patients. J Neurosurg 71: 316–326

[68] Penfield WP, Rasmussen T (1957) The cerebral cortex of man. Macmillan, New York

[69] Price CJ, Friston KJ (1997) Cognitive conjunction: a new approach to brain activation experiments. Neuroimage 5: 261–270

[70] Raemaekers M, Vink M, Zandbelt B, van Wezel RJ, Kahn RS, Ramsey NF (2007) Test-retest reliability of fMRI activation during prosaccades and antisaccades. Neuroimage 36: 532–542

[71] Ramsey NF, Sommer IE, Rutten GJ, Kahn RS (2001) Combined analysis of language tasks in fMRI improves assessment of hemispheric dominance for language functions in individual subjects. Neuroimage 13: 719–733

[72] Rijntjes M, Weiller C (2002) Recovery of motor and language abilities after stroke: the contribution of functional imaging. Prog Neurobiol 66: 109–122

[73] Risse GL, Gates JR, Fangman MC (1997) A reconsideration of bilateral language representation based on the intracarotid amobarbital procedure. Brain Cogn 33: 118–132

[74] Roessler K, Donat M, Lanzenberger R, Novak K, Geissler A, Gartus A et al (2005) Evaluation of preoperative high magnetic field motor functional MRI (3 tesla) in glioma patients by navigated electrocortical stimulation and postoperative outcome. J Neurol Neurosurg Psychiatry 76: 1152–1157

[75] Roux FE, Boetto S, Sacko O, Chollet F, Tremoulet M (2003) Writing, calculating, and finger recognition in the region of the angular gyrus: a cortical stimulation study of Gerstmann syndrome. J Neurosurg 99: 716–727

[76] Roux FE, Boulanouar K, Lotterie JA, Mejdoubi M, LeSage JP, Berry I (2003) Language functional magnetic resonance imaging in preoperative assessment of language areas: correlation with direct cortical stimulation. Neurosurgery 52: 1335–1345

[77] Roux FE, Lubrano V, Lauwers-Cances V, Tremoulet M, Mascott CR, Demonet JF (2004) Intra-operative mapping of cortical areas involved in reading in mono- and bilingual patients. Brain 127: 1796–1810

[78] Rutten GJ, Ramsey NF (2010) The role of functional magnetic resonance imaging in brain surgery. Neurosurg Focus. 28: E4

[79] Rutten GJ, Ramsey NF, Noordmans HJ, Willems PW, van Rijen PC, Berkelbach van der Sprenkel JW et al (2003) Toward functional neuronavigation: implementation of functional magnetic resonance imaging data in a surgical guidance system for intraoperative identification of motor and language cortices. Technical note and illustrative case. Neurosurg Focus 15: E6

[80] Rutten GJ, Ramsey NF, van Rijen PC, Alpherts WC, van Veelen CW (2002) fMRI-determined language lateralization in patients with unilateral or mixed language dominance according to the Wada test. Neuroimage 17: 447–460

[81] Rutten GJ, Ramsey NF, van Rijen PC, Franssen H, van Veelen CW (2002) Interhemispheric reorganization of motor hand function to the primary motor cortex predicted with functional MRI and transcranial magnetic stimulation. J Child Neurol 17: 292–297

[82] Rutten GJ, Ramsey NF, van Rijen PC, Noordmans HJ, van Veelen CW (2002) Development of a functional MRI protocol for intraoperative localization of critical temporoparietal language areas. Ann Neurol 51: 350–360

[83] Rutten GJ, Ramsey NF, van Rijen PC, van Veelen CW (2002) Reproducibility of fMRI-determined language lateralization in individual subjects. Brain Lang 80: 421–437

[84] Rutten GJ, van Rijen PC, van Veelen CW, Ramsey NF (1999) Language area localization with three-dimensional functional magnetic resonance imaging matches intrasulcal electrostimulation in Broca's area. Ann Neurol 46: 405–408

[85] Sabsevitz DS, Swanson SJ, Hammeke TA, Spanaki MV, Possing ET, Morris GL 3rd et al (2003) Use of preoperative functional neuroimaging to predict language deficits from epilepsy surgery. Neurology 60: 1788–1792

[86] Sanes JN, Donoghue JP, Thangaraj V, Edelman RR, Warach S (1995) Shared neural substrates controlling hand movements in human motor cortex. Science 268: 1775–1777

[87] Schlosser MJ, McCarthy G, Fulbright RK, Gore JC, Awad IA (1997) Cerebral vascular malformations adjacent to sensorimotor and visual cortex: functional magnetic resonance imaging studies before and after therapeutic intervention. Stroke 28: 1130–1137

[88] Sergent J (1994) Brain-imaging studies of cognitive functions. Trends Neurosci 17: 221–227

[89] Sinai A, Bowers CW, Crainiceanu CM, Boatman D, Gordon B, Lesser RP et al (2005) Electrocorticographic high gamma activity versus electrical cortical stimulation mapping of naming. Brain 128: 1556–1570

[90] Springer JA, Binder JR, Hammeke TA, Swanson SJ, Frost JA, Bellgowan PS et al (1999) Language dominance in neurologically normal and epilepsy subjects: a functional MRI study. Brain 122: 2033–2046

[91] Teixidor P, Gatignol P, Leroy M, Masuet-Aumatell C, Capelle L, Duffau H (2007) Assessment of verbal working memory before and after surgery for low-grade glioma. J Neurooncol 81: 305–313

[92] Thiebaut de Schotten M, Urbanski M, Duffau H, Volle E, Levy R, Dubois B, Bartolomeo P (2005) Direct evidence for a parietal-frontal pathway subserving spatial awareness in humans. Science 309: 2226–2228
[93] Thiel A, Habedank B, Winhuisen L, Herholz K, Kessler J, Haupt WF, Heiss WD (2005) Essential language function of the right hemisphere in brain tumor patients. Ann Neurol 57: 128–131
[94] van Oers CA, Vink M, van Zandvoort MJ, van der Worp HB, de Haan EH, Kappelle LJ et al (2010) Contribution of the left and right inferior frontal gyrus in recovery from aphasia: a functional MRI study in stroke patients with preserved hemodynamic responsiveness. Neuroimage 49: 885–893
[95] Vigneau M, Beaucousin V, Herve PY, Duffau H, Crivello F, Houde O et al (2006) Meta-analyzing left hemisphere language areas: phonology, semantics, and sentence processing. Neuroimage 30: 1414–1432
[96] Ward NS (2004) Functional reorganization of the cerebral motor system after stroke. Curr Opin Neurol 17: 725–730
[97] Winhuisen L, Thiel A, Schumacher B, Kessler J, Rudolf J, Haupt WF, Heiss WD (2005) Role of the contralateral inferior frontal gyrus in recovery of language function in poststroke aphasia: a combined repetitive transcranial magnetic stimulation and positron emission tomography study. Stroke 36: 1759–1763
[98] Wise RJ, Scott SK, Blank SC, Mummery CJ, Murphy K, Warburton EA (2001) Separate neural subsystems within "Wernicke's area". Brain 124: 83–95
[99] Wyllie E, Lüders H, Murphy D, Morris H 3rd, Dinner D, Lesser R et al (1990) Intracarotid amobarbital (Wada) test for language dominance: correlation with results of cortical stimulation. Epilepsia 31: 156–161
[100] Yacoub E, Shmuel A, Logothetis N, Ugurbil K (2007) Robust detection of ocular dominance columns in humans using Hahn spin echo BOLD functional MRI at 7 tesla. Neuroimage 37: 1161–1177
[101] Yoo SS, Talos IF, Golby AJ, Black PM, Panych LP (2004) Evaluating requirements for spatial resolution of fMRI for neurosurgical planning. Hum Brain Mapp 21: 34–43
[102] Yousry TA, Schmid UD, Alkadhi H, Schmidt D, Peraud A, Buettner A, Winkler P (1997) Localization of the motor hand area to a knob on the precentral gyrus. A new landmark. Brain 120: 141–157
[103] Zhang D, Raichle ME (2010) Disease and the brain's dark energy. Nat Rev Neurol 6: 15–28

Preoperative magnetoencephalography

Edward F. Chang, Rodney A. Gabriel, Mitchel S. Berger, Srikantan S. Nagarajan

Introduction

Magnetoencephalography (MEG) refers to the direct recording of magnetic fields produced by the human brain and is a relatively novel clinical technique for preoperative mapping of sensory and motor cortices. MEG is a complementary alternative to other functional brain imaging modalities such as functional magnetic resonance imaging (fMRI), electroencephalography (EEG), electrocorticography (ECoG), and positron emission tomography (PET). While PET and fMRI are powerful techniques that are able to reveal functionality of cortical regions, they rely on hemodynamic changes such as blood oxygenation, blood flow, and glucose metabolism. Consequently, these functional neuroimaging modalities only indirectly measure electrical activity in the brain; although they have accurate spatial resolution up to submillimeter levels. Measuring changes in vascular dynamics is, however, a distinct disadvantage of fMRI in terms of temporal accuracy as the measured hemodynamic changes occur over seconds, whereas some brain activities are occurring at the millisecond level.

Temporal changes in brain activity can be investigated more precisely with techniques such as EEG, ECoG, or MEG. EEG refers to the measurements of scalp potentials caused by bioelectric currents that are often distorted by volume conduction through heterogeneous tissues, including neurons, glial cells, meninges, skull, and scalp with different conductivity, thereby making it difficult to interpret the underlying brain activity. This problem is further compounded in the case of brain tumors wherein the conductivity of the tissue in and around a tumor is largely unknown, and so is the distortion due to scalp potentials. ECoG refers to direct invasive recordings of electric potentials on the brain's pial surface. Apart from being invasive, disadvantages of ECoG include limited coverage by the extent of the electrode arrays placed on the brain.

Bioelectric currents produced by neurons also generate magnetic fields, whose localization, on the other hand, is not distorted by the heterogeneous environment. These magnetic fields are measured by MEG. Advances in statistical signal processing methods have enabled accurate reconstructions of underlying brain sources from MEG sensor data, often referred to as electromagnetic brain imaging or magnetic source imaging. The spatial and temporal resolution of such reconstructions of activity from the measured magnetic field can be determined with exceptional accuracy (0.1 to 1 cm and 1 ms, respectively), making this technology attractive for preoperative mapping of brain function in neurosurgery patients [41].

MEG signal processing and source modeling-Biomagnetic fields detected by MEG are extremely small, about five orders of magnitude smaller than the earth's magnetic fields, and therefore appropriate data collection necessitates a magnetically shielded room and highly sensitive detectors called superconducting quantum interference devices (SQUIDs). Signals from the brain are captured by the detection coils of a biomagnetometer, which are positioned closely to the scalp, and SQUIDs act as magnetic-field-to-voltage converters from these detection coils. The total number of sensors in modern MEG systems can be about 300, spanning the entire head surface, with more sensors allowing improved spatial resolution. Although, the maximum sampling rate for SQUID detectors can be as high as 12 kHz, in most systems, MEG data is usually recorded at a rate between 500 and 1000 Hz, and still providing excellent temporal resolution at the millisecond level.

The average of sensory activity received by the local sensors may be then calculated and evoked responses to a particular stimulus can be analyzed. For example, an auditory pure tone generates an N100m peak arising from the primary auditory cortex. The spatial characteristics of evoked cortical activity may be localized by calculating parameters of equivalent current dipoles (ECD) from sensor data. By locating the magnetic dipoles, the source location of neuronal activity is estimated and correlated with the subject's MRI, thus producing the magnetic source image (MSI).

With appropriate algorithms that can be used for time-frequency analyses of oscillatory activity, MEG is not only useful for analyzing averaged evoked potentials but also for analyses of the location, timing, and frequency band of induced oscillatory activity. Research tools such as NUTMEG can be used to perform five-dimensional imaging (three spatial, time, and frequency dimensions) from MEG data and to follow the spatiotemporal pattern of oscillatory power fluctuations across the entire cortical mantle, in relation to specific tasks such as speech production or motor tasks [47] (Fig. 1).

Current applications in preoperative neurosurgery

Epilepsy surgery

Epileptic foci in patients with intractable epilepsy often lack obvious anatomic correlates and therefore accurate localization of the epileptogenic zone is essential for seizure-free outcomes after resection. Precise localization of normal functional areas is also critical when the surgical corridor is proximal to eloquent

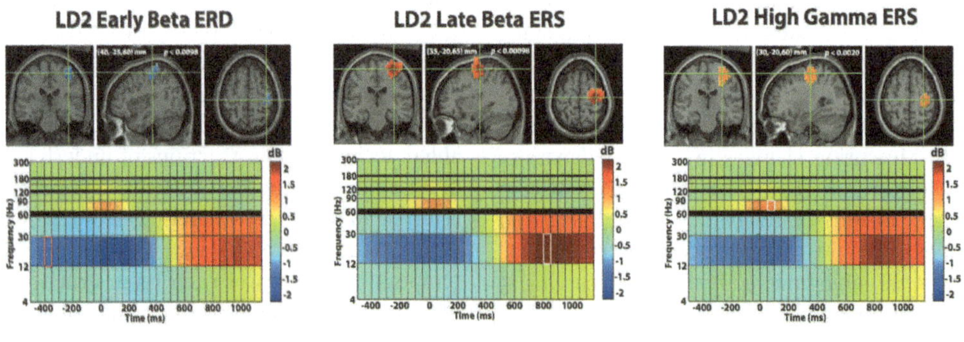

Fig. 1. Grand average reconstruction results for left index finger movement by a time-frequency beamformer, superimposed on the Montreal Neurological Institute template brain. The functional maps are superimposed on the template brain and are statistically thresholded at $p < 0.05$ (corrected for multiple comparisons). In each panel, the crosshairs mark the spatiotemporal peak for the reconstructed source, with the corresponding spectrogram shown below it. The functional map plotted on the MRI corresponds to the time-frequency window highlighted on the spectrogram

cortices. The decision on the extent and location of cortical resection for epileptic foci can be difficult and therefore a multimodal approach is preferred. Decisions on the extent of resection outside of the actual lesion must take into account the balance between preventing postoperative neurological deficits and minimizing the volume of epileptic tissue. MEG is particularly useful for both patients with nonlesional epilepsy and those with large lesions [1, 5, 27, 38], including mesial temporal lobe epilepsies [24].

Single epileptic spikes are defined as having a duration of less than 70 ms. Localization of the epileptogenic zone involves manual identification of real interictal spikes from the averaged sensor data and subsequent calculation of the location and orientation of the associated ECD [43, 48–50] (Fig. 2). One study with 455 patients demonstrated a sensitivity of 70% for identifying abnormal interictal activity [43] and several studies have also shown MEG localization of interictal spikes can lead to favorable surgical outcomes [4, 11, 14, 24].

One disadvantage of this interictal-spike localization approach is that it introduces a large amount of subjectivity from the manual visual inspection for interictal spikes (whether it is a real spike or nonepileptic origin).

Using an adaptive spatial filtering method, power changes in frequency bands associated with epileptic activity may be investigated and possibly provide a more accurate and objective methodology for preoperative evaluation with MEG. Epileptic spikes are primarily evident in the alpha, beta, and gamma frequency bands of MEG recordings. Guggisberg et al [17] describe a unique algorithm using an adaptive spatial filtering method that isolates real epileptiform activity from these high frequency bands. This technique initially involves manual identification of interictal spikes for MEG recordings followed by automated localization of the sources of the spike-locked power changes in the beta and gamma bands by an adaptive spatial filtering method. Transient increases seen in the beta and gamma power bands that were time-locked to interictal epileptic spikes

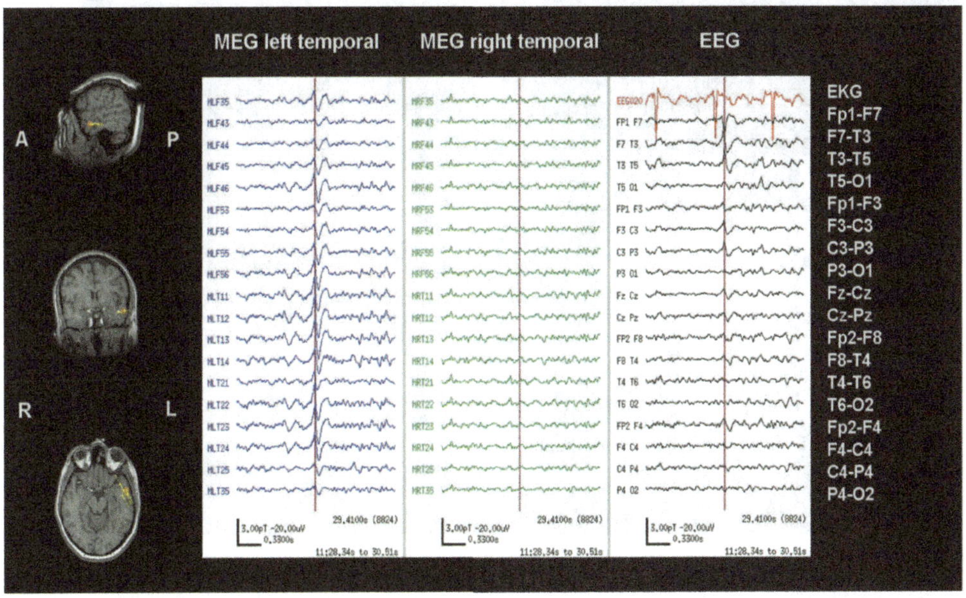

Fig. 2. MSI and MEG with simultaneous scalp EEG showing interictal unilateral temporal spikes. Single dipole source models are shown as yellow triangles with vector tails proportional to dipole strength, superimposed on the anatomical MRI. MEG and EEG show a representative left temporal spike marked with a vertical cursors. The corresponding dipole is outlined in orange on the MSI

may be used to reliably localize epileptogenic foci (Fig. 3). While still subjective, the average power changes in the frequency bands offer an alternative localization method, more robust than ECD models of single spikes. Furthermore, patients need to have a sufficient amount of interictal spikes in their recordings for effective preoperative evaluation.

Distinguishing whether generalized epileptiform discharges are from primary or secondary bilateral sources is an important decision-making factor for epilepsy surgery. Secondary bilateral synchrony is described as the rapid bilateral spread of epileptic activity from a focal cortical area of electrical abnormality. We demonstrated the usefulness of MSI in the evaluation of 16 patients with suspected secondary bilateral synchrony [8]. In this approach, interictal spikes were manually identified from MEG sensor recordings and the corresponding dipoles were fitted and visualized by MSI. This revealed a unilateral, fo-

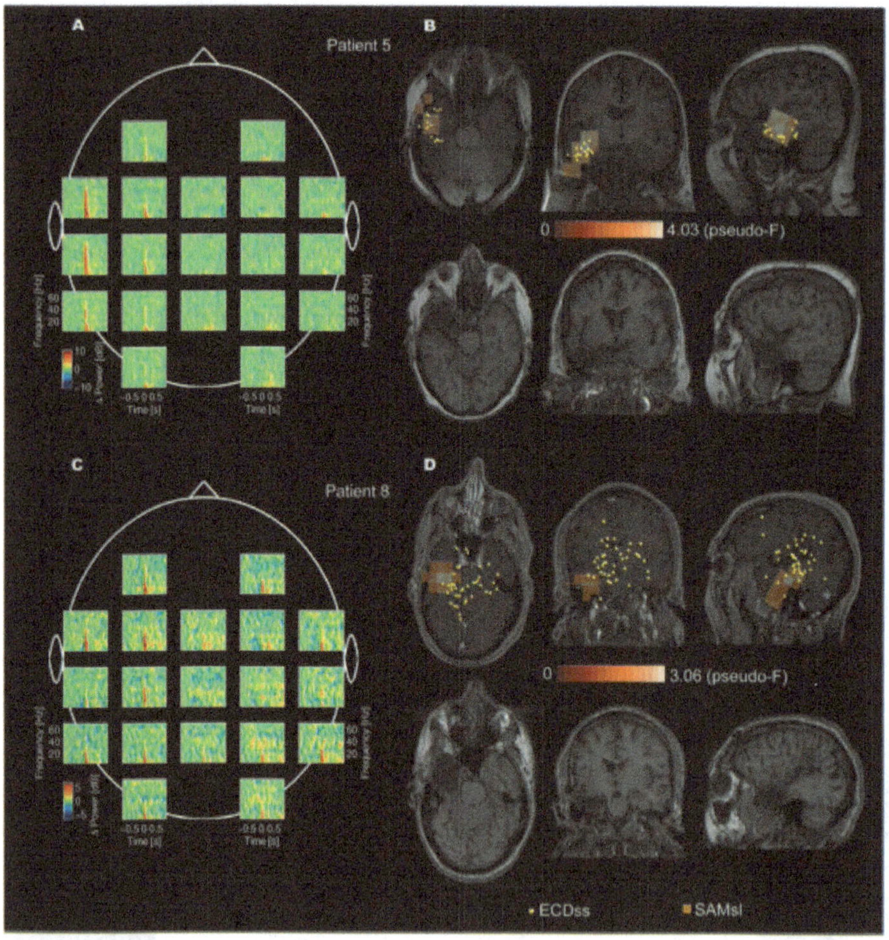

Fig. 3. MEG results in two typical patients. (**A, C**) Short-time Fourier transforms reveal transient power increases time-locked to the peak of interictal epileptic spikes at 0 seconds in a broad frequency band ranging from delta to gamma rhythms, with high-frequency changes being more focal than low-frequency changes. (**B, D**) An adaptive spatial filter localizes spike-locked beta and gamma power increases (squares in upper row) to the zone that was later surgically resected (lower row), and, in case of patient 5, to the area of clusters of corresponding ECD (yellow dots). *ECDss* Equivalent current dipoles of single spikes; *SAMsl* spike-locked dual-state synthetic aperture magnetometry

cal area of interictal spiking in the majority of patients and supported the decision to pursue resection in half of the patients.

While solitary anatomic lesions are highly prognostic of favorable seizure-free outcomes, the presence of multiple lesions can make decision-making highly complicated especially when scalp EEG is not well localizing. MSI has been applied successfully for localization in patients with multiple cerebral cavernous malformations [44]. One report found that MEG was useful in guiding the extended lesionectomy of surrounding abnormal cortex in cavernous malformations [23]. Similar results have been described for seizure localization in tuberous sclerosis patients with multiple tubers [21]. Overall, MSI appears to be an important adjunct in presurgical decision-making for epilepsy surgery, especially when there was discordance between scalp EEG and anatomic findings on MRI.

Functional mapping for preoperative planning for tumors

Gliomas often infiltrate eloquent brain areas, and it is essential to consider their proximity in the presurgical planning for patients with brain tumors. Mass lesions can frequently distort normal neuroanatomy, which consequently makes identifying eloquent cortices inaccurate with normal neuroanatomical landmarks. MEG offers reasonable accuracy for functional mapping in patients with brain tumors [13, 25, 32]. This technology has been used for localization of the sensorimotor cortex along the central sulcus [12, 30, 36, 46], as well as mapping the primary auditory [31, 41] and visual cortices [39].

The primary motor cortex and the sensory cortex are located on the anterior and posterior wall of the central sulcus, respectively. Identifying the hand region [12, 22, 30, 33, 42] and the mouth region [26, 42] of the primary sensorimotor cortex has been useful for presurgical evaluation and also confirmed with intracranial direct cortical stimulation mapping.

Motor evoked fields can be recorded by time-locking the MEG signal corresponding to movement [40], and single ECD fitting of the corresponding evoked field generated from the average sensor data [22, 26, 29, 30, 42]. Using this approach, Schiffbauer et al [42] compared MSI to intraoperative mapping in tumor patients receiving painless tactile somatosensory stimulation to the lip, hand, and foot and found that both approaches had a favorable degree of quantitative correlation. Similarly, a favorable degree of quantitative correlation was also seen from utilizing dipole fitting with MEG versus fMRI [29]. Confirmed with ECoG, dipole fitting of evoked magnetic fields to median nerve stimulation proved to be superior to fMRI for 15 patients in identifying the sensorimotor cortex [30]. Following dipole fitting of the mouth motor cortex, electrocortical stimulation sites were usually anterior and lateral to MEG localization of the lip somatosensory cortex [26].

The use of MEG spatial filtering holds promise for a more robust method for mapping the motor cortex in presurgical patients [9, 12, 33]. The use of a spatial filter beamformer while subjects perform a self-paced index finger movement can generate high-resolution imaging of the spatiotemporal patterns of premotor and motor cortex activity [9]. Peaks of the tomographic distribution of beta-band event-related desynchronization sources reliably localized the hand motor cortex in a group of 66 patients, which was confirmed with electrocortical stimulation [33] (Fig. 4).

In pediatric patients as well, the premovement motor field component in average brain response was localized. In this case, motor field time-locked to electromyography onset was successfully localized to areas corresponding to the hand region in 95% of cases (n = 10), which was confirmed with ECoG [12]. Furthermore, displacement of the sensorimotor cortex by space-occupying brain lesions does not seem to interfere with localization of the hand motor cortex [12, 33].

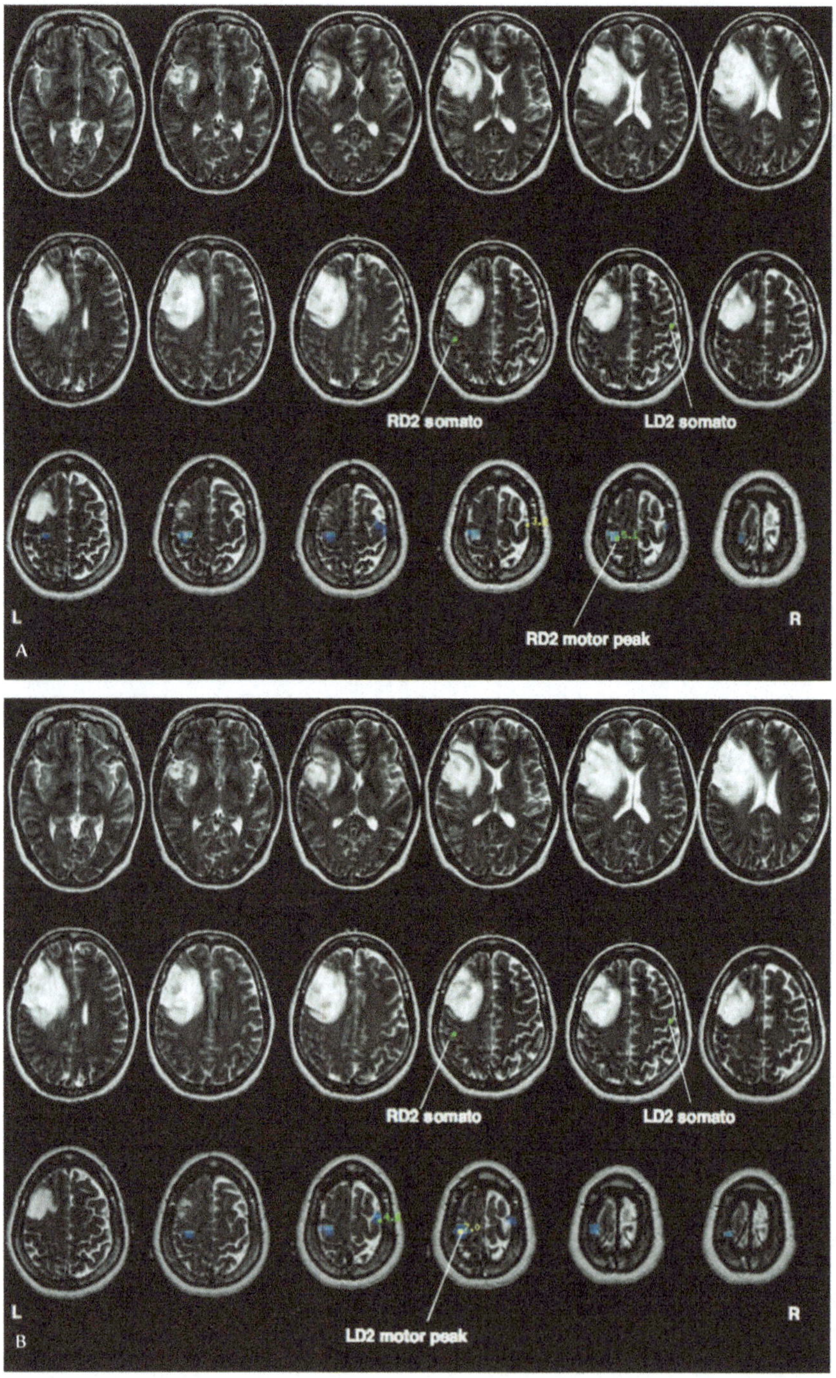

Fig. 4. **(A)** Localization of beta-band desynchronization preceding right index finger flexion for a subject with a frontal tumor. The location of hand motor cortex relative to a single dipole localization of hand somatosensory cortex is also shown. **(B)** Localization of beta-band desynchronization due to left index finger flexion in the same subject, showing contralateral hand motor cortical activation in the right hemisphere

Future applications

Lateralization of language function

Identification of a patient's dominant hemisphere is crucial when the surgical site is located near presumed language cortex. Traditionally, hemispheric language dominance is evaluated by the Wada test, which is an invasive procedure. Furthermore, crossflow of amobarbital to the contralateral hemisphere through the circle of Willis can make results inaccurate. MEG offers a noninvasive and potentially a more accurate alternative for determining hemispheric dominance.

Using MEG, language laterality can be measured by determining the asymmetry of equivalent dipole sources between both hemispheres [37]. By this approach, MSI and Wada tests were concordant in determining the dominant hemisphere in 86% of a group of 35 patients with high sensitivity and specificity [10]. Dipole sources of the late auditory evoked field components in both hemispheres can be determined while subjects undergo a recognition task for spoken words or listening to synthesized vowel sounds [37, 45]. The laterality of increased suppression of MEG activity in the 8–50 Hz range in the inferior frontal gyrus regions corresponded to the dominant hemisphere and was consistent with the Wada test among 95% of the patients [19].

Location of the language cortex (i.e., Broca's area and Wernicke's area) also holds clinical value as mass lesions can distort the anatomy and also because of interindividual anatomic variation among patients. The m100 of auditory evoked fields reside in the supratemporal auditory cortex [18], which is often surrounded by language-related cortex [34]. Grummich et al [15] compared MEG to fMRI in locating Wernicke's and Broca's areas among 172 patients. These language areas were localized in all patients; however, 4% of cases differed in MEG and fMRI and for 19% one modality showed activation while the other did not. Similarly, Kober et al [28] located Wernicke's area in the posterior part of the left superior temporal gyrus and the motor speech area in the left inferior frontal gyrus by spatially filtered MEG. Most recently, Hirata et al [20] used synthetic aperture magnetometry (an improvement over MEG dipole methods) to prospectively determine language lateralization and found high concordance with Wada testing and intraoperative cortical stimulation results. In a recent study, we have extended the approach of Hirata et al and were able to accurately characterize dynamics of language dominance by MEG, during an auditory verb generation task, a semantic association task that activates both receptive and expressive language networks. Our estimation of language lateralization based on this examination also correlates very highly with Wada test results.

Assessing functional connectivity

The term functional connectivity essentially defines the complex functional interaction between local and more remote brain areas. This concept should be considered clinically as disturbances in these networks as abnormalities in functional connecting during resting state are observed primarily in brain tumor patients when compared to healthy controls [2, 3]. Furthermore, neurocognitive effects are correlated with functional-connectivity changes in brain tumor patients, especially in patients with low-grade gliomas [6, 7]. Therefore, the mapping of functional connectivity may be an important component in surgical planning [16].

Utilizing MEG, Guggisberg et al [16] describe the changes in the time–frequency space of functional connectivity in 15 brain tumor patients compared to healthy controls (Figs. 5 and 6). Mean imaginary coherence between brain voxels was calculated as an index for functional connectivity. When compared with healthy controls, all patients with brain tumors had diffuse brain areas with decreased alpha coherence. Decreased connectivity was seen around the lesion area in patients with lesion-induced neurological deficits. In the resting connectivity in the delta and gamma frequency bands in patients with brain tumors, functional

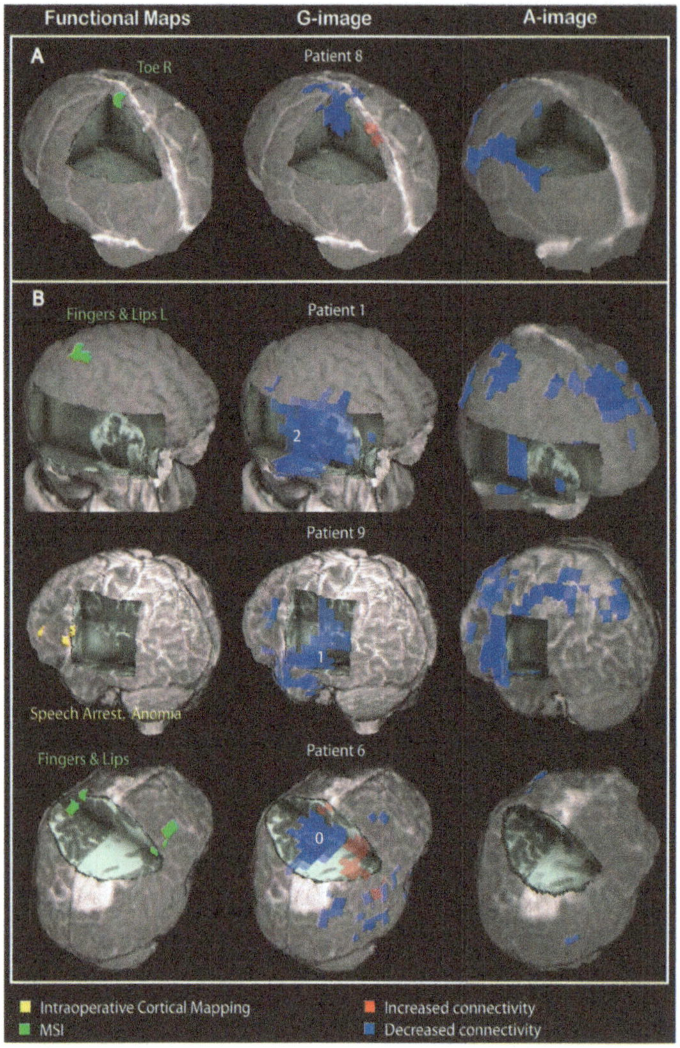

Fig. 5. Functional maps obtained by MSI or intraoperative cortical mapping as well as two different kinds (L and P) of images of functional connectivity in 4 patients with brain tumors superimposed over their three-dimensionally rendered individual brains. The L-image is a lesion-specific image of connectivity and the P-image is a patient-specific image of connectivity. (**A**) Twenty-five-year-old woman with a central paresis of the right foot due to an astrocytoma WHO grade III that infiltrated the left medial sensorimotor cortex. Note that the L-image displays a corresponding decrease in functional connectivity in the sensorimotor cortex of the right foot. (**B**) The L-images of these 3 tumor patients without presurgical functional deficits indicate functional disconnection (in blue) of different proportions of the corresponding tumor tissue (graded 0–2, with 0 indicating smallest proportion with disconnection). In agreement with the L-images and the clinical status, functional cortex was mapped outside of disconnected (blue) areas by MSI and cortical mapping in all patients. In addition, L-images predicted the functional status after radical surgery: whereas patient 6 suffered from postsurgical sensible deficits in the left arm and leg, no deficits were observed in patients 1 and 9. P-images show diffuse or scattered areas with connectivity estimates significantly lower than those of a healthy control population, but these areas are unrelated to tumor location and brain regions with functional deficits

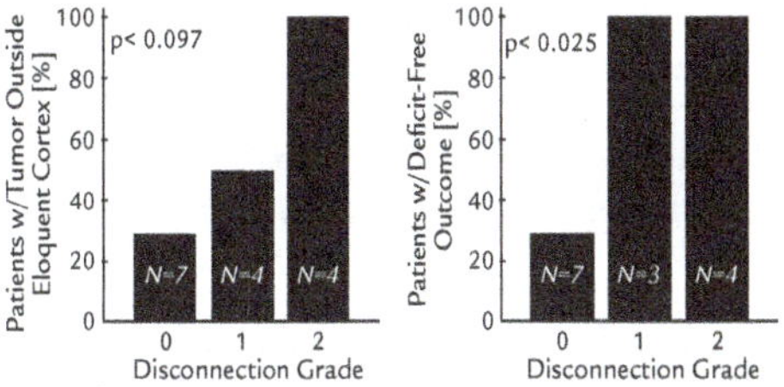

Fig. 6. Percentage of patients without critical tissue within the tumor area (**A**) and without functional deficits after tumor resection (**B**) in relation with the functional disconnection score derived from G-images

connectivity was decreased in patients with brain tumors and further decreased in left-sided versus right-sided tumors [3, 42]. Specifically, there is a decrease in high-frequency bands for long-distance connections and an increase in slower frequency bands for more local connections [2]. Another study showed that patients with low-grade gliomas had higher long-distance synchronization in the delta, theta, and lower gamma frequency bands compared with that of healthy controls [6]. The increase of relative power in the theta- and alpha-band correlated with impaired executive function, information processing, and working memory [6]. The delta- and theta-band activities were more intense in the cortex proximal to the tumor and the surrounding edematous tissues, and patients with increased volume of enhanced delta activity exhibited poor recovery of function in the early postoperative period [35]. In a recent follow-up study, we have found that these measures of functional connectivity can potentially be used to guide intraoperative electrical stimulation mapping.

Conclusion

MEG is rapidly becoming an ideal method for preoperative evaluation of patients undergoing intracranial neurosurgery. The noninvasive ability to derive high-resolution spatial and temporal cortical activity will play an increasingly important role in clinical decision-making for neurosurgery.

References

[1] Barkley GL, Baumgartner C (2003) MEG and EEG in epilepsy. J Clin Neurophysiol 20: : 163–178

[2] Bartolomei F, Bosma I, Klein M, Baayen JC, Reijneveld JC, Postma TJ et al (2006) Disturbed functional connectivity in brain tumour patients: evaluation by graph analysis of synchronization matrices. Clin Neurophysiol 117: 2039–2049

[3] Bartolomei F, Bosma I, Klein M, Baayen JC, Reijneveld JC, Postma TJ et al (2006) How do brain tumors alter functional connectivity? A magnetoencephalography study. Ann Neurol 59: 128–1386

[4] Bast T, Oezkan O, Rona S, Stippich C, Seitz A, Rupp A et al (2004) EEG and MEG source analysis of single and averaged interictal spikes reveals intrinsic epileptogenicity in focal cortical dysplasia. Epilepsia 45: 621–631

[5] Baumgartner C, Pataraia E (2006) Revisiting the role of magnetoencephalography in epilepsy. Curr Opin Neurol 19: 181–186

[6] Bosma I, Douw L, Bartolomei F, Heimans JJ, van Dijk BW, Postma TJ et al (2008) Synchronized brain activity and neurocognitive function in patients with low-grade glioma: a magnetoencephalography study. Neuro Oncol 10: 734–744

[7] Bosma I, Stam CJ, Douw L, Bartolomei F, Heimans JJ, van Dijk BW et al (2008) The influence of low-grade glioma on resting state oscillatory brain activity: a magnetoencephalography study. J Neurooncol 88: 77–85

[8] Chang EF, Nagarajan SS, Mantle M, Barbaro NM, Kirsch HE (2009) Magnetic source imaging for the surgical evaluation of electroencephalography-confirmed secondary bilateral synchrony in intractable epilepsy. J Neurosurg 111: 1248–1256

[9] Cheyne D, Bakhtazad L, Gaetz W (2006) Spatiotemporal mapping of cortical activity accompanying voluntary movements using an event-related beamforming approach. Hum Brain Mapp 27: 213–229

[10] Doss RC, Zhang W, Risse GL, Dickens DL (2009) Lateralizing language with magnetic source imaging: validation based on the Wada test. Epilepsia 50: 2242–2248

[11] Fischer MJ, Scheler G, Stefan H (2005) Utilization of magnetoencephalography results to obtain favourable outcomes in epilepsy surgery. Brain 128: 153–157

[12] Gaetz W, Cheyne D, Rutka JT, Drake J, Benifla M, Strantzas S et al (2009) Presurgical localization of primary motor cortex in pediatric patients with brain lesions by the use of spatially filtered magnetoencephalography. Neurosurgery 64: 177–185; discussion 186

[13] Gallen CC, Schwartz BJ, Bucholz RD, Malik G, Barkley

GL, Smith J et al (1995) Presurgical localization of functional cortex using magnetic source imaging. J Neurosurg 82: 988–994

[14] Genow A, Hummel C, Scheler G, Hopfengartner R, Kaltenhauser M, Buchfelder M et al (2004) Epilepsy surgery, resection volume and MSI localization in lesional frontal lobe epilepsy. Neuroimage 21: 444–449

[15] Grummich P, Nimsky C, Pauli E, Buchfelder M, Ganslandt O (2006) Combining fMRI and MEG increases the reliability of presurgical language localization: a clinical study on the difference between and congruence of both modalities. Neuroimage 32: 1793–1803

[16] Guggisberg AG, Honma SM, Findlay AM, Dalal SS, Kirsch HE, Berger MS et al (2008) Mapping functional connectivity in patients with brain lesions. Ann Neurol 63: 193–203

[17] Guggisberg AG, Kirsch HE, Mantle MM, Barbaro NM, Nagarajan SS (2008) Fast oscillations associated with interictal spikes localize the epileptogenic zone in patients with partial epilepsy. Neuroimage 39: 661–668

[18] Hari R, Aittoniemi K, Jarvinen ML, Katila T, Varpula T (1980) Auditory evoked transient and sustained magnetic fields of the human brain. Localization of neural generators. Exp Brain Res 40: 237–240

[19] Hirata M, Kato A, Taniguchi M, Saitoh Y, Ninomiya H, Ihara A et al (2004) Determination of language dominance with synthetic aperture magnetometry: comparison with the Wada test. Neuroimage 23: 46–53

[20] Hirata M, Goto T, Barnes G, Umekawa Y, Yanagisawa T, Kato A et al (2010) Language dominance and mapping based on neuromagnetic oscillatory changes: comparison with invasive procedures. J Neurosurg 112: 528–538

[21] Iida K, Otsubo H, Mohamed IS, Okuda C, Ochi A, Weiss SK et al (2005) Characterizing magnetoencephalographic spike sources in children with tuberous sclerosis complex. Epilepsia 46: 1510–1517

[22] Ishibashi H, Morioka T, Nishio S, Shigeto H, Yamamoto T, Fukui M (2001) Magnetoencephalographic investigation of somatosensory homunculus in patients with peri-Rolandic tumors. Neurol Res 23: 29–38

[23] Jin K, Nakasato N, Shamoto H, Kanno A, Itoyama Y, Tominaga T (2007) Neuromagnetic localization of spike sources in perilesional, contralateral mirror, and ipsilateral remote areas in patients with cavernoma. Epilepsia 48: 2160–2166

[24] Kaiboriboon K, Nagarajan S, Mantle M, Kirsch HE (2010) Interictal MEG/MSI in intractable mesial temporal lobe epilepsy: spike yield and characterization. Clin Neurophysiol 121: 325–331

[25] Kamada K, Takeuchi F, Kuriki S, Oshiro O, Houkin K, Abe H (1993) Functional neurosurgical simulation with brain surface magnetic resonance images and magnetoencephalography. Neurosurgery 33: 269–272; discussion 272–273

[26] Kirsch HE, Zhu Z, Honma S, Findlay A, Berger MS, Nagarajan SS (2007) Predicting the location of mouth motor cortex in patients with brain tumors by using somatosensory evoked field measurements. J Neurosurg 107: 481–487

[27] Knowlton RC, Shih J (2004) Magnetoencephalography in epilepsy. Epilepsia 45 (Suppl 4): 61–71

[28] Kober H, Moller M, Nimsky C, Vieth J, Fahlbusch R, Ganslandt O (2001) New approach to localize speech relevant brain areas and hemispheric dominance using spatially filtered magnetoencephalography. Hum Brain Mapp 14: 236–250

[29] Kober H, Nimsky C, Moller M, Hastreiter P, Fahlbusch R, Ganslandt O (2001) Correlation of sensorimotor activation with functional magnetic resonance imaging and magnetoencephalography in presurgical functional imaging: a spatial analysis. Neuroimage 14: 1214–1228

[30] Korvenoja A, Kirveskari E, Aronen HJ, Avikainen S, Brander A, Huttunen J et al (2006) Sensorimotor cortex localization: comparison of magnetoencephalography, functional MR imaging, and intraoperative cortical mapping. Radiology 241: 213–222

[31] Lutkenhoner B, Krumbholz K, Lammertmann C, Seither-Preisler A, Steinstrater O, Patterson RD (2003) Localization of primary auditory cortex in humans by magnetoencephalography. Neuroimage 18: 58–66

[32] Makela JP, Kirveskari E, Seppa M, Hamalainen M, Forss N, Avikainen S et al (2001) Three-dimensional integration of brain anatomy and function to facilitate intraoperative navigation around the sensorimotor strip. Hum Brain Mapp 12: 180–192

[33] Nagarajan S, Kirsch H, Lin P, Findlay A, Honma S, Berger MS (2008) Preoperative localization of hand motor cortex by adaptive spatial filtering of magnetoencephalography data. J Neurosurg 109: 228–237

[34] Nakasato N, Kumabe T, Kanno A, Ohtomo S, Mizoi K, Yoshimoto T (1997) Neuromagnetic evaluation of cortical auditory function in patients with temporal lobe tumors. J Neurosurg 86: 610–618

[35] Oshino S, Kato A, Wakayama A, Taniguchi M, Hirata M, Yoshimine T (2007) Magnetoencephalographic analysis of cortical oscillatory activity in patients with brain tumors: synthetic aperture magnetometry (SAM) functional imaging of delta band activity. Neuroimage 34: 957–964

[36] Ossenblok P, Leijten FS, de Munck JC, Huiskamp GJ, Barkhof F, Boon P (2003) Magnetic source imaging contributes to the presurgical identification of sensorimotor cortex in patients with frontal lobe epilepsy. Clin Neurophysiol 114: 221–232

[37] Papanicolaou AC, Simos PG, Castillo EM, Breier JI, Sarkari S, Pataraia E et al (2004) Magnetocephalography: a noninvasive alternative to the Wada procedure. J Neurosurg 100: 867–876

[38] Pataraia E, Simos PG, Castillo EM, Billingsley RL, Sarkari S, Wheless JW et al (2004) Does magnetoencephalography add to scalp video-EEG as a diagnostic tool in epilepsy surgery? Neurology 62: 943–948

[39] Plomp G, Leeuwen C, Ioannides AA (2010) Functional specialization and dynamic resource allocation in visual cortex. Hum Brain Mapp 31: 1–13

[40] Rezai AR, Hund M, Kronberg E, Zonenshayn M, Cappell J, Ribary U et al (1996) The interactive use of magnetoencephalography in stereotactic image-guided neurosurgery. Neurosurgery 39: 92–102

[41] Rowley HA, Roberts TP (1995) Functional localization by magnetoencephalography. Neuroimaging Clin N Am 5: 695–710

[42] Schiffbauer H, Berger MS, Ferrari P, Freudenstein D, Rowley HA, Roberts TP (2003) Preoperative magnetic source imaging for brain tumor surgery: a quantitative comparison with intraoperative sensory and motor mapping. Neurosurg Focus 15(1): E7

[43] Stefan H, Hummel C, Scheler G, Genow A, Druschky K, Tilz C et al (2003) Magnetic brain source imaging of focal epileptic activity: a synopsis of 455 cases. Brain 126: 2396–2405

[44] Stefan H, Scheler G, Hummel C, Walter J, Romstock J, Buchfelder M et al (2004) Magnetoencephalography (MEG) predicts focal epileptogenicity in cavernomas. J Neurol Neurosurg Psychiatry 75: 1309–1313

[45] Szymanski MD, Perry DW, Gage NM, Rowley HA, Walker J, Berger MS et al (2001) Magnetic source imaging of late evoked field responses to vowels: toward an assessment of hemispheric dominance for language. J Neurosurg 94: 445–453

[46] Taniguchi M, Kato A, Ninomiya H, Hirata M, Cheyne D, Robinson SE et al (2004) Cerebral motor control in patients with gliomas around the central sulcus studied with spatially filtered magnetoencephalography. J Neurol Neurosurg Psychiatry 75: 466–471

[47] Vartiainen J, Parviainen T, Salmelin R (2009) Spatiotemporal convergence of semantic processing in reading and speech perception. J Neurosci 29: 9271–9280

[48] Vrba J, Robinson SE (2001) Signal processing in magnetoencephalography. Methods 25: 249–271

[49] Wheless JW, Castillo E, Maggio V, Kim HL, Breier JI, Simos PG et al (2004) Magnetoencephalography (MEG) and magnetic source imaging (MSI). Neurologist 10: 138–153

[50] Wu JY, Sutherling WW, Koh S, Salamon N, Jonas R, Yudovin S et al (2006) Magnetic source imaging localizes epileptogenic zone in children with tuberous sclerosis complex. Neurology 66: 1270–1272

Preoperative transcranial magnetic stimulation: Basic principles for mapping a tumorous primary motor cortex

Thomas Picht

Background: Evolution of TMS technology for pre-surgical mapping

Soon after its introduction in the mid 1980s, TMS found widespread use in the neurosciences in general and was becoming an indispensable diagnostic tool in Neurology. By contrast, the Neurosurgical world only published two scientific papers during the 1990s on the use of TMS for the preoperative work-up of motor cortex topography [2, 8]. This was partly due to the fact that early TMS systems were yet fully developed enough to be highly accurate and easy to use, but also because TMS as a method for somatotopic mapping was not well appreciated so instead fMRI was capturing everyone's attention as the exciting new technology of the day. There are two major sets of reasons why TMS is now being rediscovered for preoperative brain mapping: on the one hand the technology has been improved, and on the other hand the surgical community is beginning to better understand its merits relative to other technologies.

Earlier TMS systems were already usable, just as computers in the 1980s and 1990s were already usable, but the early TMS systems were more demanding to use and less beneficial than TMS systems today, just as computers in the 1980s and 1990s were not as useful as the laptops and netbooks of today. The technological improvements of TMS consist of both improvements of the hardware and improvements of the software. The two main improvements of the hardware have been the refinement of the stimulation coils to make them more precisely focused, and the development of more versatile navigation systems to guide the stimulation. The two main improvements of the software have been that it now takes into account all known physical factors that have an influence on the stimulation and also that it now has a much more intuitive user-interface, so the examiner can concentrate on the mapping rather than on deciphering the TMS system itself. Together all these improvements of the hardware and software have made the technology more reliable and usable.

But as mentioned, the earlier TMS systems were already sufficiently accurate and usable, so the recent rediscovery of TMS cannot be attributed only to improvements of the technology itself. The rediscovery of TMS is also due to a change in the consciousness of the neurosurgical community. On the one hand, early TMS systems were mostly overlooked because there was a lot of preoccupation with fMRI in the 1990s, as the exciting new technology of the day. And fMRI has certainly demonstrated its usefulness. But in recent years there has been growing disappointment with its accuracy for preoperative motor mapping. On the other hand, the neurosurgical community by and large did not understand TMS well in the

1990s and simply was not ready for it. And even the researchers who thought it was valid for preoperative mapping were not really envisioning all the possibilities that TMS might open up. But now there is a growing network of neurosurgeons, engineers, and other researchers who understand more clearly what this technology could really do for neurosurgery and are communicating their vision much more clearly to the broader medical world. This chapter and book serves as part of that effort to explain what TMS can do to improve the surgical outcomes of patients with brain tumors in or near the motor cortex.

Pre-surgical mapping of the motor cortex

When and why TMS is used for preoperative mapping

There are two main indications for preoperative TMS in patients eligible for surgery with rolandic tumors. First, TMS is indicated if the functional anatomy (i.e., the exact spatial relationship between the tumor and the presumed essential motor areas) remains unclear after anatomical imaging. The reasons for this can be the mass effect of the tumor or infiltrative growth. Second, TMS is indicated if there is a discrepancy between the imaging results and the clinical findings (for example, a large tumor within M1 but no noticeable motor deficits). In such cases, the functional anatomy may have changed due to tumor-induced plasticity. In either of these two scenarios (which may occur separately or together), TMS provides elucidation of the functional anatomy and the state of the motor system preoperatively. This reduces the surgeon's uncertainty and presumably reduces the risk of causing a new iatrogenic motor deficit, though this has not yet been studied. This in turn improves the quality of preoperative decision making, in terms of patient counseling and planning of the surgical procedure. And since the results are available immediately, a TMS examination can take place, need be, even just shortly before the planned procedure.

How to map the motor cortex of a brain tumor patient using TMS

Essentially, TMS mapping is performed by stimulating multiple points on the patient's brain and recording the resulting motor output. The prerequisite for navigated TMS mapping is an MRI navigational dataset, usually in 1 mm slice thickness. The patient is positioned comfortably either in an examination chair or in a hospital bed (e.g., hemiparetic patients). Surface electrodes are attached to different muscle groups and connected to the EMG. Then the patient head and the virtual 3D navigational image are co-registered. In order to perform valid peritumoral motor mapping, one must adhere to ten principles of somatotopic mapping with TMS:

(1) The number of monitored EMG channels depends on the individual case. If detailed information on the peritumoral somatotopy is needed, more EMG channels must be recorded than in cases where identification of M1 is the only question.
(2) Stimulation spots are considered motor positive when the EMG recording shows a typically shaped MEP response at a sensible latency (between 15 and 35 msec) and the peak-to-peak amplitude exceeds a predefined threshold (usually > 50µV); the resting activity of the recorded muscle must be below the amplitude threshold.
(3) Aim for the lowest stimulation intensity. This achieves the highest accuracy, because the lower the intensity, the smaller the area of excitation.
(4) First, determine the individual resting motor threshold (RMT) by stimulation at the hand knob both on the healthy and on the tumorous side. Differences in the RMT between the hemispheres must be correlated to the clinical findings and the history of the patient. Use the RMT on the tumor side for peritumoral mapping.
(5) Peritumoral mapping should always be

targeted and should answer specific questions; mapping should be initiated at 120% of the RMT.

(6) When the course of the gyri is traceable, the stimulation (i.e., the direction of the induced e-field) should primarily be orthogonal to the gyrus, and the stimulus should be aimed at the dorsal bank of the gyrus.

(7) Stimulation controlled by neuronavigation should cover the area of the presumed precentral gyrus and the adjacent gyri. Stimulation should be performed in 1–5mm steps with the stimulation being tighter closer to areas of critical interest; a predefined raster is optional.

(8) M1 is usually reliably defined after 20–50 stimuli. If the result seems inconclusive, M1 might have been excited even though the stimulation was targeted at a neighboring gyrus. In this case, two rescue strategies can be applied. First, lower the stimulation intensity to 110% RMT, 100% RMT or even below 100%. Second, perform mapping at different angles than the optimal angle orthogonal to the gyrus, in order to prevent a stimulus from "jumping" to a neighboring gyrus.

(9) When mapping around a tumor or within a tumor (especially in low-grade gliomas), mapping should always be performed at varying angles if the anatomy is distorted by the tumor.

(10) Always keep in mind, that the neural representation of motor function may have undergone plastic changes due to the tumor with the consequence that the pattern of cortical motor representation might differ from the classic M1 somatotopy.

Usually the detailed functional somatotopy becomes clear to the examiner already during the examination. Nevertheless all stimulations and MEP responses should be stored to allow for offline analysis. If doubt about motor representation in specific areas persist, the respective MEP responses or lack of responses can be double-checked offline to rule out false positive or false negative stimulations.

Output and benefits of pre-surgical TMS mapping

There have been various TMS systems over the years, and Fig. 1 shows a couple examples. The exact form of the output depends on the TMS system that is being used, but generally speaking it consists of an MRI image of the cortex, overlaid with spots that represent the points stimulated, using various color-coding schemes to show what kind of response was obtained. The TMS responses can then be transferred to the neuronavigator for intraoperative orientation.

In Fig. 2, we see the 3D navigational view of a 64 year-old male patient suffering from a mild paresis of his right hand (BMRC grade 4/5). The MRI revealed a mass of the central region in the left hemisphere, most likely a meningeoma. Displacement of the precentral gyrus frontally was suspected. In Fig. 2A, the navigated TMS system used here applied "Line Navigation". This means that the midpoint of the figure-of-eight-coil, as the presumed spot of maximum TMS impulse, is projected onto the cortical surface. In the left panel of Fig. 2A, the TMS impulse strikes the cortex surface at an angle (blue and purple line visible), which is incorrect. In the right panel of Fig. 2A, the TMS impulse strikes the cortex surface orthogonally (only blue line visible), which is correct. The software monitors the tilt of the coil and displays the stimulation trajectory, so the examiner can aim the stimulation orthogonally at the cortex, to obtain the best mapping results. But other important physical factors, such as the rotation of the coil, are not accounted for in this system. In Fig. 2B, we see the 3D navigational view after stimulation has been carried out in a 5 mm raster. All stimulated spots are displayed, whereby white spots showed no response but yellow spots showed a reproducible CMAP > 50µV peak-to-peak amplitude on the APB recording. The examination clearly confirmed the location of the anatomical handknob immediately adjacent to the fronto-lateral aspect of the tumor.

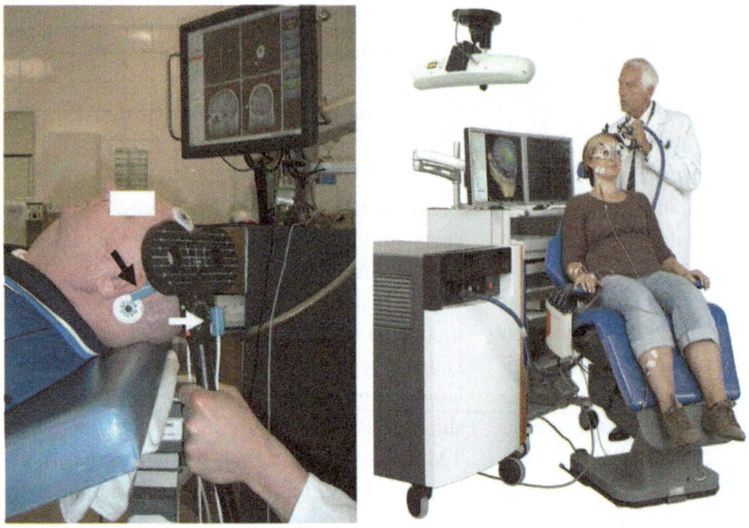

Fig. 1. Two TMS Systems. (*left panel*) "Line navigation" TMS system in use. Sensors for electromagnetic tracking are attached to the mastoid (black arrow), allowing for free positioning of the head, and to the TMS coil (white arrow). In the background, the navigation unit with an imaging display for real time navigation of the TMS coil can be seen. (*right panel*) "E-field navigation" TMS system in use. Reflective spheres are attached to the patient's head (modified glasses) and to the TMS coil. The camera for optical tracking can be seen in the upper left. The stimulator coil is shown placed against the patient's head. The motor output is recorded by surface electrodes attached to the face, arm, and leg. The anatomical map of the patient's brain is shown on the computer screen in the upper left, while the MEP output tracings are shown on the computer screen in the upper right

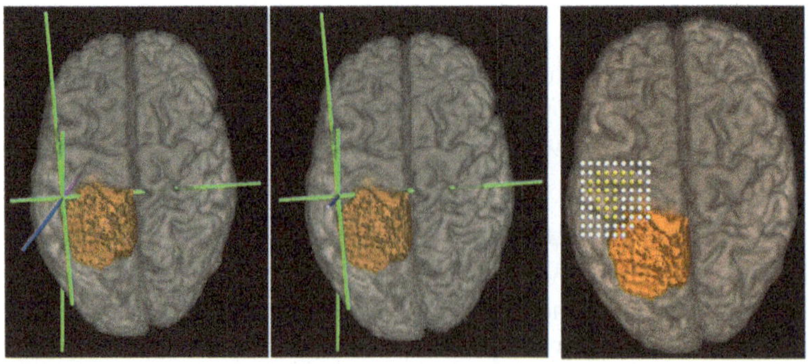

Fig. 2. TMS mapping using the system in the left panel of Fig. 1, performed on a 64 year-old male patient with a mild paresis of his right hand. (**A**) Navigational views during mapping procedure. (**B**) Mapping results (See text for further explanations)

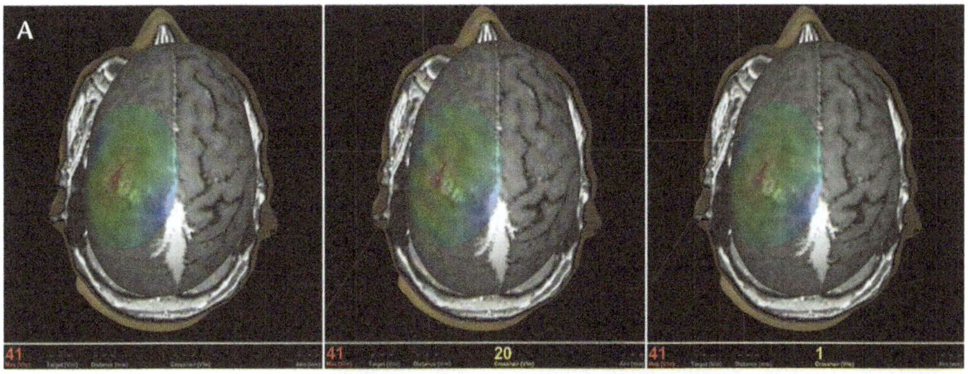

Fig. 3. TMS mapping using the system in the right panel of figure 1, performed on a 63 year-old female patient suffering from a mild hemiparesis on her right side. (A) Navigational views during mapping procedure. (B) Mapping results (See text for further explanations)

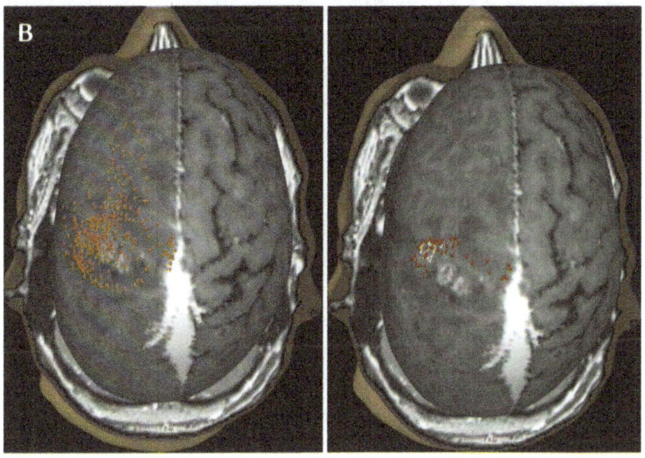

In Fig. 3, we see the 3D navigational view of a 63 year-old female patient suffering from a mild hemiparesis of her right side (BMRC grade 4/5). The MRI revealed a tumor in the central region of the left hemisphere. A high-grade glioma or metastasis was suspected. The functional anatomy of the primary motor cortex remained unclear after anatomical imaging. In Fig. 3A, the navigated TMS system used in this case applied "e-field navigation". The red arrow indicates the direction of the induced e-field. The color coding reflects the strength of the e-field (red > green > blue). The red number in the lower left corner states the maximal e-field value in volts per meter at the center of the arrow. The left yellow number states the e-field at the crosshair. The image sequence demonstrates the rapid decay of the e-field from the center to the periphery. In the left panel of Fig. 3A, we see stimulation with the e-field pointing anteriorly and slightly angled toward the midline. The e-field at the center of the arrow is 41 V/m. In the middle panel of Fig. 3A, we see the crosshair has been positioned in the green area, and the e-field at the crosshair is less than half (20 V/m) of the maximum e-field. In the right panel of Fig. 3A, we see the crosshair has been positioned in the blue area, and the e-field is now only 1 V/m. In Fig. 3B, we see the 3D navigational view showing the results after stimulation has been performed. In the left panel of Fig. 3B, all spots stimulated on the left hemisphere are displayed. The relevant area, i.e. the area adjacent to the tumor, has been stimulated in a dense raster. The premotor cortices have also been stimulated. In the right panel of Fig. 3B, the image displays only the spots where a muscle response was observed (MEP > 50 μV peak-to-peak amplitude). Three different hand muscles (APB, FDI, ADM) and one leg muscle (TA) were recorded in this case. The color coding corresponds to the intensity of the response, whereby red indicates small responses (MEP 50–500 µV), yellow indicates medium responses (MEP 500–1000 µV), and white indicates large responses (MEP > 1000 µV). This mapping makes it evident that the precentral gyrus has been displaced frontally and that the center-of-gravity for the hand muscle representation is located immediately adjacent fronto-laterally to the tumor. The responses close to the midline are from the leg (TA).

Analysis by TMS provides further information in addition to the motor topography. For example, a higher RMT on the tumor side in a neurologically intact patient can reflect an imminent risk of paresis. Or for another example, the activation pattern during M1 and M2 stimulation might reflect motor system plasticity if unusual M2 activation is observed in a patient with an M1 tumor. The synthesis of the patient's clinical status, MRI findings, TMS somatotopy, and TMS analysis of motor system excitability in primary and non-primary motor areas can improve the surgical team's ability to make a prognosis about the risk for suffering permanent motor deficits.

How to construct the TMS map image

In order to make clear and useful mapping images from the TMS examination, a few points should be kept in mind. First it must be taken into account that the excitation of the motor system takes place most probably not on the crown of the gyrus but rather at the dorsal bank of the precentral gyrus. This corresponds to a depth of about 20–30mm under the scalp, depending on individual factors such as coil-cortex-distance. So the results should be shown at a level of about 20–25 mm under the coil, to obtain the neurophysiologically most accurate visualization. This also corresponds to the depth where the gyral anatomy is most eminent, which enables the surgeon to easily understand the functional anatomy when planning his or her operation.

Second, different modes of visualization are possible. Color-coding can be used to simply distinguish motor-positive from motor-negative spots. Or it can be used to differentiate which muscles responded at which stimulation locations. Or color-coding can be used to show the strength of the motor responses. Any positive responses close to the tumor should certainly be displayed, so the surgeon does not resect those areas, but otherwise, positive spots do not necessarily all need to be shown. Identification of M1 should be self-evident, so if there are too many markers of positive responses cluttering the imaging, then some of them can be removed to make the brain anatomy visible again. Negative spots located within the tumor or immediately adjacent to it should be shown, because it is important to document the lack of function within the tumor or close to the planned resection margin. Otherwise, negative spots (outside the tumor) should not normally be shown, because they clutter the image and obscure the underlying anatomy. The mapping image(s) should be accompanied by a text file that briefly describes the mapping parameters used, the results obtained, and any interpretations or conclusions made by the examiner.

Problems and limitations of TMS mapping

In tumor patients, the routine procedure of TMS can face some special obstacles. The patients can have difficulty relaxing, which can cause a lot of noise in the EMG signal and consequently false positive responses. The patients might quickly become uncooperative. The TMS examination set-up allows for conversation with the patient before and during the procedure and can help to relax them and establish rapport with the examiner. These patient issues are also why it is important to have comfortable seating or to allow the patient to remain in the hospital bed during the procedure. In general, the procedure must be quick, easy, and flexible, because the patients are often not in top condition for enduring long tedious work-ups.

Single-pulse TMS is safe in terms of seizure induction or other adverse events. Nevertheless, experience in tumor patients is still limited. Epileptic seizures are not a contraindication for TMS, but these patients must be fully informed about the possibility of a seizure. The operator should also be consciously prepared to respond to such an eventuality.

Finally, as with all imaging modalities, reliable interpretation of TMS imaging requires an understanding of the methodology that produced that imaging. The maps cannot be viewed uncritically as the absolute truth of objective reality. The examiner must be aware of some important facts, regarding the visualization of the TMS in relation to the cortex. Due to the individual head shape and cortical anatomy as well as the neurophysiology of TMS-induced corticospinal volleys, the excitation of the pyramidal cells does not necessarily happen underneath the midpoint of the coil. The rotation and the tilting angle of the coil, the strength of the magnetic field, and the individual anatomy all have an influence on the cortical site where the excitation takes place. TMS navigation should visualize all these physical factors. The best approximation of where the excitation possibly takes place can be reached by visualizing the resulting e-field in relation to the cortex. Yet one must keep in mind that local tissue factors have an influence on the resulting electrical current. Due to all these factors, both the examiner and anyone viewing the maps later must keep in mind how the imaging is produced, in terms of the device engineering and the patient neurophysiology.

TMS compared to other mapping modalities

Other options for mapping the motor cortex

There are of course, as this book shows, several methods available for mapping the motor cortex. When used for preoperative mapping, the aim of all these methods is to identify the so-called eloquent structures, which in daily neurosurgical practice means the areas that cannot be removed or damaged without causing a permanent neurological deficit. In regard to motor function, the precentral gyrus (also called the "primary motor cortex") has, with its strict somatotopic order, historically been considered the only eloquent structure on the cortical level. Although this view has been modified by findings on overlapping motor representations within the precentral gyrus and direct corticospinal tracts originating from the superior frontal gyrus, the precentral gyrus still remains the dominant factor for assessing the risk of morbidity in rolandic tumor surgery. Studies on brain connectivity and intraoperative stimulation studies have proven that the motor system is a dynamic network that is organized hierarchically around the cortical epicenter, i.e. the precentral gyrus [12]. The various brain mapping technologies have different ways of representing this complex network, and therefore have various advantages and drawbacks.

All preoperative mapping modalities that have been used in recent years (fMRI, MEG, PET, EEG) record "brain activity" after the patient has performed certain tasks and then use biomathematical models to reconstruct the recorded data into functional information. This information has been successfully implemented into surgical planning (see previous chapters). Nevertheless, these observational methods suffer from the limitation that their biomathematical models can be inadequate to reliably determine which areas produce essential function in the vicinity of a tumor, especially when complex networks are activated during motor paradigms. There is a risk of false negatives, which can increase the surgical morbidity, and also the risk of false positives, which can leave tumors inadequately resected. The reason for this is that these observational methods also identify areas as "essential" that are part of the motor network but that are in reality merely involved in the task in inessential ways and therefore could be safely resected without causing any lasting deficit [7].

TMS is the only non-invasive method that allows for examination by stimulation, like the gold standard of intraoperative DCS. The stimulation of a precise cortical spot enables assessment without the complex biomathematical modeling involved in some other methods. Points of the brain where TMS at 120% RMT evokes MEPs within normal range latencies can be considered as "essential" with the same reliability as intraoperative stimulation mapping.

TMS vs. fMRI

Functional magnetic resonance imaging (fMRI) is currently the most widely used method for preoperative mapping of motor function, and it has been discussed extensively in the other chapters here, as well as in a vast body of journal papers. But how does it compare to TMS for preoperative mapping of the motor cortex in brain tumor patients? The first obvious advantage of fMRI is that it is already widely available at many hospitals and also that there is a vast scientific literature about its use and interpretation. Furthermore, fMRI is superior to TMS for examining functional neural *networks*. fMRI enables analysis of the whole brain over a period of time. Although the temporal resolution is low, it enables the medical team to distinguish between areas that are activated almost simultaneously versus areas that are activated separately. In other words, fMRI can provide information on the temporal sequence of all the neural areas involved in a task. Thus one may draw conclusions about distinct networks and their interconnections. Additionally, longitudinal fMRI data enables one to visualize any long-term brain plasticity, for example after a stroke or tumor resection.

On the down-side, fMRI involves a complex methodology. There are no standardized, user-independent protocols. Slight changes in task design, task performance, or data analysis can have substantial impact on the activation maps one obtains [11]. Successful identification of the precentral gyrus depends upon the quality of the data, which depends on many factors: the signal-to-noise ratio, motion and susceptibility artifacts, the chosen motor task, the subject's ability to perform the task, and an intact neurovascular coupling. All these factors can be compromised in tumor patients and negatively affect the accuracy of the fMRI mapping result. Even when the data is of good quality, the biomathematical analysis of the data requires an expert examiner who knows by experience which analysis threshold best reflects reality. And still there is a real risk of false negatives, especially if there is impaired neurovascular coupling around the tumor. There is also a risk of false positives, primarily due to activation of non-primary motor or non-motor areas, since it is very difficult to selectively activate only the primary motor cortex [19].

By contrast, TMS minimizes the risk of false negatives and positives, because it actively tests the areas that will be at risk during the operation. TMS can repeatedly target the area in doubt, in order to obtain a clear picture of the functional topography (see Fig. 3B). Furthermore, the stimulation result is obtained immediately; whereas, the analysis of functional imaging procedures requires further processing time after data acquisition, so any lingering uncertainty cannot be resolved without bringing the patient back in for another examination session. Moreover, TMS has, in theory, a high accuracy for identification of the cortical entry gate to the motor network. The more important the entry gate, the higher the motor output. For surgical planning this means that TMS has the benefit of reliably identifying the classic primary motor cortex representation. fMRI can also identify it, but the analysis is much more complex and demanding.

fMRI and TMS findings do not necessarily agree completely since they use different entry gates to the motor system and measure different outputs. So far, only one small (n=15) study from the gray literature has compared fMRI and TMS to DCS in the same patients; they found a mean (SD) distance of 10.5 (5.67) mm between the TMS and DCS hotspots and 15.0 (7.6) mm between the fMRI and DCS hotspots [5]. Further peer-reviewed comparisons of the topographic precision of fMRI vs. TMS are still needed, but regardless, the strong point of fMRI is the visualization of complex networks and the possible changes in brain activation patterns due to tumor-induced plasticity. By contrast, TMS is especially well-suited for clarification of cortical functional anatomy. Thus they are complementary methods, each best suited for answering different kinds of questions.

TMS vs. MEG

Magnetoencephalography (MEG) has demonstrated its ability to reliably identify the precentral gyrus in healthy subjects and also in tumor patients [13]. Its main strong point is its superior temporal resolution compared to both fMRI and TMS. MEG measures electromagnetic changes from neuronal activity on a timescale of milliseconds immediately before the onset of voluntary movement; whereas, fMRI is based on the measurement of hemodynamic responses to neuronal activation, but these hemodynamic responses are much longer than the underlying neuronal activity, lasting up to 15 seconds after a stimulus. And TMS does not measure brain activity at all, but EMG motor output, which does not allow for analysis of the temporal activation pattern of distinct cortical areas. MEG's superior temporal resolution enables the medical team to analyze each step of motor planning and performance in millisecond steps [9]. This ultrafine temporal analysis of the motor network is not possible with any other mapping modality, including TMS.

But like fMRI, MEG involves biomathematical analysis of the raw data that is time consuming and requires expert knowledge. By contrast, TMS does not require post-exam analysis, because the stimulation result is immediately evident. Thus, TMS can be performed easily by any trained medical personnel, if the guidelines mentioned above are taken into account, and the results are available right away. Moreover, there is a major financial barrier to the adoption of MEG technology. Its very high cost – for both initial acquisition and ongoing maintenance – ensures that it is available only at highly specialized centers.

Preoperative mapping by both MEG and TMS would not necessarily agree completely, since they use different entry gates to the motor system and different outcome measures. But so far, there have been no studies comparing MEG and TMS, undoubtedly because neither technology is widespread yet and only a few centers in the whole world have them both.

TMS vs. DCS

Intraoperative direct cortical stimulation (DCS) is still considered the gold standard for functional mapping of the primary motor cortex. When DCS is performed correctly, its sensitivity for detection of eloquent structures is 100% [10]. The major advantage of TMS over DCS is that TMS is conducted preoperatively. Clearly this allows a more timely and thorough examination of motor topography with TMS, especially if the operation is challenging (e.g., due to a severe tumor mass effect). Moreover, if intraoperative complications are encountered, these can lead to aborting DCS mapping altogether.

The results currently available suggest that navigated TMS motor mapping has a similar accuracy to DCS. In a recent study, the topographic median (range) distance between the APB hotspots of nTMS and DCS was 8.43 (0.83–15.59) mm [16]. Two earlier papers also reported a good correspondence between nTMS and DCS [6, 8], though each paper was based on only two patient cases each, so the findings cannot be taken as conclusive. In a previous report, exactly the same spots were stimulated with TMS and DCS in a 5 mm raster. The median (range) distance between nTMS and DCS hotspots in that study was 5 (0–7) mm [15]. For both TMS and DCS, the exact extent of the stimulated cortical area still remains unclear, and discrepancies between the mapping results reflect methodological differences rather than "inaccuracies" of either method.

For DCS, it has been demonstrated that during bipolar cortical stimulation the current peaks in the region directly below the bipolar electrodes; whereas, current density decreases much less rapidly with depth during monopolar anodal stimulation [14]. Consequently, suprathreshold anodic stimulation of the motor cortex primarily leads to direct stimulation of the pyramidal cells [17]; whereas, responses after bipolar stimulation are also mediated by intracortical connections that conduct the stimulus into adjacent neurons [4]. Single-pulse TMS is likely to involve both tangential cortical fibers and direct corticospinal axonal bundles [1, 18]. Depending on the e-field direction and the stimulation strength, TMS of M1 will preferentially activate the pyramidal cells indirectly (i.e., transsynaptically) or directly at their axon hillock. These observations imply that DCS and TMS stimulate preferentially the same population of neurons. Nevertheless, the exact stimulation path remains unknown in the individual case, especially around a tumor with possible changes of conductivity.

Conclusion: the future of TMS in pre-surgical motor mapping

Navigated TMS has proven to be a reliable and accurate method for preoperative peritumoral motor mapping. In comparison to functional neuroimaging methods, TMS has the benefit of being the only painless preoperative method that establishes a causal link between the stimulation of an area and the observed motor output, in a fashion similar to DCS. TMS can be performed quickly and easily by a non-expert user; the results are available immediately; and the interpretation of the results is straight-forward. Moreover, TMS can be conducted on patients who are not able to perform movement tasks for functional imaging, due to hemiparesis for example.

TMS analysis will in the near future surely take on further roles in motor cortex neurosurgery, besides topographical mapping to improve surgical planning and patient counseling. TMS is already starting to be used to evaluate the status of the motor system, by analyzing the patterns of M1 and non-primary motor area activations, the necessary stimulation strengths, and the latencies and resulting MEP amplitudes. These and other parameters provide a detailed assessment of the *status* of the motor system that far exceeds the mere clarification of topography. Researchers are also currently working on ways to use TMS as the basis for fiber-tracking, which will allow TMS to support surgical planning at subcortical levels. TMS analysis might also be useful to predict the outcomes of surgery in terms of potential brain plasticity. Similar tumors with similar clinical findings may call for different treatment strategies if the TMS findings point to different potentials for plastic reorganization and thus different potentials for recovery after resection of a lesion located in an area that would normally be regarded as a surgical no-touch area. Postoperative TMS examinations can help to further specify the individual prognosis in regard to motor function and can improve the planning of patient-specific rehabilitation. Indeed, TMS even has the potential to *promote* rehabilitation by induction of brain plasticity [3]. The use of TMS mapping in neurosurgery is just beginning to take off now in the 21st century. Undoubtedly many new discoveries and applications for TMS in neurosurgery will be found as more medical teams adopt the technology, spend time exploring its potential, and sharing their findings with the rest of the scientific and medical community.

Acknowledgments

I would like to thank Michael Hanna, PhD, (Mercury Medical Research & Writing) for providing feedback and revision of the manuscript.

References

[1] Amassian VE, Stewart M, Quirk GJ, Rosenthal JL (1987) Physiological basis of motor effects of a transient stimulus to cerebral cortex. Neurosurgery 20: 74–93

[2] Asakura T, Tokimura H, Hirahara K, Baba K (1994) Identification of the cerebral motor cortex by focal magnetic stimulation: Clinical application to neurosurgical patients. Stereotact Funct Neurosurg 63: 177–181

[3] Bolognini N, Pascual-Leone A, Fregni F (2009) Using non-invasive brain stimulation to augment motor training-induced plasticity. J Neuroeng Rehabil 17,6: 8

[4] Day BL, Rothwell JC, Thompson PD, Dick JP, Cowan JM, Berardelli A, Marsden CD (1987) Motor cortex stimulation in intact man. 2. Multiple descending volleys. Brain 110: 1191–1209

[5] Forster MT, Szelényi A (2010) Integration neuronavigierter transkranieller Magnetstimulation in die Resektionsplanung zentral gelegener Tumore. 17. Tagung der Sektion Neurophysiologie der Deutschen Gesellschaft für Neurochirurgie 19.–20. 02. 2010, Frankfurt

[6] Kantelhardt SR, Fadini T, Finke M et al (2010) Robot-assisted image-guided transcranial magnetic stimulation for somatotopic mapping of the motor cortex: a clinical pilot study. Acta Neurochir 152: 333–343

[7] Krainik A, Duffau H, Capelle L, Cornu P, Boch AL, Mangin JF, Le Bihan D, Marsault C, Chiras J, Lehéricy S (2004) Role of the healthy hemisphere in recovery after resection of the supplementary motor area. Neurology 62: 1323–1332

[8] Krings T, Buchbinder BR, Butler WE, Chiappa KH, Jiang HJ, Rosen BR, Cosgrove GR (1997) Stereotactic transcranial magnetic stimulation: correlation with direct electrical cortical stimulation. Neurosurgery 41: 1319–1326

[9] Mäkelä JP, Forss N, Jääskeläinen J, Kirveskari E, Korvenoja A, Paetau R (2006) Magnetoencephalography in neurosurgery. Neurosurgery 59: 493–511

[10] Mandonnet E, Winkler PA, Duffau H (2010) Direct electrical stimulation as an input gate into brain functional networks: Principles, advantages, and limitations. Acta Neurochir 152: 185–193

[11] McGonigle DJ, Howseman AM, Athwal BS, Friston KJ, Frackowiak RS, Holmes AP (2000) Variability in fMRI: an examination of intersession differences. Neuroimage 11: 708–734

[12] Mesulam MM (2000) Principles of behavioral and cognitive neurology. Oxford University Press, New York

[13] Nagarajan S, Kirsch H, Lin P, Findlay A, Honma S, Berger MS (2008) Preoperative localization of hand motor cortex by adaptive spatial filtering of magnetoencephalography data. J Neurosurg 109: 228–237

[14] Nathan SS, Sinha SR, Gordon B, Lesser RP, Thakor NV (1993) Determination of current density distributions generated by electrical stimulation of the human cerebral cortex. Electroencephalogr Clin Neurophysiol 86: 183–192

[15] Picht T, Mularski S, Kuehn B, Vajkoczy P, Kombos T, Suess O (2009) Navigated transcranial magnetic stimulation for preoperative functional diagnostics in brain tumor surgery. Neurosurgery 65 (Suppl 6): 93–99

[16] Picht T, Schmidt S, Brandt S, Frey D, Hannula H, Neuvonen T, Karhu J, Vajkoczy P, Suess O (2010) Preoperative funcational mapping for rolandic brain tumor surgery: Comparison of navigated transcranial magnetic stimulation to direct cortical stimulation. [under review]

[17] Ranck JB Jr (1975) Which elements are excited in electrical stimulation of mammalian central nervous system: a review. Brain Res 98: 417–440

[18] Ruohonen J, Ilmoniemi RJ (2002) The handbook of magnetic stimulation: Physical principles for TMS. Arnolds Publishers, London, pp 18–29

[19] Rutten GJ, Ramsey NF (2010) The role of functional magnetic resonance imaging in brain surgery. Neurosurg Focus 28(2): E4

Navigated repetitive Transcranial Magnetic Stimulation (TMS) for language mapping: a new tool for surgical planning

Josep M. Espadaler and Gérardo Conesa

Introduction

When a resection in dominant perisylvian areas is needed, language function is a major concern [1, 2, 3, 4]. Different methods, both invasive and non-invasive, have been used to address this issue, such as functional MRI (fMRI), MEG (magnetoencephalography), event related potentials, PET scanning, cortico-cortical potentials and DCS (direct cortical stimulation) – the latest considered up to now as the gold standard both for cortical and subcortical mapping as well as for intraoperative monitoring [5, 6, 7, 8]. Furthermore, different specific language tasks have been used to segment language function into distinct subsets of cerebral organizational resources for all the methods previously mentioned.

FMRI is the non invasive method the most frequently used, with an exponentially increasing number of papers published in the last years [9, 10, 11, 12, 13]. There is also extensive literature concerning the use of this tool for presurgical language planning and the correlation between fMRI and DCS [14]. This issues are more extensively discussed in other chapters in this book, but the core idea is that fMRI is defining participating areas in a given language task, but do not elucidate which of them are actually essential and therefore should not be resected. In spite of this dubious intrahemispheric localization, it is considered a useful tool for language lateralization, with a good correlation with the Wada test. Anyhow fMRI discloses important information about related cortical areas for a given task and thus leads to information on anatomo-functional connectivity that can be attained by using these areas as ROIs for DTI tractography. Furthermore, longitudinal fMRI gives insight to neural plasticity of language function [7].

Interestingly, navigated TMS (nTMS) uses a quite similar concept than DCS for language function localization. When a patient is performing a language task, an electrical (DCS) or a magnetic (nTMS) current can interfere with the test by synchronously altering the neurons included in the field. Indeed, if these neurons are crucial for the specific task, the function being studied will show an alteration such as speech arrest, paraphasia, anomia, etc. Thus, it can be considered that the method simulates a situation in which the cortical areas and subcortical connecting tracts are virtually and transitorily damaged. Therefore nTMS arises as an alternative non invasive method for presurgical language mapping and may play an important role in designing tailored surgical strategies.

Speech arrest induced by repetitive TMS (rTMS) has been previously reported [15–21]. The use of a navigation system allows precise location of the stimulus delivered by TMS in the cortex of the individual by showing it onto a 3D model rendered from an MRI previously loaded into the system. As shown in the previous chapter, nTMS with a single stimulus has been used to perform primary motor cortical maps by recording responses from the muscles [22]. In this setting the estimated accuracy of the cortical area stimulated is less than 10 mm [22–24].

Papers studying the correlation of nTMS with DCS language mappings are scarce. In our experience, we have explored language using nTMS in 18 patients with a different subset of lesions such as gliomas, cavernomas, AVMs, meningiomas and also for epilepsy surgery cases. DCS has been correlated in all of these patients. Here, we will show some illustrative cases of this series.

In addition, diffusion tensor imaging (DTI) tractography allows the visualization of subcortical pathways both in healthy volunteers and in patients. Different methods have been applied to define these tracts and all of them have several limitations. Therefore, they have to be addressed carefully for case planning, especially in relation to false positives and false negatives with regard to the different possible tract interpretations provided by distinct softwares. Nevertheless DTI tractography is currently used in surgical planning and translated to the operating room in neuronavigation systems. There is also a recent literature which attempted to correlate data obtained during intraoperative DCS of the subcortical language pathways with data provided by DTI tractography [25–29]. A different and appealing approach may consist to use the nTMS language mapping areas as ROIs for DTI tract localization in order to further understand connectivity at the individual level and therefore to delineate safer surgical resections. This has also been done in our series.

Finally, since 2004, we use an augmented reality stereoscopic planning system, namely Dextroscope® (Volume Interactions). It has a full range of visualization, segmentation, registration and reporting tools, and all of non emergent cases are previously planned in this setting. We can include data on tumour metabolism with methionine PET scanning as well as information on neural function with MEG, fMRI and nTMS, allowing the performance of a 5D stereoscopic tool (3D stereoscopic + physiology + function). In this stereoscopic setting, the more complex is the image, the more advantage for easier understanding is provided by a 3D system. This is particularly true for DTI and functional interpretation. Indeed, it is difficult to interpret the data put into a statistical analysis, but it is very easy to see them just by pressing the monoscopic or stereoscopic button in the virtual keyboard with the natural stereoscopic human vision. This system may also be helpful in longitudinal analysis for neural plasticity studies and as a neuroimaging database tool.

Methodology

We have used two repetitive butterfly stimulation coils, MagPro Series Magstim (Medtronic, USA) and Nexstim TMS (Nexstim, eXimia, Finland), coupled to a nTMS system (NBS: Navigation Brain System, Nexstim, eXimia, Finland) to perform the presurgical language mapping. Two kinds of localizing glasses have been used in order to be able to fully explore the temporal, parietal and frontal areas.

For the mapping of speech areas, we use repetitive nTMS (rnTMS). After setting the threshold of tenar muscles at rest, we raise the rnTMS to a 110 to 120% over this motor threshold and rnTMS train of 20 pulses at 10 Hz during 2 seconds is delivered while the subject or patient is performing a language task.

Counting, reading and naming tasks have been used in these patients. No synchronization device between object presentation and stimulation has been used. Some other tasks such as verb generation and spontaneous speech have also been used in specific patients. A language area has been assigned onto the 3D MRI model when a language disturbance has been produced at least three times at the same location.

The aim has always been to explore a wide perisylvian area, comprising the inferior-posterior frontal cortex, first and second temporal gyri, supramarginal and angular gyri. When a patient has experienced pain or has not been comfortable with the magnetic stimulation, the area causing this problem has not been explored.

A language map has been depicted with language areas shown as cubic volumes of 5 mm sides, and exported by means of a DICOM export program to the Dextroscope® and to the surgical navigation system (stealth station, Medtronic, USA). This is helpful for DTI tractography and it also allows intraoperative DCS and nTMS comparison. Intraoperative pictures of DCS maps have also been taken for every patient.

DCS language mapping has been performed by using square biphasic 0.2 msec stimuli in 50 Hz trains of 2 seconds duration at increasing intensities starting at 1 mA up to 16 mA. A stimulation area is considered crucial for language when eliciting a language problem in at least 2 out of 3 stimulations – performed just below the afterdischarges threshold determined by electrocorticography.

Correlation is made by comparing the coordinates between DCS with nTMS in the Dextroscope® and by comparing intraoperative pictures with 3D stereoscopic renderings. A good concordance is defined when the distance between both methods is under 10 mm.

A 3.0 T Philips and a 1,5 GE MR systems have been used for DTI sequence acquisition. A deterministic eigenvector paradigm and software have been used to define tractography using the Dextroscope® planning system.

Results

A global map of the areas of speech arrest elicited by nTMS in our first 7 patients (with tumours within the left fronto-temporal region), further confirmed by intraoperative DCS, is shown in Fig. 1.

Illustrative cases

Case 1

A 30-year-old bilingual right-handed female presented with a generalized seizure. A MRI showed a left extra-axial tumor, typical for a meningioma. The nTMS study showed different cortical areas for Euskera and Spanish (Figs. 2 and 3). Being the patient a medical doctor and after discussion with her, it was agreed to operate on her with a conscious craniotomy, and a language mapping was made both for L1 (Spanish) and L2 (Euskera) (Fig. 4). A good correlation between both methods was confirmed in this case.

Case 2

A 35-year-old bilingual right-handed male presented with a 20 months history of increasing seizures. These were mainly partial (motor and aphasic types). He was on multiple antiepileptic drugs with persistent daily crisis, and he developed at least two episodes of epilepticus status in the 6 months period before surgery. MRI showed an inferior and posterior left frontal brain tumour, typical for a low grade glioma. This lesion, involving the inferior motor strip, steadily grew in subsequent scans.

The rnTMS study showed language areas in the pars opercularis and pars triangularis for Spanish (L2) (Fig. 5). Discomfort was elicited when stimulating in the first temporal gyrus so no further stimulations were performed in the rest of perisylvian areas. An intraoperative DCS language mapping study was performed, which allowed the detection of areas predicted by the rnTMS, that is, language sites within the inferior frontal gyrus and within the superior temporal gyrus (Fig. 6).

Discussion

From our own experience and from the literature, it has been shown that language disturbances and particularly speech arrest can be elicited by nTMS stimulation, and therefore depicted onto a 3D model of a patient. When

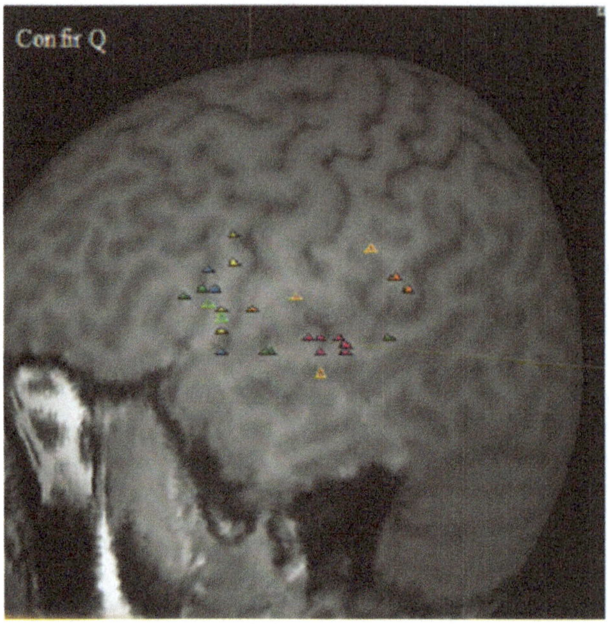

Fig. 1. The color triangles show the location of the language sites in the first 7 patients of this series

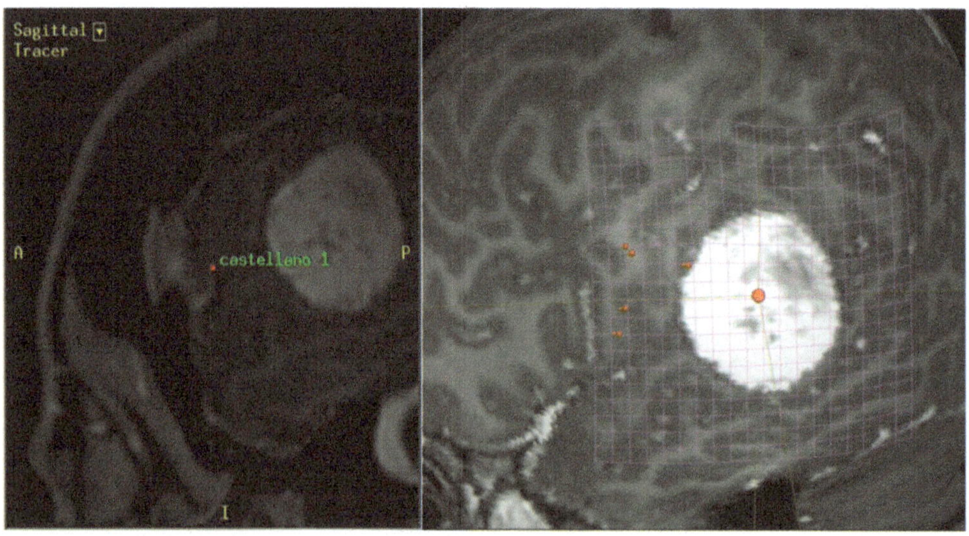

Fig. 2. Cortical areas involved in Spanish detected by preoperative nTMS in a patient with meningioma (Case 1)

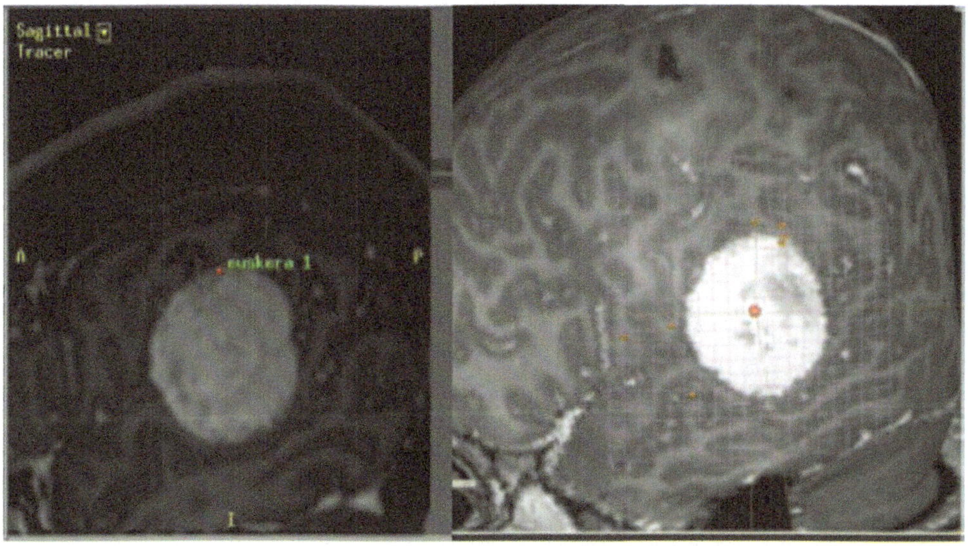

Fig. 3. Cortical areas involved in Euskera detected by preoperative nTMS in the same patient with meningioma

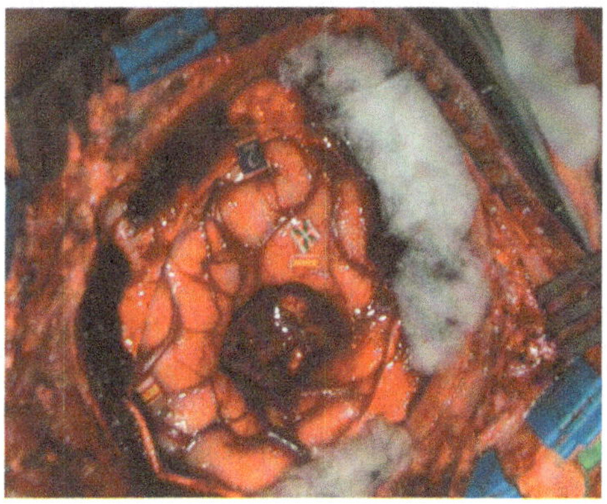

Fig. 4. Intraoperative language mapping using DCS in the same patient with meningioma, demonstrating a good correlation between nTMS and DCS in this case

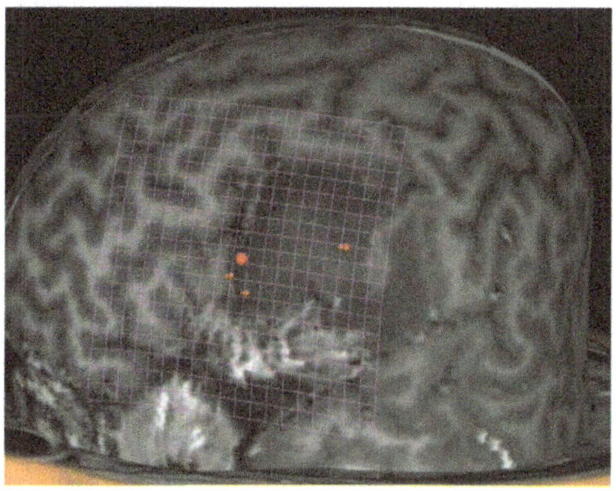

Fig. 5. Cortical areas crucial for Spanish detected by preoperative nTMS in a patient with left low-grade glioma involving the rolandic operculum (Case 2)

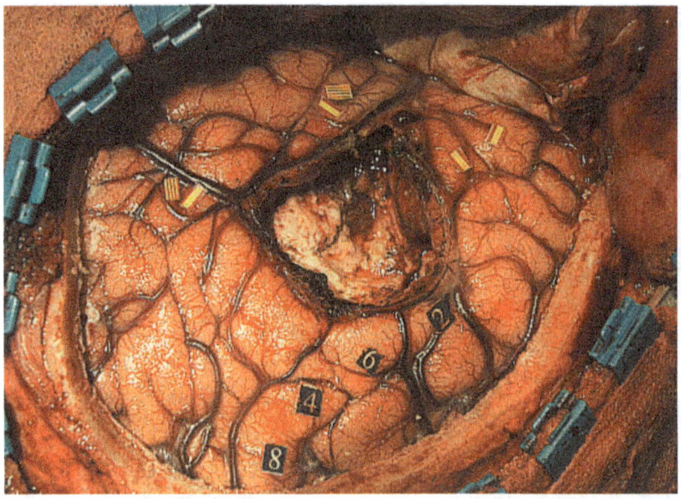

Fig. 6. Intraoperative language mapping using DCS in the same patient with left glioma, demonstrating a good correlation between nTMS and DCS

finding these nTMS areas and comparing them to the DCS intraoperative mapping, we have found a good agreement with DCS language mapping. Although a comprehensive report is not yet given here, some positive trends are clearly shown in our preliminary analysis. Therefore, these first data deserve further discussion, especially with regard to the sensibility, specificity and accuracy of the method, but also disclosure of present methodological problems and insight into further improvements.

Indeed, one of the more important problems is the sensitivity of the technique. There are several circumstances that make difficult to evaluate speech arrest responses. Coil positioning can interfere with the reference localization glasses of the patient, making difficult to achieve a correct positioning of the stimulating coil to cover all the perisylvian area. A new reference glasses with a different attachment to the patient head to free the temporal area has been used in the last cases, allowing minimization of this problem.

In addition, TMS can also induce tetanisation due to peripheral stimulation of temporal and masseter muscles. The facial and trigeminal nerves can also be stimulated. All of these structures can cause pain, eliciting a stop of the task that can be misinterpreted as a language disturbance. Another cause of misinterpretation in this setting is due to direct muscle stimulation without pain, which can also be interpreted as a positive speech arrest. The patient presents a muscle contraction that follows the rhythm of the magnetic stimulation and that creates a difficulty in keeping the fulfilment of the task. Video recording of the patient study is important to review these studies, and can also benefit of adding a later analysis by different specialists. All of the aforementioned problems are more important when trying to stimulate temporal rather than frontal or parietal areas, so the first trend is that nTMS is more comprehensive and useful for localizing language areas in these regions.

The idea to compare rnTMS with intraoperative DCS language mapping lead us to use an object naming task as previously described by Berger et al [11]. For rnTMS we did not used a synchronizing device, starting the magnetic stimulation when the image was visually presented to the patient. We have found that a continuous language task such as reading or counting has been somewhat more efficient in this setting.

Nonetheless, we have to admit that we have not been able to produce speech arrest in all the subjects studied. Furthermore, in some subjects, we have produced speech arrest in one task (reading), while it has been impossible to induce disturbances during another type of language task (e.g., naming). This can be specifically interesting for discrimination of different language networks.

It is clear now that speech arrest or other language problems elicited by DCS need a certain degree of charge density to produce a virtual lesion, and that this virtual lesion is cortically limited using the usual stimulation paradigms. It is not clear however how wide and deep the area of cortical and subcortical stimulation is, when using a repetitive magnetic stimulation, since it is most probably spreading wider and deeper with every repetition of the pulse. The use of a certain intensity of the magnetic field is therefore also a matter of discussion. A 110–120% intensity over the motor threshold is used on the assumption that the language areas have a similar threshold. This can be more probably the case in volunteers but not necessarily be the same for patients with different types of lesions, different lesion locations and different drugs such as antiepileptic medications.

In our practice, we have found efficient language stimulation with short term (2 seconds) high rate frequency trains, and no more that 10 Hz were needed. Although we have used 110–120% intensities over the motor threshold, sometimes we have needed to rise up to 130% to obtain a speech arrest. Of note, the shape of the coil, and the magnetic field volume formed is also influencing the result of the stimulation.

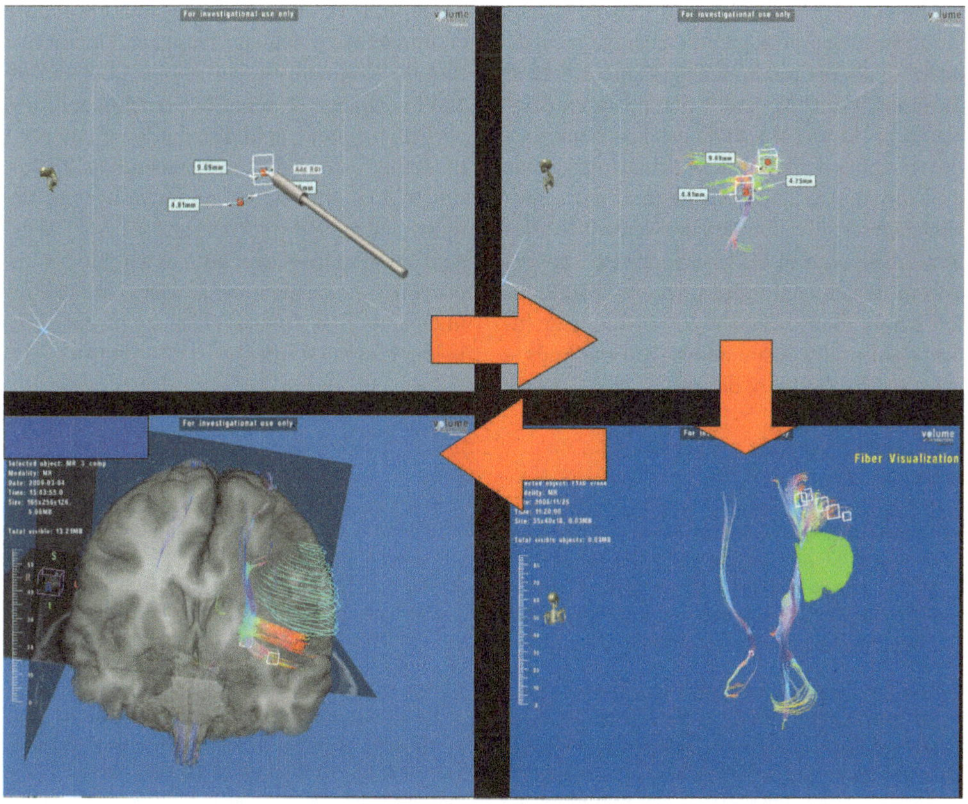

Fig. 7. The step by step process of fiber tracking acquisition using rnTMS ROIs is shown in this figure

When finding rnTMS language area, this site has also been shown by DCS intraoperative in more than 80% of the cases. For example, it was the case with L1 and L2 specific tasks in bilingual patients, with a precision of correlations less than 10 mm. So this leads us to think that nTMS is a precise and a specific method. However, once again, pain, movements, muscle artefacts and the tracking glasses problems previously mentioned make the method much less sensitive for temporal language areas compared to frontal and parietal ones, as also shown in case 2.

Finally, the use of rnTMS as volume ROIs for language fiber tracking may also give insight into the connectivity of the depicted areas and may help to understand functional significance for the different tracts (Fig. 7). The cubic volume ROIs of the system are initially meant for motor stimulation which uses single pulses. In repetitive stimulation the area will not be in any case smaller than showed by single pulse cube. Although we can only speculate on the depth of the stimulated magnetic field in the brain, it can be possible that the tracts originated in the rnTMS ROIs are less numerous than the ones depicted in the actual image. This probably makes again this method more specific than sensitive. In the same sense is the deterministic type of DTI tracking used by our system, which could give us more reliable information about tractography in the near future than probabilistic and Q-ball methods – even is still a matter of debate.

Conclusions

Although with certain limitations, especially in relation to stimulation of temporal areas, nTMS shows up as an accurate and specific method for preoperative language mapping, with a good agreement with the intraoperative DCS language mapping (even in bilingual patients)

The link between this rnTMS data detecting functional mosaics and tractography (by using TMS results as volume ROIs) gives a unique opportunity to better understand functional connectivity, by providing functional meaning to these anatomical tracts. Indeed, these pathways can be further investigated when assigning specific language functions to the rnTMS ROIs, since rnTMS may discriminate "sub-functions", such as reading and naming. Nonetheless, both magnetic stimulation conditions and language tasks paradigms need to be improved in further research for a more comprehensive language mapping. Finally, further developments in coil cooling, coil design and knowledge on repetitive cortical and subcortical magnetic field distribution are needed to further extend this novel technique. The ultimate goal is to define non-invasively at the individual scale the cortical and subcortical organization of language, both as a guide for improving surgical planning as well as for a better understanding of neural plasticity mechanisms in longitudinal studies.

References

[1] Duffau H, Capelle L, Denvil D, Sichez N, Gatignol P, Lopes M, Mitchell MC, Sichez JP, Van Effenterre P (2003) Functional recovery after surgical resection of low grade gliomas in eloquent brain: hypothesis of brain compensation. J Neurol Neurosurg Psychiatry 74: 901–907

[2] Peraud A, Meschede M, Eisner W, Ilmberger J, Reulen H (2002) Surgical resection of grade II astrocytomas in the superior frontal gyrus. Neurosurgery 50: 966–975

[3] Mandonnet E, Jbabdi S, Taillandier L, Galanaud D, Benali H, Capelle L, Duffau H (2007) Preoperative estimation of residual volume for WHO grade II glioma resected with intraoperative functional mapping. Neuro-Oncology 9: 63–69

[4] Robert E (2005) Linguistic procedure in "awake neurosurgery". Stem-, Spraak- en Taalpathologie 32.8310/01/1305-54: 13(1) 54–64

[5] Duffau H, Lopes M, Arthuis F, Bitar A, Sichez JP, Van Effenterre R, Capelle L (2005) Contribution of intraoperative electrical stimulations in surgery of low grade gliomas: a comparative study between two series without (1985–96) and with (1996–2003) functional mapping in the same institution. J Neurol Neurosurg Psychiatry 76: 845–851

[6] Duffau H, Capelle L, Denvil D, Gatignol P, Sichez N, Lopes M, Sichez JP, Van Effenterre R (2003) The role of dominant premotor cortex in language: a study using intraoperative functional mapping in awake patients. NeuroImage 20: 1903–1914

[7] Duffau H (2006) New concepts in surgery of WHO grade II gliomas: functional brain mapping, connectionism and plasticity – a review. J Neurooncol 79: 77–115

[8] Duffau H (2007) Contribution of cortical and subcortical electrostimulation in brain glioma surgery: Methodological and functional considerations. Clin Neurophysiol 37: 373–382

[9] Andoh J, Artiges E, Pallier C, Rivière D, Mangin JF, Cachia A, Plaze M, Paillère-Martinot ML, Martinot JL (2006) Modulation of language areas with functional MR image-guided magnetic stimulation. NeuroImage 29: 619–627

[10] Munk S, Forchhammer HB, Brennum J, Hansen AE, Larsson HB (2007) Presurgical functional MR imaging in the mapping of language function. Ugeskr Laeger 169: 3571–3574

[11] Sanai N, Mirzadeh Z, Berger MS (2008) Functional outcome after language mapping for glioma resection. N Engl J Med 358: 18–27

[12] Petrella JR, Shah LM, Harris KM, Friedman AH, George TM, Sampson JH, Pekala JS, Voyvodic JT (2006) Preoperative functional MR imaging localization of language and motor areas: effect on therapeutic decision making in patients with potentially resectable brain tumors. Radiology 240: 793–802

[13] Quiñones-Hinojosa A, Ojemann SG, Sanai N, Dillon WP, Berger MS (2003) Preoperative correlation of intraoperative cortical mapping with magnetic imaging landmarks to predict localization of Broca area. J Neurosurg 99: 311–318

[14] Roux FE, Boulanouar K, Lotterie JA, Mejdoubi M, LeSage JP, Berry I. (2003) Language functional magnetic resonance imaging in preoperative assessment of language areas: correlation with direct cortical stimulation. Neurosurgery 52: 1335–1347

[15] Pascual-Leone A, Tormos-Muñoz JM (2008) Transcranial magnetic stimulation: the foundation and potential of modulating specific neuronal networks. Rev Neurol 46: S3–S10

[16] Ruohonen J, Karhu J (2010) Navigated transcranial magnetic stimulation. Clin Neurophysiol 40: 7–17

[17] Devlin JT, Watkins KE (2007) Stimulating language: insights from TMS. Brain 130: 610–622

[18] Devlin JT, Watkins KE (2008) Investigation language organization with TMS. The Oxford Handbook of Trancraneal Stimulation. Oxford University Press, New York pp 479–499

[19] Epstein CM, Meador KJ, Loring DW, Wright RJ, Weissman JD, Sheppard S, Lah JJ, Puhalovich F, Gaitan L, Davey KR (1999) Localization and characterization of speech arrest during transcranial magnetic stimulation. Clin Neurophysiol 110: 1073–1079

[20] Lo Gerfo E, Oliveri M, Torriero S, Salerno S, Koch G, Caltagirone C (2008) The influence of rTMS over prefrontal and motor areas in a morphological task: Grammatical vs. semantic effects. Neuropsychologia 46: 764–767

[21] Sparing R, Mottaghy FM (2008) Noninvasive brain stimulation with transcranial magnetic or direct current stimulation (TMS/tDCS) – from insights into human memory to therapy of its dysfunction. Methods 44: 329–337

[22] Picht T, Mularski S, Kuehn B, Vajkoczy P, Kombos T, Suess O (2009) Navigated transcranial magnetic stimulation for preoperative functional diagnostic in brain tumor surgery. Neurosurgery 6 (Suppl 1): 93–99

[23] 23. Hannula H, Ylioja S, Pertovaara A, Korvenoja A, Ruohonen J, Ilmoniemi RJ, Carlson S (2005) Somatotopic blocking of sensation with navigated transcranial magnetic stimulation of the primary somatosensory cortex. Hum Brain Mapp 26: 100–109

[24] Hannula H, Neuvonen T, Savolainen P, Tukiainen T, Salonen O, Carlson S, Pertovaara A (2008) Navigated transcranial magnetic stimulation of the primary somatosensory cortex impairs perceptual processing of tactile temporal discrimination. Neurosci Lett 437: 144–147

[25] Catani M, Ffytche DH (2005) The rises and falls of disconnection syndromes. Brain 128: 2224–2239

[26] Catani M, Jones DK, Ffytche DH (2005) Perisylvian language networks of the human brain. Ann Neurol 57: 8–16

[27] Catani M, Mesulam M (2008) The arcuate fasciculus and the disconnection theme in language and aphasia: History and current state. Cortex 44: 953–961

[28] Bello L, Gallucci M, Fava M, Carrabba G, Giussani C, Acerbi F, Baratta P, Songa V, Conte V, Branca V, Stocchetti N, Papagno C, Gaini SM (2007) Intraoperative subcortical language tract mapping guides surgical removal of gliomas involving speech areas. Neurosurgery 60: 67–82

[29] Glasser MF, Rilling JK (2008) Feature article DTI tractography of the human brain's language pathways. Cereb Cortex 18: 2471–2482

Preoperative Diffuson Tensor Imaging (DTI): contribution to surgical planning and validation by intraoperative electrostimulation

Lorenzo Bello, Antonella Castellano, Enrica Fava, Giuseppe Casaceli, Marco Riva, Andrea Falini

Introduction

Surgical resection of tumors located within the so-called eloquent areas requires the pre- and intraoperative identification of cortical and subcortical functional sites to achieve the goal of a satisfactory tumor resection associated with a full preservation of the patient's integrity.

Diffusion tensor imaging (DTI) and fiber tractography (FT) are magnetic resonance techniques based on the concept of anisotropic water diffusion in myelinated fibers, which enable three-dimensional reconstruction and visualization of white-matter tracts. Three-dimensional visualization of the functional fibers and their relationship with brain lesions is helpful for preoperative evaluation and intraoperative navigation, combining DTI information with those of direct electrical stimulation (DES).

In this chapter, we will describe the contribution of DTI-FT to surgical planning and, when combined with DES, to safety in the performance of surgical removal of tumors involving functional brain areas.

DTI data acquisition and FT processing

DTI is an innovative magnetic resonance imaging (MRI) technique, introduced in the mid 1990s [37, 38], that allows to assess the axonal organization of the brain, using the anisotropic water diffusion. DTI uses the translational motion of water molecules to obtain anatomic information [1]. Water molecules are supposed to move more easily along the axonal bundles rather than perpendicular to these bundles because there are fewer obstacles to prevent movement along the fibers. By the characterization of the anisotropic diffusion of water, an entirely new image contrast is provided which is based on structural orientation [2, 12, 16].

To identify the course of white matter tracts, DTI tractography methods require the delineation of regions of interest (ROIs) as starting seed points for tracking [25]. ROIs can be delinated automatically or manually. Manual delineation of ROIs requires a priori anatomical knowledge and it is very helpful when the anatomy is distorted such as in the case of tumors. In our institute we use the method described by Catani and Thiebaut de Schotten [13]: a ROIs is defined around areas of white matter that represent "obligatory passages" along the course of each tract. If the ROI representing an obligatory passage contains only fibers of the tract of interest, a single ROI approach is used. A one-ROI approach is used for the arcuate fasciculus, cingulum, corpus callosum, anterior commissure, and fornix. When a tract shares its obligatory passages with one or more other tracts, a two-ROI approach is used. The second ROI is defined such that it contains at least a section of the desired fasciculus but does not contain any fibers of the undesired fasciculi that pass through the first ROI. The two-ROI approach is used for the corticospinal tract and the uncinate, inferior longitudinal, and inferior fronto-occipito fasciculi. A second ROI can also be used to exclude undesired streamlines.

At our institute, DTI data are obtained with a 3 T MR scanner using a single-shot echo planar imaging sequence (repetition time, 8986 ms; echo time, 80 ms) with parallel imaging (sensitivity encoding factor, R = 2.5) [3–6]. Diffusion gradients are applied along 32 axes, using a b-value of 0 and 1000 mm^2/s. A field of view of 240 by 240 mm and a data matrix of 96 by 96 are used and these lead to isotropic voxel dimensions (2.5 by 2.5 by 2.5 mm). The data are interpolated in-plane to a matrix of 256 by 256 leading to a voxel size of 0.94 by 0.94 by 2.5 mm. Fifty-six slices are obtained, with a thickness of 2.5 mm, with no gap. The sequence is repeated 2 consecutive times and data are averaged off-line to increase the signal-to-noise ratio; thus, the total time for diffusion-tensor MR imaging is 646 s. Three-dimensional fast field echo T1-weighted imaging (repetition time, 8 ms; echo time, 4 ms; image resolution equal to DTI) was performed for anatomic guidance. Alternatively, a 1.5 T machine can be used [23]. At least 6 directions should be used.

DTI datasets are realigned off-line on a PC workstation by the AIR (automatic image registration) software to correct artifacts due to rigid-body movement during scan acquisition [7].

Usually, deterministic tractography is performed for clinical use with patients [14, 33, 38], as well as in our series [5, 6]. Reconstruction is performed by various software; we use DTI Studio v2.4.01 software (H. Jiang, S. Mori, Radiology Department, Johns Hopkins University, Baltimore, HMD, USA), by which we obtain main eigenvector and fractional anisotropy (FA) maps [8]. Subcortical connections were reconstructed by the "fiber assignment by continuous tracking" method [9, 10, 31, 32]. An FA threshold of 0.1 and a turning angle of >55° are used as criteria to start and stop tracking. Seeding ROIs for tractography were defined around areas of white matter that represented the brain regions that all the fibers of each tract must pass through in order to reach their cortical or subcortical terminations. ROIs were chosen on the basis of previous anatomic knowledge, in sections perpendicular to the main course of the tracts [49], and were adapted to intraoperative neurophysiological findings.

By DTI-FT we reconstructed various motor tracts, such as corticospinal tract (CST) and supplementary motor area (SMA) fibers, and language tracts, involved in the phonologic and semantic components of language: the superior longitudinal fascicle (SLF) is the basic tract involved in the phonologic component of language, the inferior fronto-occipital fascicle (IFO) is the basic tract involved in the semantic component of language [6]. Additional tracts such as the following can be reconstructed: the uncinate and inferior longitudinal (ILF) fascicles, which provide information on the semantic and phonologic component of language in the frontal and temporal lobes, or the subcallosal fascicle, involved in the phonologic component of language and located in the lateral border of the lateral ventricle [6].

To reconstruct the CST, a ROI should be placed on an axial section at the level of subcortical white matter of the precentral gyri. For the IFOF, ILF, and uncinate fascicle, a ROI is placed on a coronal section at the level of the anterior part of the external capsule at the junction of the frontal and temporal lobes, where the two tracts run in contiguity.

To reconstruct the SLF, a first ROI should be placed on a coronal section at the level of a high-anisotropy region laterally to the central part of the lateral ventricle; a second ROI is placed in a peritrigonal site at the level of the descending branch of the fascicle. For all the tracts reconstructed, any contaminating fibers have to be removed.

After tract reconstruction, the following step is to superimpose the reconstruction to volumetric morphological MR images. Volumetric precontrast T1-weighted or FLAIR (fluid attenuated inversion recovery) images are usually coregistered to the mean of all diffusion-weighted images by the SPM2 software to obtain the superimposition of the white-matter tracts on T1-weighted anatomical images. This allows to compare the trajectories of the tracts in the involved hemisphere with those of the controlateral unaffected hemisphere and to evaluate the anatomical relationship between the tract and the tumor mass as well as the effect exerted by the tumor on the tract of interest. Finally, DTI-FT data are saved as a compatible format (DICOM) to be transferred to the neuronavigation system, to be loaded and being available for surgical planning and intraoperative guidance.

DTI-FT techniques allow to depict the functional white-matter fibers around and inside a tumor [24], producing an impressive amount information concerning the anatomical and functional boundaries of the lesion to be resected. In the following section, we will report our experience on the combined use of DTI-FT and direct electrical stimulation during surgical removal of gliomas.

Use of DTI during surgery of gliomas: pitfalls

DTI-FT images are fused with morphologic volumetric postcontrast T1-weighted and FLAIR images and fMRI images [7, 11, 21]. These images are loaded into the neuronavigation system and are used preoperatively for surgical planning by localizing the tumor and defining its relationship with subcortical and cortical functional areas and the main vasculature [11, 15]. In addition, the same images, being loaded into the neuronavigation system, are available during surgery for guidance.

Two critical points appear as important for the correct use of DTI-FT data in glioma surgery: the transfer of the data to the neuronavigation system [40] and the use of technical adjustments during surgery to maintain a global accuracy of the information and to reduce the problem of brain shift [9, 10, 26, 39].

We save DTI data as a compatible format (DICOM) by the Medx software (Medical Numerics, Inc., Germantown, MD, USA). This allows the images to be transferred and loaded into the neuronavigational system. The neuronavigational system (Brain Lab, Munich, Germany, in our experience) performs an automatic coregistration between DTI-FT datasets and the preoperative MR images acquired with references applied on the skull of the patient by a voxel-by-voxel intensity matching nonlinear algorithm.

As estimate for the clinical navigation accuracy, the target registration error localizing a separate fiducial, which is not used for registration, is usually performed at the beginning of surgery. The target registration error should be less than 2 mm to be considered as acceptable.

The brain shift is the main limitation of the use of neuronavigation, particularly in case of large tumors; it occurs already at the beginning of surgery, when the dura is opened, and increases with the progress of tumor removal.

In order to reduce the problem of brain shift, it is critical that the resection is performed with repeated landmark checks during surgery, the craniotomy is restricted to the minimum necessary to expose the tumor area and a limited portion of the surrounding brain, resection should proceed so that the maintenance of accuracy of the information is maximally ensured. For frontal tumors located close to CST, the resection should be started from the posterior border where the CST is located, and after its identification, the tract is followed inside the tumor mass. Afterwards the remaining anterior part of the tumor is removed. Similarly, for parietal tumors, the resection is started from the anterior border following the same principle.

Another way to overcome these problems is to update the preoperative images intraoperatively. The use of intraoperative MR is the gold standard [34, 35], but its main limitation is the cost of the system. Ultrasound is another imaging option used for intraoperative visualization of low-grade gliomas. Advances in ultrasound technology have made the image quality of the ultrasound comparable to intraoperative MR. Recent studies [39] showed that the integration of intraoperative ultrasound with neuronavigation represents an efficient and inexpensive tool for intraoperative imaging and surgical guidance. Brain shift detected with intraoperative ultrasound could be used to update preoperative image data such as fMRI and DTI-FT in order to increase the value of this information throughout the operation. Unfortunately, the ability of these methods to reveal overlooked tumor remnants is lower than that of intraoperative MR systems.

Correlation between DES and DTI-FT: intraoperative use of DTI-FT

DTI-FT data allow depicting the relationship of subcortical tracts with the neoplastic mass: the tract can be unchanged, dislocated, infiltrated, or interrupted by the tumor mass and its effect [1]. This information can be influenced by some technical factors, linked to the tract reconstruction, such as the FA threshold used to commence and cease tracking [32] or the characteristics of the tumor (histology, edema, location) or the location of ROI placement [20, 46 ,48]. To better define the intraoperative use of DTI-FT information, we have routinely correlated DTI-FT data with subcortical DES data during surgery [6]. This correlation has been performed for more than 350 cases, in which motor (CST) or language (IFOF or SLF) tracts were reconstructed and their images were loaded into the neuronavigation system and available during surgery for correlation with DES data. At the time of surgery, the correlation was performed by registering the location of each positive or negative motor or language response into the neuronavigation system in which the DTI-FT data were loaded and fused with morphological MR images. In addition, the correlation was also performed postoperatively, by transferring postoperative MR images on a workstation and fusing them with preoperative MR images and DTI-FT ones. The distance between the point of response at DES and the tract border was measured on transverse navigation images and was graded as corresponding (at the tract border or within the tract), close (within 1 cm), or distant (>1 cm). In addition, each subcortical site identified during subcortical mapping was marked with a sterile numbered tag and a digital picture of the surgical cavity was taken at the end of the resection. The anatomical location of the subcortical pathways, i.e., the periphery of the surgical cavity, where the resection was stopped according to the functional responses elicited by intraoperative stimulation [18], was evaluated on an immediate postoperative MR scan. Technically this was accomplished by transferring immediate postoperative MR images of each patient to a Brain Lab workstation and fusing them with preoperative MR images on which DTI-FT for each fascicle were merged. The existence of a correspondence between the location of a subcortical site identified by DES and that of the tract in DTI-FT images was considered only when the distance between the location of the subcortical site evoked by DES and identified intraoperatively by the use of neuronavigation system was graded as adjacent to the tract in transverse neuronavigation images. In all the other cases, the correspondence was considered null.

DES for cortical and subcortical mapping [6, 7, 30] is usually performed with a bipolar hand-held stimulator equipped with 1 mm electrodes, tips 5 mm apart, connected to an Ojemann cortical stimulator (Integra, Plainsboro, NJ, USA) or an Osiris or Isis stimulator (Inomed, Emmendingen, Germany), which is delivering biphasic square wave pulses, each phase lasting 1 ms, at 60 Hz in trains lasting 1–2 s for cortical mapping and 1–4 s for subcortical mapping. Subcortical mapping is alternated with the resection in a back and forth fashion. Subcortical mapping is performed with the same current threshold as applied for cortical mapping. Alternatively, a monopolar stimulation can be used, either cortically or subcortically, by delivering a single- or double-pulse stimulus, according to the train-of-five technique.

For motor mapping, the patient is connected to a multichannel electromyography system which allows simultaneous monitoring of different and distant muscles. Once determined, the same intensity of the current for stimulation is used in most of the cases throughout the procedure. Initially, a low current intensity (2 mA) is used, which then is progressively increased till a movement is induced. A stimulus duration of 1 or 2 s is usually enough to generate a motor response. At this point, it is good practice to stimulate the areas close to that in which the current induced the movement, to map them, and to check if the current is able to evoke motor responses also in these zones. If not, the current intensity may be increased and adjusted to evoke appreciable motor responses. When a response was induced at a subcortical level, it is recommended to generate an intensity–response curve, to assess the maintenance of the response either at very low current intensity levels. This can help in estimating the distance between the point of stimulation and the functional tract. Usually, if a response is maintained till the intensity is decreased to 5–3 mA, the location of the stimulation is very close or can be considered as being in the tract.

For language mapping [6], counting is used as initial test. The current is usually applied onto the premotor cortex related to the face, and the test aims to check if the current is able to stop the patient's counting. In order to be reliable, this has to be repeated several times and the counting to be stopped at least three times. Then, the same current is applied during the other testing (object, verb naming, word or sentences comprehension) to locate the site involved in the semantic and phonologic component of language. As before, when a response is found at subcortical level, a intensity–curve is also generated to assess the distance between the site of stimulation and the tract.

During surgery and all the time of stimulation, it is also recommended to check by electrocorticography if the applied current may induce afterdischarges in the nearby brain areas. Only a current which is immediately below that which is inducing afterdischarges has to be used for mapping. If afterdischarges are seen, the current should be set up at an intensity of at least 0.5 mA less than the previous one. By doing so, electrocorticography recording is used to keep the test reliable. In fact, only the responses evoked in absence of afterdischarges are considered to be trustworthy.

We will present data according to motor or language tracts. Motor tracts were reconstructed in 294 patients with lesions involving motor pathways and language tracts in 305 patients with lesions involving language pathways.

Motor tracts

For routine clinical use, the CST should be always reconstructed [22, 27]. For selected patients, additional tracts running in the premotor areas can be also reconstructed. For patients with frontal tumors on the dominant side, the course of the motor fibers belonging to the premotor face areas should be also reconstructed. For reason of simplicity, their findings have been reported in the language tract section.

In all the cases of precentral tumors, the CST was shown either as unchanged (66% of cases) or as dislocated posteriorly (34% of tumors, in case of large lesions). In both cases, subcortical stimulation located the tract at the posterior border of the tumor mass. Motor responses appeared as focal (few muscles) when the tract was stimulated in close vicinity to the surface, while multiple muscle groups were affected by deep stimulation. The cortical or near-cortical stimulation always induced evident movements, while subcortical stimulation frequently induced at the beginning some muscle activations which were detected only by amplified electromyography while the resection was approaching the subcortical tracts and which became overt movements when the CST was reached. In all cases, subcortical DES located the CST where it was shown by DTI-FT and therefore it was graded as corresponding.

In cases of rolandic tumors, DTI reconstructed the CST mainly inside the tumor mass

(98% of cases). In the majority of bulky tumors (92%), the tract was displaced either anteriorly (22%) or more frequently posteriorly (78%) and highly infiltrated by the tumor mass (Fig. 1). Less frequently, and in case of highly infiltrating and diffuse low-grade gliomas, the tract was shown inside the tumor mass and as highly infiltrated. In the first group of tumor, subcortical DES located the tract in the same position where it was shown by DTI-FT (Fig. 1). Some discrepancies were observed only in the superior portion of the tract, close to the cortical surface, where DTI-FT failed to reconstruct fibers but DES located motor responses. As we previously reported, even the placement of additional ROI in those regions did not improve the fiber reconstruction [5]. More problematic are the cases of highly diffuse low-grade gliomas, where DTI-FT usually reconstructed the tract as highly infiltrated and inside the tumor mass. Particularly in those cases with a long history of seizures, and at the beginning of the resection, when 60 Hz stimulation was applied over the regions of the tumor where DTI-FT depicted the location of the upper portion of tract, it usually failed to locate overt motor responses. When the current intensity was progressively increased to induce responses, this usually resulted in seizures, without overt movements. In these cases, the electrical identification of the CST required the use of different modalities of stimulation, such as the monopolar one, or alternatively, the initial resection could be performed under DTI-FT guidance. However, as in the previous cases, DTI-FT failed to show fibers close to the more lateral portion of the homunculus, probably due to the presence of crossing fibers that cannot be visualized by the simple tensor model here used for tractography [48], where DES (generally with monopolar stimulation) induced laryngeal or upper or lower face responses. When a portion of the tumor was removed, and the CST partially decompressed, the 60 Hz stimulation started again to identify motor responses, usually in the same location where DTI-FT reconstructed the deeper portion of the CST.

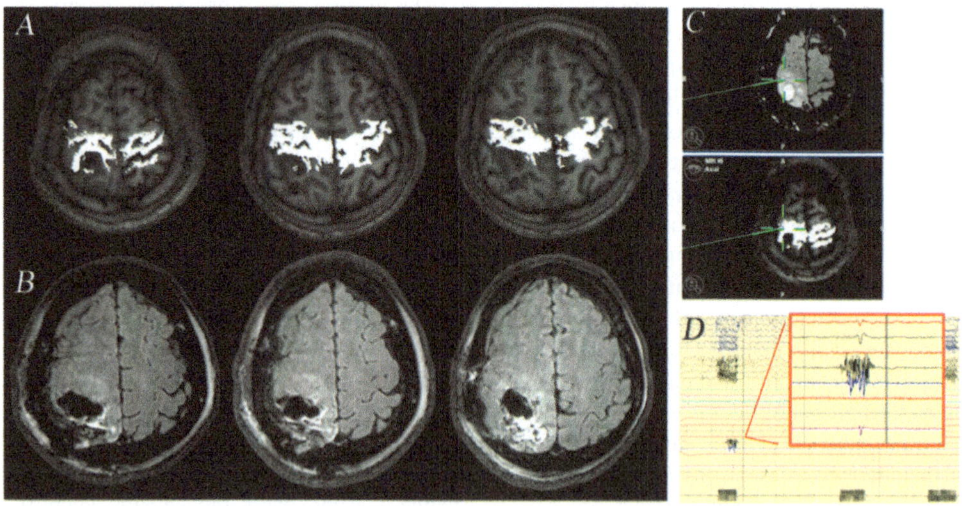

Fig. 1. (A–D). CST reconstruction in a case of right parietal low-grade oligodendroglioma. **(A)** The CST (white bundle) was reconstructed by DTI-FT and superimposed on morphologic T1-weighted images; the CST was dislocated and infiltrated by the tumor mass. **(C)** The CST was identified intraoperatively by DES where the CST was located by DTI FT. **(D)** Example of motor response at subcortical level (hand). **(B)** Postoperative FLAIR images show that the margin of resection was coincident with the CST

In the cases of parietal tumors, DTI-FT usually located the CST at the anterior border of the tumor and showed it as either unchanged when the tumor volume was small (16.7% of cases) or dislocated anteriorly when the tumor was larger (83.3%). DES located the tract in a similar position.

In cases of insular or temporal tumors, the CST was located at the medial and posterior border of the tumor and shown as either unchanged or dislocated. The concordance was tested in a subgroup of patients in which the tract was located close to the border of the tumor, where DES during surgery identified motor responses.

Globally considered, these data indicate that there is usually a very high concordance between DTI-FT data for CST and subcortical mapping, the CST being located in the same position where it was shown by DTI FT. Some pitfalls may occur in a limited subgroup of highly diffuse low-grade gliomas located in the rolandic areas, in which DTI-FT may fail in reconstructing portions of CST, particularly in the areas of extensive infiltration. In addition, in these particular cases, also the electrical identification of the CST may be problematic and require the use of alternative modalities of stimulation [7, 11].

Language tracts

The SLF and IFOF were reconstructed in all patients with either temporal or frontal dominant tumors, these two tracts being essential for the functional preservation of language [17, 19].

The SLF is a large tract [28] running from the parietal to the frontal lobe, mediating the phonemic component of language. It was shown by DTI-FT as dislocated or infiltrated by the tumor, generally depending on histology, size, and location of the tumor mass (Fig. 2). The tract was usually unchanged or dislocated in high-grade tumors, and infiltrated in case of low-grade gliomas. Large tumors dislocated or infiltrated the tract, smaller ones left the tract unchanged. When the SLF was found during

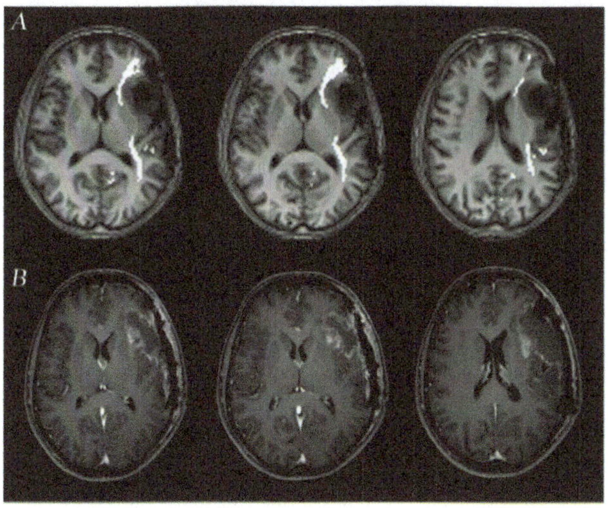

Fig. 2. (A, B). IFOF reconstruction in a case of left frontal recurrent low-grade oligodendroglioma. The IFOF (white bundle) was reconstructed by DTI-FT and superimposed on morphologic T1-weighted images; the tract was dislocated and infiltrated by the tumor mass (**A**); the tract was identified intraoperatively by DES (onset of semantic paraphasia) in the same position where it was located by DTI-FT, as demonstrated by the postoperative postcontrast T1-weighted images (**B**)

surgery and stimulated with 60 Hz current, its stimulation induced phonemic paraphasias in the same places where the SLF was located by DTI FT. As previously reported, the anatomic extension of this tract is usually considerably larger than the functional one when subcortical language mapping is performed. Therefore, a large part of the tract can be safely resected because it is not functional in terms of language. This is particular so for frontal and temporal tumors. In addition, in case of low-grade gliomas involving the F3 gyrus, an additional ROI should be placed in the subcortical area of F3 to visualize the portion of the SLF tract whose stimulation induced phonemic paraphasia. This portion usually constituted the anterior and superior border of the tumor.

The IFOF is a discrete tract running from the occipital to the frontal lobe and mediating the semantic component of language [5–7]. The relationship between the tract and a tumor depends on the tumor's location, size, and histology. The IFOF was shown as unchanged or dislocated in case of high-grade tumors, or as dislocated and/or infiltrated in case of low-grade gliomas. Again, large tumors dislocated or infiltrated the tract, small tumors left the tract unchanged. It is to be stressed that the anatomic extension of this tract is small and usually corresponds to the functional one shown by subcortical mapping (Fig. 3). In fact, DES evoked no language disturbances in areas of lacking fibers when the tract was shown as interrupted inside the tumor mass. Only in a few occasional cases of frontal low-grade gliomas (11 cases out of 96), where the tract was located inside the more anterior portion of the tumor mass and shown as highly dislocated and infiltrated, a limited portion of tract can be safely removed because it is not functional. In addition, some problems may occur for F3 low-grade gliomas in which DTI-FT may fail in reconstructing the more superior part of the tract at the inferior border of the tumor when

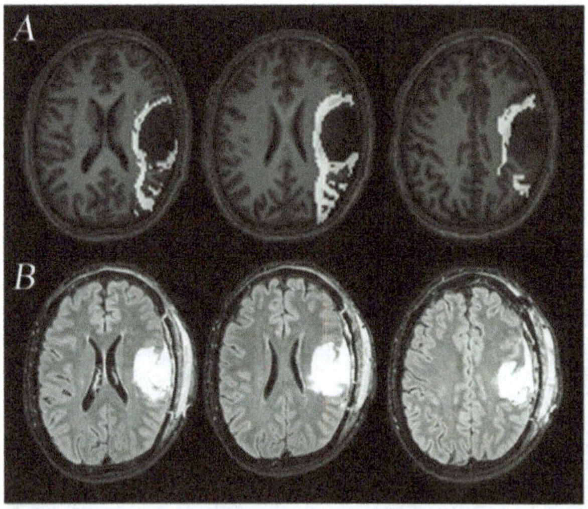

Fig. 3. (A, B). SLF reconstruction in case of left frontoparietal low-grade oligodendroglioma. The SLF (white bundle) was reconstructed by DTI-FT and superimposed on morphologic T1-weighted images; the tract was dislocated and slightly infiltrated by the tumor mass (**A**); the tract was identified intraoperatively by DES (onset of phonemic paraphasia) in the same position where it was located by DTI-FT, as shown by the postoperative postcontrast T1-weighted images, where the margin of resection was coincident with the location of the SLF (**B**). Part of the tract was removed at the time of surgery because found not functional

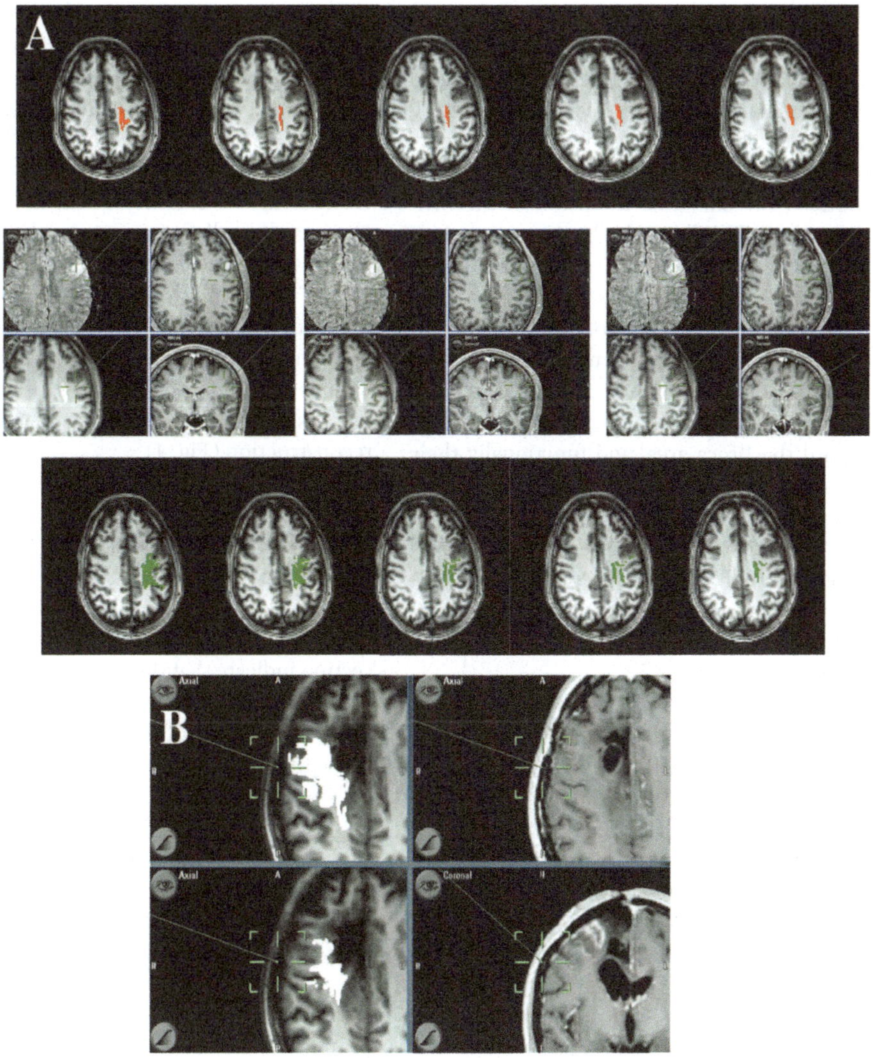

Fig. 4. (**A**) Importance of correct placement of ROI for a correct reconstruction of tracts. In the initial set of images, the upper ROI for the CST was located just behind M1, and the resulting reconstruction of CST is reported in the first row of images as a red bundle superimposed on T1-weighted images. The tract apparently is not in relationship with the tumor mass (left frontal oligodendroglioma). These images were loaded into the neruonavigation system and available during surgery for correlation. The patient was operated in awake anesthesia. Motor reponses (face and hand) were located at the posterior border of the tumor and the location of the responses registered into the neuronavigation system (intraoperative screenshots, second row of images), indicating a null concordance with DTI-FT reconstruction of CST. Postoperatively, the ROI was located more inferiorly to the previous position, and the resultant CST tract (visualized as a green bundle, fused with T1-weighted images, lower row) was then correctly reconstructed and being in relationship with the posterior portion of the tumor mass, where motor responses were induced intraoperatively by DES. (**B**) Reconstruction of tract depends on choice FA level. In a case of left frontal recurrent oligodendroglioma, the CST was reconstructed with two different FA (0.1 for upper left image, and 0.2 for lower left image). These images were both loaded into the neuronavigation system and available during surgery for correlation with DES data. The intraoperative screenshot shows the location where motor (pharynx) response was induced by DES. A good concordance was found only with the tract reconstructed with the FA of 0.1 (upper left image), whereas the tract reconstructed with the FA of 0.2 failed to show fibers in this position

the tumor infiltration in this area is quite extensive.

Globally considered, these data indicate that the reconstruction of the IFOF by DTI-FT is of particular use, because it strictly was correlated with the functional mapping by DES. In contrast, the anatomic extension of the SLF shown by DTI-FT is quite large, and particularly inside the tumor mass, a large part of the tract can be safely resected.

In case of F3 or F2 frontal or insular tumors, we found it useful to reconstruct the course of the face premotor fibers. These are fibers running from the dorsal premotor cortex toward the Broca area, and functionally their stimulation during surgery induced anarthrya. As for IFOF and SLF, the maintenance of these fibers is essential for the preservation of language. DTI-FT showed the course of these fibers at the posterior border of F3, F2, or insular tumors, where they were identified by DES.

Factors critical for tract reconstruction

As previously stated, the reconstruction of the tracts by DTI-FT is also influenced by some technical factors such as the FA threshold used to commence and cease tracking or the characteristics of the tumor (histology, edema, location) or the location of ROI placement. The characteristics of the tumor, specifically the degree of infiltration and edema, strongly influenced the ability to reconstruct fibers. The ability to reconstruct fibers depends on the FA. In a tumor setting, particularly for low-grade gliomas or those with a high level of edema, the preferential FA to reconstruct fibers is 0.1. When this is done, a good concordance between DTI-FT data and DES findings is usually found. Unfortunately, also in this setting, DTI may fail to show fibers in the regions of highly infiltrating tumors where DES induces responses. In addition, DTI-FT may show fibers in some areas of a tumor which during surgery turned out to be not functional as determined by DES. Furthermore, a small change in the FA may significantly alter the reconstruction of tracts. In a series of preliminary cases with CST, we showed that small variations in the FA may significantly change the reconstruction of the CST particularly at the upper part of the tract. When these data were combined with those of DES at the time of surgery, it turned out that in the upper portion of the tract, changes in the FA may result in significant changes for adjacent areas, and that these changes were not clearly predictable by the degree of tumor infiltration or edema. However, in the lower portion of the tract, no significant differences were observed according to the various FA used. Finally, a correct ROI placement is also important to obtain a good tract reconstruction (Fig. 4).

Significance of the use of DTI in surgery of gliomas

As a whole, the data presented in the previous section indicate that the reconstruction of specific tracts, such as CST, IFOF, SLF and face premotor fibers, by DTI-FT and its availability at the time of surgery are of particular help, especially for resection of low-grade gliomas. Our experience with a large number of patients showed that the combined use of DTI-FT and DES is a feasible approach that can be effectively and safely applied in routine clinical activity. DTI-FT, when available and loaded into the neuronavigation system, can allow to decrease the time of surgery, by helping the surgeon in designing the surgical approach and in finding at the time of the resection the location of the tract where to start the subcortical stimulation and proceed with a careful resection. This may result in a reduction of the number of stimulations needed to safely locate a tract, of the occurrence of seizures, and of patient fatigue. In some cases, such as in diffuse low-grade gliomas, this combined approach may help at the beginning of the resection when the use of 60 Hz current may be problematic. In addition, in our experience it also helped in a few cases of surgery performed under awake anesthesia, when at the end of a long procedure the level of collaboration of the patient became suboptimal, making direct language guidance not useful. In these few cases, the resection was improved under the guidance of DTI-FT images till the position of subcortical tracts was encountered, in order to maintain the patient's functional integrity.

Identification of subcortical tracts during surgery by the combined use of DTI-FT and DES resulted in the appearance of immediate postoperative deficits in 72.2% and 84% of patients with lesions involving motor or language pathways, respectively, which lasted for one week on average. Most of deficits were most likely to develop in those cases in which preoperative DTI showed the tracts as dislocated or infiltrated. At one-month follow-up, 94% of patients with a motor lesion had a normal motor exam and 96.8% of those with lesions involving speech areas or pathways had a normal language.

Conclusions

Globally considered, our data indicate the uselfulness of the routine combined use of DTI-FT and subcortical mapping, particularly in patients with low-grade gliomas. These tumors display an infiltrative modality of growth, along short and long connecting fibers and in which visualizing the trajectory of the tracts is important for planning and performing surgery. When used in combination with subcortical mapping, DTI-FT offers the opportunity to quickly find the fibers associated with motor or language functions during surgery. The clinical relevance of this combined approach comes from the fact that it further enhances surgical safety, maintaining a high rate of functional preservation. A careful use requires the knowledge of its limitations, mainly the occurrence of brain shift, the dependence of tract reconstruction on FA changes, and correct ROI placement.

References

[1] Basser PJ, Pajevic S, Pierpaoli C, Duda J, Aldroubi A (2000) *In vivo* fiber tractography using DTI MRI data. Magn Reson Med 44: 625–632

[2] Beaulieu C (2002) The basis of anisotropic water diffusion in the nervous system – a technical review. NMR Biomed 15: 435–455

[3] Bello L, Acerbi F, Giussani C, Baratta P, Taccone P, Songa V et al (2006) Intraoperative language localization in multilingual patients with gliomas. Neurosurgery 59: 115–125

[4] Bello L, Gallucci M, Fava M, Carrabba G, Giussani C, Acerbi F et al (2007) Intraoperative subcortical language tract mapping guides surgical removal of gliomas involving speech areas. Neurosurgery 60: 67–80; discussion 80–82

[5] Bello L, Gambini A, Castellano A, Carrabba G, Acerbi F, Fava E et al (2008) Motor and language DTI fiber tracking combined with intraoperative subcortical mapping for surgical removal of gliomas. Neuroimage 39: 369–382

[6] Bello L, Castellano A, Fava E, Casaceli G, Riva M, Scotti G et al (2010) Intraoperative use of diffusion tensor imaging fiber tractography and subcortical mapping for resection of gliomas: technical considerations. Neurosurg Focus 28(2): E6

[7] Bello L, Fava E, Carrabba G, Papagno C, Gaini SM (2010) Present day's standards in microsurgery of low-grade gliomas. Adv Tech Stand Neurosurg 35: 113–157

[8] Berman J (2009) Diffusion MR tractography as a tool for surgical planning. Magn Reson Imaging Clin N Am 17: 205–214

[9] Berman JI, Berger MS, Mukherjee P, Henry RG (2004) Diffusion-tensor imaging-guided tracking of fibers of the pyramidal tract combined with intraoperative cortical stimulation mapping in patients with gliomas. J Neurosurg 101: 66–72

[10] Berman JI, Berger MS, Chung SW, Nagarajan SS, Henry RG (2007) Accuracy of diffusion tensor magnetic resonance imaging tractography assessed using intraoperative subcortical stimulation mapping and magnetic source imaging. J Neurosurg 107: 488–494

[11] Bertani G, Fava E, Casaceli G, Carrabba G, Casarotti A, Papagno C et al (2009) Intraoperative mapping and monitoring of brain functions for the resection of low-grade gliomas: technical considerations. Neurosurg Focus 27(4): E4

[12] Burgel U, Amunts K, Hoemke L, Mohlberg H, Gilsbach JM, Zilles K (2006) White matter fiber tracts of the human brain: three-dimensional mapping at microscopic resolution, topography and intersubject variability. Neuroimage 29: 1092–1105

[13] Catani M, Thiebaut de Schotten M (2008) A diffusion tensor imaging tractography atlas for virtual *in vivo* dissection. Cortex 44: 1105–1132

[14] Clark CA, Barrick TR, Murphy MM, Bell BA (2003) White matter fiber tracking in patients with space-occupying lesions of the brain: a new technique for neurosurgical planning? Neuroimage 20: 1601–1608

[15] Coenen VA, Krings T, Mayfrank L, Polin RS, Reinges MH, Thron A, Gilsbach JM (2001) Three-dimensional visualization of the pyramidal tract in a neuronavigation system during brain tumor surgery: first experiences and technical note. Neurosurgery 49: 86–92; discussion 92–93

[16] Conturo TE, Lori NF, Cull TS, Akbudak E, Snyder AZ, Shimony JS et al (1999) Tracking neuronal fiber pathways in the living human brain. Proc Natl Acad Sci USA 96: 10422–10427

[17] Duffau H (2008) The anatomo-functional connectivity of language revisited. New insights provided by electrostimulation and tractography. Neuropsychologia 46: 927–934

[18] Duffau H, Capelle L, Sichez N, Denvil D, Lopes M, Sichez JP et al (2002) Intraoperative mapping of the subcortical language pathways using direct stimulations. An anatomo-functional study. Brain 125: 199–214

[19] Ellmore TM, Beauchamp MS, O'Neill TJ, Dreyer S, Tandon N (2009) Relationships between essential cortical language sites and subcortical pathways. J Neurosurg 111: 755–766

[20] Goebell E, Paustenbach S, Vaeterlein O, Ding XQ, Heese O, Fiehler J et al (2006) Low-grade and anaplastic gliomas: differences in architecture evaluated with diffusion-tensor MR imaging. Radiology 239: 217–222

[21] Hendler T, Pianka P, Sigal M, Kafri M, Ben-Bashat D, Constantini S et al (2003) Delineating gray and white matter involvement in brain lesions: three-dimensional alignment of functional magnetic resonance and diffusion-tensor imaging. J Neurosurg 99: 1018–1027

[22] Holodny AI, Schwartz TH, Ollenschleger M, Liu WC, Schulder M (2001) Tumor involvement of the corticospinal tract: diffusion magnetic resonance tractography with intraoperative correlation. J Neurosurg 95: 1082

[23] Hunsche S, Moseley ME, Stoeter P, Hedehus M (2001) Diffusion-tensor MR imaging at 1.5 and 3.0 T: initial observations. Radiology 221: 550–556

[24] Jellison BJ, Field AS, Medow J, Lazar M, Salamat MS, Alexander AL (2004) Diffusion tensor imaging of cerebral white matter: a pictorial review of physics, fiber tract anatomy, and tumor imaging patterns. Am J Neuroradiol 25: 356–369

[25] Jones DK (2003) Determining and visualizing uncertainty in estimates of fiber orientation from diffusion tensor MRI. Magn Reson Med 49: 7–123

[26] Kamada K, Todo T, Masutani Y, Aoki S, Ino K, Takano T et al (2005) Combined use of tractography-integrated functional neuronavigation and direct fiber stimulation. J Neurosurg 102: 664–672

[27] Keles GE, Lundin DA, Lamborn KR, Chang EF, Ojemann G, Berger MS (2004) Intraoperative subcortical stimulation mapping for hemispherical perirolandic gliomas located within or adjacent to the descending motor pathways: evaluation of morbidity and assessment of functional outcome in 294 patients. J Neurosurg 100: 369–375

[28] Makris N, Kennedy DN, McInerney S, Sorensen AG, Wang R, Caviness VS Jr, Pandya DN (2005) Segmentation of subcomponents within the superior longitudinal fascicle in humans: a quantitative, *in vivo*, DT-MRI study. Cereb Cortex 15: 854–869

[29] Mikuni N, Okada T, Enatsu R, Miki Y, Hanakawa T, Urayama S et al (2007) Clinical impact of integrated functional neuronavigation and subcortical electrical stimulation to preserve motor function during resection of brain tumors. J Neurosurg 106: 593–598

[30] Mori S, van Zijl PC (2002) Fiber tracking: principles and strategies – a technical review. NMR Biomed 15: 468–480

[31] Mori S, Crain BJ, Chacko VP, van Zijl PC (1999) Three-dimensional tracking of axonal projections in the brain by magnetic resonance imaging. Ann Neurol 45: 265–269

[32] Mori S, Frederiksen K, van Zijl PC, Stieltjes B, Kraut MA, Solaiyappan M, Pomper MG (2002) Brain white matter anatomy of tumor patients evaluated with diffusion tensor imaging. Ann Neurol 51: 377–380

[33] Mori S, Kaufmann WE, Davatzikos C, Stieltjes B, Amodei L, Fredericksen K et al (2002) Imaging cortical association tracts in the human brain using diffusion-tensor-based axonal tracking. Magn Reson Med 47: 215–223

[34] Nimsky C, Ganslandt O, Hastreiter P, Wang R, Benner T, Sorensen AG, Fahlbusch R (2005) Preoperative and intraoperative diffusion tensor imaging-based fiber tracking in glioma surgery. Neurosurgery 56: 130–137; discussion 138

[35] Nimsky C, Ganslandt O, Hastreiter P, Wang R, Benner T, Sorensen AG, Fahlbusch R (2005) Intraoperative diffusion-tensor MR imaging: shifting of white matter tracts during neurosurgical procedures – initial experience. Radiology 234: 218–225

[36] Okada T, Mikuni N, Miki Y, Kikuta K, Urayama S, Hanakawa T et al (2006) Corticospinal tract localization: integration of diffusion-tensor tractography at 3-T MR imaging with intraoperative white matter stimulation mapping – preliminary results. Radiology 240: 849–857

[37] Pajevic S, Pierpaoli C (1999) Color schemes to represent the orientation of anisotropic tissues from diffusion tensor data: application to white matter fiber tract mapping in the human brain. Magn Reson Med 42: 526–540; Erratum in Magn Reson Med 43: 921 (2000)

[38] Pierpaoli C, Jezzard P, Basser PJ, Barnett A, Di Chiro G (1996) Diffusion tensor MR imaging of the human brain. Radiology 201: 637–648

[39] Rasmussen IA Jr, Lindseth F, Rygh OM, Berntsen EM, Selbekk T, Xu J et al (2007) Functional neuronavigation combined with intra-operative 3D ultrasound: initial experiences during surgical resections close to eloquent brain areas and future directions in automatic brain shift compensation of preoperative data. Acta Neurochir (Wien) 149: 365–378

[40] Roux FE, Ibarrola D, Tremoulet M, Lazorthes Y, Henry P, Sol JC, Berry I (2001) Methodological and technical issues for integrating functional magnetic resonance imaging data in a neuronavigational system. Neurosurgery 49: 1145–1156; discussion 1156–1157

[41] Schiffbauer H, Ferrari P, Rowley HA, Berger MS, Roberts TP (2001) Functional activity within brain tumors: A magnetic source imaging study. Neurosurgery 49: 1313–1320; discussion 1320–1321

[42] Schiffbauer H, Berger MS, Ferrari P, Freudenstein D, Rowley HA, Roberts TP (2002) Preoperative magnetic source imaging for brain tumor surgery: a quantitative comparison with intraoperative sensory and motor mapping. J Neurosurg 97: 1333–1342

[43] Schonberg T, Pianka P, Hendler T, Pasternak O, Assaf Y (2006) Characterization of displaced white matter by brain tumors using combined DTI and fMRI. Neuroimage 30: 1100–1111

[44] Skirboll SS, Ojemann GA, Berger MS, Lettich E, Winn HR (1996) Functional cortex and subcortical white matter located within gliomas. Neurosurgery 38: 678–684; discussion 684–685

[45] Stadlbauer A, Ganslandt O, Buslei R, Hammen T, Gruber S, Moser E et al (2006) Gliomas: histopathologic evaluation of changes in directionality and magnitude of water diffusion at diffusion-tensor MR imaging. Radiology 240: 803–810

[46] Stadlbauer A, Nimsky C, Buslei R, Salomonowitz E, Hammen T, Buchfelder M et al (2007) Diffusion tensor imaging and optimized fiber tracking in glioma patients: histopathologic evaluation of tumor-invaded white matter structures. Neuroimage 34: 949–956

[47] Wakana S, Jiang H, Nagae-Poetscher LM, van Zijl PC, Mori S (2004) Fiber tract-based atlas of human white matter anatomy. Radiology 230: 77–87

[48] Wiegell MR, Larsson HB, Wedeen VJ (2000) Fiber crossing in human brain depicted with diffusion tensor MR imaging. Radiology 217: 897–903

[49] Wieshmann UC, Clark CA, Symms MR, Franconi F, Barker GJ, Shorvon SD (1999) Reduced anisotropy of water diffusion in structural cerebral abnormalities demonstrated with diffusion tensor imaging. Magn Reson Imaging 17: 1269–1274

Multimodal functional neuronavigation and intraoperative imaging

Oliver Ganslandt, Peter Grummich, Christopher Nimsky

Introduction

Neurosurgical resection techniques in glioma surgery have evolved from simple tumor debulking to a holistic concept which includes anatomy, function and metabolism both of tumor and adjacent brain parenchyma. Driven by the dichotomic aim not to induce neurological deterioration while attempting to achieve gross tumor resection, neurosurgeons have sought to improve surgical outcome with the help of modern image guided techniques. In this chapter the implementation of multimodal functional neuronavigation and intraoperative imaging in glioma surgery is described.

Multimodal neuronavigation

While electrical stimulation is still regarded as the "gold standard" for neurosurgical functional brain localization this invasive technique is not available for preoperative decision making and surgical planning. A different approach to make brain function accessible for the neurosurgeon is the use of brain imaging techniques and its implementation into the neurosurgical workspace. In recent years, two noninvasive techniques have been found especially suitable for the presurgical localization of the eloquent cortex: magnetoencephalography (MEG) and functional magnetic resonance imaging (fMRI). Studies using these techniques successfully localized functional activity. One of the most interesting applications was the merge of functional brain imaging with frame-based and frameless stereotaxy, also known as functional neuronavigation. There is initial evidence that the use of functional neuronavigation for lesions adjacent to eloquent brain areas may favour clinical outcome [10].

If surgery near eloquent brain areas is planned, a detailed knowledge about the topographic relation of a lesion to the adjacent functional brain area is crucial to avoid postoperative neurological deficits. In neurosurgery, the primary sensorimotor cortex and the cortical areas subserving language comprehension and production are considered to be the main risk structures. These structures can usually not be depicted from conventional structural imaging techniques. Other reasons that warrant a detailed evaluation are the individual representation of these eloquent areas and the phenomenon of cortical reorganization of these areas from their original positions [6]. Furthermore, normal sulcal anatomy is often not discernible because of a space occupying lesion. These situations require methods for localizing functional areas prior to surgery for decision making, planning and avoiding crippling postoperative results.

Functional MRI has become indispensable in neurosurgery to easily gain knowledge about the topographic relation of a given lesion to the functional brain area at risk and thus to plan the surgical approach. Furthermore, fMRI-derived information about the extent of cortical involvement in function can be used in conjunction with image-guided surgery during resection of lesions adjacent to eloquent brain areas under general anesthesia for navigation. In addition, its non-invasiveness gives the opportunity to repeat the examinations and conduct follow-up studies on reorganization of cortical function. Advances in MRI technology, such as the introduction of higher field strengths or the development of multi-channel coils, have undoubtedly improved signal acquisition and processing. Over the last years, a substantial number of publications have described the usefulness of clinical fMRI for neurosurgical applications [43]. The use of fMRI for the presurgical localization of the sensorimotor cortex is now widely appreciated and has been investigated by several groups, which also performed comparisons with direct motor stimulation. Language fMRI has been found to be an alternative to the invasive Wada test for language lateralization [4, 40]. Furthermore, fMRI has been used to predict memory localization. Concerning the reliability of fMRI-localization of speech areas in the fronto-temporal cortex, as compared to direct electrical stimulation, there is not enough data that support its application as a substitute for electrical stimulation [34]. However, more studies are needed to fully understand the mechanisms of language activation and the use of fMRI adds complementary information that is essential for neurosurgeons. Besides advances in fMRI-techniques another MRI based modality has been established in neurosurgery to visualize subcortical white matter fibers: Diffusion tensor imaging (DTI) and fiber tracking. Fiber tracking is a reconstruction technique of diffusion tensor (DT) imaging data to visualize neuronal pathways in the human brain. This technique is increasingly used to plan and guide neurosurgical resection of intracranial tumors close to functionally important brain areas [32]. The ability to reconstruct fiber bundles out of MRI diffusion parameters has opened new possibilities in surgery of deep seated gliomas.

Functional imaging

In surgery close to functional brain areas it is critical to have information about the relation of eloquent cortex or important white matter tracts and the lesion.

Problems may be:

(1) Individual localization of functional brain areas (in particular language areas)
(2) Gyral displacement caused by the lesion
(3) Transfer of functions in new brain areas caused by the lesion may have occurred (brain plasticity)
(4) Knowledge about hemispheric dominance

Functional imaging serves in this situation in two ways:

(1) Determination of the risk of surgery and presurgical planning of approach and extent of surgery.
(2) Use of functional imaging during surgery with neuronavigation (functional neuronavigation.

Methods of functional imaging

In our department we perform MEG (Magnetoencephalography) measurements simultaneously over both hemispheres with a dual-sensor 74-channel (2 x 37 channels) biomagnetometer system (Magnes II, 4-D Neuroimaging, San Diego, CA, USA). For source localization we use a single equivalent-current dipole fit_algorithm based on a least-square search, when there is a single focal predominant pattern of activity. We discard dipole localizations with a correlation smaller than a minimum value of 0.94.

If results of the single moving dipole algorithm for the assumpted language area (Broca or Wernicke) has low correlation values of the calculated field distribution compared to the measured field distribution of less than 0.94 we apply a current-density reconstruction approach.

Our current-density reconstruction approach is a minimum variance beamformer using a spatial filter algorithm [8, 9]. A detailed description of the methods is found elsewhere [10].

MEG results are superimposed on T1-weighted three-dimensional MR images (3D MPRAGE, FOV 256 mm x 256 mm with a 256 x 256 matrix, slice thickness 1 mm). To obtain the transformation matrix, we digitize the surface of the patient's head with an electromagnetic 3D digitizer (Polhemus Navigation Sciences Inc., Colchester, VT, USA), both before and after data acquisition and fitting it to the head shape reconstructed from the patient's MR data set using an edge detection algorithm [19] (co-registration accuracy better than 2 mm).

FMRI (Functional Magnetic Resonance Imaging) measurements are acquired on a 1.5 T MR scanner (Magnetom Sonata, Siemens, Erlangen, Germany) using an echo-planar imaging (EPI) sequence.

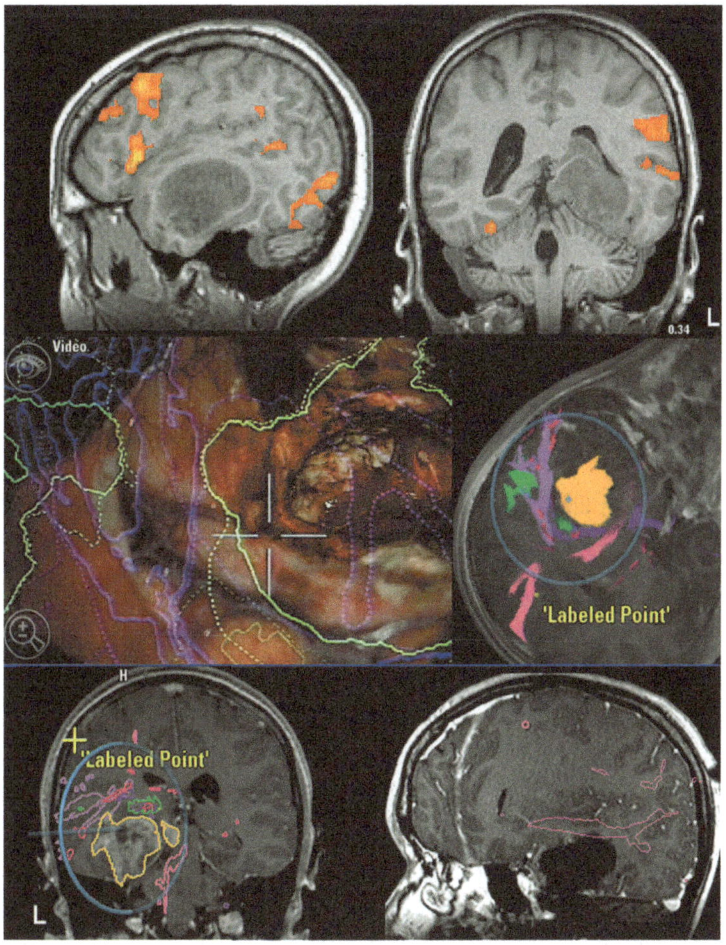

Fig. 1. *First row:* FMRI activity during language task (build sentence from a given noun). *Second row:* microscopic view with Wernicke area in light green on the left, DTI-reconstruction of the fasciculus occipito-frontalis (purple) during navigation. *Third row left:* Focus of microscope shown during surgery in coronal slice. *Right:* Final MR showing resection result with overlaid preoperative fiber tract of Fasciculus occipito- frontalis showing resection touching the fiber tract. The patient was operated upon a glioblastoma

Measurements for language are done with 25 slices of 3mm thickness, a TR= 2470, a TE= 60. Stimulation is done in a block paradigm ("boxcar") with 180 stimulus presentations in 6 blocks. We present 30 visual stimuli in an activation condition during which the patient is instructed to perform a language task; we alternate these with 30 presentations in a resting condition

Motion correction is performed by an image-based prospective acquisition correction applying interpolation in the k-space [44]. Activation maps are calculated by the correlation between signal intensity and a square wave reference function for each pixel according to the course of stimulation. Pixels exceeding a significance threshold (typical correlations above a threshold of 0.4 are displayed, if at least 6 contiguous voxels constitute a cluster, to eliminate isolated voxels. We align the functional slices to magnetization prepared rapid acquisition gradient echo (MPRAGE) images using 160 slices of 1mm slice thickness.

Functional imaging of language functions

So far we localized language activity in 283 patients. Of these patients 225 underwent surgery (196 resections, 29 biopsies).

In Fig. 1, a typical distribution of activated areas during a silent sentence building task is displayed, which includes receptive and expressive language functions. Activity is found in the Wernicke area in the supratemporal sulcus and at the end of the Sylvian fissure in the supramarginal gyrus or the planum temporale. In the frontal lobe activity can be found at the bottom and top of the inferior frontal gyrus, operculum frontale, at the pars triangularis of the inferior frontal gyrus and at the upper frontal corner of the insula (not visible in the figure). Further activity is found in the premotor cortex.

Comparison FMRI-MEG

There are several differences between fMRI and MEG. MEG shows the brain activity at a certain time after the stimulus. FMRI shows the activity during the whole task. So it enhances those activities, which are active the most time, whereas MEG enhances the activity, which occurs at a certain time after the stimulus.

FMRI measures brain function not directly but by changes in the blood supply of the active cortex areas. There may occur errors if this coupling of brain activity to blood supply, the hemodynamic response, is not working normally. This seems to happen especially in fMRI studies of glioblastoma. Schreiber described a reduction of the BOLD effect in the vicinity of gliomas but not in the vicinity of non-glial tumors [35].

In our own series we found in 53 % of patients with high grade glioma, the language areas are not clearly detectable by fMRI [10]. Therefore we find it useful to compare MEG and fMRI measurements to increase reliability of language mapping (Fig. 2).

Outcome of surgery

54 (24%) of the 225 patients who underwent surgery close to language centers experienced transient language disorders, which resolved after few weeks. Slight permanent postoperative deterioration of speech was observed in only one patient. No patient suffered from permanent global aphasia after surgery. Distance of the resection to the functional areas and fiber tracts were 5 mm to 20 mm.

Two of our patients showed an amelioration of language function after surgery. One patient, who was not able to talk before surgery, was able to talk afterwards. Another patient, who had severe naming problems, showed an improvement after surgery.

Marked transient language disorders occurred, when surgery was done in the medial temporal lobe close to the fasciculus occipitofrontalis inferior.

The fact of having no severe permanent speech disorder in our patients indicates that our language mapping protocol is reliable. The transitory language deficits indicate that resection was conducted very close to the boundaries of functional areas.

It should be mentioned that we found that certain functional areas can be compensated for, if they are resected. These were the supplementary motor area (SMA) and the area in the fusiform gyrus for word recognition.

DTI navigation

While fMRI and MEG enables identification of cortical eloquent brain areas also major white matter tracts have to be preserved to avoid neurological sequelae when resecting a brain tumor. Major white matter tracts can be identified by diffusion weighted imaging visualizing differences in tissue anisotropy by measuring the self-diffusion properties, i.e. the Brownian motion of water molecules. Diffusion is anisotropic, i.e. orientation-dependent, in areas with a strong aligned microstructure, including cell membranes and the myelin sheath surrounding myelinated white matter, causing a slight impediment of the water motion, so that a differentiation between white and grey matter becomes possible [1]. DTI is based on measuring multiple diffusion weighted images in different gradient directions to resolve the orientation of the white matter tracts. The dominant fiber orientation in each voxel element is resolved by DTI, representing the mean longitudinal direction of axons in major white matter tracts. DTI provides information about the normal course, the displacement, or interruption of white matter tracts around a tumor, as well as a widening of fiber bundles due to edema or tumor infiltration can be measured. Fiber tracking is probably the most appealing and understandable technique for representing major white matter tracts and has been investigated by various groups. Fiber tracking algorithms which compare local tensor field orientations

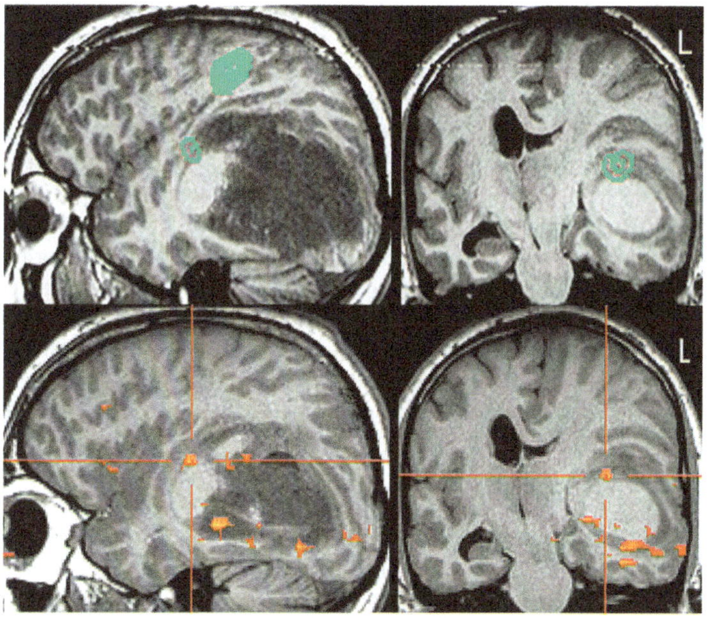

Fig. 2. Comparison of MEG beamformer localization with FMRI during picture naming in a patient with glioblastoma. The bold effect is suppressed so that FMRI does not show clear activity of language areas

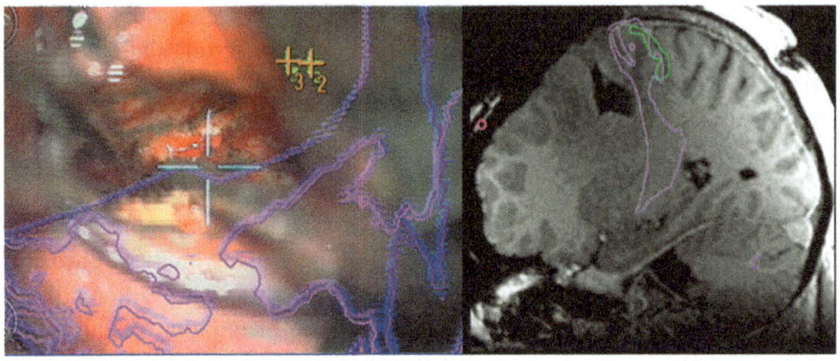

Fig. 3. *Left:* Microscopic view showing the pyramidal fibers overlaid on the exposed brain surface. *Right:* Resection result with pyramidal tract reconstructed by DTI images. Patient with glioblastoma

measured by DTI from voxel to voxel have been implemented, allowing a non-invasive tracing of large fiber tract bundles in the human brain [2, 20, 21, 39]. The first applications integrating tractography data into a stereotactic coordinate system were standalone applications developed for individual clinical sites [15, 18, 42, 46]. A broad application for routine clinical use, as well as a standardization was possible by the implementation of a fiber tracking algorithm into a standard navigation system, allowing routine usage and broad availability [24]. Registration with standard anatomical image data greatly facilitates the generation and selection of the fibers of interest. The implemented approach allows a straightforward definition of volumes of interest for selection of the fiber tracts of interest. Only two parameters, the FA (fractional anisotropy) threshold and the minimum length of the fibers that will be computed have to be selected by the user. The total generation of the fiber tracts, including image transfer, registration of the diffusion data with the standard anatomical image data, tensor calculation, fiber tracking, and the final generation of a 3D object needs less than 10 minutes, depending to some extent on the individual strategy how the different seed volumes of interest during initiation of the tracking algorithm are selected.

Clinical data applying pyramidal tract navigation in a routine setting for glioma resection proved that the occurrence of unwanted neurological sequelae could be reduced. In 11% of a series of 70 patients where the pyramidal tract was visualized in the surgical field a new or aggravated postoperative paresis could be observed, which was transient in 6.8% of them. Thus, only in 3 patients (4.2%) there was a new permanent neurological deficit [31]. These data clearly support the concept of functional neuronavigation, i.e. adding functional information to 3D anatomical datasets to reduce postoperative morbidity in surgery close to eloquent brain areas (Fig. 3). Maximal safety may require combining electrophysiological brain mapping with functional navigation that integrates fMRI/MEG-data and DTI-based fibertracking acquired before or during surgery. Like cortical eloquent brain areas can be identified by intraoperative electrophysiological mapping, subcortical electrical stimulation helps to identify major white matter tracts intraoperatively [5, 47]. It should be emphasized that functional neuronavigation and subcortical stimulation are complementary methods that facilitate the preservation of pyramidal tracts.

Integrating DTI data into a neuronavigation environment offers the possibility to correlate DTI findings in a multimodal setup, e.g. correlations with MR spectroscopy and PET become possible. Besides preservation of neurological deficits DTI neuronavigation also enables to directly correlate histological findings with DTI results, so tumor invasion of major white matter tracts can be detected and quantified [7, 17, 36–38].

Intraoperative imaging

Neuronavigation systems have a decreasing accuracy during the course of a surgical procedure if the system is based on preoperative data only. This is due to brain shift, i.e. an intraoperative deformation of the brain, caused by tumor removal, brain swelling, the use of brain retractors, and cerebrospinal-fluid drainage [13, 22]. The introduction of intraoperative imaging offers a possibility to compensate for the effects of brain shift. Intraoperative imaging depicts a virtual reproduction of the actual intraoperative physical reality, on how the brain is deformed and on the extent of tumor removal [14, 22, 25, 45]. The ability to objectively determine the extent of tumor removal during surgery is of great value. If a resection is incomplete, tumour residues that were initially missed can be further removed during the same operation. Intraoperative imaging allows an objective evaluation of the actual intraoperative situation, thus serving as quality control during surgery [3, 11, 12, 23, 30, 41] (Fig. 4).

Intraoperative high-field MRI [28] is the most sophisticated possibility for intraoperative imaging compared to the alternatives such as intraoperative computed tomography and intraoperative ultrasound. Combining high-field MRI and microscope-based navigation enables an intraoperative possibility to compensate for the effects of brain shift by intraopertive updating the image information [30]. For the update procedure the intraoperative images have to be registered either by a calibrated registration matrix that can be attached to the upper part of the imaging coil and tracked by the navigation system [33], or by direct registration of pre- and intraoperative images, assuming hat there was no positional shift, i.e. that the initial patient registration was not erroneous, so that the coordinate system of the initial patient registration can be also applied for the intraoperative images. The third alternative, applying bone fiducial markers, which was used in low-field MRI setups, is a cumbersome procedure and does no longer play a role in clinical routine. Neuronavigation with updated intraoperative image data allows a reliable identification of tumor remnants. Microscope-based heads-up displays with the direct visualization of the segmented tumor remnant in the surgical field play a crucial role in the precise localization and orientation in the resection cavity.

Intraoperative high-field MRI does not only allow for standard anatomical imaging such as T1- and T2-weighted sequences, however, also intraoperative fMRI and more important clinically intraoperative DTI is feasible. Still intraoperative MRS has its limitations due to the brain air interface, so that there remain some distinct challenges for updating also MRS information intraoperatively. Intraoperative DTI depicted a great variability in fiber tract shifting [26, 27, 29], emphasizing the need for an intraoperative update also regarding fiber tract data. Reports comparing subcortical electrical stimulation investigations and preoperative DTI data showed some inconsistencies, which were probably due to the effects of brain shift [16, 18]. An update visualizing the actual course of major white matter tracts based on intraoperative DTI data, is a prerequisite for a reliable electrophysiological validation of these data.

Conclusion

Multimodal functional navigation enables removing a tumor close to eloquent brain areas with low postoperative deficits, while additional intraoperative imaging ensures that the maximum extent of the resection can be achieved, as well as it allows to update the image data compensating for the effects of brain shift.

While these methods are basically used to render information where not to resect tumor tissue to avoid postoperative neurological deficits another approach is to seek information where to resect tumor tissue that can not be seen in conventional MRI imaging or by microscope vision. We use proton MR spectroscopy and FET-PET to better delineate tumor infiltration otherwise not to be seen in conventional MR imaging.

References

[1] Basser PJ, Mattiello J, LeBihan D (1994) MR diffusion tensor spectroscopy and imaging. Biophys J 66: 259–267

[2] Basser PJ, Pajevic S, Pierpaoli C, Duda J, Aldroubi A (2000) *In vivo* fiber tractography using DT-MRI data. Magn Reson Med 44: 625–632

[3] Black PM, Moriarty T, Alexander III E, Stieg P, Woodard EJ, Gleason PL, Martin CH, Kikinis R, Schwartz RB, Jolesz FA (1997) Development and implementation of intraoperative magnetic resonance imaging and its neurosurgical applications. Neurosurgery 41: 831–845

[4] Branco DM, Suarez RO, Whalen S, O'shea JP, Nelson AP, da Costa JC, Golby AJ (2006) Functional MRI of memory in the hippocampus: Laterality indices may be more meaningful if calculated from whole voxel distributions. Neuroimage 32: 592–602

[5] Duffau H, Capelle L, Denvil D, Sichez N, Gatignol P, Taillandier L, Lopes M, Mitchell MC, Roche S, Muller JC, Bitar A, Sichez JP, van Effenterre R (2003) Usefulness of intraoperative electrical subcortical mapping during surgery for low-grade gliomas located within eloquent brain regions: functional results in a consecutive series of 103 patients. J Neurosurg 98: 764–778

[6] Duffau H, Denvil D, Capelle L (2002) Long term reshaping of language, sensory, and motor maps after glioma resection: a new parameter to integrate in the surgical strategy. J Neurol Neurosurg Psychiatry 72: 511–516

[7] Goebell E, Fiehler J, Ding XQ, Paustenbach S, Nietz S, Heese O, Kucinski T, Hagel C, Westphal M, Zeumer H (2006) Disarrangement of fiber tracts and decline of neuronal density correlate in glioma patients – a combined diffusion tensor imaging and 1H-MR spectroscopy study. Am J Neuroradiol 27: 1426–1431

[8] Grummich P, Kober H, Vieth J (1992) Localization of multi source magnetic brain activity using a spatial filter reconstruction. In: Dittmar A, Froment J (eds) Proc Satellite Symp on Neuroscience and Technology of the 14th Ann Int Conf of the IEEE Eng Med Biol Soc, Lyon, France

[9] Grummich P, Kober H, Vieth J (1992) Localization of the underlying current of magnetic brain activity using spatial filtering. Biomed Eng 37: 158–159

[10] Grummich P, Nimsky C, Pauli E, Buchfelder M, Ganslandt O (2006) Combining fMRI and MEG increases the reliability of presurgical language localization: a clinical study on the difference between and congruence of both modalities. Neuroimage 32: 1793–1803

[11] Hall WA, Kowalik K, Liu H, Truwit CL, Kucharezyk J (2003) Costs and benefits of intraoperative MR-guided brain tumor resection. Acta Neurochir (Suppl 85): 137–142

[12] Hall WA, Liu H, Martin AJ, Pozza CH, Maxwell RE, Truwit CL (2000) Safety, efficacy, and functionality of high-field strength interventional magnetic resonance imaging for neurosurgery. Neurosurgery 46: 632–642

[13] Hastreiter P, Rezk-Salama C, Nimsky C, Lürig C, Greiner G, Ertl T (2000) Registration techniques for the analysis of the brain shift in neurosurgery. Comput Graph 24: 385–389

[14] Hastreiter P, Rezk-Salama C, Soza G, Bauer M, Greiner G, Fahlbusch R, Ganslandt O, Nimsky C (2004) Strategies for brain shift evaluation. Med Image Anal 8: 447–464

[15] Kamada K, Sawamura Y, Takeuchi F, Kawaguchi H, Kuriki S, Todo T, Morita A, Masutani Y, Aoki S, Kirino T (2005) Functional identification of the primary motor area by corticospinal tractography. Neurosurgery 56: 98–109

[16] Kamada K, Todo T, Masutani Y, Aoki S, Ino K, Takano T, Kirino T, Kawahara N, Morita A (2005) Combined use of tractography-integrated functional neuronavigation and direct fiber stimulation. J Neurosurg 102: 664–672

[17] Kinoshita M, Hashimoto N, Goto T, Kagawa N, Kishima H, Izumoto S, Tanaka H, Fujita N, Yoshimine T (2008) Fractional anisotropy and tumor cell density of the tumor core show positive correlation in diffusion tensor magnetic resonance imaging of malignant brain tumors. Neuroimage 43: 29–35

[18] Kinoshita M, Yamada K, Hashimoto N, Kato A, Izumoto S, Baba T, Maruno M, Nishimura T, Yoshimine T (2005) Fiber-tracking does not accurately estimate size of fiber bundle in pathological condition: initial neurosurgical experience using neuronavigation and subcortical white matter stimulation. Neuroimage 25: 424–429

[19] Kober H, Nimsky C, Vieth J, Fahlbusch R, Ganslandt O (2002) Co-registration of function and anatomy in frameless stereotaxy by contour fitting. Stereotact Funct Neurosurg 79: 272–283

[20] Mori S, Crain BJ, Chacko VP, van Zijl PC (1999) Three-dimensional tracking of axonal projections in the brain by magnetic resonance imaging. Ann Neurol 45: 265–269

[21] Mori S, Van Zijl PC (2002) Fiber tracking: principles and strategies – a technical review. NMR Biomed 15: 468–480

[22] Nabavi A, Black PM, Gering DT, Westin CF, Mehta V, Pergolizzi RS, Jr., Ferrant M, Warfield SK, Hata N, Schwartz RB, Wells WM, 3rd, Kikinis R, Jolesz FA (2001) Serial intraoperative magnetic resonance imaging of brain shift. Neurosurgery 48: 787–798

[23] Nimsky C, Ganslandt O, Fahlbusch R (2005) Comparing 0.2 tesla with 1.5 tesla intraoperative magnetic resonance imaging analysis of setup, workflow, and efficiency. Acad Radiol 12: 1065–1079

[24] Nimsky C, Ganslandt O, Fahlbusch R (2006) Implementation of fiber tract navigation. Neurosurgery 58: ONS-292–304

[25] Nimsky C, Ganslandt O, Hastreiter P, Fahlbusch R (2001) Intraoperative compensation for brain shift. Surg Neurol 56: 357–365

[26] Nimsky C, Ganslandt O, Hastreiter P, Wang R, Benner T, Sorensen AG, Fahlbusch R (2005) Intraoperative diffusion tensor imaging: shifting of white matter tracts during neurosurgical procedures - initial experience. Radiology 234: 218–225

[27] Nimsky C, Ganslandt O, Hastreiter P, Wang R, Benner T, Sorensen AG, Fahlbusch R (2005) Preoperative and intraoperative diffusion tensor imaging-based fiber tracking in glioma surgery. Neurosurgery 56: 130–138

[28] Nimsky C, Ganslandt O, Keller v B, Romstöck J, Fahlbusch R (2004) Intraoperative high-field strength MR imaging: implementation and experience in 200 patients. Radiology 233: 67–78

[29] Nimsky C, Ganslandt O, Merhof D, Sorensen AG, Fahlbusch R (2006) Intraoperative visualization of the pyramidal tract by diffusion-tensor-imaging-based fiber tracking. Neuroimage 30: 1219–1229

[30] Nimsky C, Ganslandt O, Von Keller B, Romstock J, Fahlbusch R (2004) Intraoperative high-field-strength MR imaging: implementation and experience in 200 patients. Radiology 233: 67–78

[31] Nimsky C, Ganslandt O, Weigel D, Keller v B, Stadlbauer A, Akutsu H, Hammen T, Buchfelder M (2008) Intraoperative tractography and neuronavigation of the pyramidal tract. Jpn J Neurosurg 17: 21–26

[32] Nimsky C, Grummich P, Sorensen AG, Fahlbusch R, Ganslandt O (2005) Visualization of the pyramidal tract in glioma surgery by integrating diffusion tensor imaging in functional neuronavigation. Zentralbl Neurochir 66: 133–141

[33] Rachinger J, von Keller B, Ganslandt O, Fahlbusch R, Nimsky C (2006) Application accuracy of automatic registration in frameless stereotaxy. Stereotact Funct Neurosurg 84: 109–117

[34] Roux FE, Boulanouar K, Lotterie JA, Mejdoubi M, LeSage JP, Berry I (2003) Language functional magnetic resonance imaging in preoperative assessment of language areas: correlation with direct cortical stimulation. Neurosurgery 52: 1335–1345; discussion 1345–1337

[35] Schreiber A, Hubbe U, Ziyeh S, Hennig J (2000) The influence of gliomas and nonglial space-occupying lesions on blood-oxygen-level-dependent contrast enhancement. Am J Neuroradiol 21: 1055–1063

[36] Stadlbauer A, Ganslandt O, Buslei R, Hammen T, Gruber S, Moser E, Buchfelder M, Salomonowitz E, Nimsky C (2006) Gliomas: histopathologic evaluation of changes in directionality and magnitude of water diffusion at diffusion-tensor MR imaging. Radiology 240: 803–810

[37] Stadlbauer A, Nimsky C, Buslei R, Salomonowitz E, Hammen T, Buchfelder M, Moser E, Ernst-Stecken A, Ganslandt O (2007) Diffusion tensor imaging and optimized fiber tracking in glioma patients: Histopathologic evaluation of tumor-invaded white matter structures. Neuroimage 34: 949–956

[38] Stadlbauer A, Nimsky C, Gruber S, Moser E, Hammen T, Engelhorn T, Buchfelder M, Ganslandt O (2007) Changes in fiber integrity, diffusivity, and metabolism of the pyramidal tract adjacent to gliomas: a quantitative diffusion tensor fiber tracking and MR spectroscopic imaging study. Am J Neuroradiol 28: 462–469

[39] Stieltjes B, Kaufmann WE, van Zijl PC, Fredericksen K, Pearlson GD, Solaiyappan M, Mori S (2001) Diffusion tensor imaging and axonal tracking in the human brainstem. Neuroimage 14: 723–735

[40] Stippich C, Rapps N, Dreyhaupt J, Durst A, Kress B, Nennig E, Tronnier VM, Sartor K (2007) Localizing and lateralizing language in patients with brain tumors: feasibility of routine preoperative functional MR imaging in 81 consecutive patients. Radiology 243: 828–836

[41] Sutherland GR, Kaibara T, Louw D, Hoult DI, Tomanek B, Saunders J (1999) A mobile high-field magnetic resonance system for neurosurgery. J Neurosurg 91: 804–813

[42] Talos I, O'Donnell L, Westin CF, Warfield SK, Wells WM, 3rd, Yoo S, Panych L, Golby A, Mamata H, Maier S, Ratiu P, Guttmann C, Black PM, Jolesz F, Kikinis R (2003) Diffusion tensor and functional MRI fusion with anatomical MRI for image-guided neurosurgery. In: Ellis R, Peters T (eds) MICCAI 2003. Springer, Berlin Heidelberg, pp 407–415

[43] Tharin S, Golby A (2007) Functional brain mapping and its applications to neurosurgery. Neurosurgery 60: 185–201; discussion 201–182

[44] Thesen S, Heid O, Mueller E, Schad LR (2000) Prospective acquisition correction for head motion with image-based tracking for real-time fMRI. Magn Reson Med 44: 457–465

[45] Wirtz CR, Bonsanto MM, Knauth M, Tronnier VM, Albert FK, Staubert A, Kunze S (1997) Intraoperative magnetic resonance imaging to update interactive navigation in neurosurgery: method and preliminary experience. Comput Aided Surg 2: 172–179

[46] Wu JS, Zhou LF, Hong XN, Mao Y, Du GH (2003) Role of diffusion tensor imaging in neuronavigation surgery of brain tumors involving pyramidal tracts. Zhonghua Wai Ke Za Zhi 41: 662–666

[47] Yingling CD, Ojemann S, Dodson B, Harrington MJ, Berger MS (1999) Identification of motor pathways during tumor surgery facilitated by multichannel electromyographic recording. J Neurosurg 91: 922–927

Intraoperative neurophysiological monitoring under general anesthesia

Andrea Szelényi

Introduction

The early steps in determining cortical functional organization during neurosurgical procedures were performed in awake patients and date back into the early 1930s [12, 30]. At that time, the identification and assessment of cortical functional organization within the vicinity of a brain pathology (e.g., tumor, epileptic foci) was possible only by direct electrical stimulation of the cerebral cortex. The observation of the elicited interference with the awake patient's behavior, movement, and language performance served as guidance for the surgical tumor resection. Only in the late 1970s monitoring of somatosensory evoked potentials (SEP) and in the early 1990s monitoring of motor evoked potentials (MEPs) were introduced into the operating room. It was utilized in spine and spinal cord surgery and then for neurovascular procedures, before it was finally implemented into brain tumor surgery.

Overall, the development of reliable application of anesthetics, microsurgical tools and commercially available neuromonitoring equipment did allow for routine and standardized intraoperative neurophysiological monitoring in the anesthetized patient. The intraoperative methods should help to achieve the aim of a maximal tumor resection and a minimal – if any – permanent morbidity. The neurophysiological methods should be sensitive and specific towards the neuronal pathways assessed and easy and safe to perform and should provide real-time information and online analysis. An essential prerequisite for successful neurophysiological monitoring is the potential to reverse changes indicating neurological injury and thereby to prevent injury to the nervous tissue, as well as high re-test reliability.

Methods

In the anesthetized patient, intraoperative neurophysiological monitoring for brain tumor surgery combines localizing and monitoring methods. For localization, the phenomenon of phase-reversed SEP – recorded across the central sulcus – and direct cortical stimulation (DCS) with low-intensity electrical pulses to elicit MEPs are being used. Monitoring methods are SEPs and MEPs, as well as – for brainstem surgery – auditory evoked potentials . The complementary use of the quoted potentials allows for safe tumor surgery in the vicinity of the central and insular region.

Somatosensory evoked potentials

Somatosensory evoked potentials were first described in 1947 [6], but it took about 30 years of further technical development before successful intraoperative utilization was reported [26].

To elicit SEPs, a peripheral nerve, commonly the median nerve at the wrist and the posterior tibial nerve at the medial malleolus are stimulated at a frequency of 3.1 to 5.8 Hz. The response is recorded either directly at exposed cortex or at the scalp of the primary somatosensory cortex at C3', Cz', and C4' (according to the international 10-20 EEG system). The simultaneous recording of the responses generated at the upper cervical level allows for excluding general effects to the SEPs such as temperature, peripheral nerve conduction block due to malpositioning of a limb, or anesthesia.

Because of the near linear correlation between the cortical SEP amplitudes and the cerebral perfusion when decreased below 15 ml per 100 g of brain parenchyma, SEPs are used in neurovascular procedures. The loss of SEP amplitude correlates to cortical infarcts in the territories of the middle cerebral artery and the internal carotid artery [17]. In contrast, the relative insensitivity of SEPs to subcortical ischemia gave rise to concerns and might limit the use of SEPs in indicating ischemia resulting from injury of perforating arteries supplying the internal capsule. The lack of publications about pure SEP losses in intracranial surgery should be seen in the light of the introduction of MEPs.

As the SEPs reflect the activity of the leminscal pathways and somatosensory cortex, reports of false-negative SEPs – i.e., not predicting motor deficit – have been driven by the expectation to monitor the motor pathways alike. As neurosurgical outcome assessment tends to focus on postoperative motor status, even studies analyzing the predictive value of SEPs relate postoperative motor deficits instead of sensory deficits with intraoperative SEP alterations. In intracranial neurosurgical procedures, the sensitivity of SEPs to predict minor postoperative deficits was 64% and the negative predictive value was 95%; regarding severe postoperative deficits, the sensitivity was 81% and the negative predictive value 98% [39].

Motor evoked potentials

Transcranial electric stimulation (TES) and transcranial magnetic stimulation (TMS) became routinely used clinically in the 1980s. Those devices generated single-pulse outputs and did not reliably elicit MEPs intraoperatively. The breakthrough was the technical modification towards a short train of stimuli in 1993 [38]. The applied train contains a short series of pulses at high frequency (mostly five pulses with a duration of 0.5 ms each, 250–500 Hz). The pulses activate preferentially fast-conducting axons of the corticospinal tract. Those fast-conducting neurons are essential for executing voluntary movements. Studies of monkeys with direct recordings from the corticospinal tract demonstrated that the direct activation of the corticospinal tract (D-wave) occurred after a single pulse was applied to the motor cortex. Those results were later confirmed with human patients during intramedullary tumor surgery [4, 8, 9, 29]. The multipulse stimulation activates a series of descending volleys which activate spinal alpha-motoneurons and thus evoke muscle responses. When the stimulation is applied transcranially, the site of neuronal activation within the white matter is critical (Fig. 1). With increasing stimulation intensity a shorter latency of the D-wave is recorded, indicating an activation of fibers with increasing depth within the white matter. Only when high-voltage stimulation (1000 V) is used, current penetrates as deep as to the level of the foramen magnum [18, 31]. On the other hand, there is evidence that moderate anodal suprathreshold TES, as well as anodal direct cortial stimulation, activates the corticospinal tract close to the axon hillock [10]. Thus, MEPs are elicited within the white matter, in contrast to the cortical SEPs, which are generated within the grey matter.

This knowledge and the distribution of the vascular territories are of importance for the intraoperative interpretation of SEP and MEP data. In the absence of a preexisting motor deficit and after establishing the reliable MEP technique, monitorability is achieved in 95–

99% of the patients. In insular glioma and central-region tumor surgery, MEP alteration might occur in up to 44% of the patients [27, 28]. Unaltered MEPs with regard to MEP configuration, amplitude, and stimulation threshold correlated with no new motor deficit. This has to be distinguished from lesions of the supplementary motor area, for which the intraoperative preservation of MEPs predicts the full or nearly complete recovery of a patient's voluntary-movement abilities [32, 42]. The irreversible loss of MEPs is always followed by a severe motor deficit, disabling 42% of the patients severely and permanently (Fig. 2 shows an exemplary MEP loss). MEP alterations ranging from reversible deterioration over reversible loss to irreversible deterioration are followed by a range of unchanged, transiently deteriorated to moderate permanent motor deficits [27, 28, 32]. A critical approach to the value of MEPs might therefore conclude that intraoperative MEP alteration is not sensitive enough to predict postoperative motor outcome. Whereas for spine surgery, the presence or disappearance of MEP amplitude criteria is commonly accepted, it became evident soon

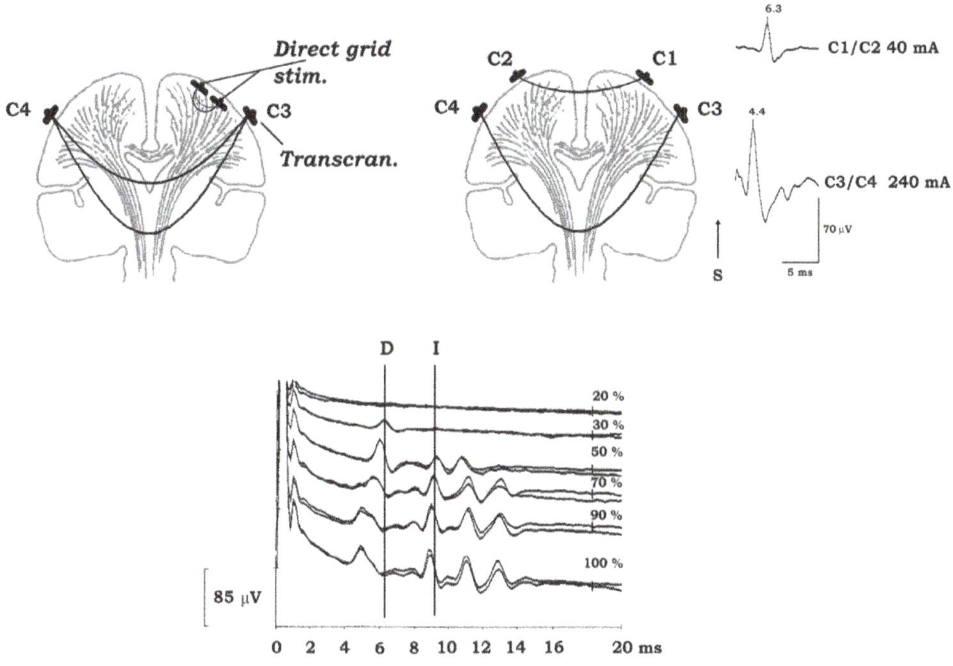

Fig. 1. *Top left*: Current flow during TES and direct brain stimulation via grid electrode are presented schematically. During strong TES, current penetrates deep in the brain, activating both corticospinal tracts. During direct brain stimulation, using a grid electrode current flow is restricted to a single corticospinal tract if one uses low current intensity and activates only restricted motoneuron pools from selective cortical areas (upper or lower extremities depending on the position of the stimulating electrode). *Top right*: Difference in amplitude and latencies of the D-waves record epidurally over the upper thoracic spinal cord in a patient undergoing surgery for a spinal cord tumor. Note the 1.9 ms difference between latencies of the D-waves when elicited with low intensity of current and stimulating montage C1/C2 versus high intensity of current and montage C3/C4. Note the higher amplitude of the D-wave when more axons of the corticospinal tract are recruited and current penetrates deep in the brain (C3/C4 montage and 240 mA stimulating current). *Bottom*: D- and I-waves recorded after single electrical stimulus delivered transcranially (anode at Cz, cathode 6 cm anterior) in a 14-year-old patient with idiopathic scoliosis. As a result of increasing the intensity of the stimulus, the electrical current activates the corticospinal tract deeper within the brain and the latency of the D-wave becomes shorter. As current becomes stronger, more I-waves are induced (100% corresponds to 750 V of stimulator output). Note that at the bottom, the three traces of D-waves have a double peak as a result of corticospinal tract activation at different depths within the brain. (Reproduced from [7] with permission, © Elsevier)

after the first experiences that this has to be refined for supratentorial surgery [25]. Even amplitude deterioration and prolonged transient losses were related with postoperative motor deficit. This empirical experience has led to a 50% amplitude decrement criterion, also being used for SEPs. This is supported by a data analysis of 29 patients experiencing only MEP alteration during the course of intracranial tumor resection. Irreversible MEP alteration was significantly more often correlated with postoperative motor deficit than was reversible MEP alteration ($p < 0.0001$) [35]. In those patients, irreversible MEP alteration was more often associated with postoperative new signal alteration in MRI than was reversible MEP alteration ($p = 0.018$). Further, MEP loss was significantly more often associated with subcortically located new signal alteration ($p = 0.006$). MEP deterioration was significantly more often followed by new signal alterations located in the precentral gyrus ($p = 0.036$). This supports the findings of previous studies by Neuloh et al [27].

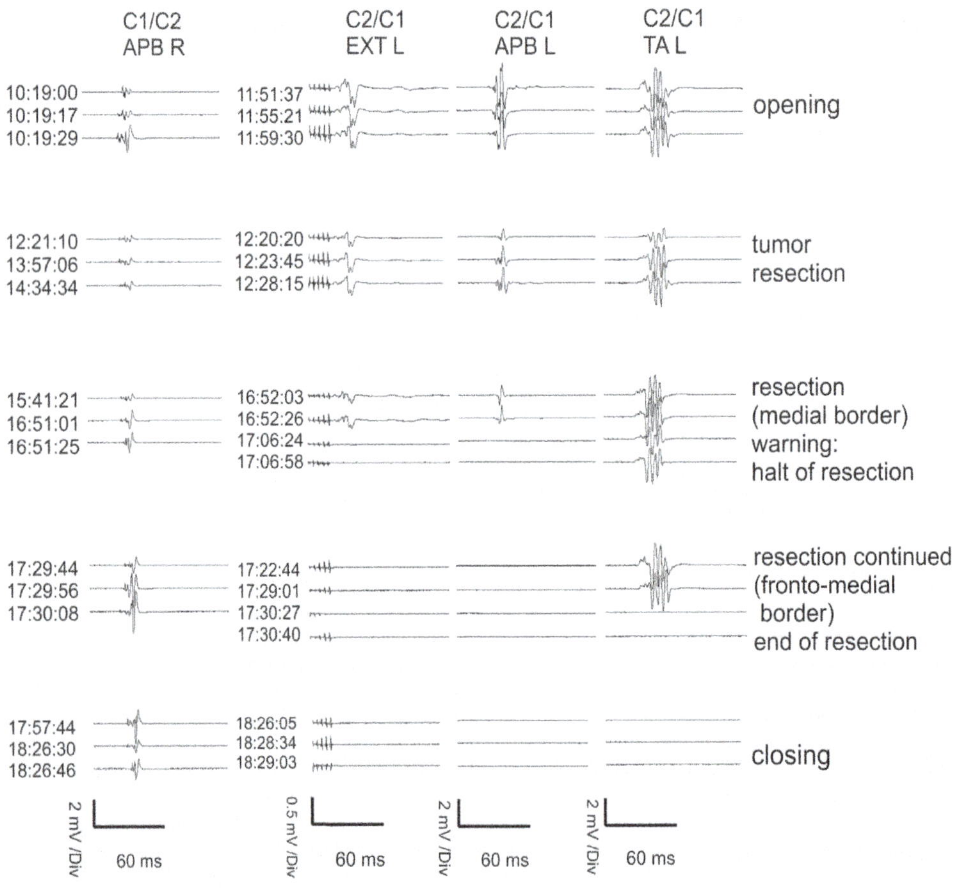

Fig. 2. In the course of a resection of an insular glioma, a subsequent MEP loss was observed, which was followed by a dense hemiplegia and a capsular infarct in this patient. The MEPs of the unaffected hemisphere remained unchanged, thus serving as control. (Reproduced from [35] with permission, © Lippincott Williams & Wilkins)

Visual evoked potentials

Despite some enthusiastic reports about the intraoperative monitoring of VEPs and good correlation with postoperative visual function in surgery around the orbita, the method is not widely used [15, 34]. For surgeries adjacent to the visual pathways and occipital lobe, there is no strong evidence for a correlation between amplitude or latency changes and postoperative visual field defects. The flash stimulation method to elicit VEPs does not seem to be appropriate with regard to the functional organization of the visual pathways. This, in combination with the sensitivity of VEPs to anesthetics, remains a problem to be solved.

Intraoperative mapping in the anesthetized patient

Due to tumor-related distorted anatomy, anatomical landmarks for the identification of the central sulcus are not always helpful. Imaging data such as 3 T MRI, special projection techniques (Mercator projection [16]), and fMRI enable the identification of the precentral gyrus. But studies of anatomy and function may demonstrate differing results [22]. This underlines that intraoperative DCS remains the gold standard for intraoperative verification of the motor cortex.

Phase reversal

The phenomenon of a phase-reversed cortical SEP potential recorded over the precentral gyrus has been implemented in epilepsy surgery [13] and was transferred to brain tumor surgery. A strip or grid electrode consisting of plane or spherical electrodes is placed tangentially over the hand knob and presumably the central sulcus. The reliability to identify the central sulcus at the hand knob ranges from 90 to 94% [5, 20, 40]. In the absence of a somatosensory deficit, the method is reliable, but it may encounter problems in the presence of somatosensory deficits. It further might be difficult to obtain a phase-reversed potential from tibial nerve SEPs, which is of importance in parasagittally located tumors.

Direct cortical stimulation

DCS subsumes two techniques. First, the 60 Hz technique was introduced to the broader neurosurgical community by Penfield in 1937 [30]. For this, a stimulation at 50 to 60 Hz is applied with a bipolar probe. The technique is commonly performed for mapping of cortical areas representing motor and language function in surgery of awake patients. In 1991, LeRoux et al [23] demonstrated its application in anesthetized patients. As minimal movement remains easily unrecognized in anesthetized patients, the additional recording of evoked muscle activity with a multichannel electromyogram proved advantageous [41].

Second, the technique with a train of five stimuli, a technical modification as described by Taniguchi et al in 1993 [38], can be applied for mapping and continuous monitoring of cortical and subcortical motor pathways. This method is predominantly applied with a monopolar anodal stimulation cortically and cathodal stimulation subcortically. Just recently the application for language testing was described, although the routine clinical application will need further development [2]. Performing DCS according to Taniguchi with a short series of high-frequency pulses, the stimulation parameters are the same as for TES except the limitation of the maximum stimulation intensity to 25 mA (see below for side effects).

In a comparison of the two methods, three major differences have to be highlighted: (1) the duration of stimulation, (2) the frequency of stimulation, and (3) the type of probe being used. The 50 Hz technique has to be applied for more than 0.5 s in order to observe an effect of the stimulation and thus it is commonly applied for 1–4 s. The resulting charge (stimulus) exceeds the one necessary to elicit MEPs by the train of five pulses. This might well explain the higher incidence of seizures compared to that with the train of five pulses [36]. The induced tonic movement and the high probability of seizures limit the use the 50 Hz technique as a continuous monitoring method and thus it is used only for mapping. Stimulation with a bipolar probe creates a more focal electric field compared to that with a monopolar probe. It is thought that the stimulation with the bipolar probe provides more precise results in localizing. With the monopolar probe 69% of all motor responses were elicited by DCS in the precentral gyrus and 23% in the premotor area compared to 54% and 38%, respectively, when stimulation was done with the bipolar probe. By stimulation of the motor cortex with the monopolar probe, 96% of all stimulation sites elicited MEPs compared to 95% with the bipolar probe; when the premotor cortex was stimulated with the monopolar probe, that rate was only 15%, whereas it was 27% when the bipolar probe was used [21].

Side effects and safety

TES has a low incidence of side effects. For about 1% of patients, seizures are reported [37]. With the application of bite blocks and moderate stimulation techniques, serious tongue bite injuries and airway obstruction can easily be avoided [24].

When directly applying electric current to the brain, side effects and safety have to be addressed. The most likely side effect is the occurrence of a focal or secondary generalized seizure. This is usually self-terminating. As the bolus administration of sedatives reduces the excitability of the nervous system, this alters the further mapping and monitoring procedure. This is avoided by the administration of cold Ringer solution directly onto the cortex, which will terminate the seizure [33]. Long-term stimulation in animal experiments revealed that the application of a charge of 40 μC per phase is safe without introducing kindling or lesioning of brain tissue [1]. There are no reports for humans which relate intraoperative DCS with histopathological findings or kindling, although in daily practice the applied charge per phase exceeds 40 μC [14]. This might well be explained by the fact that intraoperatively the duration and frequency of stimulation is timely limited.

Principles of clinical application

The intraoperative workflow in functional-monitoring-guided resection of central-region tumors utilizes intraoperative monitoring with transcranially elicited MEPs and recorded SEPs. The unaffected hemisphere serves as control and helps to judge intraoperative signal alteration. Comparable to testing in awake procedures, the first step after dura opening is the localization. The central sulcus is determined by the phase reversal, which is followed by a targeted mapping of the motor cortex. A strip electrode (disc electrodes embedded in silicone) is placed over the cortex for a stimulation at the intensity of the lowest motor threshold. The selection of muscles is determined by the tumor location and should cover the area with the greatest risk of damage. Electrodes not being used for stimulation might be used for electrocorticography or SEP. Alternatively, the exposed motor cortex is mapped first with a stimulating probe, and the strip electrode is being placed parallel over the precentral gyrus thereafter.

During tumor resection, SEPs and MEPs are recorded in an alternating fashion, providing real-time information in less than a minute about the functional integrity of the somatosensory and motor cortex and their related pathways. In the Frankfurt setting, for dissections close to the pyramidal tract any deterioration of MEPs is indicated. Comparable to those groups using only mapping methods, tumor resection is further guided by intermittent cortical and subcortical stimulation to delineate the extension of the resection. The decision when to stop resection in the presence of MEPs is highly dependent on the stimulation parameters being used, the aspect of the resection cavity, and the surgeon's experience. When clinical outcome and imaging results of experienced groups are compared, especially lesions resulting from subcortical ischemia due to perforator injury appear to contribute to permanent morbidity [3, 11, 19, 27, 28].

Conclusion

The combined use of mapping and monitoring techniques in supratentorial surgery and especially in tumor surgery in the vicinity of the motor cortex, corticospinal tract, somatosensory cortex, and lemniscal pathway allows for tumor resection in brain areas previously being considered unresectable. Further studies analyzing not only functional outcome but also tumor recurrence and progression in comparison with their history in patients not being operated on will help for treatment decision.

References

[1] Agnew WF, McCreery DB (1987) Considerations for safety in the use of extracranial stimulation for motor evoked potentials. Neurosurgery 20: 143–147

[2] Axelson HW, Hesselager G, Flink R (2009) Successful localization of the Broca area with short-train pulses instead of "Penfield" stimulation. Seizure 18: 374–375

[3] Berger MS, Deliganis AV, Dobbins J, Keles GE. The effect of extent of resection on recurrence in patients with low grade cerebral hemisphere gliomas. Cancer 74: 1784–1791

[4] Burke D, Hicks RG, Stephen JPH (1990) Corticospinal volleys evoked by anodal and cathodal stimulation of the human motor cortex. J Physiol 425: 283–299

[5] Cedzich C, Taniguchi M, Schafer S, Schramm J (1996) Somatosensory evoked potential phase reversal and direct motor cortex stimulation during surgery in and around the central region. Neurosurgery 38: 962–970

[6] Dawson GD (1947) Investigations on a patient subject to myoclonic seizures after sensory stimulation. J Neurol Neurosurg Psychiatry 10: 141–162

[7] Deletis V, Sala F (2008) Corticospinal tract monitoring with D- and I-waves from the spinal cord and muscle MEPs from the limb muscles. In: Nuwer MR (ed) Intraoperative monitoring of neural function. Handbook of clinical neurophysiology, vol 8. Elsevier, Amsterdam, pp 235–251

[8] Deletis V, Isgum V, Amassian VE (2001) Neurophysiological mechanisms underlying motor evoked potentials in anesthetized humans. Part 1. Recovery time of corticospinal tract waves elicited by pairs of transcranial electrical stimulation. Clin Neurophysiol 112: 438–444

[9] Deletis V, Rodi Z, Amassian VE (2001) Neurophysiological mechanisms underlying motor evoked potentials in anesthetized humans. Part 2. Relationship between epidurally and muscle recorded MEPs in man. Clin Neurophysiol 112: 445–452

[10] Di Lazzaro V, Oliviero A, Profice P, Saturno E, Pilato F, Insola A, Mazzone P, Tonali P, Rothwell JC (1998) Comparison of descending volleys evoked by transcranial magnetic and electric stimulation in conscious humans. Electroencephalogr Clin Neurophysiol 109: 397–401

[11] Duffau H (2009) A personal consecutive series of surgically treated 51 cases of insular WHO grade II glioma: advances and limitations. J Neurosurg 110: 696–708

[12] Foerster O, Penfield W (1930) The structural basis of traumatic epilepsy and results of radical operations. Brain 53: 99–119

[13] Goldring S (1978) A method for surgical management of focal epilepsy, especially as it relates to children. J Neurosurg 49: 344–356

[14] Gordon B, Lesser RP, Rance NE, Hart J, Jr., Webber R, Uematsu S, Fisher RS (1990) Parameters for direct cortical electrical stimulation in the human: histopathologic confirmation. Electroencephalogr Clin Neurophysiol 75: 371–377

[15] Harding GF, Bland JD, Smith VH (1990) Visual evoked potential monitoring of optic nerve function during surgery. J Neurol Neurosurg Psychiatry 53: 890–895

[16] Hattingen E, Hattingen J, Clusmann H, Meyer B, Koenig R, Urbach H (2004) Planar brain surface reformations for localization of cortical brain lesions. Zentralbl Neurochir 65: 75–80

[17] Holland NR (1998) Subcortical strokes from intracranial aneurysm surgery: implications for intraoperative neuromonitoring. J Clin Neurophysiol 15: 439–446

[18] Katayama Y, Tsubokawa T, Maejima S, Hirayama T, Yamamoto T (1988) Corticospinal direct response in humans: identification of the motor cortex during intracranial surgery under general anesthesia. J Neurol Neurosurg Psychiatry 51: 50–59

[19] Keles GE, Lundin DA, Lamborn KR, Chang EF, Ojemann G, Berger MS (2004) Intraoperative subcortical stimulation mapping for hemispherical perirolandic gliomas located within or adjacent to the descending motor pathways: evaluation of morbidity and assessment of functional outcome in 294 patients. J Neurosurg 100: 369–375

[20] King RB, Schell GR (1987) Cortical localization and monitoring during cerebral operations. J Neurosurg 67: 210–219

[21] Kombos T, Suess O, Kern BC, Funk T, Hoell T, Kopetsch O, Brock M (1999) Comparison between monopolar and bipolar electrical stimulation of the motor cortex. Acta Neurochir (Wien) 41: 1295–1301

[22] Krishnan R, Raabe A, Hattingen E, Szelenyi A, Yahya H, Hermann E, Zimmermann M, Seifert V (2004) Functional magnetic resonance imaging-integrated neuronavigation: correlation between lesion-to-motor cortex distance and outcome. Neurosurgery 55: 904–914

[23] LeRoux PD, Berger MS, Haglund MM, Pilcher WH, Ojemann GA (1991) Resection of intrinsic tumors from nondominant face motor cortex using stimulation mapping: report of two cases. Surg Neurol 36: 44–48

[24] MacDonald DB (2002) Safety of intraoperative transcranial electrical stimulation motor evoked potential monitoring. J Clin Neurophysiol 19: 416–429

[25] MacDonald DB (2006) Intraoperative motor evoked potential monitoring: overview and update. J Clin Monit Comput 20: 347–377

[26] Nash CL Jr, Lorig RA, Schatzinger LA, Brown RH (1977) Spinal cord monitoring during operative treatment of the spine. Clin Orthop Relat Res 126: 100–105

[27] Neuloh G, Pechstein U, Cedzich C, Schramm J (2004) Motor evoked potential monitoring in supratentorial surgery. Neurosurgery 54: 1061–1072

[28] Neuloh G, Pechstein U, Schramm J (2007) Motor tract monitoring during insular glioma surgery. J Neurosurg 106: 582–592

[29] Patton HD, Amassian VE (1954) Single and multiple-unit analysis of cortical state of pyramidal tract activation. J Neurophysiol 17: 345–363

[30] Penfield W, Boldrey E (1937) Somatic motor and sensory representation in the cerebral cortex of man as studied by electric stimulation. Brain 60: 389–443

[31] Rothwell J, Burke D, Hicks R, Stephen J, Woodforth I, Crawford M (1994) Transcranial electrical stimulation of the motor cortex in man: further evidence for the site of activation. J Physiol 481: 243–250

[32] Sala F, Lanteri P (2003) Brain surgery in motor areas: the invaluable assistance of intraoperative neurophysiological monitoring. J Neurosurg Sci 47: 79–88

[33] Sartorius CJ, Berger MS (1998) Rapid termination of intraoperative stimulation-evoked seizures with application of cold Ringer's lactate to the cortex. Technical note. J Neurosurg 88: 349–351

[34] Sasaki T, Itakura T, Suzuki K, Kasuya H, Munakata R, Muramatsu H, Ichikawa T, Sato T, Endo Y, Sakuma J, Matsumoto M (2010) Intraoperative monitoring of visual evoked potential: introduction of a clinically useful method. J Neurosurg 112: 273–284

[35] Szelenyi A, Hattingen E, Weidauer S, Seifert V, Ziemann U (2010) Intraoperative motor evoked potential alteration in intracranial tumor surgery and its relation to signal alteration in postoperative magnetic resonance imaging. Neurosurgery 67: 302–313

[36] Szelenyi A, Joksimovic B, Seifert V (2007) Intraoperative risk of seizures associated with transient direct cortical stimulation in patients with symptomatic epilepsy. J Clin Neurophysiol 24: 39–43

[37] Szelenyi A, Kothbauer K, de Camargo AB, Langer D, Flamm ES, Deletis V (2005) Motor evoked potential monitoring during cerebral aneurysm surgery: technical aspects and comparison of transcranial and direct cortical stimulation. Neurosurgery 57 (Suppl 4): 331–338

[38] Taniguchi M, Cedzich C, Schramm J (1993) Modification of cortical stimulation for motor evoked potentials under general anesthesia: technical description. Neurosurgery 32: 219–226

[39] Wiedemayer H, Sandalcioglu IE, Armbruster W, Regel J, Schaefer H, Stolke D (2004) False negative findings in intraoperative SEP monitoring: analysis of 658 consecutive neurosurgical cases and review of published reports. J Neurol Neurosurg Psychiatry 75: 280–286

[40] Wood CC, Spencer DD, Allison T, McCarthy G, Williamson PD, Goff WR (1988) Localization of human sensorimotor cortex during surgery by cortical surface recording of somatosensory evoked potentials. J Neurosurg 68: 99–111

[41] Yingling CD, Ojemann S, Dodson B, Harrington MJ, Berger MS (1999) Identification of motor pathways during tumor surgery facilitated by multichannel electromyographic recording. J Neurosurg 91: 922–927

[42] Zentner J, Hufnagel A, Pechstein U, Wolf HK, Schramm J (1996) Functional results after resective procedures involving the supplementary motor area. J Neurosurg 85: 542–549

Brain mapping in epilepsy surgery

George Ojemann

Introduction

The planning of surgical resections for medically refractory epilepsy is based primarily on two sets of data: One set indicates what ideally would be resected to provide a high probability of postoperative seizure control. This data set includes electrophysiologic findings of epileptiform activity, from preoperative interictal and ictal onset electroencephalograms (EEG), but may be supplemented by direct brain mapping of epileptic abnormalities in the electrocorticogram (ECoG), through chronic intracranial electrode or intraoperative recordings. Also included is preoperative imaging evidence of any structural lesion, usually derived from magnetic resonance imaging (MRI). It is the convergence of the electrophysiologic abnormalities with any imaging findings, supplemented by neuropsychologic assessment of cognitive function and the clinical features of the patient's seizures, to one brain region that provides the indication for this type of surgery and the extent of "epileptogenic" zone (focus) likely generating the seizures. The greater the convergence the more likely is postoperative seizure control. What constitutes an adequate amount of preoperative converging data, particularly when there is an electrophysiologically identified focus without imaging changes, is beyond the scope of this chapter but has been addressed by the author elsewhere [1].

The other data set indicates what can be safely resected without a risk of a major functional deficit. The extent of any such assessment depends on the location of the focus. Whether the resection is in the language-dominant hemisphere is usually established by the preoperative evaluation, based on intracarotid amobarbital perfusion testing [2] or functional MRI (fMRI) [3], if there are any unusual preoperative clinical features such as left handed, or language deficits with right hemisphere focus. One common location for the focus is in the anterior temporal lobe. Whether any functional mapping is indicated in those resections is controversial. One comparative study indicated that language mapping in the language- dominant hemisphere was not necessary [4] although the conclusions of that study have been questioned [5–6]. On the other hand, functional mapping of language with the electrical stimulation mapping technique is standard in resections of foci in posterior frontal or temporal lobe of dominant hemisphere, as it is the general consensus that fMRI localization of language does not identify the crucial language areas to be avoided in resections there [7–8]. Mapping of Rolandic cortex is usually performed only in cases with evidence of a focus there, usually based on a clinical pattern of simple partial motor or sensory seizures or the presence of a structural lesion. Except in research settings, mapping of other functions is not usually done.

Techniques of brain mapping in epilepsy surgery

Brain mapping for epilepsy surgery is performed through either acutely placed intraoperative electrodes or chronic subdural or intracerebral electrodes. The major limitations of the acute intraoperative approach are the reliance on interictal epileptiform discharges to identify the focus, and that the patient needs to be awake for a portion of the operation if language is to be mapped. With the chronic approach, ECoG patterns associated with seizure onsets can be recorded, as well as interictal activity and functional mapping can be performed in a more leisurely extraoperative setting, but this technique carries additional risks, including those of the extra craniotomy to place electrodes in addition to the one where they are removed and any indicated resection performed. Some Epilepsy Centers utilize chronic electrodes in all patients. Others individualize the approach depending on whether the presurgical evaluation of that patient has defined the focus with sufficient clarity that recording of seizure onsets is not indicated [9]. Cases with complex partial seizures, unilateral temporal interictal and ictal activity on scalp EEG recordings associated with imaging changes of mesial temporal sclerosis in the same temporal lobe often can be managed with the acute intraoperative approach (though some centers manage such cases without any mapping), as can many cases with refractory seizures associated with intrinsic brain tumors. Cases with extratemporal foci usually require ictal onset recording and thus are managed with the chronic approach. When the focus appears to be close to functionally important areas, the author has often used both approaches, as the acute intraoperative approach provides more detailed mapping than that obtained through chronic electrodes, and also allows for functional testing during the resection.

The extent of cortex to be mapped, whether by acute or chronic techniques, is based on the presurgical evaluation, usually including the site of the suspected focus, any structural lesion and any nearby areas likely to be functionally significant. Language mapping during epilepsy surgery has usually been directed at establishing both where interference in the task occurs as well as establishing the absence of interference in the area of proposed resection using similar stimulation parameters. In contrast to what has been reported for language mapping during tumor resections [10], it has not been established that an entirely negative mapping indicates cortex that can be safely resected during epilepsy surgery. Because patients with refractory epilepsy would seem to be particularly prone to evoking seizures with stimulation, the ECoG is usually recorded during stimulation with currents kept below the threshold for afterdischarges.

Both techniques require that the lateralization of the focus has already been determined. Although strips of subdural electrodes can be placed bilaterally, usually through burr holes, and are valuable for establishing lateralization of the focus, they tend to be of limited value for functional stimulation mapping due to the spotty coverage of cortex. The larger two-dimensional subdural arrays that are suitable for detailed electrophysiologic identification of the focus and functional mapping are rarely placed bilaterally as that requires bilateral craniotomies. Although subcortical stimulation can be done through intracerebral electrodes, mapping of subcortical pathways has not been part of most resections for epilepsy, again in contrast to mapping for tumor resections.

Awake craniotomies for acute mapping in the epilepsy surgery population use the same technique as that used for acute mapping of tumor patients [11]. Indeed the propofol technique for awake craniotomies was initially developed in epilepsy patients [12]. The author routinely maintained the patients' on their antiepileptic drugs during the surgery. Under that situation there was no increased risk of seizures in these patients. After a field block of 0.25% lidocaine and 0.25% bupivacaine in scalp and along the middle meningeal artery in the dura and the craniotomy under propofol neuroleptanelgesia, the patient can be awakened and be wide awake and reasonably comfortable for several hours of ECoG and functional stimulation mapping.

Electrocorticographic mapping

The intraoperative ECoG rarely records seizure onsets; its value in identifying the focus depends on localizing interictal epileptiform spikes (IIS). The value of IIS in determining the extent of resections that control seizures is highly controversial. Their use to plan resections for epilepsy was initially developed by Penfield and his associates at the Montreal Neurologic Institute. Their early reports indicated that in patients undergoing temporal resections, the presence or absence of lateral cortical IIS on the post-resection ECoG was a predictor of postoperative seizure control [13]. Subsequent studies have both supported this and found no predictive value [14]. The current consensus view is that lateral temporal cortical interictal ECoG is of little value in planning an anterior temporal resection. However, the distribution of IIS recorded from the surface of the hippocampus by intraoperative intraventricular electrodes has been shown to be highly predictive of the extent of a medial temporal resection that includes variable portions of anterior hippocampus that must be resected to control seizures [15] (Fig. 1). This may be an important issue in dominant hemisphere temporal lobe resections, where postoperative verbal memory deficits remain a significant problem, as the extent of hippocampal resection has been shown to correlate with the severity of the verbal memory deficit, at least in those patients with temporal lobe foci who have imaging-normal hippocampi. Use of the ECoG to determine extent of hippocampal resection then provides a mechanism to select patients

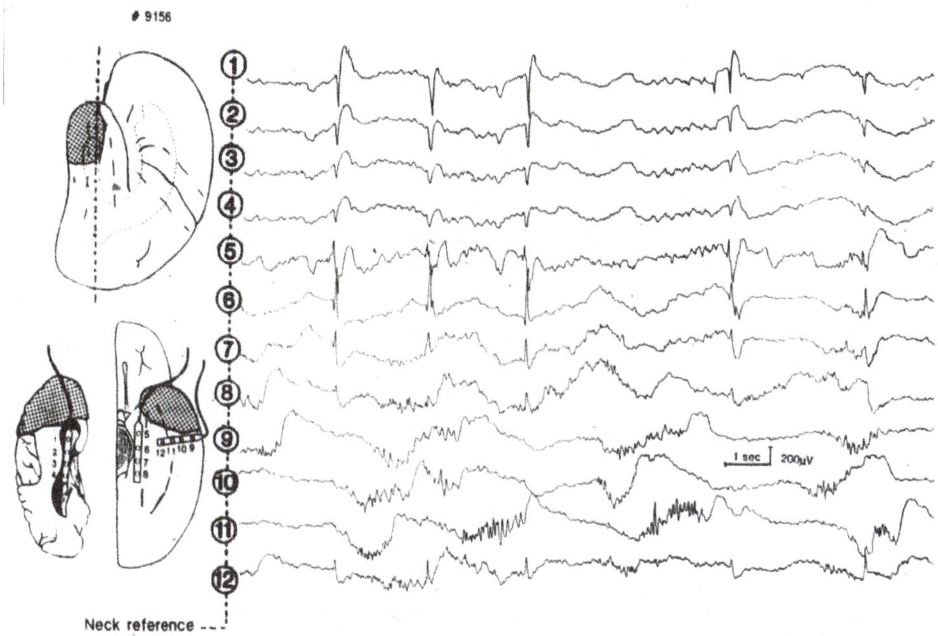

Fig. 1. Electrocorticogram recorded from surface of the hippocampus in patient with medically refractory epilepsy and a left temporal focus. 1–4 are recordings from a strip electrode on the hippocampal surface, 5–8 on the parahippocampal gyrus, 9–12 on fusiform gyrus. Interictal spikes are confined to contacts 1 and 2 in hippocampal recording with only volume conduction of those spikes in more posterior recordings, so that the medial resection need extend no further posterior than electrode 2. This resection will also include the portion of parahippocampal gyrus with interictal spikes, contacts 5 and 6 [15]

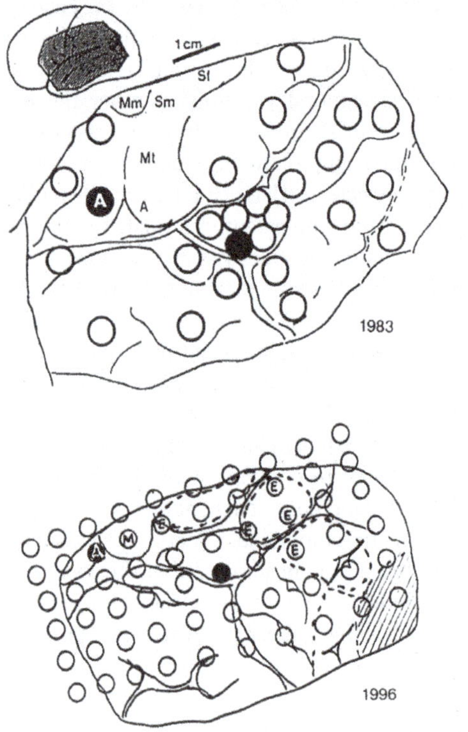

Fig. 2. Stimulation mapping of interference effects on object naming in patient with medically refractory epilepsy and a left temporal focus. Circles are sites of stimulation. Filled circles sites of repeated naming errors. Errors at site identified by "A" were arrest of speech; at site without a letter, failure to name correctly with retained ability to speak. Note the very focal nature of the temporal site of interference with naming. The patient was seizure free for a decade. Seizures then recurred, with repeat mapping during object naming through a chronic subdural electrode array performed 13 years after the initial mapping. The sites of interference at the repeat mapping are in the same location as those at the first mapping, in relation to the sylvian fissure, motor cortex and the cortical surface veins. The resection at the first operation is indicated by the crosshatched area on the second operation map. "M" identify sites with evoked movements, "S" sites of evoked sensation. "E" are sites of interictal spikes and seizure onsets recorded through the chronic subdural grid

who are likely to be seizure free with smaller hippocampal resections that carry a lower risk to verbal memory.

Functional mapping

The techniques of functional mapping now widely used in resections of intrinsic brain tumors were initially developed for epilepsy surgery, particularly by Penfield and his colleagues [16–17]. They extended earlier studies identifying primary motor and sensory cortices by the evoked effects of stimulation and were the first to report the interference effects of high-frequency (30Hz and higher) stimulation in association cortex on language. These observations were extended by the author and his colleagues to address the extent and stability of localization in individual patients [18–19] with the finding that interference effects on an object naming measure of language were frequently localized to multiple focal areas of cortex both frontally and temporally, with this localization often stable over periods of years (Fig. 2). The relation of these sites to the effects of resections was examined. Anterior temporal resections that came within 2cm along a continuous gyrus of one of these sites were associated with a significantly greater deficit on an aphasia battery administered 1 month postoperatively than resections that avoided those sites by a larger margin, an effect that was not related to the extent of the resection, degree of postoperative seizure control or preoperative verbal ability as reflected in the verbal IQ (VIQ) [20]. A similar relationship to sites of evoked interference in naming was found for clinically evident aphasias in temporal resections for tumors [21]. Curiously, although the resections encompass both surface cortex and that buried in sulci, it is surface stimulation that predicts the effects on language.

The variability in the localization of sites of stimulation interference with naming in dominant hemisphere across a population of 117 adult epilepsy surgery patients was also assessed (Fig. 3), with substantial variation such that the only region with interference sites in more than half the patients was the posterior inferior frontal cortex. Variability was particularly great in temporal lobe, where even across the entire superior temporal gyrus, which encompasses most of the traditional Wernicke language area, only 65% of the patients had a site where stimulation interfered with naming. Moreover, such sites were found in a number of regions that are classically considered to be uninvolved with language including anterior temporal cortex in front of the coronal plane through the central fissure. This variability is also evident in Penfield's data, but was not commented on [17]. It is this combination of localized sites of interference with language in an individual patient, that avoiding these sites reduces the chances of a postoperative language disturbance and that these sites are in variable anatomic locations across patient populations that justifies using stimulation mapping data to plan resections rather than relying only on anatomy.

The individual patterns of localization of the sites with interference in naming have relationships to several patient characteristics including presurgical verbal ability as reflected in the VIQ and gender [19]. Patients with lower VIQs had significantly larger total language areas and were more likely to have sites in superior temporal gyrus than in middle temporal gyrus. Females were overrepresented in the small group of patients in whom only frontal sites could be identified, despite extensive temporal mapping. Males with low VIQs were particularly likely to have inferior parietal sites. The author's group has mapped sites of interference with naming in epilepsy patients as young as 4, though extraoperatively through chronic subdural electrodes rather than acutely intraoperatively as in the above adult study. Children younger than 9 had fewer sites than

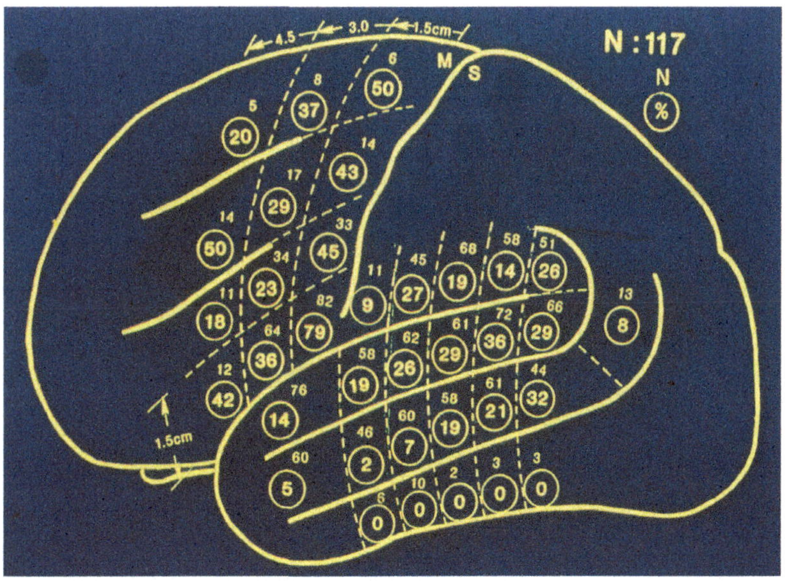

Fig. 3. Variability in location of sites of stimulation interference in object naming in the left hemisphere of 117 patients with medically refractory epilepsy with language dominance to the left hemisphere established by preoperative intracarotid amobarbital perfusion testing. Cortex divided into arbitrary zones, based on central sulcus, posterior end of the sylvian fissure and the major sulci. In each zone, number without a circle is the number of patients with stimulation in that zone, the number in the circle the percentage of those patients with significant interference in naming. The substantial variability, especially in temporal cortex, is evident [19]

those 9–16, who in turn had fewer sites than were identified in the subjects of the adult study, suggesting that during maturation the area of cortex related to a language task, naming, may increase [22]. There was no relation to age (18–80) within the adult study.

The previous studies all were concerned with naming of visually presented objects. More recently, Hamberger et al [23–24] investigated naming of auditory material – oral descriptions of concrete items. They found some separation of sites where stimulation interfered with naming of the different types of material, with sites of interference in auditory naming generally more anterior in temporal lobe from sites with interference in visual naming. They presented evidence that sites of interference with auditory naming provided more useful information in planning dominant-hemisphere anterior temporal resections that avoided language deficits than sites with interference in visual naming.

When measures of other language functions have been assessed with cortical stimulation mapping, the general pattern is that interference is often evoked from separate sites. Mapping of naming the same items in multiple languages in polyglot subjects has demonstrated sites in nearly all subjects where interference in only one language was evoked (Fig. 4), an effect that was more prominent with temporal than frontal stimulation, but was independent of the type of language, romance or oriental [25–27]. Mapping during visual naming and sentence reading also shows some degree of dissociation of sites of interference, also more often seen in temporal cortex [28], with the different patterns of localization having a particularly robust relation to high or low VIQs. These findings have some practical application to preserving function in multiple languages and reading with temporal resections.

On a research basis, the author's group has studied cortical mapping of multiple other functions in epilepsy patients. Within the dominant hemisphere, these include mimicry of speech gestures and detection of speech

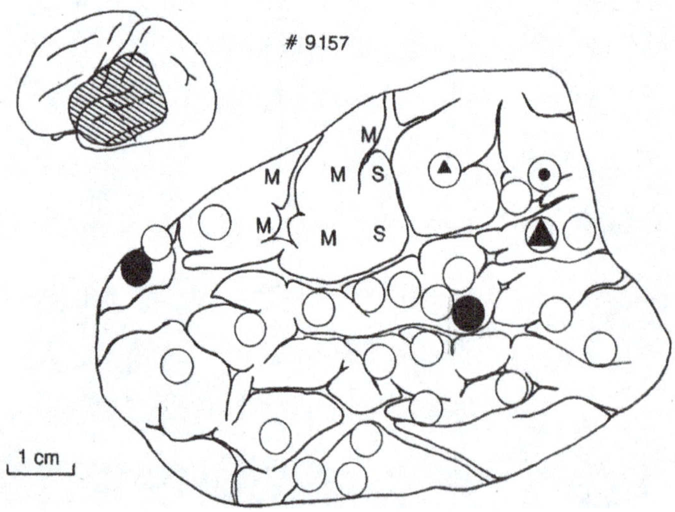

Fig. 4. Stimulation mapping of naming of the same objects in two languages, English and Spanish, in a patient with medically refractory epilepsy and a left temporal focus, who taught English to native Spanish speakers. English was this patient's native language. Circles are sites of stimulation. Filled circles are sites of repeated interference in English but not Spanish naming. Circle filled with a triangle is site of repeated interference in naming in Spanish but not English. Circles with small symbols had single errors in English only (small circle) or Spanish only (small triangle) [46]

sounds, identifying frontal and temporal perisylvian cortex common to speech production and perception [29–30], manual communication systems [31–32], verb generation including changes with learning [33], action and object naming [34] and patterns of naming in patients with preexisting aphasias, a study that reinforces the findings from remapping after periods of years, that the adult cortex shows little plasticity for the sites of interference with naming [35]. Complex figure- and face matching and facial emotion recognition have been investigated in the nondominant hemisphere [36].

Of particular relevance for epilepsy patients has been the mapping of temporal cortical sites in the language dominant hemisphere related to recent verbal memory [37–40]. Our studies used an encoding-storage-retrieval paradigm, with stimulation applied during only one phase of the memory measure. With object names as the material to be retained in memory, temporal cortical stimulation was significantly more likely to interfere with later retrieval when applied during encoding or storage than when retrieval occurred. Frontal lobe stimulation interference was more likely during retrieval. Identification of the object to be remembered was intact during encoding stimulation that interfered with later retrieval (which occurred in the absence of stimulation). By contrast, temporal cortical sites where stimulation interfered with naming usually had no effect on memory performance even when the current was applied during encoding. Sites of stimulation interference with memory also show substantial individual variation in their location, but were particularly likely in anterior temporal cortex and supramarginal gyrus, surrounding sites related to naming. There was evidence that dominant hemisphere anterior temporal resections that encroached on the sites where stimulation during encoding or storage interfered with memory were more likely to have postoperative memory disturbances than those that did not. This suggested that the extent of lateral cortical resection, as well as the medial resection might be another factor in the postoperative verbal memory deficit after dominant resections.

What physiologically characterizes the sites where stimulation interferes with a language measure? It is generally thought that high-frequency stimulation produces interference effects by depolarization blockade of local neuron activity [41]. However, those localized sites do not represent the only regions of cortex with neuron networks for language. For example, single neuron activity statistically related to object naming can be recorded widely through dominant temporal lobe, in areas with no interference effects of stimulation, as well as in the nondominant temporal cortex [42]. The proportion of neurons with significant changes was the same in either hemisphere, with dominant activity characterized by early neuron activity changes that are most often relative inhibition. Nondominant activity was later excitation. ECoG changes during object naming localized to sites where stimulation interfered with object naming have been identified [43–44]. In temporal lobe this change was local "desynchronization", loss of low-frequency potentials with replacement by low-voltage fast activity, which has been subsequently shown to be activity in the high gamma range (50–70Hz and above) [45]. In our study this change occurred in parallel with a slow potential that characterized frontal sites where stimulation interfered with naming, electrophysiologic evidence supporting the stimulation-derived evidence of cortex common to both language production and perception.

References

[1] Holmes MD, Born DE, Kutsy RL et al (2000) Outcome after surgery in patients with refractory temporal lobe epilepsy and normal MRI. Seizure 9: 407–411

[2] Wada J, Rasmussen T (1960) Intracarotid injections of sodium amytal for the lateralization of cerebral speech dominance. J Neurosurg 17: 266–282

[3] Benson RR, FitzGerald DB, LeSueur LL et al (1999) Language dominance determined by whole brain functional MRI in patients with brain lesions. Neurology 52: 798–809

[4] Hermann BP, Wyler AR, Somes G (1991) Language function following anterior temporal lobectomy. J Neurosurg 74: 560–566

[5] Barbaro NM, Walker JA, Laxer KD (1991) Temporal lobectomy and language function. J Neurosurg 75: 830–831

[6] Ojemann GA (2006) Language mapping is necessary for language-dominant temporal resections. In: Miller JW, Silbergeld DL (eds) Epilepsy surgery: Principles and controversies. Taylor and Francis, New York, pp 423–425

[7] Roux FE, Boulanouar K, Lotterie JA et al (2003) Language functional magnetic resonance imaging in preoperative assessment of language areas: correlation with direct cortical stimulation. Neurosurgery 52: 1335–1345; discussion 1345–1337

[8] Talacchi A, Gerosa M (eds) (2009) Awake surgery and cognitive mapping. Top Med 15 (special issue 1)

[9] Holmes MD, Kutsy RL, Ojemann GA et al (2000) Interictal, unifocal spikes in refractory extratemporal epilepsy predict ictal origin and postsurgical outcome. Clin Neurophysiol 111: 1802–1808

[10] Sanai N, Mirzadeh Z, Berger MS (2008) Functional outcome after language mapping for glioma resection. N Engl J Med 358: 18–27

[11] Ojemann GA (1995) Awake operations with mapping in epilepsy. In: Schmidek H, Sweet W (eds) Operative neurosurgical techniques, 3rd edn. Saunders, Philadelphia, pp 1317–1322

[12] Trop D (1986) Conscious-sedation analgesia during the neurosurgical treatment of epilepsies-practice at the Montreal Neurological Institute. Int Anesthesiol Clin 24: 175–184

[13] Bengzon A, Rasmussen T, Gloor P (1968) Prognostic factors in the surgical treatment of temporal lobe epileptics. Neurology 18: 717–731

[14] Miles AN, Ojemann GA (2004) Awake surgery for epilepsy. In: Shorvon S, Perucca E, Fish D, Dodson WE, (eds) The treatment of epilepsy, 2nd edn. Blackwell, Oxford, pp 824–832

[15] McKhann GM, 2nd, Schoenfeld-McNeill J, Born DE et al (2000) Intraoperative hippocampal electrocorticography to predict the extent of hippocampal resection in temporal lobe epilepsy surgery. J Neurosurg 93: 44–52

[16] Penfield W, Jasper H (1954) Epilepsy and the functional anatomy of the human brain. Little, Brown & Co, Boston

[17] Penfield W, Roberts L (1959) Speech and brain mechanisms. Princeton University Press, Princeton, NJ

[18] Ojemann GA, Whitaker HA (1978) Language localization and variability. Brain Lang 6: 239–260

[19] Ojemann GA, Ojemann JG, Lettich E et al (1989) Cortical language localization in left, dominant hemisphere. An electrical stimulation mapping investigation in 117 patients. J Neurosurg 71: 316–326

[20] Ojemann GA (1983) Electrical stimulation and the neurobiology of language. Behav Brain Sci 2: 221–226

[21] Haglund MM, Berger MS, Shamseldin M et al (1994) Cortical localization of temporal lobe language sites in patients with gliomas. Neurosurgery 34: 567–576; discussion 576

[22] Ojemann SG, Berger MS, Lettich E et al (2003) Localization of language function in children: results of electrical stimulation mapping. J Neurosurg 98: 465–470

[23] Hamberger MJ, Goodman RR, Perrine K et al (2001) Anatomic dissociation of auditory and visual naming in the lateral temporal cortex. Neurology 56: 56–61

[24] Hamberger MJ, Seidel WT, McKhann GM, 2nd et al (2005) Brain stimulation reveals critical auditory naming cortex. Brain 128: 2742–2749

[25] Ojemann GA, Whitaker HA (1978) The bilingual brain. Arch Neurol 35: 409–412

[26] Lucas TH, McKhann GM, Ojemann GA (2004) Functional separation of languages in the bilingual brain: a comparison of electrical stimulation language mapping in 25 bilingual patients and 117 monolingual control volunteers. J Neurosurg 101: 449–457

[27] Roux FE, Tremoulet M (2002) Organization of language areas in bilingual patients: a cortical stimulation study. J Neurosurg 97: 857–864

[28] Ojemann GA (1989) Some brain mechanisms for reading. In: Von Euler C, Lundberg I, Lennerstrand G (eds) Brain and reading. MacMillan, Hampshire (London), pp 47–59

[29] Ojemann G, Mateer C (1979) Human language cortex: localization of memory, syntax, and sequential motor-phoneme identification systems. Science 205: 1401–1403

[30] Ojemann GA (1983) Brain organization for language from the perspective of electrical stimulation mapping. Behav Brain Sci 6: 189–206

[31] Mateer CA, Polen SB, Ojemann GA et al (1982) Cortical localization of finger spelling and oral language: a case study. Brain Lang 17: 46–57

[32] Haglund MM, Ojemann GA, Lettich E et al (1993) Dissociation of cortical and single unit activity in spoken and signed languages. Brain Lang 44: 19–27

[33] Ojemann JG, Ojemann GA, Lettich E (2002) Cortical stimulation mapping of language cortex by using a verb generation task: effects of learning and comparison to mapping based on object naming. J Neurosurg 97: 33–38
[34] Corina D, Gibson E, Martin RF et al (2005) Dissociation of action and object naming: Evidence from cortical stimulation mapping. Hum Brain 24: 1–10
[35] Lucas TH, 2nd, Drane DL, Dodrill CB et al (2008) Language reorganization in aphasics: an electrical stimulation mapping investigation. Neurosurgery 63: 487–497; discussion 497
[36] Fried I, Mateer C, Ojemann G et al (1982) Organization of visuospatial functions in human cortex. Evidence from electrical stimulation. Brain 105: 349–371
[37] Ojemann GA (1978) Organization of short-term verbal memory in language areas of human cortex: evidence from electrical stimulation. Brain Lang 5: 331–340
[38] Ojemann GA, Dodrill CB (1985) Verbal memory deficits after left temporal lobectomy for epilepsy. Mechanism and intraoperative prediction. J Neurosurg 62: 101–107
[39] Ojemann GA, Dodrill CB (1987) Intraoperative techniques for reducing language and memory deficits with left temporal lobectomy. Adv Epileptology 16: 327–330
[40] Perrine K, Devinsky O, Uysal S et al (1994) Left temporal neocortex mediation of verbal memory: evidence from functional mapping with cortical stimulation. Neurology 44: 1845–1850
[41] Ranck JB, Jr. (1975) Which elements are excited in electrical stimulation of mammalian central nervous system: a review. Brain Res 98: 417–440
[42] Schwartz TH, Ojemann GA, Haglund MM et al (1996) Cerebral lateralization of neuronal activity during naming, reading and line-matching. Cogn Brain Res 4: 263–273
[43] Fried I, Ojemann GA, Fetz EE (1981) Language-related potentials specific to human language cortex. Science 212: 353–356
[44] Ojemann GA, Fried I, Lettich E (1989) Electrocorticographic (ECoG) correlates of language. I. Desynchronization in temporal language cortex during object naming. Electroencephalogr Clin Neurophysiol 73: 453–463
[45] Crone NE, Sinai A, Korzeniewska A (2006) High-frequency gamma oscillations and human brain mapping with electrocorticography. Prog Brain Res 159: 275–295
[46] Ojemann GA (1994) Cortical stimulation and recording in language. In: Kertesz A (ed) Localization and neuroimaging in neuropsychology. Academic Press, SanDiego, CA, chap 2, pp 35–55

Awake mapping and tumor surgery

Hugues Duffau

Introduction

The first goal of brain tumor surgery is to optimize the extent of resection (EOR) of the lesion. Indeed, maximal resection of glioma, when possible, is currently the first therapeutic option to consider for low-grade gliomas (LGG) (as recommended by the European Guidelines [48]) as well as for high-grade gliomas [49]. In recent series measuring objectively the EOR on repeated postoperative MRI, all of them supported EOR as a statistically significant predictor of overall survival. With WHO grade II gliomas, when no signal abnormality was visible on control MRI, especially on FLAIR-weighted imaging (i.e., the so-called "complete resection"), patients had a significantly longer overall survival compared with patients having any residual abnormality [1,8,19,41,47]. Interestingly, even in cases of incomplete tumor removal, patients with a greater percentage of resection had a significantly longer overall survival. In addition to the percentage of resection, the postoperative tumor volume is also a predictor of survival, with a significantly longer overall survival when the residue is less than 10 ml ("subtotal resection") compared with residues greater than 10 ml ("partial resection") [47]. For glioblastomas, it was also shown that the complete removal of the contrast-enhanced part of the tumor detected on postsurgical MRI increased the median survival to around 17 months, instead of only 12 months if a residual enhancement remained [49].

Therefore, the dilemma of cerebral surgery in neurooncology is to maximize the EOR while preserving brain functions. Nonetheless, due to the frequent location of supratentorial gliomas near or within the so-called eloquent areas, and due to their infiltrative feature (poorly demarcated), it was for a long time considered that the chances to perform an extensive glioma removal were low, whereas the risk to generate postoperative sequelae was high. Indeed, many surgical series have reported a rate of permanent and severe deficit between 13 and 27.5% following removal of intra-axial tumors (for reviews, see [9] and [19]).

As a consequence, to optimize the benefit-to-risk ratio of surgery, an increased use of functional mapping methods was observed in the last decade. Indeed, considerable interindividual anatomo-functional variability was demonstrated for healthy volunteers [54] as well as for epileptic patients [42]. Furthermore, this variability is increased in cases of gliomas, due to cerebral plasticity, explaining why many patients have no or only mild deficits before surgery, especially with slowly growing tumor such as LGG [11]. Therefore, many arguments support the unpredictability of functional eloquence on the basis of anatomical features alone and the fact that patients should not be considered ineligible for surgical intervention on the basis of anatomical considerations alone (for a review, see ref. 44). Rather, neurosurgeons need to take advantage of modern technology and mapping techniques to create individualized maps and management plans. Indeed, it is now mandatory to study for each patient the individual anatomo-functional organization, in order to tailor the resection according not only to oncological but also to cortico-subcortical functional boundaries.

The goal of this chapter is to review how, in addition to functional neuroimaging, the method of intraoperative electrostimulation mapping (IEM), especially in awake patients, has enabled a significant improvement of the results of glioma surgery, with (i) an increase of the surgical indications for tumors located within eloquent areas classically considered as "inoperable", (ii) an optimization of the EOR with an increased impact on the natural history of the glioma, and (iii) preservation or even improvement of the quality of life.

Presurgical functional assessment: advances and limitations

Preoperative neurocognitive examination

Gliomas, especially LGG, are usually revealed by inaugural seizures in young patients who enjoy a normal social and professional life, with no or only mild neurological deficit. However, recent extensive neuropsychological examinations have demonstrated that most patients had cognitive disturbances, especially concerning working memory and executive functions [51]. This is the reason why a systematic preoperative assessment of higher functions and health-related quality of life is now recommended [40]: (i) to search the possible neuropsychological deficit not identified by a classical neurological examination, (ii) to adapt the surgical methodology to the results of this assessment, e.g., to perform functional mapping under local anesthesia even in the right hemisphere [22], (iii) to benefit from a presurgical baseline allowing a comparison with the postsurgical evaluation, and (iv) to plan a specific functional rehabilitation following the resection, which can induce a transient neurological worsening [26]. As a consequence, standardized examination of neurocognitive outcome has recently been proposed (see the chapter by Klein and de Witt Hamer).

It is nonetheless puzzling to note that these deficits are not more pronounced, despite the frequent location of LGG in the so-called eloquent areas. This can be explained by mechanisms of brain reshaping allowing functional compensation in cases of slowly growing lesions. Indeed, it was shown that cerebral remapping was possible, with a recruitment of perilesional or remote areas within the ipsilesional hemisphere and/or recruitment of contrahemispheric homologous areas. The recent integration of these concepts into the therapeutic strategy has resulted in dramatic changes in the surgical management of LGG patients, with an increase of surgical indications in functional areas classically considered as inoperable [11].

Preoperative functional neuroimaging: a necessary baseline

In this context, advances in functional neuroimaging (FNI), e.g., functional MRI (fMRI), magnetoencephalography, and transcranial magnetic stimulation, have enabled to perform a noninvasive cortical mapping of the whole brain and is currently a standard before resection of gliomas. FNI gives an estimation of the location of the eloquent areas (e.g., regions involved in sensorimotor, language, visual and even higher cognitive function – see the chapter by Champod et al) in relation to the tumor and provides information with regard to the hemispheric language lateralization. Thus, these methods may be useful for (i) the surgical indications, partly depending on the location of the tumor and its relationships with eloquent areas detected by FNI (allowing an estimation of the tumor resectability), (ii) the surgical planning, namely, the selection of the surgical approach and the delineation of the limits of resection, (iii) the selection of the surgical technique, especially the decision to wake the patient intraoperatively if the glioma is close to somatosensory, language, or cognitive areas – even in the right hemisphere,on the basis of the laterality index on FNI for language in addition to the handedness of the patient provided by the neuropsychological examination [52].

However, it is worth noting that FNI methods are not yet reliable enough at the individual scale, despite constant efforts for their improvement, mainly because the results depend on the biomathematical models used for the reconstruction. Concerning fMRI (see chapter by Ramsey and Rutten), correlations with intraoperative electrophysiology have recently demonstrated that the sensitivity of fMRI was currently only around 71% for movement [2], and from 59 to 100% for language (specificity from 0 to 97%) [29]. Such discrepancies can be explained by a neurovascular decoupling in cases of glioma (BOLD [blood oxygenation level-dependent] response in the vicinity of gliomas does not reflect the neuronal signal as accurately as it does in healthy tissue), by inadequate tasks (not adapted to the location of the glioma and/or to the neurological status of the patient), or to methodological problems (e.g., selection of the threshold). As a consequence, there is a risk of false-negative and then to operate a patient without intraoperative mapping, although the glioma is actually located in crucial areas for the function – but not detected by preoperative FNI – thus with a high risk to induce a permanent deficit. Moreover, an erroneous interpretation of brain reshaping ("pseudo-reorganization") can be made. In addition, these methods are not able to differentiate the structures essential for the function, which should be surgically preserved, from those which can be functionally compensated and so potentially resected without permanent deficit.Therefore, there is a double risk (i) to not select a patient for surgery, although the tumor was operable, or (ii) to stop prematurely the resection with a lower impact on the natural history of the glioma. Finally, these techniques allow a mapping of the grey-matter but not of the white-matter connectivity.

Interestingly, the recent development of diffusion tensor imaging (DTI) now allows the identification of the main bundles, that is, their tractography (see chapter by Catani and Dell'Acqua), as well as their location in relation to a tumor (see chapter by Bello et al). However, this new method needs to be validated, especially by intraoperative electrophysiological techniques, before it can be used routinely for surgical planning, especially due to the fact that results of DTI, as those of FNI, strongly depend on the biomathematical models used for the fiber tracking. Indeed, comparison between distinct fiber tracking software tools found different results, showing that neurosurgeons have to be cautious about applying tractography results intraoperatively, especially when dealing with an abnormal or distorted fiber tract anatomy [6]. Furthermore, correlations between DTI and intrasurgical subcortical stimulation demonstrated that, despite a good correspondence, DTI is not yet optimal to map language tracts in patients. Negative tractography does not rule out the persistence of a fiber tract, especially when invaded by a glioma [34]. Moreover, DTI enables the study of only the anatomy of the subcortical pathways, but not their function.

With the aim to overcome these pitfalls, one can currently consider to perform longitudinal studies based on pre-, intra-, and postoperative mapping rather than to content oneself with static information based on a unique presurgical functional neuroimaging analysis [28].

Furthermore, despite the progress in FNI, its current limitations make the additional use of invasive electrophysiological investigations highly recommendable for surgery in eloquent structures.

Intrasurgical functional brain mapping

Intraoperative functional neuroimaging

Intraoperatively, the integration of multimodal imaging into frameless stereotactic surgery was extensively used in the past decade and referred to as "functional neuronavigation" (see the chapter by Ganslandt et al). However, the sole randomized trial failed to demonstrate a significant impact of navigation on postoperative results [56]. This can be explained by the limitations of the presurgical FNI and DTI detailed above, as well as to the high risk of intraoperative brain shift due to surgical retraction, mass effect, gravity, extent of resection (especially for voluminous tumors), and cerebrospinal fluid leakage. Several technical improvements have been proposed to reduce the effects of this shift, but their reliability has still to be optimized: combination with intraoperative ultrasound, producing real-time imaging; use of mathematical models based on data from ultrasonography or digital images that track cortical displacement; and intraoperative MRI. Nevertheless, their actual value on the improvement of EOR and preservation of quality of life remains to be demonstrated.

As a consequence, invasive electrophysiological investigations currently remain the "gold standard" when operating on eloquent brain structures.

Evoked potentials and electrocorticography

First, the technique of somatosensory and motor evoked potentials was extensively used in the past decades for intraoperative identification of the central region (see chapter by Szelenyi). However, its reliability regarding the localization of the rolandic sulcus is not optimal, with accurate localization of the central sulcus reported for only between 91 and 94%. Estimation of the overall sensitivity and negative predictive value of this method were evaluated at about 79 and 96%, respectively [55]. Moreover, phase reversal recording identifies only the central sulcus itself but offers no direct information on the particular distribution of motor function on the adjacent exposed cerebral structures. Also, although the method of motor evoked potentials was improved, when recording compound muscle action potentials, only the monitored muscles can be controlled, that is, this method does not allow detection and possible avoidance of motor deficits in nonmonitored muscles. In addition, monitoring of muscle action potentials does not mean monitoring of complex movements, action adapted to the environment, and intention to act, which is nonetheless the ultimate goal for the patient [33]. Above all, intraoperative evoked potentials cannot currently be used to map langague, memory, or other higher functions crucial for the quality of life of the patients (for a review, see [18]).

Numerous authors have also suggested the use of extraoperative electrophysiological recordings (electrocorticography) and stimulations through the implantation of subdural grids (see the chapter by Winkler). By this method, the patient is in optimal conditions, in his room, to perform the tasks, this point is particularly important for children. Moreover, recent advances in the interpretation of the electrophysiological signal, such as electrocorticographic spectral analysis evaluating the event-related synchronization in specific bands of frequency, have allowed a better understanding of the organization of the functional cortex and the study of the connectivity, in particular by the recording of "cortico-cortical evoked potential". However, extraoperative electrophysiological mapping usually uses grids with electrodes spaced 1 cm apart, thus with a limited accuracy. Also, it is necessary to perform two surgical procedures, one to implant the grids and a second to remove the lesion. In addition, there is still a risk of infectious complications due to the presence of subdural grids for several days. Above all, although this method was extensively advocated in epilepsy surgery because it also allows detection of the seizure foci, only the cortex can be mapped, it provides no information about the axonal connectivity, i.e., it is not possible to map the subcortical structures. Thus, this technique is not suited for neurooncology, since it is well known that gliomas migrate along the white-matter bundles [35].

Intrasurgical cortical and subcortical electrostimulation mapping

Taking into account the advantages and the limits of these different mapping techniques, an increased number of neurosurgeons currently advocates the additional use of IEM during surgery in eloquent areas, possibly under general anesthesia but more and more frequently under local anesthesia [18, 23, 30, 32, 45, 46]. Indeed, except for tumors located within the motor structures, the mapping is performed in awake patients. However, as previously mentioned, since movements and action are more complex than single muscle contractions, it is also currently proposed to map the motor function under local anesthesia with active participation of the patient [16]. The principle is to use IEM as a focal and transitory virtual lesion, to obtain an invidual functional mapping at both cortical and subcortical levels, and to test if a structure involved by a lesion is still crucial for the function, which is observed in 15 to 20% of cases in LGG. Stimulation of an essential area generates a transient disruption of the task performed by the patient, and this area should be preserved. An individual cortical mapping is thus obtained before the resection, which can be tailored according to functional boundaries (Fig. 1). In practice, a bipolar electrode with tips spaced 5 mm apart and delivering a biphasic current (pulse frequency of 60 Hz, single-pulse phase duration of 1 ms) is applied to the brain. The current intensity adapted to each patient is determined by progressively increasing the amplitude by 1 mA increments from a baseline of 2 mA until a functional response is elicited, with 6 mA as the upper limit under local anesthesia, and with 16 mA as the upper limit under general anesthesia, so that the generation of seizures will be avoided. The patient is never informed when the brain is stimulated. No site is stimulated twice in succession, in order to avoid seizures. Each cortical site of the entire cortex exposed by the bone flap is tested three times. Indeed, it is admitted nowadays that three trials are sufficient to assure if an area is crucial for brain function, by generating disturbances during its three stimulations and with normalization of the function as soon as the stimulation is stopped. This limitation of trials and tasks is required by the timing of the surgical procedure because the patient is awake and may become tired at the end of the resection (see chapter by Ojemann).

Interestingly, recent series showed that the surgical procedure could be simplified by avoiding the use of intraoperative electrocorticography because electrical mapping is equivalently reliable and does not increase the rate of seizures [23]. However, in cases of stimulation-induced seizures, the use of cold Ringer's lactate is recommended to abrogate the seizure activity. Moreover, some authors emphasized the value of "negative mapping" (no identification of eloquent sites) in the setting of a tailored cortical exposure [32, 45]. Although such recommendation is acceptable for high-grade gliomas, since the surgical goal is mainly to remove the enhanced part of the tumor, a negative mapping can be dangerous in surgery of diffuse LGG, especially in nonexpert hands. Indeed, due to the fact that LGG is poorly delineated, the limit of the resection will be essentially guided according to functional criteria. Because negative mapping can be due to false-negative for methodological reasons, it does not guarantee the absence of eloquent sites. In the experience reported by Sanai et al, all 4 of the patients with permanent postoperative deficits had no positive sites detected prior to their resections [45]. Therefore, other authors continue to recommend a wider bone flap, in order to obtain a systematic positive mapping before performing the resection [15, 18, 23]. Moreover, a positive mapping might also allow an optimization of the EOR, since the resection can be pursued until eloquent areas are encountered, i.e., with no margin

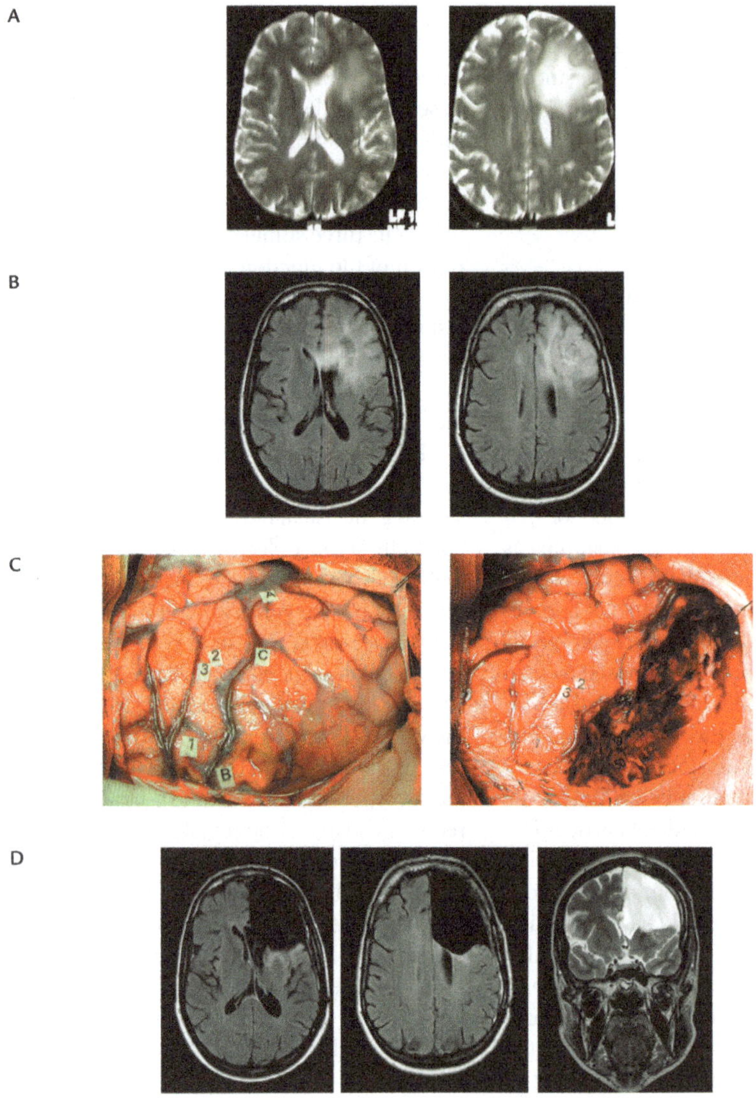

Fig. 1. (A) Axial T2-weighted MRI in a right-handed young patient with inaugural seizures. No oncological treatment was administrated. **(B)** Axial FLAIR-weighted MRI four years later. The patient met several neurosurgeons who said that the tumor was stable and impossible to operate due to the invasion of Broca's area. In fact, volumetric measurement showed that the tumor had doubled, then with an involvement of the corpus callosum. An awake surgery was thus proposed at our institution. **(C)** Intraoperative views before (*left*) and after (*right*) glioma resection delineated by letter tags. Intrasurgical electrical mapping showed a reshaping of the eloquent maps within the precentral gyrus (*1–3*), with a recruitment of perilesional language sites located behind the glioma. There was no crucial site within the left inferior frontal gyrus. Thus, an extensive resection of Broca's area was possible and the subcortical connectivity in the depth of the cavity (49 and 50, corresponding to language pathways) was preserved. **(D)** Postoperative axial FLAIR- and coronal T2-weighted MRI demonstrating a near-complete resection of the glioma with removal of the corpus callosum, in a patient with neither neurological nor neurocognitive deficit, leading a normal socio-professional life. It is worth noting that a FLAIR-hyperintensity is visible in the depth of the cavity, i.e., within the deep gray nuclei and white matter tracts, still functional (Reproduced from [11])

around the functional structures (Fig. 1). A recent study demonstrated that, in a consecutive and homogeneous series of 115 LGG in the left dominant hemisphere, the rate of permanent deficit remained lower than 2% despite the absence of a margin around the language sites [23]. Indeed, Gil Robles and Duffau. [26] showed that the EOR could be dramatically reduced by avoiding the preservation of 5 to 10 mm around the functional areas, as usually proposed in the classical literature (Fig. 2). Interestingly, Gil Robles and Duffau [26] also showed that it was not logical to leave a small amount of tumor involving the cortex when the resection was performed at the subcortical level in contact with white-matter pathways (see below), because that means that the cortical area not removed was in fact disconnected and then no longer functional.

IEM allows the mapping of motor function (possibly under general anesthesia, by inducing unvoluntary motor response, but also in awake patient by eliciting a disturbance of the movement), somatosensory function (by eliciting dysesthesias described by the patient himself intraoperatively), visual function (by eliciting phosphenes and/or visual field deficit described by the patient), auditory-vestibular function (by inducing vertigo), language (spontaneous speech, counting, object naming, comprehension, writing, reading, bilingualism, language switching from one language to another), and also the mapping of higher-order functions such as calculation, memory, spatial cognition, cross-modal judgement or even emotional processing – by generating transient disturbances if the electrical stimulation is applied at the level of a functional "epicenter". It is crucial that a speech therapist, neuropsychologist, or neurologist be present in the operative room, in order to interpret accurately the kind of disorders induced by IEM, for instance, speech arrest, anarthria, speech apraxia, phonological disturbances, semantic paraphasia, perseveration, anomia, syntactic errors [20, 27, 53]. Thus, IEM is able to identify in real time the cortical sites essential for the function before the beginning of the resection, in order both to select the best surgical approach and to define the cortical limits of the lesion removal.

Another major issue is the use of subcortical mapping throughout the resection, in addition to the cortical mapping before the lesion removal [3, 10, 18, 23]. Brain lesion studies have taught that damage of the white-matter pathways generated more severe deficit than cortical injury. Therefore, the subcortical tracts subserving motor, somatosensory, visual, auditory-vestibular, language, and cognitive functions must be detected during the lesion removal, in order to preserve the anatomo-functional connectivity while optimizing the EOR, namely, to pursue the resection until eloquent pathways are detected. Interestingly, according to the same principle as that described at the cortical level, IEM can also identify eloquent subcortical structures. It allows the study of the anatomo-functional connectivity by directly and regularly stimulating the white-matter tracts and deep gray nuclei throughout the resection and by eliciting functional response when in contact with deep crucial areas (Figs. 1 and 3). Furthermore, IEM enabled a better understanding of the brain connectivity, showing that dynamic cerebral processing is underlain by parallel distributed and interactive networks, the so-called hodotopy [12]. This connectionist view also opens the door to the concept of cerebral plasticity, crucial in LGG surgery.

One of the major advantages of IEM for brain mapping in adult patients is that they intrinsically do not cause any false-negatives, in any case if the methodology is rigorously applied, as detailed above. Indeed, IEM is highly sensitive for detecting the cortical and axonal eloquent structures, and it also provides a unique opportunity to study brain connectivity, since each area responsive to stimulation is in fact an input gate into large-scale network rather than an isolated discrete functional site. IEM, however, has also a limitation: its specificity is suboptimal. Indeed, IEM may lead to an interpretation of a structure as being crucial due to the induction of a transient functional

A

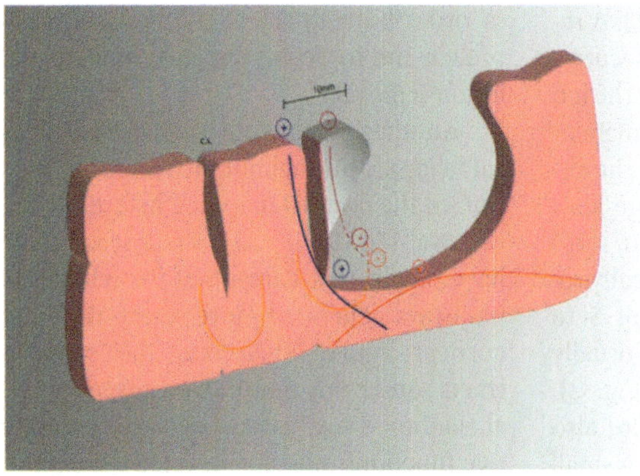

B

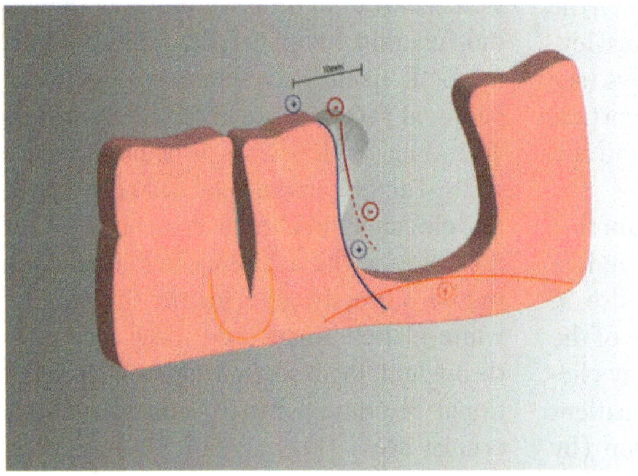

Fig. 2. Schematic drawings of exemplary resections with margin. **(A)** First, a tumor coming into contact with a sulcus removed by leaving a 10 mm security margin from the positive cortical stimulation site (blue +) (i.e., with no subpial dissection). If subcortical stimulation is performed, the "vertical connectivity"of projections fibers will be identified at the bottom of the sulcus (blue +; e.g., pyramidal tracts). The U fibers coming from the other side of the sulcus will also be tested throughout the resection (orange –). At the bottom of the cavity, the "horizontal connectivity" or long-distace assocation fibers (e.g., arcuate fascicle) will represent the deep functional subcortical boundary (red +). Interestingly, the cortex invaded by the tumor (in grey) is functionally useless, since the resection has been pursued until the subcortical pathways have been encountered (blue and red +). Thus, the fibers arising from this cortex have been disrupted, as they do not respond to stimulation (brown –). As a consequence, this cortex was disconnected and can therefore be removed with no functional risk. **(B)** Example of an intragyral dissection (Reproduced from [26])

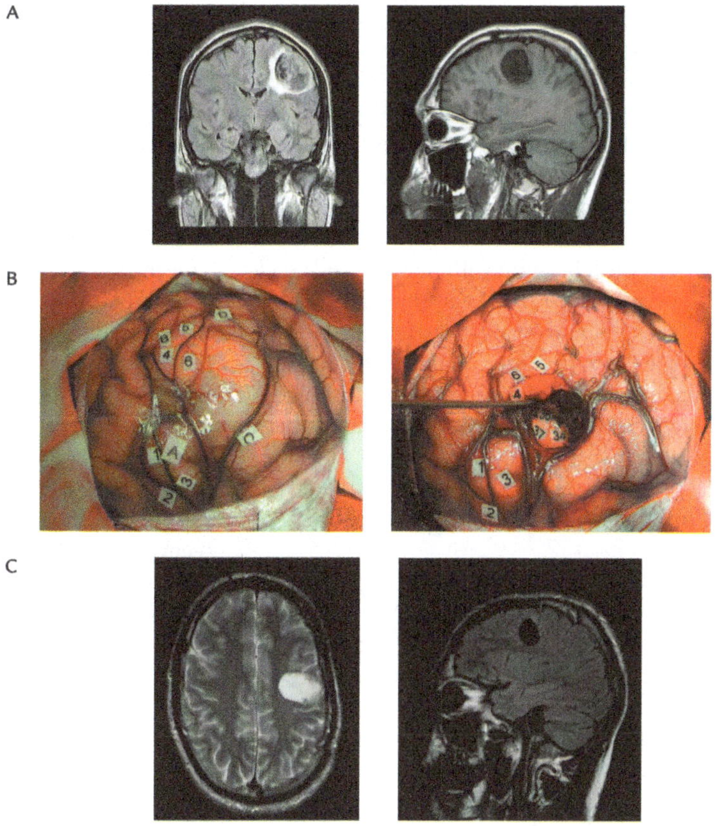

Fig. 3. **(A)** Preoperative coronal FLAIR-weighted MRI (left) and sagittal T1-weighted MRI (right) showing a left precentral glioma involving the middle frontal gyrus. **(B)** Intraoperative cortical mapping: tags *A*, *B*, and *C* are tumor boundaries as shown by intraoperative ultrasonography. *1–3* Primary motor area of the hand, *4* speech arrest with facial movements, *5* and *6* speech arrest in the ventral premotor cortex. On the right, subcortical mapping shows the resection cavity with no margin between functional limits and tumor, thanks to a subpial dissection along the precentral sulcus (spatula). The subcortical functional boundaries were: the pyramidal fiber track of the hand at the bottom of the sulcus (*37*) and fibers coming from the ventral premotor cortex inducing anarthria (*36*), both constituting the so-called "vertical connectivity" of projection fibers, as well as phonemic paraphasia elicited by stimulation over the arcuate fascicle, i.e., long-distance association pathways, or "deep connectivity" or "horizontal connectivty" (*34*). **(C)** Postoperative axial T2-weighted MRI (*left*) shows surgical cavity, which is directly in contact with the precentral sulcus, where subpial dissection was performed with no residual tumor at this level and no margin. Sagittal FLAIR-weighted MRI (*right*) shows inferior limit of the surgical cavity directly in contact with the arcuate fascicle, which lies just above the insula (Reproduced from [26])

response when stimulated, whereas (i) this effect is caused by the backward spreading of the electrostimulation along the network to an essential area and/or (ii) the stimulated region can be functionally compensated thanks to long-term brain plasticity mechanisms. In brief, although IEM is still the gold standard for brain mapping, due to the risk of false-positives, its combination with new methods such as perioperative FNI and biomathematical modeling is now mandatory, to clearly differentiate those networks that are actually indispensable to function from those that can be compensated [38] (see chapter by Mandonnet).

Functional and oncological results

A dramatic improvement of the surgical results was provided by advances in IEM.

First, it has been demonstrated that the use of IEM has allowed to significantly increase the surgical indications in eloquent areas which were classically considered as inoperable, such as the central region, the insular lobe, or Broca's area [4, 11, 13, 19, 24, 28].

In addition, despite a frequent transitory neurological worsening in the immediate postoperative period due to the attempt to perform a maximal tumor removal according to cortico-subcortical functional limits determined by IEM [26], leading to a specific functional rehabilitation, more than 98% of patients recovered the same status as before glioma resection within eloquent brain areas guided by functional mapping and returned to a normal socio-professional life [23,45]. Even more, for patients with LGG it was observed that 15 to 20% can improve in comparison with their preoperative neurological and neuropsychological assessment [51], and 80% of patients with presurgical intractable epilepsy can benefit from a relief of their seizures [5, 17]. In other words, LGG surgery presently not only is able to preserve brain functions but also may improve the quality of life of patients, as demonstrated by extensive neurocognitive assessment performed after surgical resection. These data support the existence of additional brain plasticity mechanisms occurring after the operation, likely facilitated by a systematic and adapted rehabilitation. Interestingly, the rate of less than 2% of sequelae is highy reproducible among the teams using IES worlwide. In comparison, in series which did not use IES, the rate of sequelae ranged from 13 to 27.5%, with a mean of about 19% (for a review, see [19]).

Moreover, a comparative study between LGG resection performed without or with IEM showed that the EOR was significantly increased thanks to IEM, with better, rather than impaired; functional results following resection within eloquent areas [19]. Indeed, since IESM allows identification of the cortical and subcortical eloquent structures individually, it seems logical to perform a resection according to functional boundaries. The resection is continued until, rather than ceased before, the functional regions are detected by IESM, without a margin, in order to optimize the EOR without increased risk of permanent deficit [26]. Interestingly, a recent study reported for the first time a series of patients who underwent 2 consecutive surgeries without and with awake mapping. Under general anesthesia, 9 patients underwent surgery for a low-grade glioma in functional sites in other institutions. The resection was subtotal in 3 cases and partial in 6 cases. There was a postoperative worsening in 3 cases. A second surgery was then performed in the awake condition with intraoperative electrostimulation (resection according to functional boundaries at both the cortical and subcortical levels). Postoperative MRI showed that the resection was complete in 5 cases and subtotal in 4 cases (no partial removal) and that it was improved in all cases compared with the first surgery ($p = 0.04$). There was no permanent neurological worsening. Of the 9 patients, 3 improved compared with the presurgical status. All patients returned to normal professional and social lives. These original results demonstrate that awake surgery, known to preserve the quality of life in patients with LGG, is also able to significantly improve the extent of resection for lesions located in functional regions [9]. Even more recently, the Chang et al [7] demonstrated for the first time in a series of 281 patients that the use of functional-mapping-guided resection of LGG in presumed eloquent areas, thanks to a reliable delineation of true functional and nonfunctional regions, allowed not only a maximization of tumor resection but also a signitificant improvement of long-term survival.

Interestingly, reoperation was also demonstrated to have a significant impact on the overall survival of patients with LGG [1], including tumors within eloquent areas [39], without a higher risk to induce permanent deficit, thanks to mechanisms of brain plasticity which can occur between two surgeries [11, 28].

Conclusions and perspectives

Brain surgery may now benefit from important technical developments in the field of functional mapping, using complementary noninvasive methods of FNI and invasive IEM. Such recent advances have enabled a better understanding of the organization of the eloquent brain areas for each patient, in order to integrate the concept of interindividual anatomo-functional variability in the surgical strategy. Indeed, intraoperative real-time cortical and subcortical stimulations make possible on-line correlations between discrete and transient "virtual" lesions, which can be performed at each site of a distributed network (each cortical and subcortical site being perfectly identified anatomically by three-dimensional MRI), and their functional consequences (accurately analyzed by a speech therapist or neuropychologist along the surgical procedure). Therefore, IEM allows to perform tumor resection according to functional boundaries, leading to an optimization of the benefit-to-risk ratio of surgical removal of cerebral gliomas, as it allows (i) to significantly extend the indications of resection in eloquent areas such as Broca's area, the insula – even in the left dominant hemisphere –, the central area, or the left posterior temporal regions, (ii) to significantly decrease the rate of permanent deficit at less than 2%, instead of 13 to 27% (mean of 19%) without mapping, (iii) even to improve the quality of life, thanks to seizure control (in around 80% of cases, especially for insular and/or temporal LGG [17, 24, 50]) and thanks to cognitive rehabilitation, and (iv) to significantly increase the EOR compared with series without brain mapping, thus with a significant improvement of the overall survival.

The next step is now to discuss the possible significance of a "supercomplete" resection performed by continuing the resection until, rather than ceasing it before, functional areas are encountered even for LGG located within noneloquent areas, in order to resect a marginal zone around the MRI-defined abnormalities. Indeed, a recent study using biopsy samples taken within and beyond MRI signal abnormalities demonstrated that conventional MRI underestimated the actual spatial extent of LGG because tumor cells were present beyond the MRI signal abnormalities, at distances up to 20 mm, even when gliomas were well defined in MRI [43]. As a consequence, perspectives could be to select the optimal indications of such supercomplete resection, on the basis notably of MRI advances, as some LGG are more "invasive", whereas some other LGG are more "proliferative". To this end, it has been suggested that spectroscopy and perfusion MRI could be more efficient than conventional MRI to evaluate the actual LGG infiltration. The metabolic imaging might be more sensitive than morphologic MRI and could be closer to the neuropathological tumor infiltration, as abnormal metabolic areas exceed areas of MRI-defined abnormalities on T2-weighted sequences [25]. However, these techniques still lack reliability and need to be validated. Another way could be to use the new biomathematical models of proliferation and diffusion with at least two MRIs recorded with an interval of 3 to 6 months before any treatment [31, 37]. Finally, a probabilistic atlas allowing preoperative estimation of the residual volume for LGG resection with intrasurgical functional mapping can also be built [36].

This strong link between cognitive neurosciences and neurooncology will enable to improve both the overall survival and the quality of life, leading to a new concept of "functional neurooncology" [14]. Indeed, although it was recently noted that "the most important part of our work as neurosurgeons: the disease" (see comment on [16]), one can also suggest that the most important part of our work as neurosurgeons is the patient. In other words, neurosurgeons have to see the brain first but not the tumor. Interestingly, such philosophy based on a better understanding of individual dynamic brain organization – hodotopy and plasticity – not only could be applied to the treatment of tumors but also could be used as a paradigm in view of other neurological pathologies such as multiple sclerosis (which can be considered as a "disconnection syndrome" due to damages of the white-matter pathways), stroke, or even degenerative diseases.

References

[1] Ahmadi R, Dictus C, Hartmann C, Zürn O, Edler L, Hartmann M et al (2009) Long-term outcome and survival of surgically treated supratentorial low-grade glioma in adult patients. Acta Neurochir (Wien) 151: 1359–1365

[2] Bartos R, Jech R, Vymazal J, Petrovicky P, Vachata P, Hejcl A et al (2009) Validity of primary motor area localization with fMRI versus electric cortical stimulation: a comparative study. Acta Neurochir (Wien) 151: 1071–1080

[3] Bello L, Gallucci M, Fava M, Carraba G, Giussani C, Acerbi F et al (2007) Intraoperative subcortical language tract mapping guides surgical removal of gliomas involving speech areas. Neurosurgery 60: 67–80

[4] Benzagmout M, Gatignol P, Duffau H (2007) Resection of World Health Organization grade II gliomas involving Broca's area: methodological and functional considerations. Neurosurgery 61: 741–752

[5] Brogna C, Gil Robles S, Duffau H (2008) Brain tumors and epilepsy. Expert Rev Neurother 8: 941–955

[6] Bürgel U, Mädler B, Honey CR, Thron A, Gilsbach J, Coenen VA (2009) Fiber tracking with distinct software tools results in a clear diversity in anatomical fiber tract portrayal. Cent Eur Neurosurg 70: 27–35

[7] Chang EF, Clark A, Smith JS, Polley MY, Chang SM, Barbaro NM et al (2010) Functional mapping-guided resection of low-grade gliomas in eloquent areas of the brain: improvement of long term survival. J Neurosurg, DOI 10.3171/2010.6.JNS091246

[8] Claus EB, Horlacher A, Hsu L, Schwartz RB, Dello-Iacono D, Talos F (2005) Survival rates in patients with low-grade glioma after intraoperative magnetic resonance image guidance. Cancer 103: 1227–1233

[9] de Benedictis A, Moritz-Gasser S, Duffau H (2010) Awake mapping optimizes the extent of resection for low-grade gliomas in eloquent areas. Neurosurgery 66: 1074-1084

[10] Duffau H (2005a) Intraoperative cortico-subcortical stimulations in surgery of low-grade gliomas. Exp Rev Neurother 5: 473–485

[11] Duffau H (2005b) Lessons from brain mapping in surgery for low-grade glioma: insights into associations between tumour and brain plasticity. Lancet Neurol 4: 476–486

[12] Duffau H (2008) The anatomo-functional connectivity of language revisited: new insights provided by electrostimulation and tractography. Neuropsychologia 4: 927–934

[13] Duffau H (2009a) A personal consecutive series of surgically treated 51 cases of insular WHO grade II glioma: advances and limitations. J Neurosurg 110: 696–708

[14] Duffau H (2009b) Surgery of low-grade gliomas: towards a "functional neurooncology". Curr Opin Oncol 21: 543–549

[15] Duffau H (2010a) Brain mapping in neuro-oncology: what is the future? Future Neurol 5: 433–448

[16] Duffau H (2010b) Awake surgery for non-language mapping. Neurosurgery 66: 523–528

[17] Duffau H, Capelle L, Lopes M, Bitar A, Sichez JP, van Effenterre R (2002) Medically intractable epilepsy from insular low-grade gliomas: improvement after an extended lesionectomy. Acta Neurochir (Wien) 144: 563–572

[18] Duffau H, Capelle L, Denvil D, Sichez N, Gatignol P, Taillandier L et al (2003) Usefulness of intraoperative electrical subcortical mapping during surgery for low-grade gliomas located within eloquent brain regions: functional results in a consecutive series of 103 patients. J Neurosurg 98: 764–778

[19] Duffau H, Lopes M, Arthuis F, Bitar A, Sichez JP, van Effenterre R et al (2005) Contribution of intraoperative electrical stimulations in surgery of low-grade gliomas: a comparative study between two series without (1985–96) and with (1996–2003) functional mapping in the same institution. J Neurol Neurosurg Psychiatry 76: 845–851

[20] Duffau H, Gatignol P, Mandonnet E, Peruzzi P, Tzourio-Mazoyer N, Capelle L (2005) New insights into the anatomo-functional connectivity of the semantic system: a study using cortico-subcortical electrostimulations. Brain 128: 797–810

[21] Duffau H, Kujas M, Taillandier L (2006) Episodic nocturnal wandering in a patient with epilepsy due to a right temporoinsular low-grade glioma: relief following resection. J Neurosurg 104: 436–439

[22] Duffau H, Leroy M, Gatignol P (2008) Cortico-subcortical organization of language networks in the right hemisphere: an electrostimulation study in left-handers. Neuropsychologia 46: 3197-3209

[23] Duffau H, Gatignol P, Mandonnet E, Capelle L, Taillandier L (2008) Contribution of intraoperative subcortical stimulation mapping of language pathways: a consecutive series of 115 patients operated on for a WHO grade II glioma in the left dominant hemisphere. J Neurosurg 109: 461–471

[24] Duffau H, Moritz-Gasser S, Gatignol P (2009) Functional outcome after language mapping for insular World Health Organization grade II gliomas in the dominant hemisphere: experience with 24 patients. Neurosurg Focus 27(2): E7

[25] Ganslandt O, Stadlbauer A, Fahlbusch R, Kamada K, Buslei R, Blumcke I et al (2005) Proton magnetic resonance spectroscopic imaging integrated into image-guided surgery: correlation to standard magnetic resonance imaging and tumor cell density. Neurosurgery 56: 291–298

[26] Gil Robles S, Duffau H (2010) Surgical management of World Health Organization grade II gliomas in eloquent areas: the necessity of preserving a margin around functional structures. Neurosurg Focus 28(2): E8

[27] Gil Robles S, Gatignol P, Capelle L, Mitchell MC, Duffau H (2005) The role of dominant striatum in language: a study using intraoperative electrical stimulations. J Neurol Neurosurg Psychiatry 76: 940–946

[28] Gil Robles S, Gatignol P, Lehéricy S, Duffau H (2008) Long-term brain plasticity allowing multiple-stages surgical approach for WHO grade II gliomas in eloquent areas: a combined study using longitudinal functional MRI and intraoperative electrical stimulation. J Neurosurg 109: 615–624

[29] Giussani C, Roux FE, Ojemann J, Sganzerla EP, Pirillo D, Papagno C (2010) Is preoperative functional magnetic resonance imaging reliable for language areas mapping in brain tumor surgery? Review of language functional magnetic resonance imaging and direct cortical stimulation correlation studies. Neurosurgery 66: 113–120

[30] Ilmberger J, Ruge M, Kreth FW, Briegel J, Reulen HJ, Tonn JC (2008) Intraoperative mapping of language functions: a longitudinal neurolinguistic analysis. J Neurosurg 109: 583–592

[31] Jbabdi S, Mandonnet E, Duffau H, Capelle L, Swanson KR, Pélégrini-Issac M et al (2005) Simulation of anisotropic growth of low-grade gliomas using diffusion tensor imaging. Magn Reson Med 54: 616–624

[32] Kim SS, McCutcheon IE, Suki D, Weinberg JS, Sawaya R, Lang FF et al (2009) Awake craniotomy for brain tumors near eloquent cortex: correlation of intraoperative cortical mapping with neurological outcomes in 309 consecutive patients. Neurosurgery 64: 836–846

[33] Lafargue G, Duffau H (2008) Awareness of intending to act following parietal cortex resection. Neuropsychologia 46: 2662–2667

[34] Leclercq D, Duffau H, Delmaire C, Capelle L, Gatignol P, Ducors M et al (2010) Comparison of diffusion tensor imaging tractography of language tracts and intraoperative subcortical stimulations. J Neurosurg 112: 503–511

[35] Mandonnet E, Capelle L, Duffau H (2006) Extension of paralimbic low-grade gliomas toward an anatomical classification based on white matter invasion patterns. J Neurooncol 78: 179–185

[36] Mandonnet E, Jbabdi S, Taillandier L, Galanaud D, Benali H, Capelle L, Duffau H (2007) Preoperative estimation of residual volume for WHO grade II glioma resected with intraoperative functional mapping. Neuro Oncol 9: 63–69

[37] Mandonnet E, Pallud J, Clatz O, Taillandier L, Konukoglu E, Duffau H et al (2008) Computational modeling of the WHO grade II glioma dynamics: principles and applications to management paradigm. Neurosurg Rev 31: 263–269

[38] Mandonnet E, Winkler PA, Duffau H (2010) Direct electrical stimulation as an input gate into brain functional networks: principles, advantages and limitations. Acta Neurochir (Wien) 152: 185–193

[39] Martino J, Taillandier L, Moritz-Gasser S, Gatignol P, Duffau H (2009) Re-operation is a safe and effective therapeutic strategy in recurrent WHO grade II gliomas within eloquent areas. Acta Neurochir (Wien) 151: 427–436

[40] Mauer ME, Bottomley A, Taphoorn MJ (2008) Evaluating health-related quality of life and symptom burden in brain tumour patients: instruments for use in experimental trials and clinical practice. Curr Opin Neurol 21: 745–753

[41] McGirt MJ, Chaichana KL, Attenello FJ, Weingart JD, Than K, Burger PC et al (2008) Extent of surgical resection is independently associated with survival in patients with hemispheric infiltrating low-grade gliomas. Neurosurgery 63: 700–707

[42] Ojemann G, Ojemann J, Lettich E, Berger MS (1989) Cortical language localization in left, dominant hemisphere. An electrical stimulation mapping investigation in 117 patients. J Neurosurg 71: 316–326

[43] Pallud J, Varlet P, Devaux B, Geha S, Badoual M, Deroulers C et al (2010) Diffuse low-grade oligodendrogliomas extend beyond MRI-defined abnormalities. Neurology 74: 1724–1731

[44] Pouratian N, Bookheimer SY (2010) The reliability of neuroanatomy as a predictor of eloquence: a review. Neurosurg Focus 28(3): E3

[45] Sanai N, Mirzadeh Z, Berger MS (2008) Functional outcome after language mapping for glioma resection. N Engl J Med 358: 18–27

[46] Serletis D, Bernstein M (2007) Prospective study of awake craniotomy used routinely and nonselectively for supratentorial tumors. J Neurosurg 107: 1–6

[47] Smith JS, Chang EF, Lamborn KR, Chang SM, Prados MD, Cha S et al (2008) Role of extent of resection in the long-term outcome of low-grade hemispheric gliomas. J Clin Oncol 26: 1338–1345

[48] Soffietti R, Baumert B, Bello L, von Deimling A, Duffau H, Frenay M et al (2010) Guidelines on management of low grade gliomas: report of an EFNS-EANO task force. Eur J Neurol 17: 1124–1133

[49] Stummer W, Pichlmeier U, Meinel T, Wiestler OD, Zanella F, Reulen HJ et al (2006) Fluorescence-guided surgery with 5-aminolevulinic acid for resection of malignant glioma: a randomised controlled multicentre phase III trial. Lancet Oncol 7: 392–401

[50] Taillandier L, Duffau H (2009) Epilepsy and insular grade II gliomas: an interdisciplinary point of view from a retrospective monocentric series of 46 cases. Neurosurg Focus 27(2): E8

[51] Teixidor P, Gatignol P, Leroy M, Masuet-Aumatell C, Capelle L, Duffau H (2007) Assessment of verbal working memory before and after surgery for low-grade glioma. J Neurooncol 81: 305–313

[52] Vassal M, Le Bars E, Moritz-Gasser S, Menjot N, Duffau H (2010) Crossed aphasia elicited by intraoperative cortical and subcortical stimulation in awake patients. J Neurosurg 113: 1251–1258

[53] Vidorreta JG, Garcia R, Moritz-Gasser S, Duffau H (2011) Double dissociation between syntactic gender and picture naming processing: a brain stimulation mapping study. Hum Brain Mapp 32: 331–340

[54] Vigneau M, Beaucousin V, Herve PY, Duffau H, Crivello F, Houdé O et al (2006) Meta-analyzing left hemisphere language areas: phonology, semantics, and sentence processing. Neuroimage 30: 1414–1432

[55] Wiedemayer H, Sandalcioglu IE, Armbruster W, Regel J, Schaefer H, Stolkke D (2004) False negative findings in intraoperative SEP monitoring: analysis of 658 consecutive neurosurgical cases and review of published reports. J Neurol Neurosurg Psychiatry 75: 280–286

[56] Willems PW, Taphoorn MJ, Burger H, Berkelbach van der Sprenkel JW, Tulleken CA (2006) Effectiveness of neuronavigation in resecting solitary intracerebral contrast-enhancing tumors: a randomized controlled trial. J Neurosurg 104: 360–368

Perspectives

Indications of awake mapping and selection of intraoperative tasks

Hugues Duffau

Introduction

Interindividual anatomo-functional variability of the human brain is now well demonstrated by both functional neuroimaging in healthy volunteers [108] and intraoperative brain stimulation in patients with neurological diseases such as epilepsy [77]. As a consequence, a growing number of surgical series has reported awake mapping as a valuable tool to optimize the extent of resection of lesions located within eloquent areas [17, 31], allowing an increase of the surgical impact on their natural history, e.g., significant improvement of the survival of patients with low-grade gliomas [12], while minimizing the risk of permanent deficit and even improving the health-related quality of life (HRQoL) [32, 91]. Indeed, in contrast to extraoperative electrical mapping using subdural grids, awake surgery offers the opportunity to map not only the cortex but also the subcortical structures, especially white-matter pathways, and thus to study the individual axonal connectivity [21, 22]. However, the majority of awake craniotomies, although demonstrated to be well tolerated by the patients [15, 114], have been performed almost exclusively in cases of lesions involving presumed language areas, that is, the perisylvian regions in the left hemisphere, in order to avoid postsurgical permanent oral or written language disturbances [51, 82, 91]. Surprisingly, the use of awake mapping for lesions located in brain areas a priori not implicated in language (on the basis of single anatomic criteria), especially the right hemisphere, is rare. Furthermore, mapping of nonlanguage functions received less attention, despite the possible consequences of neurological disorders other than aphasia on the daily life. Indeed, when objective neuropsychological and HRQoL assessment have been performed, visuospatial, memory, attention, planning, learning, emotional, motivational and behavioral deficits have regularly been observed after brain (epilepsy and/or tumor) surgery [2, 3, 8, 9, 14, 20, 35, 47, 49, 50, 52, 58, 64, 66, 67, 69, 70, 78, 79, 81, 89, 96, 98, 101, 107, 111]. However, in spite of these numerous reports, nonlanguage disorders have largely been underestimated by neurosurgeons. This is explained by the fact that the identification of such “subtle” deficits is not possible by a single “standard clinical examination”. Unfortunately, extensive neurocognitive evaluation was very rarely performed after glioma surgery in the classical literature, especially when the lesion was located outside the so-called language areas. Therefore, it is likely that more patients than those reported in the literature experienced this kind of disorders.

In this chapter, the aim is to provide new insights into the indications of awake craniotomies to map and preserve language structures for lesions located outside the classical presumed language areas and to map and preserve brain networks not directly involved in language but crucial for sensorymotor, visual, and vestibular functions, spatial awareness, as well as cognitive functions such as calculation, memory, understanding and even judgement, in cases of surgery for lesions invading areas a priori not related to language processing. To this end, a wider range of tasks which could be used intraoperatively should be available and their selection should be improved, i.e., adapted to each patient according to several criteria: the life of the patient (job, hobby, habits), the results of the presurgical neuropsychological assessment (especially the possible existence of a presurgical cognitive deficit - even slight - due to the lesion, the handedness), the findings provided by the preoperative neurofunctional imaging, and the characteristics of the lesion (location, size, behavior). In other words, the optimization of intraoperative tests is based on the better understanding of the individual anatomy, physiology, HRQoL, and its interactions with the natural course of the disease.

Awake language mapping for surgery outside classical language areas

Since the seminal works by Ojemann et al [77], awake surgery is essentially performed to map language (generally speech, counting, picture naming, reading, writing and multilingualism) for lesions involving the lateral cortex at the level of the perisylvian regions within the left dominant hemisphere, that is, the temporal lobe (especially the superior and posterior part), the inferior parietal lobule, and the lateral part of the frontal gyrus. However, many reports support the unpredictability of functional eloquence on the basis of anatomical features alone [84]. Thus, the surgical method (under general anesthesia versus local anesthesia) cannot be decided with respect to anatomical considerations only. Below, recent findings which show that language structures could be encountered outside the classical anatomical language areas will be reviewed, pleading in favor of the fact that neurosurgeons need to take advantage of neurocognitive evaluation and mapping techniques to create individualized maps and management plans.

Less-considered language areas in the left dominant hemisphere

Although the insular lobe was poorly studied for more than a century, recent lesion and functional neuroimaging works provided strong arguments in favor of a role of the dominant insula in language fuctioning, especially in complex planning of speech articulation [1, 19, 48, 75]. As a consequence, it was suggested to perform awake craniotomies for resection of lesions located within the insula, such as cavernomas [26] or gliomas [23, 27]. Interestingly, intrasurgical articulatory disturbances have been elicited by electrostimulation of the left insula in approximately 20% of patients, even in cases of invasion by a tumor as low-grade glioma. Such positive language mapping has enabled both a better selection of the surgical approach and a better delineation of the limits of resection, thus avoiding permanent postoperative speech disorders [34].

Although it is traditionally recognized that patients do not necessarily become permanently disabled after resections in the left dominant supplementary motor area done with the patients under general anesthesia, accurate assessment showed possible persistent mild but objective language deficits after surgery. Indeed, it is well known that resection of the anterior part of this area induces transient speech disorders (possibly a complete mutism) during the immediate postsurgical period, though with a secondary recovery [37, 60]. Accurate examination one year after surgery was able to identify mild word-finding difficulties [60]. In addition, a recent study reported several cases of permanent agraphia after resection of this structure [92]. Therefore, some authors proposed to perform awake mapping for lesions involving the left supplementary motor area, in order to improve the accurate determination of the functional (language) boundaries and to optimize the HRQoL [32].

It is worth noting that surgery within other areas classically presumed as not implicated in language networks may nevertheless generate permanent deficits, especially concerning written language. For example, resection of the left superior parietal lobule can induce a definitive spatial agraphia [92].

Finally, neurosurgeons should keep in mind that the subcortical connectivity of language can be encountered even in brain regions *a priori* not expected to participate in language. For instance, during surgery near the left intraparietal sulcus or superior parietal lobule, that is, remote from the sylvian fissure, even if no cortical language sites are detected, there is a high risk to encounter the superior longitudinal fascicle in the depth (lateral to the ventricle). If this white-matter tract is not identified and not preserved, its damage will generate permanent conduction aphasia. Therefore, long-association pathways should be mapped in a systematic manner to avoid a "disconnection syndrome". The sole way is to test the patient on-line throughout the resection under local anesthesia, especially in the depth [32]; extraoperative mapping is not sufficient because it is only able to map the cortical surface.

Language areas in the right hemisphere

As a consequence of the seminal lesional works reported in the second half of the 19th century by Broca [10], who discovered that a damage in the left inferior frontal gyrus induced a reduced capacity for articulate speech, by Wernicke [112], who associated speech comprehension with the left superior posterior temporal gyrus, and by Dejerine [18], who described alexia with agraphia subsequent to a left angular gyral injury, the preeminent role of the left hemisphere in language processing was definitively established. Moreover, in 1965, Geschwind [41] summarized the anatomic findings from aphasic patients in a model that outlined regions in the left hemisphere, and in particular their connections, that were critical for language. Consequently, these studies led to the concept of left hemispheric "dominance" or "specialization" for language, and for more than a century it was believed that the right hemisphere (RH) had little or no potential for processing language – despite a theory developed by Jackson as early as in 1874 in which the RH was thought to be responsible for the automatic, involuntary use of words in speech [54].

However, because of the refinement of clinical evaluation tools which allowed the detection of certain language deficits after an RH lesion (for a review, see [55]), the development of noninvasive functional imaging, which showed activations within the RH during language tasks (for a recent meta-analysis, see [109]), and the evolution of appropriate conceptual frameworks, the role of the RH in language functioning began to be considered. Indeed, several studies in the last decades provided strong arguments in favor of the likely implication of the RH in semantic processing [39] and spoken language comprehension [62], particularly in figurative meaning of verbal material such as metaphors [7], as well as in word recognition – with best processes for nouns, especially those for which an image can be retrieved [100] or with an emotional content [11]. RH seems also to be involved in action word processing [73], in lexical decision [99], during processing of linguistic context [59], in reading [100], and even in phonological processing [104]. In summary, the RH seems to have a more important role in language than previously thought, especially in prosody, the semantic processing of words and discourse, the processing of context and in pragmatic abilities, in addition to its involvement in executive functions, in particular, attention and working memory.

In a clinical setting, another criterion must be taken into account, namely, the concept of hemispheric dominance. Indeed, to make the decision to perform an awake procedure for language mapping in the right hemisphere, three parameters are crucial: the handedness, the results of the neurocognitive examination, and the calculation of the lateralization index on functional neuroimaging. Interestingly, the three parameters can be coherent, i.e., left-handed patients with presurgical language disturbances and right dominance on neuroimaging. Dissociations are nonetheless possible. For instance, using the Wada intracarotid Amytal test, Rasmussen and Milner [85] showed that in non-right-handed patients, speech was represented in the left cerebral hemisphere in nearly a third of the group, in the right hemisphere in half the group, and bilaterally in the remainder. They also suggested the possibility of a functionally asymmetric participation of the two hemispheres in the language processes of some normal left-handers. Later, Annett [4] offered a genetic model, the right shift theory, based on the existence of a gene with two alleles, one of which influences the distributions of asymmetries related to language and manual activities by favoring the left hemisphere. In homozygous subjects who do not have this allele, asymmetries are distributed independently and in random fashion, leftward asymmetries being as frequent as rightward asymmetries. This would explain why right-hemispheric specialization for language is far from being systematically associated with left manual preference [13]. By functional neuroimaging, the notion of interindividual variability of language lateralization was reinforced [105], a factor currently known to exert a crucial influence on language processing and being at least partly related to handedness [56]. It seems that left-handed subjects present less asymmetrical language lateralization [97].

On the basis of these data, some neurosurgeons proposed to perform awake language mapping in non-right-handed patients, that is, in left-handed and ambidextrous patients [33, 53]. In a recent study, language was mapped by cortico-subcortical electrostimulation in nine patients with a right low-grade glioma. In frontal regions, cortical stimulation elicited articulatory disorders (ventral premotor cortex), anomia (dorsal premotor cortex), speech arrest (pars opercularis), and semantic paraphasia (dorsolateral prefrontal cortex). Insular stimulation generated dysarthria, parietal stimulation induced phonemic paraphasias, and temporal stimulation generated semantic paraphasias. Subcortically, the superior longitudinal fascicle (phonological disturbances), inferior occipitofrontal fascicle (semantic disturbances), subcallosal fascicle (control disturbances), and common final pathway (articulatory disorders) were identified. These structures were preserved, avoiding permanent aphasia. These results support the major role of the RH in language in left-handers and provide new insights into the anatomo-functional cortico-subcortical organization of the language networks in the RH, suggesting a "mirror" configuration in comparison with the left hemisphere [33].

Even more recently, crossed aphasia, that is, aphasia resulting from a right-hemispheric lesion in right-handers, has been induced by intraoperative stimulation in awake patients harboring a right low-grade glioma [106]. These original findings highlight the possibility to find crucial cortico-subcortical language networks in the RH in a subgroup of atypical right-handers. Thus, the challenge is to detect the patients with a risk of crossed aphasia. To this end, the most important criterion is the existence of a preoperative language deficit, even if subtle. Such deficit means, in essence, that the RH has a crucial role in language in this patient at this moment, and therefore that intrasurgical language mapping is mandatory to avoid postoperative aphasia. Therefore, in right-handed patients, if language disturbances are detected during seizures or on presurgical neuropsychological assessment, awake craniotomy with intraoperative language mapping should be considered. This is particularly true when right activations are observed on language-functional MRI, independently of the lateralization index. Indeed, it is possible that, despite a left lateralization index, a single right activation corresponds to a crucial site which should be preserved intraoperatively as its injury could lead to a permanent aphasia. As a consequence, presurgical functional neuroimaging is not enough to make the decision whether awake mapping should be performed: a preoperative cognitive examination is essential.

Intraoperative mapping of movement and visuospatial cognition

For a long time, motor function was solely considered as the ability to move, i.e., to generate muscle contraction. However, movement is actually a more complex and integrated function, coordinating not only the mobilization itself but also the control of adapted action thanks to somatosensory, visual, and vestibular feedback, as well as the awareness and intention to act [61]. Consequently, even if several authors reported the possibility to detect and preserve the pyramidal structures in asleep patients by intraoperative monitoring (motor evoked potentials) and/or direct electrostimulation at both cortical and subcortical levels, for lesions within or near motor pathways (as premotor and insular lesions) [57, 74], others have proposed to perform awake mapping to task movement throughout the resection [16, 31, 36, 93, 94, 113]. Such strategy enables to check the real action, its accuracy, its adaptability, its reaction time, and the absence of a possible occurrence of ataxia or apraxia, rather than only the muscle contraction, all along the resection. Furthermore, in addition to the primary motor area, it is possible to detect the negative motor area under local anesthesia, which was shown to play a significant role in the planning of voluntary movement and generated negative motor phenomena when stimulated [65].

Concerning retrocentral lesions, awake surgery also enables the mapping of the somatosensory function with greater accuracy than do recordings of somatosensory evoked potentials under general anesthesia. Indeed, by phase reversal recording, somatosensory evoked potentials can identify the central sulcus but they are unable to provide details regarding the precise organization of the somatosensory cortex (for a review, see [29]). In contrast, during an awake procedure, the patient can accurately describe both superficial (dysesthesia) and deep (proprioceptive) somatoesthetic sensations throughout the electrical mapping and the resection [25]. It is therefore possible to map very precisely the distribution of the eloquent sites both within the somatoesthetic cortex and at the level of the thalamo-cortical pathways [29]. Interestingly, while new permanent deficits after glioma resection within nonlanguage parietal regions under general anesthesia were recently reported for 4 out of 28 patients [90], no permanent worsening was observed after awake surgery in the same regions [25, 102].

Furthermore, visual feedback is crucial for adapted movement and HRQoL. For example, with a homonymous hemianopsia driving is forbidden in many countries. Interestingly, for posterior lesions located within the temporo-parieto-occipital junction, intraoperative stimulation has been demonstrated to enable the detection of visual pathways during surgical resection in awake patients by inducing transient visual disturbances such as the perception of a shadow or phosphenes in a specific visual quadrant (i.e., detection of optic radiation) or even the perception of visual illusion such as metamorphopsia (i.e., detection of associative structures involved in higher-order visual processing) [30].

Moreover, vestibular processing is also essential for movement. The vestibular system is a part of the high-order multisensory associative system that coordinates different inputs (visual, somatosensory, auditory) to control the position of the body itself and its relation to the surrounding space. Neural foundations underlying vestibular function, especially the postero-inferior parieto-insular cortex and the superior longitudinal fascicle within the hemisphere nondominant for language, can be mapped and preserved during surgery for temporo-parietal lesions by intraoperative cortico-subcortical stimulation in awake patients which generates reproducible vertigo [95]. Interestingly, extraoperative electrostimulation has also induced experiential responses such as complex vestibulo-somatosensory sensations creating an out-of-body sensory illusion induced by stimulation of the right angular gyrus and the superior temporal gyrus [6, 103].

Finally, motion accurately adapted to the environment is not possible without awareness of action. A deficit in cognition of movement induces hemineglect as a result of a failure of the coordination of all outcoming inputs into an egocentric reference system, which may have important consequences on HRQoL. Direct electrostimulation of the superior temporal gyrus and the supramarginal gyrus in the RH may generate a significant deficit in visual search [42] or rightward deviations on line bisection [5] (Fig. 1). The same transient unilateral spatial neglect can also be induced by stimulation of the superior longitudinal fascicle because of a transient disconnection between the different functional units, showing that both the cortical areas as well as the subcortical pathway subserving spatial awareness can be identified and preserved during resection within the right parieto-occipital junction [102] (Fig. 2).

In summary, movement is a complex multimodal function which integrates multiple unimodal processings (muscle contraction, somatosensory, visual, and vestibular functions) as well as supramodal control. Preservation of

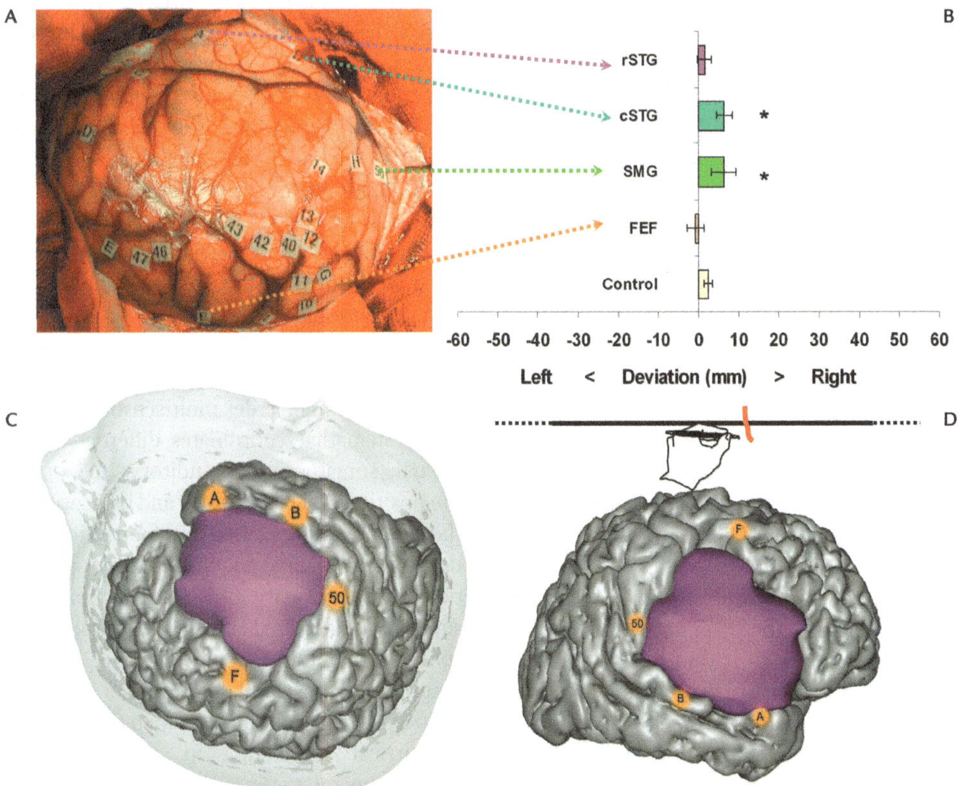

Fig. 1. (A) Surgical field in a patient with a low-grade glioma centered on the right inferior parietal lobule who performed a line bisection task during electrical mapping. **(B)** Mean deviation with 95% confidence intervals during stimulation of the rostral part of the superior temporal gyrus (*rSTG*, label A), of the caudal part of the superior temporal gyrus (*cSTG*, label B), of the supramarginal gyrus (*SMG*, label 50), of the frontal eye field (*FEF*, label F) and of control neighboring regions. A significant rightward deviation during line bisection performance was elicited during direct cortical stimulation of the supramarginal gyrus and caudal part of the superior temporal gyrus; *$p < 0.05$ (two-tailed). **(C)** Three-dimensional reconstruction of the low-grade glioma (in purple) and of the stimulated regions (in yellow). **(D)** Lateral view (From [102])

normal motor function is not adequately grasped by avoiding any damage of cortico-spinal structures but also implies to better understand the anatomo-functional organization of this whole distributed network at the individual scale. To this end, awake mapping can add valuable information to preserve a high level of somatosensory-vestibulo-visuo-spatial processing allowing awareness of the intention to act [61], anticipation, planning, execution, and on-line control of complex action, essential for a normal HRQoL.

Intraoperative mapping of higher cognitive functions: towards a better selection of tasks

In contrast to the wide use of intraoperative mapping for language, higher cognitive functions were poorly studied before, during, and after brain surgery except by a few authors who have regularly tested memory in addition to language during awake procedures in the past decades, especially for temporal epilepsy [76]. Nonetheless, by more accurate neuropsycho-

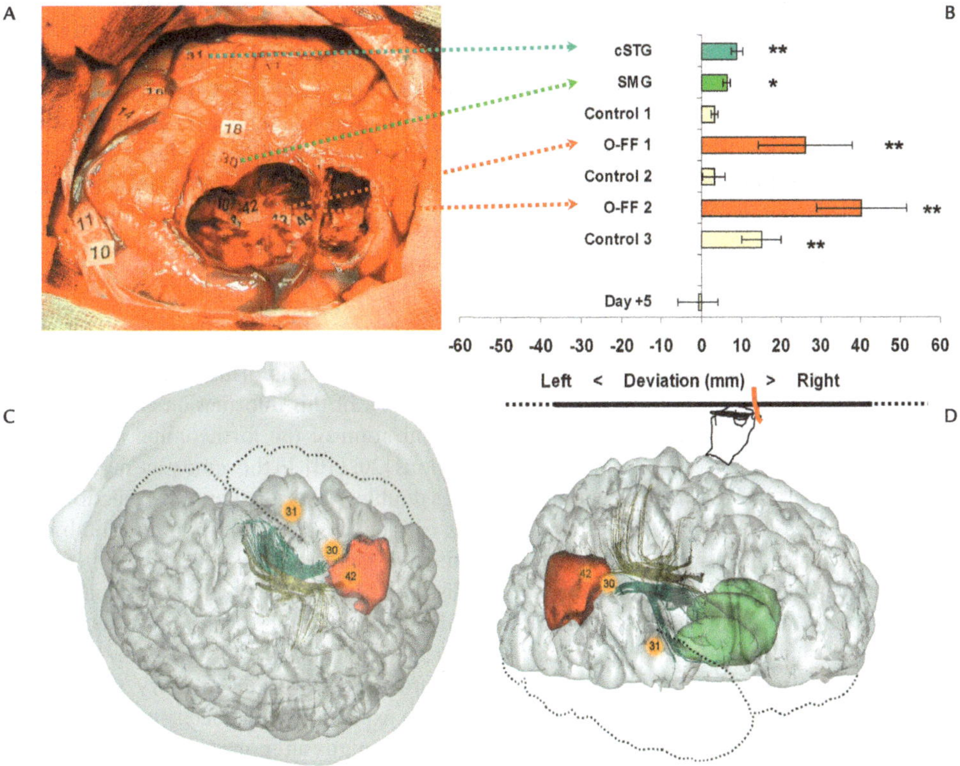

Fig. 2. (A) Surgical field in a patient harboring a low-grade glioma centered on the right inferior parietal lobule who performed a line bisection task throughout the surgical resection. **(B)** Mean deviation with 95% confidence intervals during stimulation of the caudal part of the superior temporal gyrus (*cSTG*, label 31), of the supramarginal gyrus (*SMG*, label 30), of the superior longitudinal fascicle (part II) (label 42) during (*O-FF 1*) and after tumor resection (*O-FF 2*), and of control neighboring regions before resection (*control 1*), during resection (*control 2*), and after resection (*control 3*). Performance 5 days (*day +5*) after surgery is also shown; *$p < 0.05$, **$p < 0.01$ (two-tailed). **(C)** Three-dimensional reconstruction of the surgical resection (in red) and of the stimulated regions, showing their relationships with the superior occipito-frontal fascicle (in yellow) and the superior longitudinal fascicle (in blue). The head of the caudate nucleus and the putamen are shown in green. **(D)** Lateral view (From [102])

logical examination, it was recently observed that patients could experience postsurgical neurocognitive deficit, in particular concerning working memory, even after surgery performed under local anesthesia with language mapping [20, 101]. Such results mean that an awake procedure by itself does not prevent cognitive worsening if only intraoperative assessment of language (e.g., speech, naming, reading, writing, bilinguism) and/or complex movement (see above) is done, but that specific testing should be added to map and preserve higher-order functions. Therefore, it is now mandatory to select tasks adapted to the individual patient (according to her or his job, hobby, possible preoperative neurocognitive disturbances, results of the presurgical functional neuroimaging) and to the specific lesion (location, side, nature), also taking into account the fact that there is a limitation of time for intraoperative awake mapping.

For example, mapping of calculation (e.g., multiplication and subtraction) was reported especially for a school teacher with a lesion involving the left parietal lobe [28]. Identification and preservation of the calculation sites allowed the patient to work normally after surgery, without dyscalculia and anarithmetia. Similar results were reported by other authors, notably for lesions located near or within the left angular gyrus [86, 88]. Recently, singing was also tested intraoperatively in patients who were amateur singers and had undergone surgery for brain tumors; dissociations between speech and singing showed that in some cortical stages these two functions used different cerebral pathways [87]. Double dissociation between language tested by picture naming and nonverbal comprehension tested by a task of semantic association was also observed [38]. These findings support the need to incorporate specific tasks, e.g., the Pyramids and Palm Tree test, in addition to the usual naming test during surgery for lesions involving the left posterior temporal areas [38].

Concerning executive functions regularly disturbed after brain surgery [20, 101], recent memory can be tested intraoperatively by performing stimulation during encoding, storage, or retrieval, in patients harboring a lesion within temporo-parietal or frontal sites [76]. Indeed, short-term memory errors were observed by intra- and extraoperative stimulation of the left temporal neocortex [80]. Interestingly, beyond electrical mapping, the hippocampus has also been cooled by rinsing with cold saline intraoperatively to evaluate memory and learning performance and to determine the risk of postoperative memory disorder [63]. In addition, intraoperative mapping of the dominant frontal premotor area and anterior temporal lobe has identified specific areas involved in famous-face recognition [45].

Furthermore, ocular saccades, which are known to be closely related to attention, were tested in awake patients with lesions involving the frontal eye field by electrical stimulation which induced ocular deviation or saccade suppression [68]. Indeed, the resection of that premotor area was demonstrated to induce executive dysfunction, especially concerning working memory, namely, fundamental processes which include the short-term maintenance of relevant information, the mental manipulation of this information and the mental organization of the forthcoming sequence of actions [20, 110]. In the same way, performance of a double task throughout the resection, such as regular movement of the upper limb combined to a language task (e.g., naming), may be disturbed in awake patients. It is nonetheless difficult to determine accurately whether such decline is due to damages of networks underlying attention or working memory or whether it can be explained by the occurrence of a physiological fatigue after one to two hours of intrasurgical work.

For tumors located within the left dominant prefrontal cortex, a task of crossmodal (visual-verbal) congruent and incongruent judgment has been performed by awake patients. Visual and auditory stimuli were presented simultaneously, either referring to the same item (congruence condition) or to different items (semantic or phonemic incongruent condition). It was demonstrated that stimulation of brain areas not involved in naming processing elicited reproducible deficits of incongruent judgment, especially at the level of the left dorso-lateral prefrontal region – even though an interindividual variability was observed as for other functions [83]. Preservation of such executive functions is essential for the daily life, in particular with regard to decision-making and planning of complex strategy.

More recently, beyond the classical intraoperative mapping of multiple languages frequently reported in the literature (for a review, see [44]), the neural bases of language switching have also been mapped, at both the cortical and the subcortical level [71]. It seems that this neurocognitive process of switching might be subserved by a large-scale cortico-subcortical network with an executive system (prefrontal cortex, anterior cingulum, caudate nucleus) controlling a more dedicated language subcircuit which involves postero-temporal areas, supramarginal and angular gyri, Broca's area, and the superior longitudinal fascicle [72]. The involvement of such a striato-cortical executive loop in cognitive control may explain why perseverations are reproducibly elicited when the head of the left caudate nucleus is stimulated [43].

Finally, concerning emotional aspects, a recent study mapped the capability of recognizing the expressions of facial emotions in awake patients. The authors advocated offering an intrasurgical brain mapping of facial emotion recognition ("mirror of the soul") to patients with right posterior perisylvian tumors [46].

Conclusions and perspectives

In the era of "functional neurosurgery", it is necessary for neurosurgeons to reinforce the link between the improved understanding of the processing underlying neurocognition and the clinical applications of that knowledge, with the aim to better test the patients operated for a brain lesion before, during, and after the surgical resection. Although intraoperative language mapping under local anesthesia has without any doubt minimized the risk of permanent aphasia during surgery in so-called eloquent areas, other higher functions such as visuospatial cognition, executive functions, and emotional components have been neglected. Recent extensive neuropsychological studies have demonstrated a significant rate of cognitive disorders following brain surgery even when language mapping was performed, which indicates a (partial) dissociation between language and cognitive neural networks. New insights provided by the present review advocate the more systematic use of awake procedure also for lesions not located within presumed language regions – even if some of them have classically been considered as noneloquent [24]. Clearly, the goal is to task intrasurgically additional high-level functions whose preservation is essential for an optimal HRQoL. Beyond the field of cognition, perspectives might be to map networks involved in emotion and behavior, underestimated but crucial for the daily personal, familial, social, and professional life. Specific intraoperative tasks for awake patients should be elaborated to meet this goal. Furthermore, with the aim to make the results of different teams comparable, a common general protocol should be established for pre-, intra-, and postoperative application. However, in addition to a common core, specific tests should be tailored to the individual patients' particular needs and preoperative findings, e.g., for spatial cognition in dancers with lesions invading the right parietal region [102], for calculation in school teachers with lesions located within the left parietal lobe [28], or for judgement in managers with a lesion involving the prefontral cortex [83].

On the other hand, it is important to keep in mind that the goal of brain surgery, especially for epilepsy and in neuro-oncology, is to optimize the extent of resection. Thus, the role of the neurosurgeon is to find the optimal balance between the surgical impact on the natural history of the disease and the preservation of the HRQoL, namely, to adapt the tasks to the indiviadual patient rather than to perform a standardized resection according exclusively to anatomical boundaries in a localisationist view of cerebral organization. Most importantly, whatever the lesion, the surgical modalities have to be adapted to the individual benefit-to-risk ratio of the surgery. While neurosurgeons must not neglect the main goal of surgery, which is to increase the median survival, they also have to preserve or even to improve the functional status. The matter of deliberation is no longer one of median survival versus HRQoL but of both median survival and HRQoL. To this end, awake mapping has been demonstrated to allow both a significant increase of the extent of resection and a significant decrease of the rate of permanent deficit [17, 31]. Regarding preoperative neurofunctional imaging, even if these noninvasive methods can be valuable for surgical planning, they are not yet reliable enough at the individual scale. These techniques can be very useful to select the optimal intraoperative tasks for the individual patient and to better understand mechanisms of plasticity by performing longitudinal fMRI after surgery, but awake mapping is still the gold standard and should be more developed in clinical practice. Only an interactive work between neurosurgeons, neurologists, neuropsychologists, speech therapists, neurophysiologists, and neuroscientists would improve our understanding of the neural basis of cognition and emotion and the benefit-to-risk ratio of surgery tailored to the individual patient, that is, with intraoperative tasks adapted to the individual HRQoL, for which the criteria can change from patient to patient. New specific programs of cognitive rehabilitation have also to be built, not only to allow a functional recovery after surgery but also to try to improve the quality of life in comparison to the presurgical status [40].

References

[1] Ackermann H, Riecker A (2004) The contribution of the insula to motor aspects of speech production: a review and a hypothesis. Brain Lang 89: 320–328

[2] Anderson SI, Taylor R, Whittle IR (1999) Mood disorders in patients after treatment for primary intracranial tumours. Br J Neurosurg 13: 480–485

[3] Andrewes DG, Kaye A, Murphy M, Harris B, Aitken S, Parr C, Bates L (2003) Emotional and social dysfunction in patients following surgical treatment for brain tumour. J Clin Neurosci 10: 428–433

[4] Annett M (1996) In defence of the right shift theory. Percept Mot Skills 82: 115–137

[5] Bartolomeo P, Thiebaut de Schotten M, Duffau H (2007) Mapping of visuospatial functions during brain surgery: a new tool to prevent unilateral spatial neglect. Neurosurgery 61: E1340, DOI 10.1227/01.neu. 0000306126. 46657.79

[6] Blanke O, Ortigue S, Landis T, Seeck M (2002) Stimulating illusory own-body perceptions. Nature 419: 269–270

[7] Bottini G, Corcoran R, Sterzi R, Paulesu E, Schenone P, Scarpa P et al (1994) The role of the right hemisphere in the interpretation of the figurative aspects of language. A positron emission tomography activation study. Brain 117: 1241–1253

[8] Braun CM, Denault C, Cohen H, Rouleau I (1994) Discrimination of facial identity and facial affect by temporal and frontal lobectomy patients. Brain Cogn 24: 198-212

[9] Braun V, Albrecht A, Kretschmer T, Richter HP, Wunderlich A (2006) Brain tumour surgery in the vicinity of short-term memory representation – results of neuronavigation using fMRI images. Acta Neurochir (Wien) 148: 733–739

[10] Broca P (1861) Nouvelle observation d'aphémie par une lésion de la moitié postérieure des deuxième et troisième circonvolutions frontales gauches. Bull Soc Anat 36: 398–407

[11] Buchanan TW, Lutz K, Mirzazade S, Specht K, Shah NJ, Zilles K, Jäncke L (2000) Recognition of emotional prosody and verbal components of spoken language: an fMRI study. Brain Res Cogn Brain Res 9: 227–238

[12] Chang EF, Clark A, Smith JS, Polley MY, Chang SM, Barbaro NM et al (2011) Functional mapping-guided resection of low-grade gliomas in eloquent areas of the brain: improvement of long-term survival. J Neurosurg 114: 566–573

[13] Corballis MC (2003) From mouth to hand: gesture, speech, and the evolution of right-handedness. Behav Brain Sci 26: 199–208

[14] D'Angelo C, Mirijello A, Leggio L, Ferrulli A, Carotenuto V, Icolaro N et al (2008) State and trait anxiety and depression in patients with primary brain tumors before and after surgery: 1-year longitudinal study. J Neurosurg 108: 281–286

[15] Danks RA, Rogers M, Aglio LS, Gugino LD, Black PM (1998) Patient tolerance of craniotomy performed with

the patient under local anesthesia and monitored conscious sedation. Neurosurgery 42: 28–36
[16] Danks RA, Aglio LS, Gugino LD, Black PM (2000) Craniotomy under local anesthesia and monitored conscious sedation for the resection of tumors involving eloquent cortex. J Neurooncol 49: 131–139
[17] de Benedictis A, Moritz-Gasser S, Duffau H (2010) Awake mapping optimizes the extent of resection for low-grade gliomas in eloquent areas. Neurosurgery 66: 1074–1084
[18] Déjerine J (1891) Sur un cas de cécité verbale avec agraphie, suivi d'autopsie. C R Soc Biol 43: 197–201
[19] Dronkers NF (1996) A new region for coordinating speech articulation. Nature 384: 159–161
[20] du Boisgueheneuc F, Levy R, Volle E, Seassau M, Duffau H, Kinkingnehun S et al (2006) Functions of the left superior frontal gyrus in humans: a lesion study. Brain 129: 3315–3328
[21] Duffau H (2005) Intraoperative cortico-subcortical stimulations in surgery of low-grade gliomas. Expert Rev Neurother 5: 473–485
[22] Duffau H (2007) Contribution of cortical and subcortical electrostimulation in brain glioma surgery: methodological and functional considerations. Neurophysiol Clin 37: 373–382
[23] Duffau H (2009) A personal consecutive series of surgically treated 51 cases of insular WHO Grade II glioma: advances and limitations. J Neurosurg 110: 696–708
[24] Duffau H (2010) Awake surgery for non-language mapping. Neurosurgery 66: 523–528
[25] Duffau H, Capelle L (2001) Functional recuperation after resection of gliomas infiltrating primary somatosensory fields. Study of perioperative electrical stimulation. Neurochirurgie 47: 534–541
[26] Duffau H, Fontaine D (2005) Successful resection of a left insular cavernous angioma using neuronavigation and intraoperative language mapping. Acta Neurochir (Wien) 147: 205–208
[27] Duffau H, Capelle L, Lopes M, Faillot T, Sichez JP, Fohanno D (2000) The insular lobe: physiopathological and surgical considerations. Neurosurgery 47: 801–811
[28] Duffau H, Denvil D, Lopes M, Gasparini F, Cohen L, Capelle L, Van Effenterre R (2002) Intraoperative mapping of the cortical areas involved in multiplication and subtraction: an electrostimulation study in a patient with a left parietal glioma. J Neurol Neurosurg Psychiatry 73: 733–738
[29] Duffau H, Capelle L, Denvil D, Sichez N, Gatignol P, Taillandier L et al (2003) Usefulness of intraoperative electrical subcortical mapping in surgery of low grade gliomas located within eloquent regions functional results in a consecutive series of 103 patients. J Neurosurg 98: 764–778
[30] Duffau H, Velut S, Mitchell MC, Gatignol P, Capelle L (2004) Intra-operative mapping of the subcortical visual pathways using direct electrical stimulations. Acta Neurochir (Wien) 146: 265–269
[31] Duffau H, Lopes M, Arthuis F, Bitar A, Sichez JP, Van Effenterre R, Capelle L (2005) Contribution of intraoperative electrical stimulations in surgery of low grade gliomas: a comparative study between two series without (1985–96) and with (1996–2003) functional mapping in the same institution. J Neurol Neurosurg Psychiatry 76: 845–851
[32] Duffau H, Gatignol P, Mandonnet E, Capelle L, Taillandier L (2008) Intraoperative subcortical stimulation mapping of language pathways in a consecutive series of 115 patients with grade II glioma in the left dominant hemisphere. J Neurosurg 109: 461–471
[33] Duffau H, Leroy M, Gatignol P (2008) Cortico-subcortical organization of language networks in the right hemisphere: an electrostimulation study in left-handers. Neuropsychologia 46: 3197–3209
[34] Duffau H, Moritz-Gasser S, Gatignol P (2009) Functional outcome after language mapping for insular World Health Organization grade II gliomas in the dominant hemisphere: experience with 24 patients. Neurosurg Focus 27(2): E7
[35] Dulay MF, Levin HS, York MK, Mizrahi EM, Verma A, Goldsmith I et al (2009) Predictors of individual visual memory decline after unilateral anterior temporal lobe resection. Neurology 72: 1837–1842
[36] Ebel H, Ebel M, Schillinger G, Klimek M, Sobesky J, Klug N (2000) Surgery of intrinsic cerebral neoplasms in eloquent areas under local anesthesia. Minim Invasive Neurosurg 43: 192–196
[37] Fontaine D, Capelle L, Duffau H (2002) Somatotopy of the supplementary motor area: evidence from correlation of the extent of surgical resection with the clinical patterns of deficit. Neurosurgery 50: 297–303
[38] Gatignol P, Capelle L, Le Bihan R, Duffau H (2004) Double dissociation between picture naming and comprehension: an electrostimulation study. Neuroreport 15: 191–195
[39] Gazzaniga MS, Hillyard SA (1971) Language and speech capacity of the right hemisphere. Neuropsychologia 9: 273–280
[40] Gehring K, Sitskoorn MM, Aaronson NK, Taphoorn MJ (2008) Interventions for cognitive deficits in adults with brain tumours. Lancet Neurol 7: 548–560
[41] Geschwind N (1965) Disconnection syndromes in animals and man. Part 1. Brain 88: 237–294
[42] Gharabaghi A, Fruhmann Berger M, Tatagiba M, Karnath HO (2006) The role of the right superior temporal gyrus in visual search – insights from intraoperative electrical stimulation. Neuropsychologia 44: 2578–2581
[43] Gil Robles S, Gatignol P, Capelle L, Mitchell MC, Duffau H (2005) The role of dominant striatum in language: a study using intraoperative electrical stimulations. J Neurol Neurosurg Psychiatry 76: 940–946
[44] Giussani C, Roux FE, Lubrano V, Gaini SM, Bello L (2007) Review of language organisation in bilingual patients: what can we learn from direct brain mapping? Acta Neurochir (Wien) 149: 1109–1116

[45] Giussani C, Roux FE, Bello L, Lauwers-Cances V, Papagno C, Gaini SM et al (2009) Who is who: areas of the brain associated with recognizing and naming famous faces. J Neurosurg 110: 2898–299
[46] Giussani C, Pirillo D, Roux FE (2010) Mirror of the soul: a cortical stimulation study on recognition of facial emotions. J Neurosurg 112: 520–527
[47] Goldstein B, Armstrong CL, John C, Tallent EM (2003) Attention in adult intracranial tumors patients. J Clin Exp Neuropsychol 25: 66–78
[48] Hillis AE, Work M, Barker PB, Jacobs MA, Breese EL, Maurer K (2004) Re-examining the brain regions crucial for orchestrating speech articulation. Brain 127: 1479–1487
[49] Hornak J, Bramham J, Rolls ET, Morris RG, O'Doherty J, Bullock PR, Polkey CE (2003) Changes in emotion after circumscribed surgical lesions of the orbitofrontal and cingulate cortices. Brain 126: 1691–1712
[50] Hornak J, O'Doherty J, Bramham J, Rolls ET, Morris RG, Bullock PR, Polkey CE (2004) Reward-related reversal learning after surgical excisions in orbito-frontal or dorsolateral prefrontal cortex in humans. J Cogn Neurosci 16: 463–478
[51] Ilmberger J, Ruge M, Kreth FW, Briegel J, Reulen HJ, Tonn JC (2008) Intraoperative mapping of language functions: a longitudinal neurolinguistic analysis. J Neurosurg 109: 583–592
[52] Irle E, Peper M, Wowra B, Kunze S (1994) Mood changes after surgery for tumors of the cerebral cortex. Arch Neurol 51: 164–174
[53] Jabbour RA, Hempel A, Gates JR, Zhang W, Risse GL (2005) Right hemisphere language mapping in patients with bilateral language. Epilepsy Behav 6: 587–592
[54] Jackson HJ (1874) On the nature of the duality of the brain. Med Press Circ New Ser 17: 19–21, 41–44, 63–66; reprinted in Brain 38: 80–103, 1915
[55] Joanette Y, Ansaldo AI, Kahlaoui K, Côté H, Abusamra V, Ferreres A, Roch-Lecours A (2008) The impact of lesions in the right hemisphere on linguistic skills: theoretical and clinical perspectives. Rev Neurol 46: 481–488
[56] Josse G, Tzourio-Mazoyer N (2004) Hemispheric specialization for language. Brain Res Brain Res Rev 44: 1–12
[57] Keles GE, Lundin DA, Lamborn KR, Chang EF, Ojemann G, Berger MS (2004) Intraoperative subcortical stimulation mapping for hemispherical perirolandic gliomas located within or adjacent to the descending motor pathways: evaluation of morbidity and assessment of functional outcome in 294 patients. J Neurosurg 100: 369–375
[58] Kessels RP, Postma A, Kappelle LJ, de Haan EH (2000) Spatial memory impairment in patients after tumour resection: evidence for a double dissociation. J Neurol Neurosurg Psychiatry 69: 389–391
[59] Kircher TT, Brammer M, Tous AN, Williams SC, McGuire PK (2001) Engagement of right temporal cortex during processing of linguistic context. Neuropsychologia 39: 798–809
[60] Krainik A, Lehéricy S, Duffau H, Capelle L, Chainay H, Cornu P et al (2003) Postoperative speech disorder after medial frontal surgery: role of the supplementary motor area. Neurology 60: 587–594
[61] Lafargue G, Duffau H (2008) Awareness of intending to act following parietal cortex resection. Neuropsychologia 46: 2662–2667
[62] Lattner S, Meyer ME, Friederici AD (2005) Voice perception: sex, pitch and the right hemisphere. Hum Brain Mapp 24: 11--20
[63] Lee GP, Smith JR, Loring DW, Flanigin HF (1995) Intraoperative thermal inactivation of the hippocampus in an effort to prevent global amnesia after temporal lobectomy. Epilepsia 36: 892–898
[64] Litofsky NS, Resnick AG (2009) The relationships between depression and brain tumors. J Neurooncol 94: 153–161
[65] Lüders HO, Dinner DS, Morris HH, Wyllie E, Comair YG (1995) Cortical electrical stimulation in humans. The negative motor areas. Adv Neurol 67: 115–129
[66] Mainio A, Hakko H, Niemelä A, Koivukangas J, Räsänen P (2005) Depression and functional outcome in patients with brain tumors: a population-based 1-year follow-up study. J Neurosurg 103: 841–847
[67] Mainio A, Tuunanen S, Hakko H, Niemelä A, Koivukangas J, Räsänen P (2006) Decreased quality of life and depression as predictors for shorter survival among patients with low-grade gliomas: a follow-up from 1990 to 2003. Eur Arch Psychiatry Clin Neurosci 256: 516–521
[68] Milea D, Lobel E, Lehericy S, Duffau H, Rivaud-Péchoux S, Berthoz A, Pierrot-Deseilligny C (2002) Intraoperative frontal eye field stimulation elicits ocular deviation and saccade suppression. Neuroreport 13: 1359–1364
[69] Miller LA (1992) Impulsivity, risk-taking, and the ability to synthesize fragmented information after frontal lobectomy. Neuropsychologia 30: 69–79
[70] Miotto EC, Morris RG (1998) Virtual planning in patients with frontal lobe lesions. Cortex 34: 639–657
[71] Moritz-Gasser S, Duffau H (2009) Evidence of a large-scale network underlying language switching: a brain stimulation study. J Neurosurg 111: 729–732
[72] Moritz-Gasser S, Duffau H (2009) Cognitive processes and neural basis of language switching: proposal of a new model. Neuroreport 20: 1577–1580
[73] Neininger B, Pulvermüller F (2001) The right hemisphere's role in action word processing: a double case study. Neurocase 7: 303–317
[74] Neuloh G, Pechstein U, Schramm J (2007) Motor tract monitoring during insular glioma surgery. J Neurosurg 106: 582–592
[75] Ogar J, Willock S, Baldo J, Wilkins D, Ludy C, Dronkers N (2006) Clinical and anatomical correlates of apraxia of speech. Brain Lang 97: 343–350

[76] Ojemann GA (2003) The neurobiology of language and verbal memory: observations from awake neurosurgery. Int J Psychophysiol 48: 141-146

[77] Ojemann G, Ojemann J, Lettich E, Berger MS (1989) Cortical language localization in left, dominant hemisphere. An electrical stimulation mapping investigation in 117 patients. J Neurosurg 71: 316–326

[78] Owen AM, Downes JJ, Sahakian BJ, Polkey CE, Robbins TW (1990) Planning and spatial working memory following frontal lobe lesions in man. Neuropsychologia 28: 1021–1034

[79] Peper M, Irle E (1997) Categorical and dimensional decoding of emotional intonations in patients with focal brain lesions. Brain Lang 58: 233–264

[80] Perrine K, Devinsky O, Uysal S, Luciano DJ, Dogali M (1994) Left temporal neocortex mediation of verbal memory: evidence from functional mapping with cortical stimulation. Neurology 44: 1845–1850

[81] Petrides M (1997) Visuo-motor conditional associative learning after frontal and temporal lesions in the human brain. Neuropsychologia 35: 989–997

[82] Picht T, Kombos T, Gramm HJ, Brock M, Suess O (2006) Multimodal protocol for awake craniotomy in language cortex tumour surgery. Acta Neurochir (Wien) 148: 127–137

[83] Plaza M, Gatignol P, Cohen H, Berger B, Duffau H (2008) A discrete area within the left dorsolateral prefrontal cortex involved in visual-verbal incongruence judgment. Cereb Cortex 18: 1253–1259

[84] Pouratian N, Bookheimer SY (2010) The reliability of neuroanatomy as a predictor of eloquence: a review. Neurosurg Focus 28(2): E3

[85] Rasmussen T, Milner B (1977) The role of early left-brain injury in determining lateralization of cerebral speech functions. Ann N Y Acad Sci 299: 355–369

[86] Roux FE, Boetto S, Sacko O, Chollet F, Trémoulet M (2003) Writing, calculating, and finger recognition in the region of the angular gyrus: a cortical stimulation study of Gerstmann syndrome. J Neurosurg 99: 716–727

[87] Roux FE, Borsa S, Démonet JF (2009) "The mute who can sing": a cortical stimulation study on singing. J Neurosurg 110: 282–288

[88] Roux FE, Boukhatem L, Draper L, Sacko O, Démonet JF (2009) Cortical calculation localization using electrostimulation. J Neurosurg 110: 1291–1299

[89] Rowe AD, Bullock PR, Polkey CE, Morris RG (2001) "Theory of mind" impairments and their relationship to executive functioning following frontal lobe excisions. Brain 124: 600–616

[90] Russel SM, Elliott R, Forshaw D, Kelly PJ, Golfinos JG (2005) Resection of parietal lobe gliomas: incidence and evolution of neurological deficits in 28 consecutive patients correlated to the location and morphological characteristics of the tumor. J Neurosurg 103: 1010–1017

[91] Sanai N, Mirzadeh Z, Berger MS (2008) Functional outcome after language mapping for glioma resection. N Engl J Med 358: 18–27

[92] Scarone P, Gatignol P, Guillaume S, Denvil D, Capelle L, Duffau H (2009) Agraphia after awake surgery for brain tumor: new insights into the anatomo-functional network of writing. Surg Neurol 72: 223–241

[93] Serletis D, Bernstein M (2007) Prospective study of awake craniotomy used routinely and nonselectively for supratentorial tumors. J Neurosurg 107: 1–6

[94] Shinoura N, Yamada R, Kodama T, Suzuki Y, Takahashi M, Yagi K (2005) Preoperative fMRI, tractography and continuous task during awake surgery for maintenance of motor function following surgical resection of metastatic tumor spread to the primary motor area. Minim Invasive Neurosurg 48: 85–90

[95] Spena G, Gatignol P, Capelle L, Duffau H (2006) Superior longitudinal fasciculus subserves vestibular network in humans. Neuroreport 17: 1403–1406

[96] Spiers HJ, Burgess N, Maguire EA, Baxendale SA, Hartley T, Thompson PJ, O'Keefe J (2001) Unilateral temporal lobectomy patients show lateralized topographical and episodic memory deficits in a virtual town. Brain 124: 2476–2489

[97] Szaflarski JP, Binder JR, Possing ET, McKiernan KA, Ward BD, Hammeke TA (2002) Language lateralization in left-handed and ambidextrous people: fMRI data. Neurology 59: 238–244

[98] Taphoorn MJB, Klein M (2004) Cognitive deficits in adult patients with brain tumours. Lancet Neurol 3: 159–168

[99] Taylor KI, Crelier G, Alkadhi H, Brugger P, Kollias SS, Weniger D (2001) A fMRI study of hemispheric competency for lexical and semantic processes in reading. Neuroimage 13 (Suppl 1): 615

[100] Taylor KI, Regard M (2003) Language in the right cerebral hemisphere: contributions from reading studies. News Physiol Sci 18: 257–261

[101] Teixidor P, Gatignol P, Leroy M, Masuet-Aumatell C, Capelle L, Duffau H (2007) Assessment of verbal working memory before and after surgery for low-grade glioma. J Neurooncol 81: 305–313

[102] Thiebaut de Schotten M, Urbanski M, Duffau H, Volle E, Lévy R, Dubois B, Bartolomeo P (2005) Direct evidence for a parietal-frontal pathway subserving spatial awareness in humans. Science 309: 2226–2228

[103] Tong F (2003) Out-of-body experiences: from Penfield to present. Trends Cogn Sci 7: 104–106

[104] Tremblay T, Monetta L, Joanette Y (2004) Phonological processing of words in right- and left-handers. Brain Cogn 55: 427–432

[105] Tzourio-Mazoyer N, Josse G, Crivello F, Mazoyer B (2004) Interindividual variability in the hemispheric organizatin for speech. Neuroimage 21: 422–435

[106] Vassal M, Le Bars E, Moritz-Gasser S, Menjot N, Duffau H (2010) Crossed aphasia elicited by intraoperative cortical and subcortical stimulation in awake patients. J Neurosurg 113: 1251–1258

[107] Vendrell P, Junque C, Pujol J, Jurado MA, Molet J, Grafman J (1995) The role of prefrontal regions in the Stroop task. Neuropsychologia 33: 341–352

[108] Vigneau M, Beaucousin V, Herve PY, Duffau H, Crivello F, Houdé O et al (2006) Meta-analyzing left hemisphere language areas: phonology, semantics, and sentence processing. Neuroimage 30: 1414–1432

[109] Vigneau M, Beaucousin V, Hervé PY, Jobard G, Petit L, Crivello F et al (2011) What is right-hemisphere contribution to phonological, lexico-semantic, and sentence processing? Insights from a meta-analysis. Neuroimage 54: 577–593

[110] Volle E, Kinkingnéhun S, Pochon JB, Mondon K, Thiebaut de Schotten M, Seassau M et al (2008) The functional architecture of the left posterior and lateral prefrontal cortex in humans. Cereb Cortex 18: 2460–2469

[111] Weniger G, Irle E (2002) Impaired facial affect recognition and emotional changes in subjects with transmodal cortical lesions. Cereb Cortex 12: 258–268

[112] Wernicke C (1874) Der aphasische Symptomenkomplex: eine psychologische Studie auf anatomischer Basis. Max Cohn und Wegert, Breslau

[113] Whittle IR, Borthwick S, Haq N (2003) Brain dysfunction following "awake" craniotomy, brain mapping and resection of glioma. Br J Neurosurg 17: 130–137

[114] Whittle IR, Midgley S, Georges H, Pringle AM, Taylor R (2005) Patient perceptions of "awake" brain tumour surgery. Acta Neurochir (Wien) 147: 275–277

Brain hodotopy: new insights provided by intrasurgical mapping

Hugues Duffau

Introduction

The dilemma of cerebral surgery is to optimize the extent of resection while preserving brain function [18]. This is particularly challenging when the lesion is located within eloquent areas, which is quite frequently the case, for example, for low-grade gliomas [19]. Since anatomical landmarks are crucial but definitely not sufficient to understand the individual anatomo-functional organization, brain mapping methods should now be used in a systematic manner in both the perioperative and the intraoperative period. Functional cortical mapping is the first step, as regularly reported in the recent literature. However, although detection and preserveration of the axonal connectivity are also essential, the subcortical structures have yet received little attention.

The study of both cortical and subcortical organization is mandatory to avoid postsurgical permanent deficit. Indeed, lessons from stoke studies have thaught that a damage of the white-matter pathways generated a more severe neurological worsening than lesions of the cortex. By combining cortical function and axonal connectivity, an updated model of cerebral processing has recently been proposed, moving from a classical "localizationist" view to a "hodotopical" framework [4]. In pathology, according to this new concept, a topological mechanism (from Greek *topos*, place) refers to a dysfunction of the cortex (deficit, hyperfunction of a combination of the two), whereas a hodological mechanism (from Greek *hodos*, road or path) refers to a dysfunction related to connecting pathways (disconnection, hyperconnection, or a combination of the two) [5]. In other words, it is mandatory to take into account the complex functioning of a large-scale distributed cortico-subcortical network to understand both the physiology and the functional consequences of a lesion of this circuit, with possibly different deficits depending on the location and the extent of the damage (e.g., purely cortical or purely subcortical or both).

In contrast to extraoperative electrical mapping, intraoperative direct brain stimulation enables to map not only the cortex before any surgical resection but also the white-matter bundles. Such data are very important to tailor the resection according to functional boundaries and thus to optimize the benefit-to-risk ratio of the surgery. In addition, they provide new insights into brain's processing, with fundamental implications in the field of cognitive neurosciences. The aim is to review the new findings brought by intrasurgical cortical and subcortical electrical mapping, which, in combination with functional neuroimaging, open the door to a "connectionist" view of cerebral functioning [13, 15].

Intraoperative electrostimulation: new insights into the anatomo-functional cortical organization (topos)

Anatomo-functional organization of the supplementary motor area

The supplementary motor area (SMA) – the frontomesial area located in front of the primary motor area of the inferior limb – is involved in the planning of movement. Its resection induces the classical SMA syndrome. This syndrome is characterized by a complete akinesia and even mutism in cases of lesions of the left dominant SMA and occurs approximately thirty minutes after the end of the resection as observed in awake patients [22]. It suddenly and spontaneously resolves around the tenth day after surgery, even if some rehabilitation for 1 to 3 months is often needed in order to allow a better recovery. By preoperative fMRI, it has been shown that the occurrence of this syndrome was not related to the volume of the frontal resection but was directly related to the removal of a specific structure called the SMA proper, detectable on the preoperative imaging. Thus, on the basis of the presurgical fMRI, it is now possible to predict before surgery if an SMA syndrome will or will not occur postoperatively and to inform the patient and his family [41, 42]. Moreover, by coupling preoperative fMRI, the pattern of clinical deficit after surgery, and the extent of resection on the postoperative MRI, the existence of a somatotopy within the SMA proper has been demonstrated, namely (from anterior to posterior), the representation of language (at least in the dominant hemisphere), of the face, then the superior limb, and then the inferior limb (immediately in front of the paracentral lobule) [37]. As a consequence, it is also possible to predict before SMA resection the severity and the pattern of the postoperative transient deficit (e.g., only mutism, mutism and akinesia of the superior limb, or akinesia of the entire hemibody). This has an important impact on the planning of a specific rehabilitation.

Role of the insular lobe in language and swallowing

Although tumors, particularly low-grade gliomas, frequently involve the insular lobe, this structure has long been poorly studied for technical reasons. The insula is an anatomical, cytoarchitectonic, and functional interface between the allocortex and neocortex. Recent studies have enabled to better understand the implication of this multimodal lobe in many functions (for a recent review, see [7]), particularly for language. Indeed, preoperative fMRI has regularly showed an activation of the anterior insular cortex in the dominant hemisphere during language tasks as reported for healthy volunteers. Moreover, these results were confirmed by intraoperative electrical mapping (IEM), which induced language disorders and, more specifically, articulatory disturbances when applied on the insular cortex, supporting a role of this structure in the complex planning of speech [16,20, 21, 36] as previously suggested in stroke studies [9]. These data have important implications for the neurosurgeon, since for a left dominant (fronto-temporo-)insular lesion, resection carries a high risk to be incomplete. Moreover, following resection of gliomas involving the right nondominant insulo-opercular structures, the induction of a transient Foix-Chavany-Marie syndrome can be observed, that is, a bilateral facio-linguo-pharyngo-laryngal palsy with a reversible inability of the patient to speak and swallow [25].

Anatomo-functional organization of the left inferior frontal gyrus

IEM showed that the classical "Broca area" was not basically involved in speech production, but in high-level language processing (such as language switching [51, 52]), with its posterior part (pars opercularis) being more involved in phonological processing, its superior part (pars triangularis) implied in syntatic processing [60], and its anterior part (pars orbitaris) more involved in a large semantic network underlain by the inferior fronto-occipital fascicle [31] (see below). Interestingly, these data provided by IEM are in agreement with those obtained by fMRI, as shown in a recent meta-analysis of the literature [61].

Role of the left premotor cortex in language

Although many studies have allowed a better clarification of the implication of this structure in motor function, its participation in language remains poorly understood. Interestingly, it has demonstrated by IEM that stimulation of the dominant dorsal premotor area (namely, the structure lateral to the SMA, in front of the primary motor area of the hand) induced anomia. On the other hand, stimulation of the dominant ventral premotor cortex regularly elicited anarthria [27]. These results give strong arguments in favor of (i) the involvement of the dorsal premotor cortex in the naming network, in accordance with fMRI studies which have suggested that this region could participate to lexical retrieval and that its engagement might be related to conceptual category, and (ii) the involvement of the ventral premotor cortex in the planification of articulation, explaining why lesion studies have reported that damage of the "lower motor cortex" induced speech apraxia (i.e., aphemia).

Role of the left supramarginalis gyrus in speech and language

In epilepsy surgery, IEM showed that the left supramarinal gyrus was involved in picture naming [53]. More recently, it was demonstrated that stimulation of this structure can also induce speech apraxia [26]. These findings are in agreement with tractography studies which propose that the supramarginal gyrus is a relay between frontal and temporal language sites – the so-called Geschwind territory within the "indirect pathway" [5, 6]. Interestingly, in his model of working memory, Baddeley [1] described an articulatory loop consisting of two parts: a short-term phonological store and an articulatory rehearsal component that can revitalize memorized information. Functional neuroimaging studies suggest that the phonological store involved the left supramarginal cortex, while the subvocal rehearsal system was associated to the left inferior frontal cortex and the ventral premotor cortex [54]. Such knowledge is important during surgery within the supramarginal gyrus and the fronto-parietal loop, in order to avoid postoperative verbal working memory disorders which are still very frequent [10, 57].

Anatomo-functional organization of Wernicke'sar ea

For lesions located in the dominant temporal posterior areas, tasks adapted to test comprehension during IEM have been developped. For instance, a triad of pictures is shown and the patient is asked to pair them by naming two pictures with conceptual links, e.g., a pyramid-and-palm tree test. Interestingly, during stimulation, several sites within the posterior part of the superior temporal gyrus specifically elicited an anomia without comprehension disorders, although other sites within the same gyrus elicited only comprehension disorders with preservation of the ability to name, and other areas generated only phonological disturbances [38]. These results support the complexity of the functional organization of Wernicke's area (in accordance with fMRI results), with its participation, but also with possible dissociation, between comprehension, naming, and phonological processing [61]. In addition, it was recently shown by IEM that the posterior part of the middle temporal gyrus in the left dominant hemisphere was involved in syntactic processing, by eliciting errors of grammatical gender [60]. Furthermore, the posterior part of the superior temporal gyrus seems also to participate in language switching [51, 52].

Role of the right hemisphere in language

This point is extensively discussed in the chapter "Indications of awake mapping and selection of intraoperative tasks".

Role of angular gyrus in calculation

The angular gyrus in the left dominant hemisphere is known to participate in complex cognitive functions, such as calculation. In patients with a left posterior parietal lesion, both multiplication and subtraction can be tested by IEM. Interestingly, functional epicenters more involved in arithmetic facts such as rote multiplication, with tables learned by heart, were found to be located immediately above the posterior end of the sylvian fissure, thus very close to the language sites. On the other hand, actual calculation such as subtraction recruited functional sites located in the superior part of the angular gyrus immediately below the intraparietal sulcus, namely, close to the areas involved in working memory. These results support the existence of a "calculotopy" within the angular gyrus. There was a transient dyscalculia following surgery, but the patients recovered. In addition, these results helped corroborate the "triple code theory" [24].

Involvement of frontal eye field and cingulate eye field in oculumotor behavior

The functional anatomy of the frontal eye field was studied both by preoperative fMRI and by IEM. This region, located laterally and in front of the primary motor area of the face, is implied in the regulation of the voluntary and unvoluntary ocular saccades. Indeed, IEM over this area evoked contraversive smooth eye movements recorded electro-oculographically. In addition, stimulation of an anterior subregion of this electrically determined frontal eye field both disclosed smooth eye movement and interfered with oculomotor behavior, supressing self-paced saccades in awake patient [49]. It is worth noting that the posterior part of the anterior cingulum, namely, the cingulate eye field, also plays a role in suppression of unwanted saccades (antisaccades), thus in attentional processing [50].

Role of the right supramarginal gyrus and posterior temporal areas in spatial awareness

The use of a line bisection task during awake surgery in patients with a lesion involving the right parieto-temporal junction enables the mapping of the areas involved in spatial awareness. A significant rightward deviation is usually observed during the stimulation of the antero-inferior part of the supramarginal gyrus and the caudal part of the superior temporal gyrus [2]. In other words, a transient and reproducible left neglect is induced by electrical inactivation of cortical sites essential for the visuospatial integration, over the right parieto-temporal junction. If these eloquent areas are preserved, the patients show no signs of neglect a few days after surgery. These findings demonstrate that the supramarginal gyrus and the caudal part of the superior temporal gyrus, at least in the right hemisphere, are critical for the symmetrical processing of the visual scene in humans [58].

Role of the left prefrontal dorsolateral cortex in judgment

For lesions located within the left dominant prefrontal cortex, a task of crossmodal (visual-verbal) congruent and incongruent judgment has been performed in awake patient. Visual and auditory stimuli were presented simultaneously, referring to either the same item (congruence condition) or to different items (semantically or phonologically incongruent condition). It was shown that stimulation of brain areas not involved in naming processing elicited reproducible deficit of incongruent judgment, especially at the level of the left dorsolateral prefrontal region, even though an interindividual variability was observed, as for other functions [55]. Preservation of such executive functions is essential for the daily life, in particular regarding the decision-making and planning of complex strategy.

Interestingly, other anatomo-functional correlations can also be made by IEM in awake patients, in particular with regard to writing, reading, memory, emotional processing, or even control of micturition. Thefore, the neurosurgeon must adapt his strategy, particularly the surgical technique (e.g., the selection of the functional tasks to optimize the reliability of the intrasurgical mapping), so as to apply the better knowledge of the functional anatomy to the individual patient.

Intrasurgical stimulation: a new door to the axonal connectivity (hodos)

Beyond cortical mapping, the study of individual anatomo-functional connectivity underlying the eloquent networks is mandatory in brain surgery, in order to avoid postoperative permanent neurological deficit [3, 12, 14, 17].

Motor pathways

For precentral lesions, after detection and preservation of the primary motor cortical areas by IEM, it is also important to detect by subcortical stimulation the corresponding descending motor pathways and their somatotopy, i.e., the different fibers of the corona radiata, with the pyramidal bundles of the lower limb medially, of the upper limb, and of the face more laterally. As at the cortical level, these subcortical motor fibers constitute the posterior and deep functional limits of the resection, until the opening of the ventricle. The pyramidal pathways may also be identified within the posterior limb of the internal capsule, particularly for (fronto-temporo-)insular tumors, in which the deep boundaries of the resection are given when subcortical stimulation induces motor responses in the inferior part of the corona radiata up to the superior part of the mesencephalic peduncles [11, 16, 28].

Somatosensory thalamo-cortical pathways

In the same way, the thalamo-cortical somatosensory pathways and their somatotopy can be identified by IEM, which induces dysesthesias in awake patients, in cases of retrocentral lesions [19].

Visual pathways

Optic radiations can be mapped in patients who undergo awake surgery for temporo-occipito-parietal lesions by the induction of a transient "shadow" (negative effect) or phosphenes (positive effect) in the controlateral visual field during stimulation of the postero-superior and deep part of the surgical cavity, sometimes also with metamorphopsia (i.e., visual illusion) [30]. Thus, if resection is stopped at this level, patients are left with only a residual quadrantanopsia without consequences on the quality of life, especially for their driving.

Language pathways: the anatomo-functional connectivity of language revisited

For left dominant precentral lesions, after identification of the motor and language cortical sites within the prerolandic and inferior frontal gyri (the so-called Broca area), IEM also enables the detection of the language pathways [34]. Medially, IEM can identify the fasciculus subcallosal medialis (running from the SMA and cingulate gyrus to the head of the caudate nucleus), whose stimulation elicits transient transcortical motor aphasia. This tract is involved in the initiation of language [23]. Posteriorly, the fibers coming from the premotor ventral cortex must be detected by a stimulation inducing anarthria. This pathway is crucial for speech production [27]. More laterally, the operculo-insular connections should also to be detected by generating a complete speech arrest during stimulation. These connections are involved in speech planning [36].

In addition to these loco-regional language pathways, subcortical IEM also detect the long-distance association pathways, with first of all, the deep part of the superior longitudinal fascicle – namely the arcuate fascicle (AF) [34] (Fig. 1). In patients with a lesion involving the left insula or the left inferior frontal gyrus, IEM can identify the anterior part of AF, located

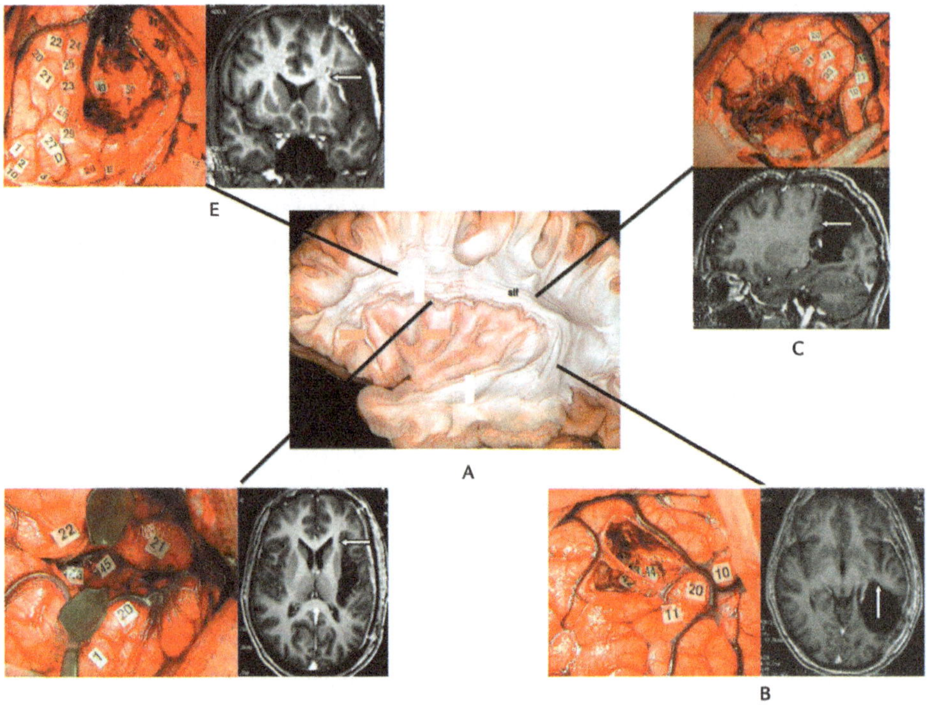

Fig. 1. Arcuate fascicle: dorsal phonological stream. **(A)** Anatomical trajectory of the white matter of the superior longitudinal fascicle (*slf*) studied by dissection. **(B–E)** Surgical field and postsurgical MRI from different patients operated on for a low-grade glioma at various brain locations: temporal **(B)**, parietal **(C)**, insular **(D)**, frontal **(E)**. In all cases, the deep functional boundary of the resection was given by a part of the arcuate fascicle identified by subcortical mapping. Electrostimulation of this tract systematically induced phonemic paraphasia. The precise locations where these language disorders were elicited were marked intraoperatively by number tags in the depth of the cavity. These sites are shown by an arrow on the postoperative anatomical imaging (Reproduced from [34])

within the anterior floor of the external capsule (under the superior part of the insula) and also under the posterior part of Broca's area (namely, the pars opercularis and pars triangularis of the inferior frontal gyrus). Stimulation induces transient symptoms observed in conduction aphasia, i.e., phonemic paraphasia and repetition disorders. In the same way, the AF must also be detected at the level of its postero-superior loop, located under the supramarginal gyrus, in patients operated on for a left parietal lesion. The same symptoms associating phonemic paraphasias and repetition disorders are induced by stimulation. Again, the AF is detected for posterior temporal lesions, the posterior part of its posterior funiculus corresponding to the anterior functional limit of the resection. Finally, the anterior part of the anterior funiculus of the AF must also been used as the posterior functional boundary of left dominant anterior and mid-temporal lobectomy [23]. Interestingly, the left AF seems also to subserve a wide network involved in language switching (from a native language to another language or vice versa): IEM can disrupt such function, crucial to detect and to preserve in bilingual patients [51, 52]. More recently, grammatical gender errors were elicited by axonal stimulation of the left AF, supporting the possible role of this pathway (connecting the middle temporal gyrus and the inferior frontal gyrus, structures whose stimulation induced the same grammatical disturbances) in syntactic processing [60].

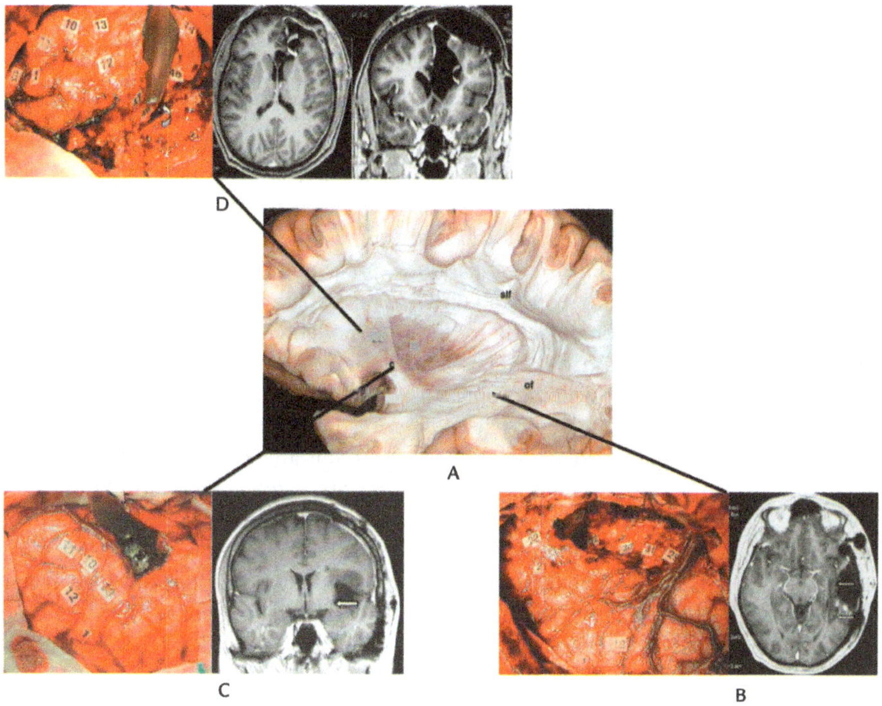

Fig. 2. Inferior fronto-occipital fascicle: ventral semantic stream. **(A)** Anatomical trajectory of the white-matter bundle of the inferior fronto-occipital fascicle (*of*) studied by dissection. **(B–D)** Surgical field and postsurgical MRI from different patients operated on for a low-grade glioma at various brain locations: temporal **(B)**, insular **(C)**, frontal **(D)**. In all cases, the deep functional boundary of the resection was given by a part of the IFOF identified by subcortical mapping. Electrostimulation of this tract systematically induced semantic paraphasia. The precise locations where these language disorders were induced were marked intraoperatively by number tags in the depth of the cavity. These sites are shown by an arrow on the postoperative anatomical imaging (Reproduced from [34])

In addition to the AF, there is a lateral part of the superior longitudinal fascicle. In patients harboring a left retrocentral suprasylvian lesion, IEM detects not only the language cortical sites at the level both of the ventral premotor cortex in front of the tumor and of the supramarginal gyrus and/or angular gyrus behind it but also a fronto-parietal subcortical network whose stimulation induces speech apraxia [26]. This operculo-opercular loop might underly the anatomo-functional connectivity of the working memory circuit. Indeed, this loop corresponds to the anterior segment of an indirect pathway of the classical superior longitudinal fascicle, which runs parallel and lateral to the AF, by connecting Broca's territory with Geschwind's territory in the inferior parietal lobe as recently shown by tractography [6]. This tract might be involved in the vocalization of semantic content. Thefore, this example illustrates well that IEM and diffusion tensor imaging can be combined in order to better understand the anatomo-functional connectivity of the brain [15, 33].

Parallel to this "dorsal phonological root", IEM supported the likely role of the inferior fronto-occipital fascicle (IFOF) in the semantic system, the "ventral semantic root" [31] (Fig. 2). In patients with a frontal lesion immediately in front and above Broca's area, i.e., within the pars orbitaris of the left inferior frontal gyrus and the dorsolateral prefrontal area, the anterior part of the IFOF has been identified under these regions by eliciting semantic paraphasias by subcortical stimulation. In the same way, the IFOF was detected in surgery for left insular lesions by stimulation in its intermediate part located in the anterior floor of the internal capsule (in front and inferior to the AF and behind and superior to the uncinate fascicle) inducing the same symptoms (semantic paraphasias). Again, the IFOF was detected for left temporal lesions by stimulation eliciting semantic disorders; it constituted the deep limit of the resection (above the roof of the temporal horn of the ventricle) [31].

Interestingly, stimulation of the anterior part of the inferior longitudinal fascicle, in front of the visual word form area (i.e., the basal part of the temporo-occipital junction, involved in high-level visual processing such as reading) [43], as well as stimulation of the uncinate fascicle [35], never generated language disturbances. In the same way, stimulation of the anterior part of the middle longitudinal fascicle (i.e., a pathway connecting the angular gyrus to the temporal pole and running under the superior temporal sulcus) never elicited language disorders [8]. Thus, these fasciculi can be removed without risk of aphasia. It seems that this indirect pathway from the temporo-occipital areas to the prefrontal region, with a relay in the temporal pole (temporo-occipital area, inferior longitudinal fascicle, temporal pole, uncinate fascicle, orbito-frontal and prefrontal areas) might be compensated by the direct pathway constituted by the IFOF [34]. It is nonetheless worth noting that the posterior part of the inferior longitudinal fascicle should be preserved, because it plays a crucial role in reading, as demonstrated by IEM which elicited reproducible visual paraphasia and dyslexia during stimulation [44].

Beyond the stimulation of the white matter, IEM also allows the mapping of the deep gray nuclei, sometimes invaded by tumors such as (low-grade) gliomas. Indeed, stimulation of the head of the dominant caudate in patients with a frontomesial lesion coming in contact with the striatum in the depth generally gener-

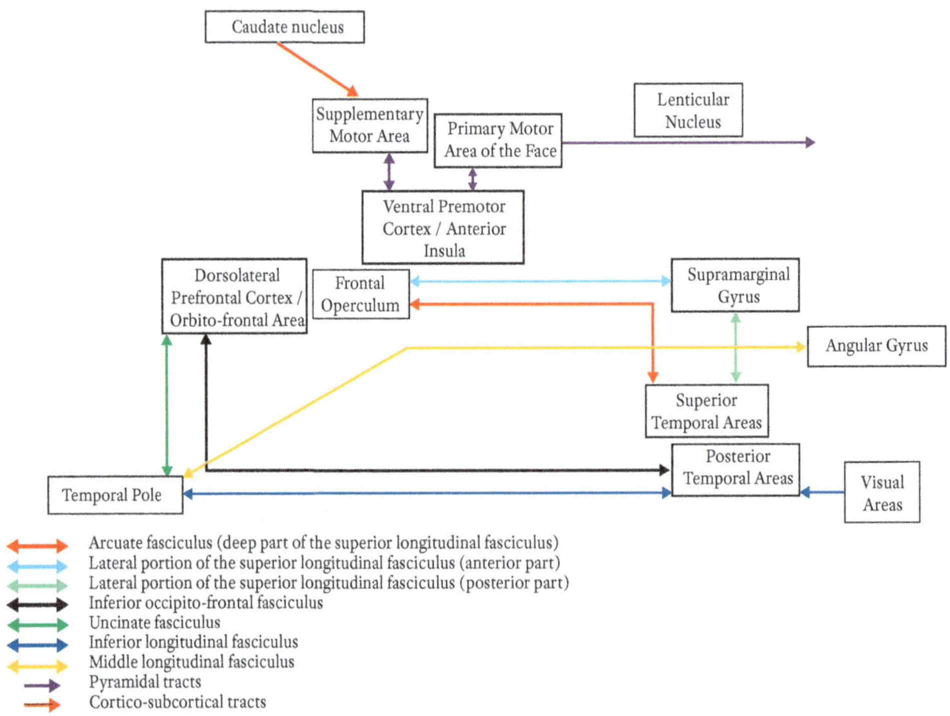

Fig. 3. Scheme of subcortical language pathways

ates perseverations, namely, the repetition of the previous item while the next item is presented to the patient. These results support an inhibitory role of the caudate in the control of cognition [40]. Equally, it is important to map the lateral part of the dominant lentiform nucleus, at the end of the resection of an insular glioma [40]. Lentiform stimulation induces anarthria, supporting the likely role of this structure in the planning of articulation, in association with the insula and ventral premotor cortex [34].

Finally, it is also important to use IEM for language mapping, at both cortical and subcortical levels, for lesions involving the right hemisphere in left-handed and ambidextrous patients [32], or even in some atypical right-handers [59], due to a possible bilateral distribution of language networks, generally, with a mirror organization of both hemispheres. In all cases, these language bundles should constitute the subcortical functional limits of the resection.

Pathways subserving spatial awareness

Using a task of line bisection during awake surgery for patients harboring a lesion within the right parieto-temporal junction (as previously described at the cortical level), IEM must also detect the white-matter tracts implied in spatial processing, in order to avoid postoperative left neglect. During the stimulation of the part II of the superior longitudinal fascicle, a significant rightward deviation is regularly observed [58]. As a consequence, it seems that this parieto-frontal pathway subserves spatial awareness and that a lesion at its level may generate a permanent left neglect.

Stimulation of the right superior longitudinal facsiculus may also induce vertigo, by disrupting a large network between the parieto-insular vestibular cortex, the visual and the sensory-motor areas [56].

These results suggest that damage to restricted regions of white matter can cause dysfunctioning of large-scale cognitive networks. Also, these data show that it is possible to adapt the intraoperative testing to each patient with the goal to map the subcortical pathway underlying cognitive functions other than language. Interestingly, although IEM of the interhemispheric white-matter pathways has been performed, no functional responses were elicited by the stimulation of the corpus callosum. Such results have allowed resection of lesions involving this structure without any consequence on the quality of life, whatever the location of the "callosectomy" [29].

Conclusions and perspectives

In summary, the vision of the neural basis of cognition begins to shift from the earlier localisationist and later associationist view towards the concept of a "hodotopical" organization (i.e., dynamic parallel large-scale networks able to compensate one another). Indeed, from Lichtheim to Geschwind [39], cognitive functions such as language were conceived in associationist terms of centers and pathways, the general assumption being that visual and auditory linguistic information were processed in localized cortical regions with a serial passage of information between regions through white-matter tracts. Presently, an alternative hodotopical account is proposed, in which language is conceived as resulting from parallel distributed processing performed by distributed groups of connected neurons rather that individual centers [15]. In contrast to the serial model of language, in which one process must be finished before another level of processing can be reached by the information, the new models of "independent networks" state that different processes can be performed simultaneously with interactive feedbacks. Interestingly, the recent methodological advances in tractography and intraoperative cortico-subcortical electrical mapping have enabled to study directly in vivo in humans the anatomo-functional connectivity that underlies cognitive functions, supporting and completing Mesulam's large-scale neural network model of language [48]. In particular, it seems that there are at least two parallel pathways, namely, the dorsal phonological stream and the ventral semantic stream, which converge into a common final tract allowing speech production (Fig. 3). Furthermore, this whole network is modulated by cortico-striato-pallido-thalamo-cortical loops. Of course, it is worth noting that the goal of this new concept is not to substitute the cortical centers (topology) with subcortical pathways (hodology) but rather to envision the common interactive processing of both grey and white matters (hodotopy). The next step to progress in the understanding of the brain connectivity might be a more accurate analysis of the interactions between the language circuit and the networks underlying the other cognitive functions, in particular, the visuospatial component, in which the role of the superior longitudinal fascicle has been emphasized, and the emotional and behavioral aspects. Such a multimodal approach seems to represent a unique opportunity to move towards an integrative model of the various functions. In this way, the

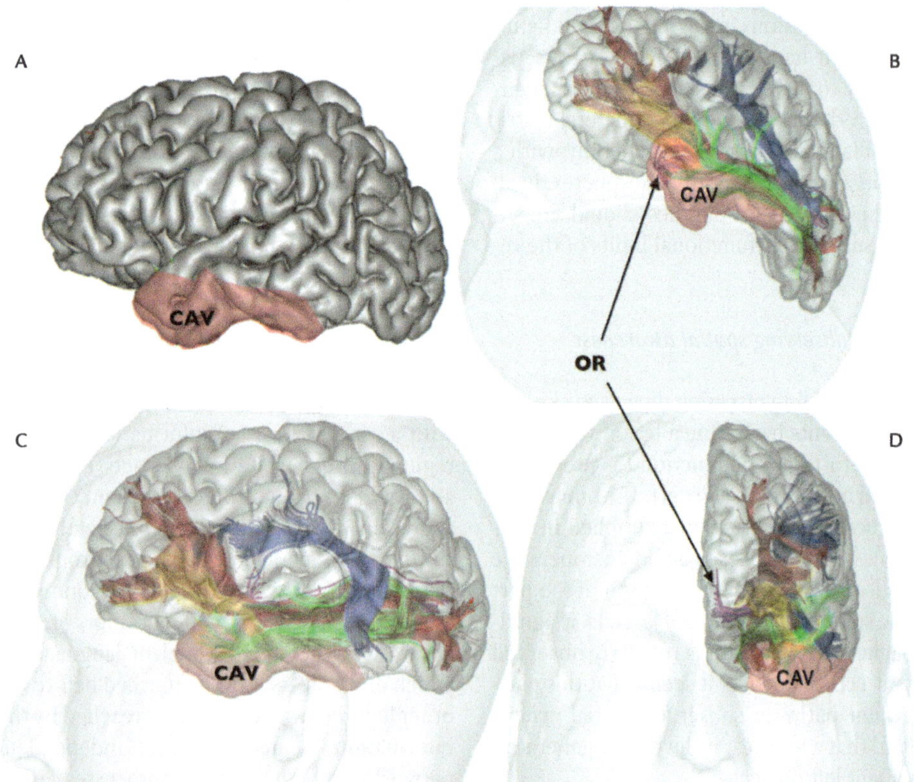

Fig. 4. (A–D). Fiber tracking of the superior longitudinal fascicle (*blue*), inferior longitudinal fascicle (*green*), inferior fronto-occipital fascicle (*red*), uncinate (*yellow*), and optic radiation (*OR*) was performed using regions of interest. A "two-region of interest" approach was used for each fascicle tracking. The procedure consisted in defining a second region of interest at such a distance from the first region of interest that it contained at least a section of the desired fascicle but did not contain any fibers of the undesired fascicle that passed through the first region. Diffusion tensor images and high-resolution three-dimensional anatomical images (**B–D**) were registered by Brainvisa 3.0.2. The derived tracts were displayed by Anatomist 3.0.2 (http://brainvisa.info). We have drawn a virtual resection cavity (*CAV*) according to essential subcortical pathways (IFOF and AF), while removing the "nonessential" tracts (uncinate, inferior longitudinal fascicle and anterior part of the optic radiation). By reporting this cavity on the three-dimensional surface reconstruction (**A**), we have obtained a resection according to cortical boundaries, similar to those classically reported in the literature (Reproduced from [33])

recent advances in biomathematical modelization of the electrophysiological and hemodynamic signals, which allow a reliable study of the activity time course within the neuronal networks via the analysis of the synchrony (the so-called chronoarchitecture), may open a new door to the effective connectivity, i.e., the influence that one neural system exerts on another.

Consequently, beyond the fundamental interest, it is also crucial for the neurosurgeon to improve her or his knowledge of the anatomo-functional connectivity in order to integrate more easily and more systematically the concept of subcortical mapping in the surgical strategy: first, because gliomas by their nature involve both cortical and subcortical structures and thus they may alter the connectivity; second, because lesions of the white matter may elicit more severe permanent deficits than do cortical damages. To this end, recent insights given by tractography are very useful (Fig. 4). In addition, new anatomic dissections of the white-matter pathways are now to be performed in the light of data provided by axonal mapping, especially with regard to the cortical terminations of the subcortical pathways still poorly known [46, 47]. Indeed, all neurocognitive models should take into account anatomic constraints, an essential point for their validation.

In conclusion, direct electrostimulation of a cerebral region (cortical or subcortical) does not correspond to the stimulation (i.e., a transient virtual lesion) of a "discrete site" but actually corresponds to the stimulation of an input gate to a complex network and thus allows the study of a large-scale distributed circuit in a connectionist framework [45]. In addition, such hodotopical view may explain why some epicenters considered as essential for language in a localisationist model, for instance, Broca's area, in certain conditions can be involved by a tumor (or even surgically removed) with no aphasia, because of a functional compensation within a large distributed network, i.e., the so-called brain plasticity.

References

[1] Baddeley A (1992) Working memory. Science 255: 556–559

[2] Bartolomeo P, Thiebaut de Schotten M, Duffau H (2007) Mapping of visuospatial functions during brain surgery: a new tool to prevent unilateral spatial neglect. Neurosurgery 61: E1340, DOI 10.1227/01.neu.0000306126.46657.79

[3] Bello L, Gallucci M, Fava M, Carraba G, Giussani C, Acerbi F et al (2007) Intraoperative subcortical language tract mapping guides surgical removal of gliomas involving speech areas. Neurosurgery 60: 67–80

[4] Catani M (2007) From hodology to function. Brain 130: 602–605

[5] Catani M, ffytche DH (2005) The rises and falls of disconnection syndromes. Brain 128: 2224–2239

[6] Catani M, Mesulam M (2008) What is a disconnection syndrome? Cortex 44: 911-913

[7] Craig AD (2009) How do you feel – now? The anterior insula and human awareness. Nat Rev Neurosci 10: 59–70

[8] De Witt Hamer P, Moritz-Gasser S, Gatignol P, Duffau H (2010) Is the human left middle longitudinal fascicle essential for language? A brain electrostimulation study. Hum Brain Mapp, DOI 10.1002/hbm.21082

[9] Dronkers NF (196) A new region for coordinating speech articulation. Nature 384: 159–161

[10] du Boisgueheneuc F, Levy R, Volle E, Seassau M, Duffau H, Kinkingnehun S, Samson Y, Zhang S, Dubois B (2006) Functions of the left superior frontal gyrus in humans: a lesion study. Brain 129: 3315–3328

[11] Duffau H (2000) Intraoperative direct subcortical stimulation for identification of the internal capsule, combined with an image-guided stereotactic system during surgery for basal ganglia lesions. Surg Neurol 53: 250–254

[12] Duffau H (2005) Intraoperative cortico-subcortical stimulations in surgery of low-grade gliomas. Expert Rev Neurother 5: 473–485

[13] Duffau H (2006) New concepts in surgery of WHO grade II gliomas: functional brain mapping, connectionism and plasticity. A review. J Neurooncol 79: 77–115

[14] Duffau H (2007) Contribution of cortical and subcortical electrostimulation in brain glioma surgery: methodological and functional considerations. Neurophysiol Clin 37: 373–382

[15] Duffau H (2008) The anatomo-functional connectivity of language revisited: new insights provided by electrostimulation and tractography. Neuropsychologia 4: 927–934

[16] Duffau H (2009) A personal consecutive series of surgically treated 51 cases of insular WHO grade II glioma: advances and limitations. J Neurosurg 110: 696–708

[17] Duffau H (2009) Does post-lesional subcortical plasticity exist in the human brain? Neurosci Res 65: 131–135

[18] Duffau H (2010) Surgery of gliomas in eloquent areas: from brain hodotopy and plasticity to functional neurooncology. Neurosurg Focus 28(2): Intro

[19] Duffau H, Capelle L (2004) Preferential brain locations of low-grade gliomas. Cancer 100: 2622–2626

[20] Duffau H, Fontaine D (2005) Successful resection of a left insular cavernous angioma using neuronavigation and intraoperative language mapping. Acta Neurochir (Wien) 147: 205–208

[21] Duffau H, Capelle L, Lopes M, Faillot T, Sichez JP, Fohanno D (2000) The insular lobe: physiopathological and surgical considerations. Neurosurgery 47: 801–811

[22] Duffau H, Lopes M, Denvil D, Capelle L (2001) Delayed onset of the supplementary motor area syndrome after surgical resection of the mesial frontal lobe: a time course study using intraoperative mapping in an awake patient. Stereotact Funct Neurosurg 76: 74–82

[23] Duffau H, Capelle L, Sichez N, Denvil D, Lopes M, Sichez JP et al (2002) Intraoperative mapping of the subcortical language pathways using direct stimulations. An anatomo-functional study. Brain 125: 199–214

[24] Duffau H, Denvil D, Lopes M, Gasparini F, Cohen L, Capelle L et al (2002) Intraoperative mapping of the cortical areas involved in multiplication and subtraction: an electrostimulation study in a patient with a left parietal glioma. J Neurol Neurosurg Psychiatry 73: 733–738

[25] Duffau H, Karachi C, Gatignol P, Capelle L (2003) Transient Foix-Chavany-Marie syndrome after surgical resection of a right insulo-opercular low-grade glioma. Neurosurgery 53: 426–431

[26] Duffau H, Gatignol P, Denvil D, Lopes M, Capelle L (2003) The articulatory loop: study of the subcortical connectivity by electrostimulation. Neuroreport 14: 2005–2008

[27] Duffau H, Capelle L, Denvil D, Gatignol P, Sichez N, Lopes M et al (2003) The role of dominant premotor cortex in language: a study using intraoperative functional mapping in awake patients. Neuroimage 20: 1903–1914

[28] Duffau H, Capelle L, Denvil D, Sichez N, Gatignol P, Taillandier L et al (2003) Usefulness of intraoperative electrical subcortical mapping during surgery for low-grade gliomas located within eloquent brain regions: functional results in a consecutive series of 103 patients. J Neurosurg 98: 764–778

[29] Duffau H, Khalil I, Gatignol P, Denvil D, Capelle L (2004) Surgical removal of corpus callosum infiltrated by low-grade glioma: functional outcome and oncological considerations. J Neurosurg 100: 431–437

[30] Duffau H, Velut S, Mitchell MC, Gatignol P, Capelle L (2004) Intra-operative maping of the subcortical visual pathways using direct electrical stimulations. Acta Neurochir (Wien) 146: 265–269

[31] Duffau H, Gatignol P, Mandonnet E, Peruzzi P, Tzourio-Mazoyer N, Capelle L (2005) New insights into the anatomo-functional connectivity of the semantic system: a study using cortico-subcortical electrostimulations. Brain 128: 797–810

[32] Duffau H, Leroy M, Gatignol P (2008) Cortico-subcortical organization of language networks in the right hemisphere: an electrostimulation study in left-handers. Neuropsychologia 46: 3197–3209

[33] Duffau H, Thiebaut de Schotten M, Mandonnet E (2008) White matter functional connectivity as an additional landmark for dominant temporal lobectomy. J Neurol Neurosurg Psychiatry 79: 492–495

[34] Duffau H, Gatignol P, Mandonnet E, Capelle L, Taillandier L (2008) Contribution of intraoperative subcortical stimulation mapping of language pathways: a consecutive series of 115 patients operated on for a WHO grade II glioma in the left dominant hemisphere. J Neurosurg 109: 461–471

[35] Duffau H, Gatignol P, Moritz-Gasser S, Mandonnet E (2009) Is the left uncinate fasciculus essential for language? A cerebral stimulation study. J Neurol 256: 382–389

[36] Duffau H, Moritz-Gasser S, Gatignol P (2009) Functional outcome after language mapping for insular World Health Organization grade II gliomas in the dominant hemisphere: experience with 24 patients. Neurosurg Focus 27(2): E7

[37] Fontaine D, Capelle L, Duffau H (2002) Somatotopy of the supplementary motor area: evidence from correlation of the extent of surgical resection with the clinical patterns of deficit. Neurosurgery 50: 297–303

[38] Gatignol P, Capelle L, Le Bihan R, Duffau H (2004) Double dissociation between picture naming and comprehension: an electrostimulation study. Neuroreport 15: 191–195

[39] Geschwind N (1970) The organization of language and the brain. Science 170: 940–944

[40] Gil Robles S, Gatignol P, Capelle L, Mitchell MC, Duffau H (2005) The role of dominant striatum in language: a study using intraoperative electrical stimulations. J Neurol Neurosurg Psychiatry 76: 940–946

[41] Krainik A, Lehéricy S, Duffau H, Vlaicu M, Poupon F, Capelle L et al (2001) Role of the supplementary motor area in motor deficit following medial frontal lobe surgery. Neurology 57: 871–878

[42] Krainik A, Lehéricy S, Duffau H, Capelle L, Chainay H, Cornu P et al (2003) Postoperative speech disorder after medial frontal surgery: role of the supplementary motor area. Neurology 60: 587–594

[43] Mandonnet E, Nouet A, Gatignol P, Capelle L, Duffau H (2007) Does the left inferior longitudinal fasciculus play a role in language? A brain stimulation study. Brain 130: 623–629

[44] Mandonnet E, Gatignol P, Duffau H (2009) Evidence for an occipito-temporal tract underlying visual recogni-

tion in picture naming. Clin Neurol Neurosurg 111: 601–605

[45] Mandonnet E, Winkler PA, Duffau H (2010) Direct electrical stimulation as an input gate into brain functional networks: principles, advantages and limitations. Acta Neurochir (Wien) 152: 185–193

[46] Martino J, Vergani F, Gil Robles S, Duffau H (2010) New insights into the anatomic dissection of the temporal stem with special emphasis on the inferior fronto-occipital fasciculus: implications in surgical approach to left mesiotemporal and temporoinsular structures. Neurosurgery 66: 4–12

[47] Martino J, Brogna C, Gil Robles S, Vergani F, Duffau H (2010) Anatomic dissection of the inferior fronto-occipital fasciculus revisited in the lights of brain stimulation data. Cortex 46: 691–699

[48] Mesulam MM (1990) Large-scale neurocognitive networks and distributed processing for attention, language, and memory. Ann Neurol 28: 597–613

[49] Milea D, Lobel E, Lehericy S, Duffau H, Rivaud-Péchoux S, Berthoz A et al (2002) Intraoperative frontal eye field stimulation elicits ocular deviation and saccade suppression. Neuroreport 13: 1359–1364

[50] Milea D, Lehéricy S, Rivaud-Péchoux S, Duffau H, Lobel E, Capelle L et al (2003) Antisaccade deficit after anterior cingulate cortex resection. Neuroreport 14: 283–287

[51] Moritz-Gasser S, Duffau H (2009) Evidence of a large-scale network underlying language switching: a brain stimulation study. J Neurosurg 111: 729–732

[52] Moritz-Gasser S, Duffau H (2009) Cognitive processes and neural basis of language switching: proposal of a new model. Neuroreport 20: 1577–1580

[53] Ojemann G, Ojemann J, Lettich E, Berger MS (1989) Cortical language localization in left, dominant hemisphere. An electrical stimulation mapping investigation in 117 patients. J Neurosurg 71: 316–326

[54] Paulesu E, Frith CD, Frackowiak RS (1993) The neural correlates of the verbal component of working memory. Nature 362: 342–345

[55] Plaza M, Gatignol P, Cohen H, Berger B, Duffau H (2008) A discrete area within the left dorsolateral prefrontal cortex involved in visual-verbal incongruence judgment. Cereb Cortex 18: 1253–1259

[56] Spena G, Gatignol P, Capelle L, Duffau H (2006) Superior longitudinal fasciculus subserves vestibular network in humans. Neuroreport 17: 1403–1406

[57] Teixidor P, Gatignol P, Leroy M, Masuet-Aumatell C, Capelle L, Duffau H (2007) Assessment of verbal working memory before and after surgery for low-grade glioma. J Neurooncol 81: 305–313

[58] Thiebaut de Schotten M, Urbanski M, Duffau H, Volle E, Lévy R, Dubois B, Bartolomeo P (2005) Direct evidence for a parietal-frontal pathway subserving spatial awareness in humans. Science 309: 2226–2228

[59] Vassal M, Le Bars E, Moritz-Gasser S, Menjot N, Duffau H (2010) Crossed aphasia elicited by intraoperative cortical and subcortical stimulation in awake patients. J Neurosurg 113: 1251–1258

[60] Vidorreta JG, Garcia R, Moritz-Gasser S, Duffau H (2011) Double dissociation between syntactic gender and picture naming processing: a brain stimulation mapping study. Hum Brain Mapp 32: 331–340

[61] Vigneau M, Beaucousin V, Herve PY, Duffau H, Crivello F, Houdé O et al (2006) Meta-analyzing left hemisphere language areas: phonology, semantics, and sentence processing. Neuroimage 30: 1414–1432

Brain plasticity: a new concept in neuroscience, a new tool in neurosurgery

Hugues Duffau

Introduction

At the beginning of the 19th century, two opposite conceptions of the functioning of the central nervous system were suggested. One was the theory of "equipotentiality", which hypothesized that the entire brain, or at least one complete hemisphere, was implied in the practice of a functional task. The other was the theory of "localizationism", which supposed each part of the brain to correspond to a specific function and was built following the seminal description of "phrenology". Progressively, the frequent reports of lesional studies led to an intermediate view, namely, a brain organized (i) in highly specialized functional areas, called "eloquent" regions (such as the central, Broca's, and Wernicke's areas, early identified), for which any lesion gives rise to major irrevocable neurological deficits, and (ii) in "nonfunctional" structures, whose lesions would be without clinical consequences. On the basis of these first anatomo-functional correlations and despite some pioneer observations of postlesional recovery, the dogma of a static functional organization of the brain, which would not be able to compensate any injury involving the so-called eloquent areas, was settled for a long time. However, through regular reports of improvement of the functional status after damage to cortical and/or subcortical structures considered as "critical", this view of a "fixed" central nervous system was called into question in the past decades. Consequently, many investigations were performed, initially in vitro and in animals and more recently, since the development of brain mapping, in humans, in order to study the mechanisms underlying these compensatory phenomena, and the concept of cerebral plasticity was born (for a review, see [4]).

Brain plasticity: definitions and mechanisms

Cerebral plasticity can be defined as the continuous processing allowing short-, medium-, and long-term remodelling of the neuronal-synaptic organization, in order to optimize the functioning of the networks of the brain during phylogeny, ontogeny, and physiological learning and after lesions involving the peripheral or the central nervous system. Several hypotheses about the pathophysiological mechanisms underlying plasticity have been considered. At a microscopic scale, these mechanisms seem to be essentially represented by synaptic efficacy modulations, unmasking of latent connections, phenotypic modifications, synchrony changes, and neurogenesis. At a macroscopic scale, diaschisis, functional redundancies, crossmodal plasticity with sensory substitution, and morphological changes are suggested to be involved. Moreover, the behavioral consequences of such cerebral phenomena have been analyzed for humans in the past decade, both in physiologic – ontogeny and learning – and in pathologic conditions. In particular, the ability to recover after a lesion of the nervous system and the patterns of functional reorganization within eloquent areas and/or within distributed networks which allow such compensation have been extensively studied [9].

In other words, brain plasticity is conceivable only in a dynamic, not in a rigid view of the organization of the central nervous system. Indeed, according to new theories, the brain is an ensemble of complex networks that form, reshape, and flush information dynamically [46, 48]. In this context, the concept of the brain "connectome" has recently emerged. Its goal is to capture the characteristics of spatially distributed dynamical neural processes at multiple spatial and temporal scales [42]. The new science of brain "connectomics" is contributing both to theoretical and computational models of the brain as a complex system [30] and, experimentally, to new indices and metrics (e.g., nodes, hubs, efficiency, modularity) to characterize and scale the functional organization of the healthy and diseased nervous system [2]. In pathology, brain plasticity is nonetheless possible only on the condition that the subcortical connectivity is preserved [13], to allow spatial communication and temporal (de)synchronization among large interconnected networks, according to the principle of hodotopy. Indeed, although different patterns of subcortical plasticity have recently been identified, namely, unmasking of perilesional latent networks, recruitment of accessory pathways, introduction of additional relays within neuronal-synaptic circuits, and involvement of parallel long-distance association pathways, the real capacity to build a new structural connectivity ("rewiring") leading to functional recovery has not yet been demonstrated in humans [13].

Cerebral plasticity and neurosurgery

Preoperative plasticity: the role of the time course of disease

Numerous patients harboring a brain tumor, especially low-grade gliomas, have usually only some mild functional deficit, in spite of the frequent invasion of eloquent structures. This means that these slowly growing lesions have likely induced progressive functional brain reshaping, as suggested by preoperative functional neuroimaging. Indeed, it was recently suggested that brain plasticity cannot be fully understood and fruitfully studied without considering the temporal pattern of the injury inflicted to the brain [4]. Thus, it was noticed for acute lesions such as stroke that, even though many patients improved within the months following the damage, only around 25% of patients totally recovered [47], while more than 90% of patients with a low-grade glioma (same location as stroke) had a normal neurological examination (independent of the slight neurocognitive deficits often diagnosed thanks to an extensive neuropsychological assessment). It is worth noting that the concept of "recovery" should be more clearly defined in the literature. Whereas this terminology should be reserved for a complete normalization of the neurological status, numerous authors speak about "recovery" in cases of partial functional improvement after brain damage. A standardization of the nomenclature is crucial to compare the results reported in the different series.

Interestingly, using a neurocomputational model based on a training of a series of parallel distributed processing neural network models, a recent work simulated acute versus slowly growing injuries [31]. The results showed a very different pattern emerging in the simulation of low-grade gliomas in comparison to the simulation of stroke, with a slow decay of the links within the same subnetwork leading to minimal performance decline, in agreement with the patient literature. Moreover, at the end of the decay regimen, the entire affected hidden layer could be "removed" on the simulation with no effect on performance, which closely matches the lack of any major impairment from low-grade glioma resection (see below). This is likely due to the fact that an abrupt stroke occasions rapid neuronal death, while a low-grade glioma initially spares neuronal tissue and thus gives time for cerebral remapping. As a consequence, it could be suggested that the functional status at the time of diagnosis may well reflect the natural history of the disease and provide a relevant insight into the behavior of the glioma, since these tumors are very heterogeneous [37].

Concerning the neural basis of such functional compensation with low-grade gliomas before any treatment, the patterns of reorganization may differ between patients, which is important to be taken into account by the neurosurgeon with the goal to optimize both indication of surgery and surgical planning [7, 8] (Fig. 1). Indeed, preoperative functional neuroimaging has shown that three kinds of preoperative functional redistribution are possible in patients without any deficit. In the first one, because of the infiltrative feature of gliomas, function still persists within the tumor, so that there is a very limited chance to perform a fair resection without inducing postoperative sequelae. In the second one, eloquent areas are redistributed around the tumor, so that there is a reasonable chance to perform at least a near-total resection despite a likely immediate transient deficit, but with secondary recovery within a few weeks to months. In the third one, there is already a preoperative compensation by remote areas within the lesional hemisphere and/or by the controlateral homologs; consequently, the chances to perform a really total resection (or even a "supercomplete" resection) of this kind of gliomas are very high, with only a slight and very transient deficit. Therefore, in cases of brain lesions involving eloquent areas, plasticity mechanisms seem to follow a hierarchical model, so that first there is an intrinsic reorganization within the injured areas (sign of favorable outcome); if this reshaping is not sufficient, then other regions implicated in the functional network are recruited in the ipsilateral hemisphere (close or more remote to the damaged area) and then in the controlateral hemisphere if necessary [10].

Intraoperative plasticity

Intraoperative stimulation mapping before any resection has allowed the confirmation of the existence of a functional reshaping induced by brain lesions, notably with a possible remapping of the sensorimotor homunculus as well as a reorganization of the language sites. Moreover, acute reorganization of functional maps was equally observed during the resection, likely due to the surgical act itself, which can generate a loco-regional hyperexcitability, as already demonstrated for head injury. Indeed, in several patients harboring a frontal lesion, stimulation of the precentral gyrus induced motor responses at the level of only a limited number of cortical sites before resection, but immediately after lesion removal, an acute unmasking of redundant motor sites was observed which were located within the same precentral gyrus and elicited the same movements as the adjacent previous sites when stimulated [5]. Acute unmasking of redundant somatosensory sites was also regularly observed within the retrocentral gyrus in patients operated on for a parietal glioma. Furthermore, it was equally possible to detect a redistribution within a larger network involving the whole rolandic region, that is, an unmasking of functional homologs located in the precentral gyrus for the first cortical representation and in the retrocentral gyrus for its redundancy (or vice versa) [16]. Finally, intraoperative mapping has also a prognostic value concerning the postoperative recovery for movement: a positive motor response elicited by cortical stimulation of the primary motor area at the end of the resection means that the patient will recover, even if the patient had a presurgical hemiplegia [6, 15].

A

B

C

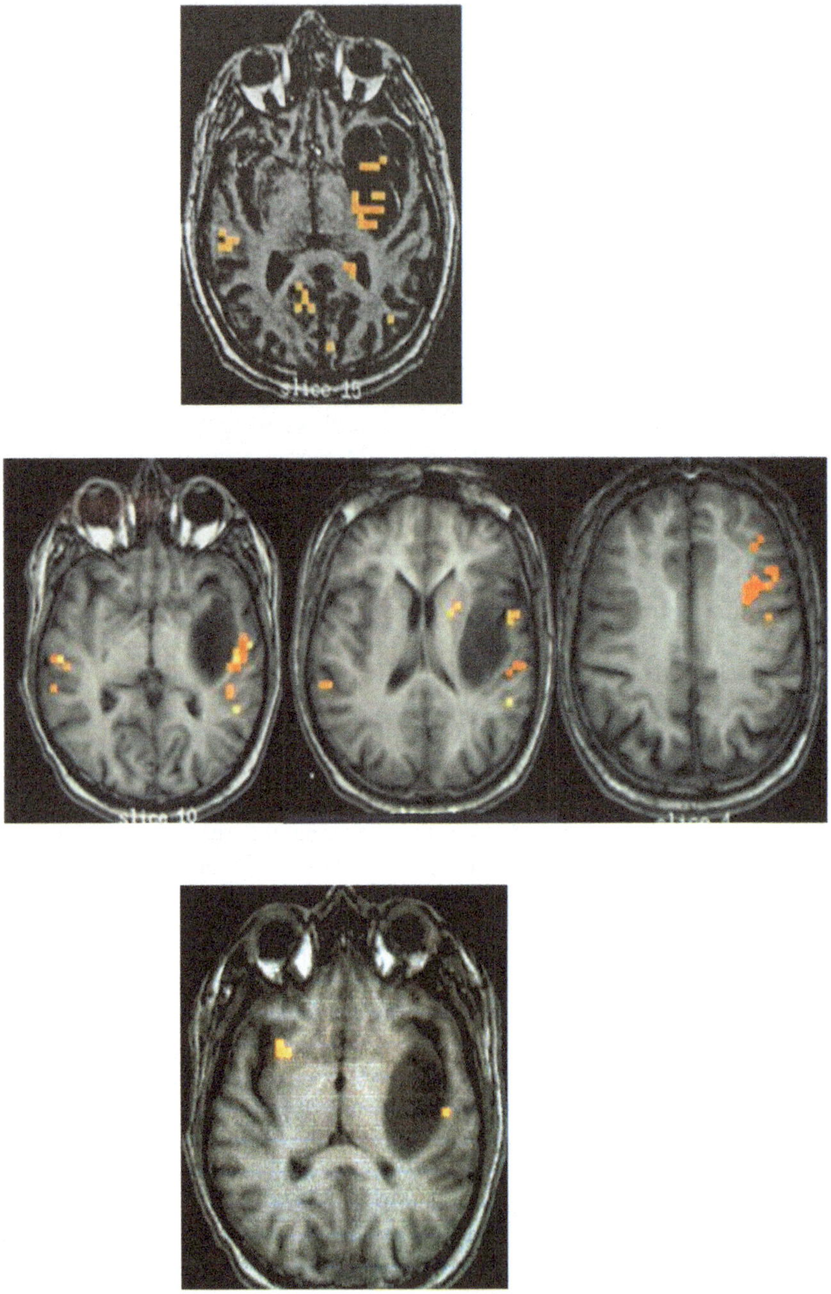

Fig. 1. (A–C). Different patterns of functional (re)organization illustrated by language functional MRI performed before treatment in three right-handed patients with an insular low-grade glioma diagnosed after inaugural seizures and with a normal neurological examination. (**A**) Activations within the tumor, due to the infiltrative feature of low-grade glioma. (**B**) Perilesional reshaping with recruitment of left temporal areas and left frontal operculum. (**C**) Contra-hemispheric remapping with recruitment of the contralateral homologous insula (Modified from [10] and [17])

Postoperative plasticity

The mechanisms of such plasticity induced by surgical resection within eloquent areas were also studied by performing postoperative functional neuroimaging once the patient recovered the preoperative functional status. In particular, several patients were examined after the resection of gliomas involving the supplementary motor area (SMA), which elicited a transient postsurgical SMA syndrome. Functional MRI showed, in comparison with the preoperative imaging, the occurrence of activations of the SMA and premotor cortex contralateral to the lesion: the contrahemispheric homologs thus participated in the postsurgical functional compensation [32].

Therapeutic implications in oncological neurosurgery

It was recently proposed to incorporate the better understanding of the individual plastic potential into the surgical strategy for gliomas, especially slowly growing low-grade gliomas, with the goal (i) to extend the indications of resection in eloquent structures so far considered as "inoperable", (ii) to maximize the extent of glioma removal by performing the resection according to (not fixed) functional boundaries with no margin, and (iii) to minimize the risk of postoperative permanent neurological deficit or even to improve the patient's quality of life [10, 20].

Consequently, several surgical series showed that it was possible to remove low-grade gliomas invading "eloquent" brain structures (Fig. 2).

Resection of the SMA induces the occurrence of an SMA syndrome. The patients recover, and postoperative fMRI has supported functional compensation by the controlateral SMA and premotor cortex as well as by the ipsilesional primary motor cortex [32].

Resection of the insula induces a hemiparesis after removal of the right insula, likely because this region is a nonprimary motor area, and transient speech disturbances after removal of the left dominant insula, but all patients recover, except in rare cases of deep stroke due to a damage of the lenticulo-striate arteries [12, 17, 22, 25]. Moreover, it was possible for right nondominant fronto-temporo-insular LGG involving the deep grey nuclei to remove the claustrum without any cognitive disorders (despite its suggested role in consciousness) [24] and also to remove the invaded striatum without inducing either motor deficit or movement disorders. This compensation can be explained by a recruitment of parallel subcortical circuits such as pallido-luyso-pallidal, strio-nigro-striate, cortico-strio-nigro-thalamo-cortical, and cortico-luysal networks [19].

Resection of the primary somatosensory area was suggested by pre- and postoperative functional neuroimaging to be possibly compensated by a recruitment of "redundant" eloquent sites around the cavity within the postcentral gyrus [36]. This is in accordance with intraoperative electrical mapping data showing an unmasking of redundant somatosensory sites during resection, likely explained by a decrease of the cortico-cortical inhibition. The recruitment of the second somatosensory area or posterior parietal cortex, primary motor area (due to strong anatomo-functional connections between the pre- and retro-central gyri), and controlateral primary somatosensory area also may explain the recovery [14].

Resection of the (dominant) parietal posterior lobe can be performed without inducing any sequelae, and even with a possible improvement in comparison with the preoperative status, especially with a pointing task [4].

Resection of nondominant primary motor area of the face usually induces transient central facial palsy, potentially with a Foix-Chavany-Marie syndrome when the insula is also involved, but these symptoms resolve. The recovery is likely to be explained by the disinhibition of the controlateral homologous sites, via the transcallosal pathways [21].

Resection of primary motor area of the upper limb could be compensated by the recruitment of parallel networks within the primary

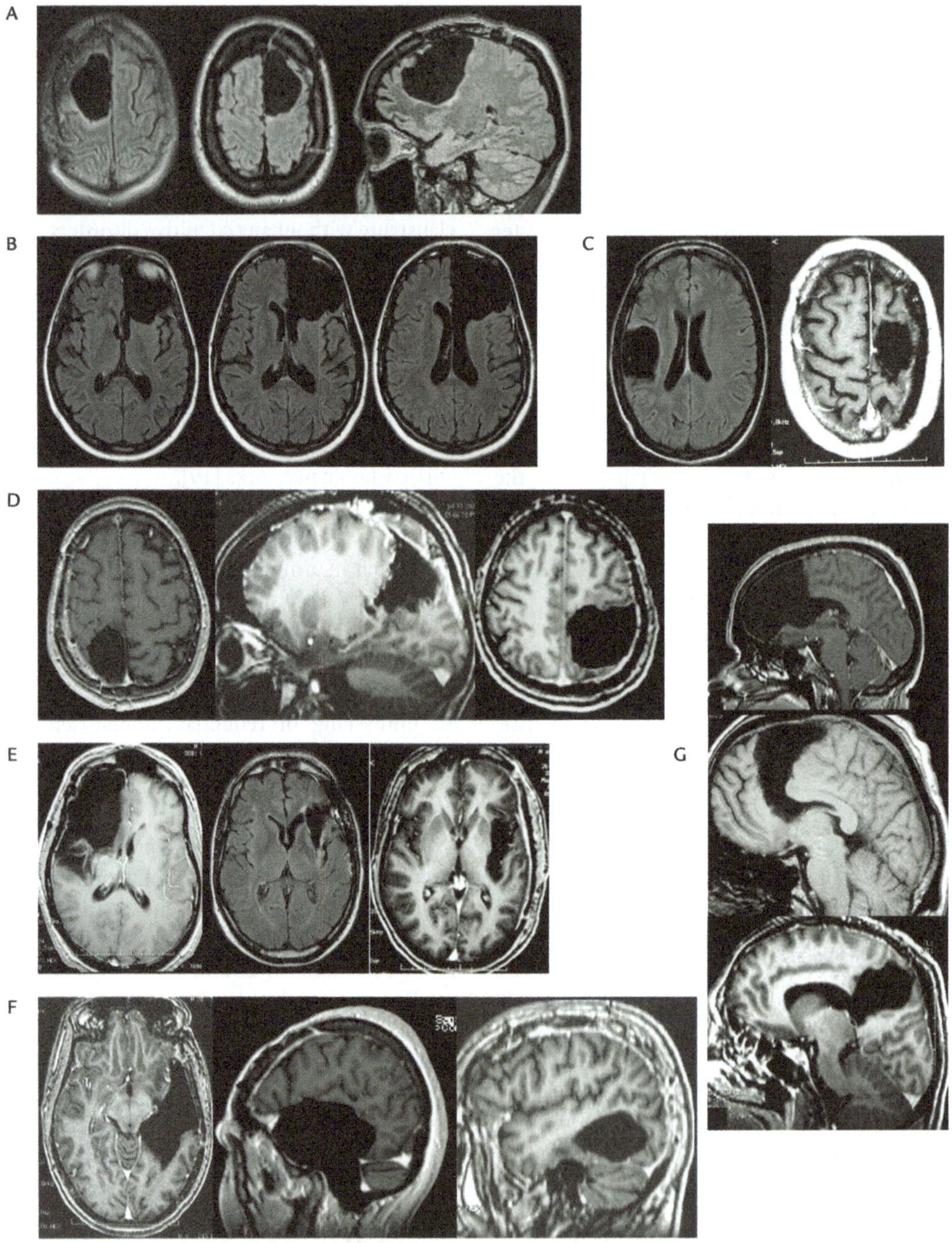

Fig. 2. (A–G). Examples of extensive glioma resection performed within the so-called eloquent areas by intraoperative electrical mapping, the quality of life preserved thanks to brain plasticity. **(A)** Right and left SMA; **(B)** entire left frontal lobe including Broca's area; **(C)** right primary sensorimotor area of the face and left primary motor area of the hand; **(D)** primary somatosensory area and parietal lobe in right and left hemispheres; **(E)** right paralimbic system and left insula; **(F)** anterior-mid and posterior left dominant temporal lobe; **(G)** corpus callosum (Modified from [10])

motor cortex because the existence of multiple cortical motor representations was shown for humans by functional neuroimaging and intraoperative stimulation mapping. Eventually two consecutive surgeries may be performed in order to induce durable remapping following the first one [18] (see below).

Resection of Broca's area in the left dominant hemisphere may take advantage of the recruitment of adjacent regions for functions of language, in particular the pars orbitaris of the inferior frontal gyrus, the dorsolateral prefrontal cortex and the insula [3]. A recent study with extensive neuropsychological examination following resection of Broca's area confirmed the complete functional recovery [40].

Resection of the left-, dominant-hemisphere temporal language area may take advantage of the fact that the complex language function seems to be organized in multiple parallel networks. Consequently, beyond the recruitment of areas adjacent to the surgical cavity, the long-term reshaping could be related to progressive involvement of remote regions within the left, dominant hemisphere – such as the posterior part of the superior temporal gyrus, the pars triangularis of inferior frontal gyrus or other left frontolateral regions – and even the controlateral, right, nondominant hemisphere because of a transcallosal disinhibition phenomenon [20, 45].

Multiple-stage surgical approach and the role of serial mapping

Sometimes, favorable functional results can be obtained only at the price of incomplete resection of a glioma when the tumor invaded areas still crucial for the function. A new concept recently proposed is the postoperative use of functional neuroimaging, which can be easily repeated because of its noninvasive nature, when the patient has perfectly recovered, in order to compare the new maps with those obtained before surgery. Indeed, even if this method has some limitations, subtraction of a pre- from a postoperative acquisition may nonetheless show a possible additional functional reshaping due to (i) the resection itself, (ii) the postsurgical rehabilitation, or (iii) the regrowth of the residual tumor (as before surgery). Such findings have led to propose the new strategy of the multiple-stage surgical approach.

Experimental observations in animals

First, the possibility that functional recovery is modulated by kinetic factors has been addressed in a series of animal studies. The main idea behind these studies was to mimic the development of slowly growing lesions by performing successive partial surgical ablations within a cerebral structure. These partial ablations were then compared to acute resections. In most experiments a control group was included. In this case several surgeries were performed but no cerebral tissue was removed ("sham" operation). Beyond some marginal disparities, the take-home message of all these studies is quite clear: The negative functional impact of large cerebral lesions is much smaller with progressive than with acute lesions. For instance, rats still showed major deficits 36 days after acute ablations of the entire somatosensory cortex . These deficits were absent when the same area was removed in two stages. In this case, the experimental rats could not be differentiated from a nonoperated control group [26]. A similar and even more spectacular report was provided by Adametz with cats [1]. The animals were submitted to a progressive (up to 8 surgeries) or acute resection of the midbrain reticular formation. In the latter case, the cats fell into deep coma and died within a few days after the surgery. In the former case, by contrast, complete recovery was found. The same type of dissociation was observed with monkeys. Acute ablations of the prefrontal cortex were found to induce functional deficits that were much more severe than those produced by serial lesions [41].

Probably the most direct demonstration that functional recovery is directly influenced by the kinetics of the lesion caused to the brain

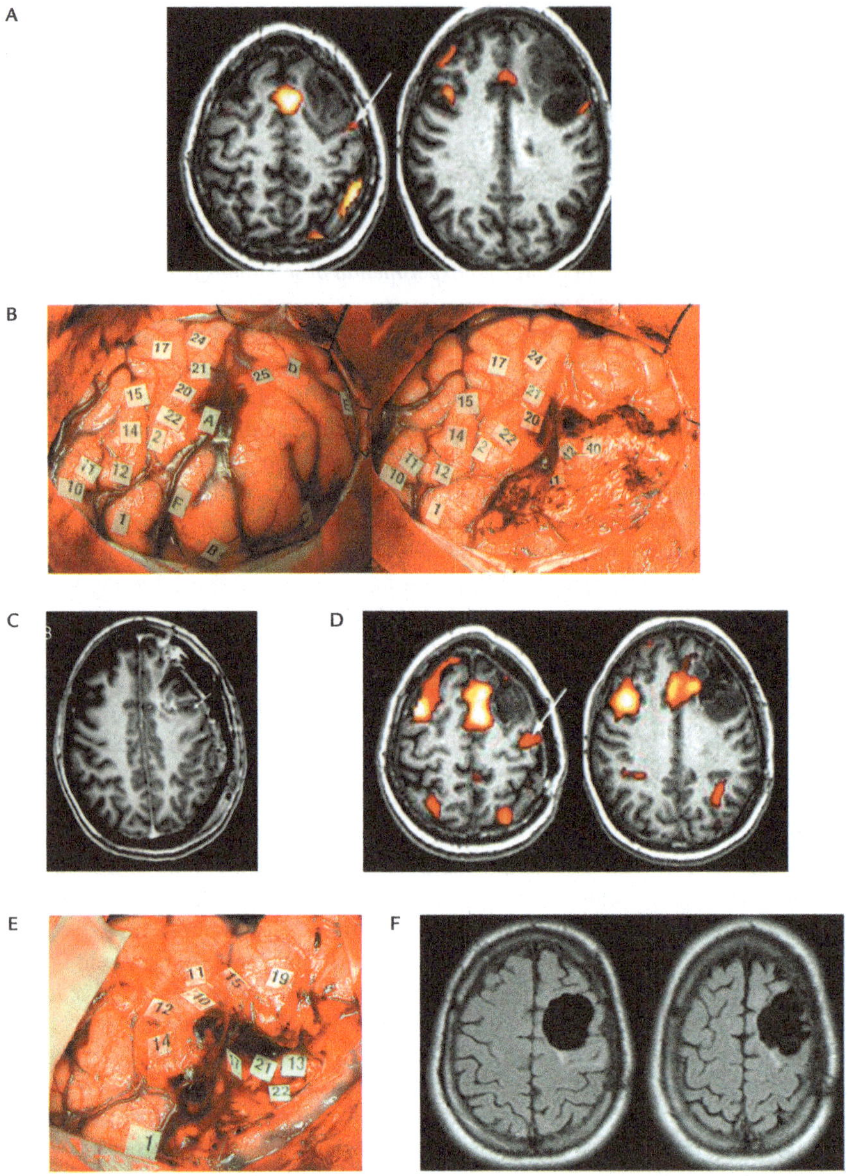

Fig. 3. (A–F). Multiple-stage surgical approach. **(A)** Preoperative language fMRI of a patient without deficit, bearing a low-grade glioma involving the left premotor area: language activation was very close to the posterior part of the tumor (arrow). **(B)** Intraoperative views before (left) and after (right) resection of the glioma, delineated by letter tags. Intraoperative electrostimulation mapping shows a reshaping of the eloquent maps, with a recruitment of perilesional language sites, allowing a subtotal resection with a posterior residue due to invasion of crucial areas (number tags). **(C)** Immediate postoperative enhanced T1-weighted MRI showing the residue (arrow). **(D)** Postoperative language fMRI 4 years after the first fMRI, showing a recruitment of the controlateral hemisphere and the posterior displacement of activation previously located at the posterior border of the tumour (arrow). **(E)** Intraoperative view during the second surgery, confirming the remapping and allowing a more extensive tumor resection with no permanent deficit. **(F)** Postoperative axial FLAIR(fluid-attenuated inversion recovery)-weighted MRI showing the improvement of the extent of resection thanks to functional reshaping (From [28])

has been provided by Patrissi and Stein [39]. They trained a group of rats to retrieve water alternatively located in the right or the left branch of a conventional T-maze. Following a period of training, the rats were divided into several subgroups which were subjected to one-stage bilateral resection of the frontal cortex or two-stage bilateral resection of the frontal cortex (one hemisphere per operation) or one- or two-stage sham operations (control group). For the two-stage groups, three interlesion intervals were considered: 10, 20, or 30 days. The rats given sequential (two-stage) frontal lesions with either a 20- or 30-day interoperative interval could not be differentiated from the sham-operated controls. Animals with two-stage lesions produced 10 days apart exhibited substantial deficits when contrasted with the sham-operated, the 20-day, or the 30-day two-stage groups. However, the 10-day two-stage animals performed significantly better than the rats operated at a single stage. Similar results were found in other studies involving resections of the frontal cortex [29] and the superior temporal gyri [44]. In all these studies, the animals were reported to show a complete recovery when the different surgeries were spaced by a sufficient interval. This interval varied from study to study but it was never smaller than 6 days. Whatever the interlesion interval, for the multi-stage surgeries, the level of recovery was always better than that for the one-stage operations.

Of course, the positive effect of sequential lesions on functional recovery depends strongly on the amount of tissue resected at each surgical stage. This was clearly shown by Stein and colleagues in a monkey study involving the resection of the sulcus principalis. In that study, the total amount of tissue resected was kept constant. It was reported that four partial lesions performed three weeks apart produced a greater level of recovery than two partial lesions performed ten weeks apart [43]. This result pleads directly for the idea that the progressiveness of neural destruction is a key predictor of functional recuperation.

Application in patients with low-grade glioma

Interestingly, recent series demonstrated that remapping was not a theoretical concept but a concrete reality in humans [28]. Postoperative functional neuroimaging performed some months or years following the surgery for low-grade glioma in patients with a complete recovery clearly showed a new recruitment of perilesional areas and/or remote regions within the ipsilesional hemisphere and/or a recruitment of controlateral structures [32]. On the basis of these data, a second surgery was proposed for patients who continued to lead a normal life, before the occurrence of new symptoms (except possible seizures) only because of an increase of the volume of the glioma [18]. The second surgery was also conducted using intraoperative cortical and subcortical mapping, in order to validate the mechanisms of brain reshaping supposed but not proven by preoperative functional neuroimaging before the additional resection was performd [28] (Fig. 3). The preliminary results have supported the efficacy and the safety of such reoperation for LGG not totally removed during a first surgery because of their location within eloquent areas. Indeed, in this recent experience, 74% of the resections were complete or subtotal (less than 10 ml of residue) after the second operation, without any additional serious neurological deficit but rather with an improvement of the neurological status in 16% of cases. Again, the seizures were reduced or disappeared in 82% of patients with epilepsy before the second operation. The median time between the two operations was 4.1 years, and all patients were still alive with a median follow-up of 6.6 years despite an initial incomplete resection. Therefore, these original data demonstrated that, thanks to mechanisms of cerebral plasticity, it is possible to reoperate patients with a low-grade glioma involving eloquent areas with a minimal morbidity and an increase of the extent of resection. However, 58% of tumors had already progressed to high-grade glioma at the second surgery, raising the problem of the timing of reoperation. It was thus suggested to "overindicate" an early reintervention, in order to antedate the second surgery before the anaplastic transformation [35].

In addition, one can currently consider performing postoperative functional neuroimaging after rehabilitation, demonstrated to induce a significant improvement in patients with brain tumors [27], and after recovery following a second surgery, in order to open the door to a possible third or even fourth resection several years after the previous operations. The goal is both to allow the patient to enjoy a normal life and to increase the overall survival. It is also possible to integrate surgeries within a dynamic therapeutic strategy including chemotherapy and radiotherapy, especially when a wide removal is not possible for functional reasons [11]. To this end, neoadjuvant chemotherapy was recently advocated for low-grade glioma, with the goal to induce a shrinking of the tumor before an operation or a reoperation [23] but also to possibly facilitate functional brain reshaping.

Conclusions and perspectives

Combination of on-line intraoperative anatomo-functional correlations (transient virtual lesion) with data provided by tractography (subcortical anatomical information), magnetoencephalography (temporal data), and serial fMRI (perioperative functional data) could enable the elaboration of individual and predictive models of functioning of neurono-synaptic circuits. Such models may lead to a better knowledge of the dynamic potential of spatio-temporal reorganization of the parallel and interactive networks, namely, the mechanisms of brain plasticity thought to play a major role in functional compensation for slowly growing lesions and their surgical resection. In practice, in order to evolve towards a multiple-stage surgical approach (i.e., second or third surgery being more extensive than the first incomplete resection within eloquent areas), a dynamic strategy has to be envisaged for functional neuroimaging. The goal is to switch from a "static" use of a unique preoperative functional neuroimaging assessment (limited technique with lack of reliability) to longitudinal studies based on the repetition of neuroimaging before and after surgical resection(s), with the goal to analyze a possible brain reshaping at the individual scale and to select the candidates for reoperation(s). The next step is now to use biomathematical models able to examine brain functional interaction through effective connectivity, in order to attempt to predict before surgery the patterns of postsurgical remapping at the individual scale on the basis of the data provided by the preoperative functional neuroimaging. The new theory of graphs could probably support the development towards such individual prediction [33, 34].

From the point of view of perspectives, it could be suggested to induce and guide brain plasticity, particularly with pharmacologic drugs, by functional rehabilitation, or even by transcranial magnetic stimulation, to promote functional recovery not only after surgery but also before surgery. One could hypothesize that such preoperative remapping might enable to increase the extent of the resection (and possibly to take a margin around the lesion) while avoiding postsurgical worsening, even in the classically so-called eloquent areas according to anatomic criteria. Furthermore, the use of plasticity could lead to the indication of surgery for asymptomatic patients. Indeed, thanks to the current development of neuroimaging, incidental discovery of tumors will progressively increase in the next future. Interestingly, concerning low-grade gliomas, it was recently demonstrated that their natural history was the same in the presymptomatic period as after the first symptom (usually seizures) [38]. Therefore, on the basis of the new neuroscientific concept of a "hodotopic and plastic brain", the next surgical goal could be to evolve towards a "preventive functional neurooncology".

References

[1] Adametz J (1959) Rate of recovery of functioning in cats with rostral reticular lesions; an experimental study. J Neurosurg 16: 85–97

[2] Bassett DS, Bullmore ET (2009) Human brain networks in health and disease. Curr Opin Neurol 22: 340–347

[3] Benzagmout M, Gatignol P, Duffau H (2007) Resection of World Health Organization grade II gliomas involving Broca's area: methodological and functional considerations. Neurosurgery 61: 741–752

[4] Desmurget M, Bonnetblanc F, Duffau H (2007) Contrasting acute and slow-growing lesions: a new door to brain plasticity. Brain 130: 898–914

[5] Duffau H (2001) Acute functional reorganisation of the human motor cortex during resection of central lesions: a study using intraoperative brain mapping. J Neurol Neurosurg Psychiatry 70: 506–513

[6] Duffau H (2001) Recovery from complete hemiplegia following resection of a retrocentral metastasis: the prognostic value of intraoperative cortical stimulation. J Neurosurg 95: 1050–1052

[7] Duffau H (2005) Lessons from brain mapping in surgery for low-grade glioma: insights into associations between tumour and brain plasticity. Lancet Neurol 4: 476–486

[8] Duffau H (2006a) New concepts in surgery of WHO grade II gliomas: functional brain mapping, connectionism and plasticity – a review. J Neurooncol 79: 77–115

[9] Duffau H (2006b) Brain plasticity: from pathophysiological mechanisms to therapeutic applications. J Clin Neurosci 13: 885–897

[10] Duffau H (2008) Brain plasticity and tumors. Adv Tech Stand Neurosurg 33: 3–33

[11] Duffau H (2009a) Surgery of low-grade gliomas: towards a "functional neurooncology". Curr Opin Oncol 21: 543–549

[12] Duffau H (2009b) A personal consecutive series of surgically treated 51 cases of insular WHO grade II glioma: advances and limitations. J Neurosurg 110: 696–708

[13] Duffau H (2009c) Does post-lesional subcortical plasticity exist in the human brain? Neurosci Res 65: 131–135

[14] Duffau H, Capelle L (2001) Functional recuperation following lesions of the primary somatosensory fields. Study of compensatory mechanisms. Neurochirurgie 47: 557–563

[15] Duffau H, Capelle L, Sichez J, Faillot T, Abdennour L, Law Koune JD et al (1999) Intra-operative direct electrical stimulations of the central nervous system: the Salpêtrière experience with 60 patients. Acta Neurochir (Wien) 141: 1157–1167

[16] Duffau H, Sichez JP, Lehéricy S (2000) Intraoperative unmasking of brain redundant motor sites during resection of a precentral angioma: evidence using direct cortical stimulations. Ann Neurol 47: 132–135

[17] Duffau H, Bauchet L, Lehéricy S, Capelle L (2001) Functional compensation of the left dominant insula for language. Neuroreport 12: 2159–2163

[18] Duffau H, Denvil D, Capelle L (2002) Long term reshaping of language, sensory, and motor maps after glioma resection: a new parameter to integrate in the surgical strategy. J Neurol Neurosurg Psychiatry 72: 511–516

[19] Duffau H, Denvil D, Capelle L (2002) Absence of movement disorders after surgical resection of glioma invading the right striatum. J Neurosurg 97: 363–369

[20] Duffau H, Capelle L, Denvil D, Sichez N, Gatignol P, Lopes M et al (2003) Functional recovery after surgical resection of low grade gliomas in eloquent brain: hypothesis of brain compensation. J Neurol Neurosurg Psychiatry 74: 901–907

[21] Duffau H, Karachi C, Gatignol P, Capelle L (2003) Transient Foix–Chavany–Marie syndrome after surgical resection of a right insulo-opercular low-grade glioma. Neurosurgery 53: 426–431

[22] Duffau H, Taillandier L, Gatignol P, Capelle L (2006) The insular lobe and brain plasticity: lessons from tumor surgery. Clin Neurol Neurosurg 108: 543–548

[23] Duffau H, Taillandier L, Capelle L (2006) Radical surgery after chemotherapy: a new therapeutic strategy to envision in grade II glioma. J Neurooncol 80: 171–176

[24] Duffau H, Mandonnet E, Gatignol P, Capelle L (2007) Functional compensation of the claustrum: lessons from low-grade glioma surgery. J Neurooncol 81: 327–329

[25] Duffau H, Moritz-Gasser S, Gatignol P (2009) Functional outcome after language mapping for insular World Health Organization grade II gliomas in the dominant hemisphere: experience with 24 patients. Neurosurg Focus 27(2): E7

[26] Finger S, Marshak RA, Cohen M, Scheff S, Trace R, Niemand D (1971) Effects of successive and simultaneous lesions of somatosensory cortex on tactile discrimination in the rat. J Comp Physiol Psychol 77: 221–227

[27] Gehring K, Sitskoorn MM, Gundy CM, Sikkes SA, Klein M, Postma TJ et al (2009) Cognitive rehabilitation in patients with gliomas: a randomized, controlled trial. J Clin Oncol 27: 3712–3722

[28] Gil Robles S, Gatignol P, Lehéricy S, Duffau H (2008) Long-term brain plasticity allowing multistage surgical approach to WHO grade II gliomas in eloquent areas: a combined study using longitudinal functional MRI and intraoperative electrical stimulation. J Neurosurg 109: 615–624

[29] Glick SD, Zimmerberg B (1972) Comparative recovery following simultaneous- and successive-stage frontal brain damage in mice. J Comp Physiol Psychol 79: 481–487

[30] Honey CJ, Kötter R, Breakspear M, Sporns O (2007) Network structure of cerebral cortex shapes functional

connectivity on multiple time scales. Proc Natl Acad Sci USA 104: 10240–10245

[31] Keidel JL, Welbourne SR, Lambon Ralph MA (2010) Solving the paradox of the equipotential and modular brain: a neurocomputational model of stroke vs. slow-growing glioma. Neuropsychologia 48: 1716–1724

[32] Krainik A, Duffau H, Capelle L, Cornu P, Boch AL, Mangin JF et al (2004) Role of the healthy hemisphere in recovery after resection of the supplementary motor area. Neurology 62: 1323–1332

[33] Marrelec G, Krainik A, Duffau H, Pélégrini-Issac M, Lehéricy S, Doyon J et al (2006) Partial correlation for functional brain interactivity investigation in functional MRI. Neuroimage 32: 228–237

[34] Marrelec G, Bellec P, Krainik A, Duffau H, Pélégrini-Issac M, Lehéricy S et al (2008) Regions, systems and the brain: hierarchical measures of functional integration in fMRI. Med Image Anal 12: 484–496

[35] Martino J, Taillandier L, Moritz-Gasser S et al (2009) Re-operation is a safe and effective therapeutic strategy in recurrent WHO grade II gliomas within eloquent areas. Acta Neurochir (Wien) 151: 427–436

[36] Meunier S, Duffau H, Garnero L, Capelle L, Ducorps A (2000) Comparison of the somatosensory cortical mapping of the fingers using a whole head magnetoencephalography (MEG) and direct electrical stimulations during surgery in awake patients. Neuroimage 11 (Suppl 1): S868

[37] Pallud J, Mandonnet E, Duffau H, Kujas M, Guillevin R, Galanaud D et al (2006) Prognostic value of initial magnetic resonance imaging growth rates for World Health Organization grade II gliomas. Ann Neurol 60: 380–383

[38] Pallud J, Fontaine D, Duffau H, Mandonnet E, Sanai N, Taillandier L et al (2010) Natural history of incidental WHO grade II gliomas. Ann Neurol 68: 727–733

[39] Patrissi G, Stein DG (1975) Temporal factors in recovery of function after brain damage. Exp Neurol 47: 470–480

[40] Plaza M, Gatignol P, Leroy M, Duffau H (2009) Speaking without Broca's area after tumor resection. Neurocase 15: 294–310

[41] Rosen J, Stein D, Butters N (1971) Recovery of function after serial ablation of prefrontal cortex in the rhesus monkey. Science 173: 353–356

[42] Sporns O,Tononi G, Kötter R (2005) The human connectome: a structural description of the human brain. PLoS Comput Biol 1: e42

[43] Stein DG, Butters N, Rosen J (1977) A comparison of two- and four-stage ablations of sulcus principalis on recovery of spatial performance in the rhesus monkey. Neuropsychologia 15: 179–182

[44] Stewart JW, JW, Ades H (1951) The time factor in reintegration of a learned habit lost after temporal lobe lesions in the monkey (Macaca mulatta). J Comp Physiol Psychol 44: 479–486

[45] Thiel A, Herholz K, Koyuncu A, Ghaemi M, Kracht LW, Habedank B et al (2001) Plasticity of language networks in patients with brain tumors: a positron emission tomography activation study. Ann Neurol 50: 620–629

[46] Varela F, Lachaux JP, Rodriguez E, Martinerie J (2001) The brainweb: phase synchronization and large-scale integration. Nat Rev Neurosci, 2: 229–239

[47] Varona JF, JF, Bermejo F, Guerra JM, Molina JA (2004) Long-term prognosis of ischemic stroke in young adults. Study of 272 cases. J Neurol 251: 1507–1514

[48] Werner G (2007) Brain dynamics across levels of organization. J Physiol Paris 101: 273–279

Resting-state brain networks in functional Magnetic Resonance Imaging (MRI)

Pierre Bellec, Arnaud Messé, David Coynel, Vincent Perlbarg, Habib Benali, Guillaume Marrelec

Introduction

Blood oxygen level-dependent (BOLD) functional magnetic resonance imaging (fMRI) has been extensively used to study how task performance modulates brain activity. Such an approach emphasizes the principle of functional segregation, in that different brain areas are characterized by their involvement in specific cognitive processes. However, this type of analysis essentially ignores that the brain maintains a constant level of spontaneous activity, i.e., activity that is not a direct consequence of environmental stimulations. Investigating these spontaneous brain fluctuations is a challenge precisely because they cannot be controlled by an experimental design. The first successful study of brain spontaneous fluctuations by BOLD fMRI was performed by Biswal and colleagues [3], who computed the correlation between the time course of a seed region and all brain voxels in the absence of any experimental task. Such resting-state functional connectivity maps, which identified distributed sets of brain regions whose spontaneous activity showed a large degree of temporal coherence, are currently known as resting-state networks (RSNs). By contrast with task-oriented studies, investigation of RSNs relies on the principle of functional integration: brain regions are characterized by the set of regions with which their signal is highly synchronized.

While the exact mechanisms at the origin of spontaneous fluctuations are still not fully elucidated, it is now firmly established that RSNs are mainly the consequence of neural activity. Approaches combining multiple neuroimaging modalities have contributed to a better understanding of their neuronal and anatomical basis. Much progress has also been made regarding methods, with the emergence of fully exploratory multivariate techniques. Beyond the mapping of RSNs, some recent approaches use a graph representation, which gives a better insight into the structure of information flow within RSNs.

Investigation of RSNs by BOLD fMRI is a recent yet quickly growing field that has already generated a wealth of applications. This chapter is a concise review of the most prominent results and challenges, as well as a discussion of existing and potential applications. References were carefully selected to point the interested readers to key works, as well as to some specialized reviews that cover the various topics presented here more thoroughly.

Mapping RSNs

In this section, we summarize the main methods used to map the spatial distributions of RSNs and we also present a typology of the maps that have been reported so far in the literature.

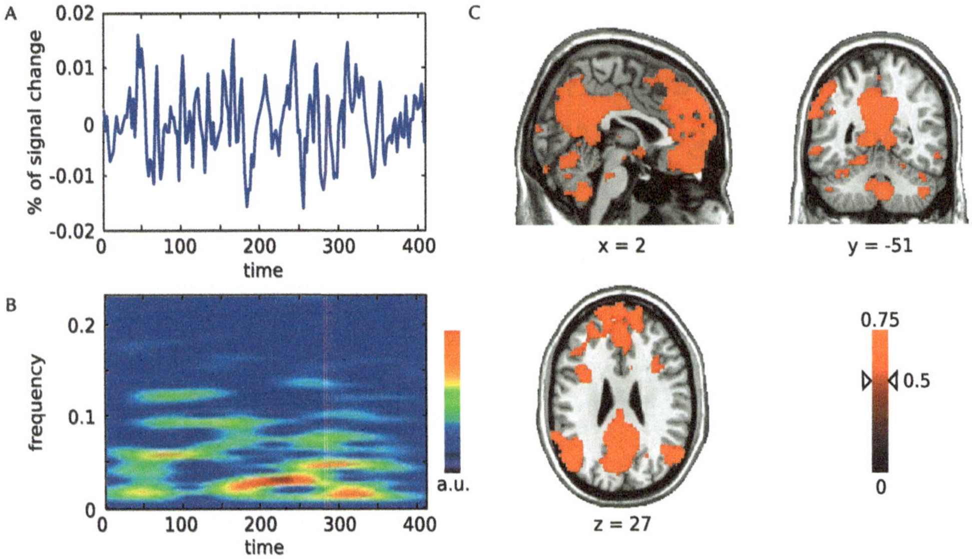

Fig. 1. (A–C). Resting-state functional connectivity maps. An individual functional connectivity map of the default-mode network (DMN) was derived with the posterior cingulate seed used by Greicius et al [16]. **(A)** and **(B)** The time series of the seed **(A)** is dominated by slow-frequency fluctuations (below 0.1 Hz), as evidenced by a window Fourier transform representing the power spectrum as a function of time **(B)**. **(C)** A correlation map (with an arbitrary threshold of 0.5) between the time series of the seed and that of all other brain voxels identifies an extended set of regions known as the DMN

Methods

Mapping RSNs consists of finding sets of voxels that exhibit coherent spontaneous BOLD fluctuations. In a seminal work on resting-state fMRI, Biswal et al [3] introduced functional connectivity maps, i.e., volumes representing the correlation between the time course of a seed region and that of every voxel in the brain (Fig. 1). A variety of fully exploratory and multivariate algorithms have then been proposed to automatically identify RSNs without having to rely on the choice of a particular seed region (see [27] for a review). The most popular of these techniques is spatial independent component analysis (ICA) (Fig. 2). The significant ICA maps delineate groups of brain regions that seem interpretable in terms of brain networks. Such identification is performed at the level of a single subject; for group-level analyses, the methodology needs to be extended to identify group-level RSNs that would summarize the distribution of individual maps across many subjects. The first and still popular approach for group ICA essentially consists of concatenating subjects in time and run a standard ICA on the concatenated time series to derive group components. Many methodological challenges remain in this area, which is still under active development (see [8] for an excellent review of recent approaches). In the perspective of clinical applications, perhaps the most important of these open challenges is to propose a framework that will be flexible enough to capture individual variability, while still allowing for a clear correspondence between group-level and individual-level RSNs.

Reliability of extracted maps

A key question is the degree of reliability of the extracted maps, which can be affected by two main factors. The first one is algorithmic: How does the choice of algorithm and parameters impact the maps? The second one is related to the particular sample of datasets that is selected for the study: How stable would the maps be if the study were to be replicated with a new, independent database? Altogether, the RSNs reported so far in the literature exhibit a fair level of agreement across studies, suggesting that RSN identification is overall reliable.

Regarding algorithmic reliability, Margulies et al [28] showed that even a small shift of the seed voxel could lead to dramatic instability of the connectivity pattern (see Fig. 4). Similar results were reported with slightly different regions of interest within the so-called DMN (see the next subsection below for definition and description of that particular RSN) [5, 10]. As to RSN extraction by spatial ICA, it was found that the algorithm used was rather robust to its initial conditions. Few studies have compared specific RSNs identified by different methods (e.g., connectivity maps and ICA). The existing results suggest a good level of agreement across methods, yet many of the alternatives to ICA have not yet been included in such investigation.

Regarding sampling reliability, an early work had concluded there to be a high degree of reproducibility of some specific RSNs identified by ICA both across subjects and across datasets acquired on the same subject. More systematic extensions of this study to test–retest databases found moderate to high reliability at the group level [44, 50]. Tests–retest stability can be approximated from a single sample of the population by resampling techniques. The work of Damoiseaux et al [12] was notably the first to claim the consistency of group RSNs by this type of approach.

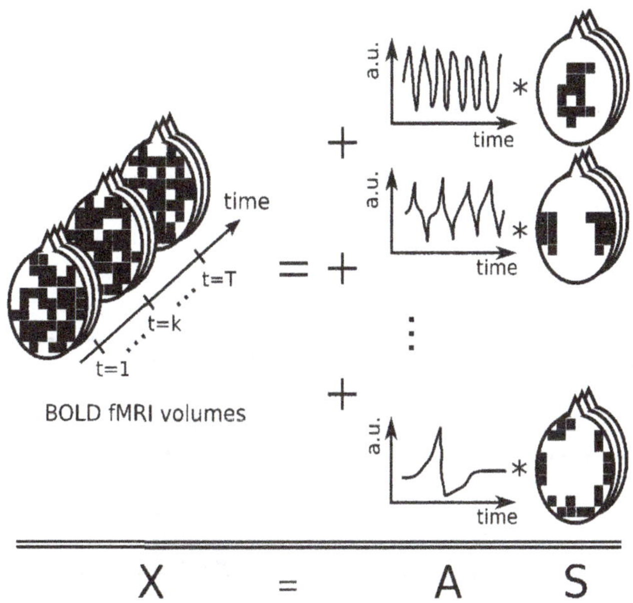

Fig. 2. Linear model for spatial ICA. ICA is based on the same linear mixture model that is used in conventional task activation paradigms. According to this model, the space-time fMRI dataset is decomposed into a number of components, each component consisting of a spatial map (in matrix S) and an associated time course (in matrix A). The spatial component describes the weight of the associated time course at each voxel. By contrast with task activation experimental designs, ICA proceeds by estimating simultaneously the spatial and temporal components. It is then possible to test for the significance of the contribution of a particular spatial component at each voxel

Typology of RSNs

The sensorimotor network was the first RSN studied for functional connectivity with a seed in the primary motor cortex [3]. A visual RSNs was identified with a seed around the calcarine fissure, and a limbic network with a seed in the amygdala. An auditory and a language RSN were later extracted, as well as a fronto-parietal network involved notably in working-memory tasks. Also the DMN was identified with a seed in the posterior cingulate cortex. Interestingly, this network was first defined as a set of regions consistently showing larger BOLD activity during rest periods than during a broad range of task-oriented behaviors and, in addition, it has the highest level of metabolic activity in the brain. These networks could also be identified in a reliable manner through ICA [12] (Fig. 3 provides a summary of findings at the group level). An interesting phenomenon which can occur in ICA is the splitting of a network in one or more subnetworks (e.g., the subdivision of the visual cortex into primary and extrastriate) when the number of extracted networks is large [38]. The anatomofunctional significance of such fine RSNs cartography largely remains to be investigated, even though it has been shown to be consistent with the organization of the monkey brain as well as the organization of white-matter tracts in some specific areas such as the precuneus [28] (Fig. 4). A more detailed typology of the RSNs along with Talairach coordinates was provided by Perlbarg and Marrelec [32]. For reviews that are specific to the DMN, see [5] and [33].

Investigating information flow within RSNs

Graph representations are perfectly suited to characterize information flow within RSNs. Graphs are composed of two objects: vertices or nodes (regions) and edges (links or connections), both of which can be defined by various techniques. Once an RSN has been modeled in the form of a graph, tools from graph theory can be used to summarize its key properties.

Defining nodes

Graph nodes are the functional units that are being connected. It is generally not feasible to identify each voxel with a specific node, because it would lead to analyses that are compu-

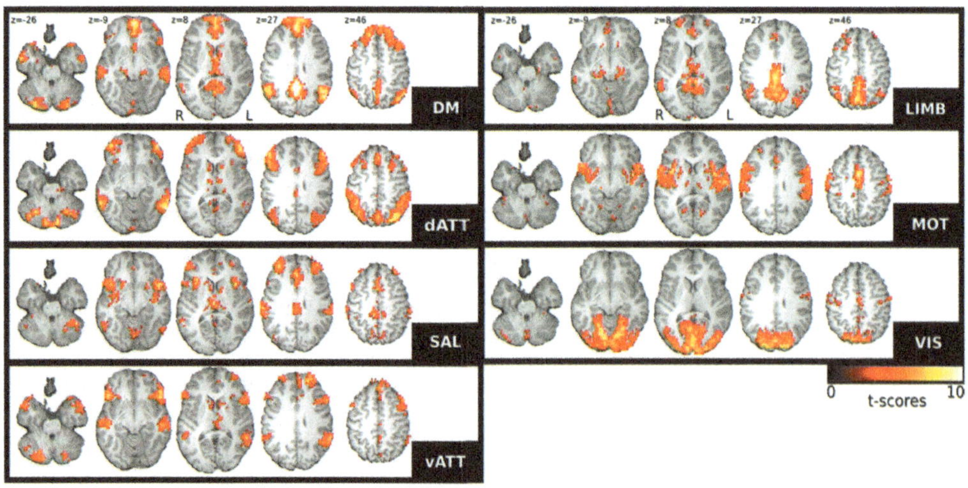

Fig. 3. Example of RSNs extracted by group ICA. Group RSNs maps from 20 subjects scanned at rest extracted by hierarchical clustering of individual maps. *DM* default mode, *dATT* dorsal attentional, *SAL* salient processing, *vATT* ventral attentional, *LIMB* limbic, *MOT* motor, *VIS* visual

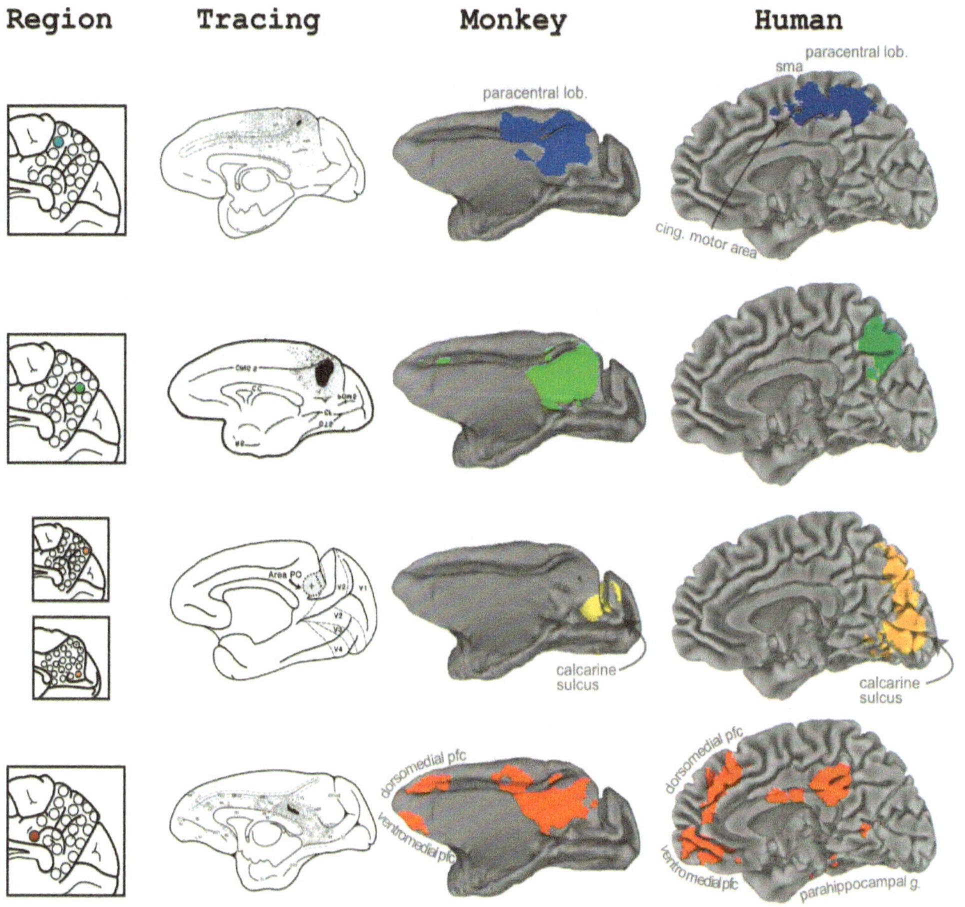

Fig. 4. Functional and anatomical parcellations of the precuneus areas in human and monkey brains. Four seed regions (*far left*) have been used to exemplify the markedly different patterns of functional connectivity found in the precuneus area for humans (*far right*) and macaque monkeys (*middle right*). Similar patterns of functional connectivity were identified for both species, which were consistent with the anatomical connections of the seed regions as identified with tracers in macaques (*middle left*). (Figure adapted from [28], courtesy of M. P. Milham and M. Petrides on behalf of the co-authors)

tationally very challenging. More fundamentally, the fMRI time courses within individual voxels are also highly redundant and noisy. It is therefore common practice to select a very limited set of regions (e.g., from 3 to 20) on the basis of prior knowledge; each region is then defined as a sphere centered around a stereotaxic coordinate. Alternatively, a full brain exploration can be performed by grouping voxels into brain regions that are then used as nodes. A common strategy is to rely on an atlas of brain regions manually segmented on the basis of anatomical landmarks in a stereotaxic space of reference. However, such an atlas usually has a limited number of regions (about 100), potentially gathering voxels with heterogeneous signals. To circumvent this issue and investigate graph properties at a finer spatial scale, one can either generate a gray-matter parcellation of arbitrary size or reduce the number of voxels by resampling the data to a coarser resolution. Using functional data, brain regions can be obtained at the individual level as sets of voxels with highly homogeneous time courses; group-level studies then require to establish a correspondence between individual brain regions across subjects. Finally, it has also been advocated to base brain regions on structural (i.e., white matter) connectivity rather than functional connectivity.

Quantifying links

Once nodes have been selected, the strength of the connection between two nodes is usually quantified by their so-called functional connectivity, i.e., the temporal correlation between the time courses of these regions. Many studies have used correlation as a way to characterize connections between regions. Other statistical measures of functional connectivity have been proposed to circumvent certain shortcomings of correlation, such as conditional and partial correlations. Measures have been used either in the temporal or the frequency domain.

When the number of regions remains limited, the strength of connection between regions can also be quantified by effective connectivity, which considers the influence that regions exert on each other. The notion of effective connectivity is strongly tied with that of causality. The two main frameworks to investigate it are structural equation modeling and dynamic causal modeling. Both approaches are model-based and strongly rely (at least to some level) on the definition of a structural model in

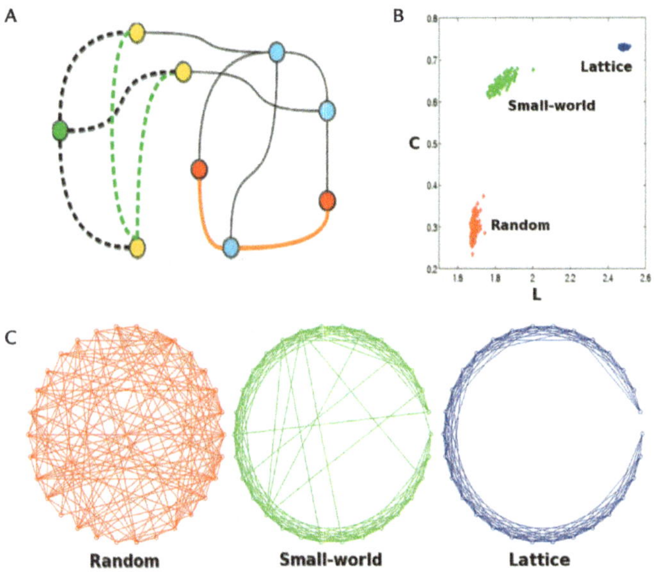

Fig. 5. (A–C). Characteristic path length and clustering coefficient as key measures of graph theory. The characteristic path length L of a network is the average length of the shortest path between any two nodes, where the path length (also called distance) between two nodes is defined as the minimum number of distinct edges required to link these nodes. The clustering coefficient C of a network is the average of the clustering coefficients of all nodes, where the clustering coefficient of a node is the ratio between the actual number of edges among the node neighbors and the largest possible number of such connections. **(A)** Example of graph with 9 nodes and 12 edges. The path length between the red nodes is $L = 2$, corresponding to the shortest path (orange) between the two nodes. The green node has 3 neighbors (yellow), connected to it by dashed edges. These 3 nodes could potentially be connected by up to $3(3 - 1)/2 = 3$ edges, but are effectively connected by only 2 edges (green) in the present graph. The clustering coefficient is therefore equal to $C = 2/3$. **(B)** Example of characteristic path length and clustering coefficient for simulated random, small-world, and lattice graphs with the same number of nodes and edges. **(C)** Examples of such graphs

the form of a directed graph prior to the analysis. While causality appears to be naturally embedded in structural equation and dynamic causal modeling in the form of arrow directions, exploratory investigation of causal relationships remains a challenge. Few methods have been proposed in fMRI, with the notable exception of the application of Granger causality. To our knowledge, only one study has investigated effective connectivity within RSNs [23]. For reviews on measures of functional and effective connectivity, see [29] and [34].

Characterizing graphs

When the number of regions N increases, the number of potential connections increases as $N(N - 1)/2$ (45 for $N = 10$, 4950 for $N = 100$) and direct interpretation of the graph and its connections quickly becomes impossible. For this reason, measures have been proposed that compute a global summary of the graph features. In statistics, such measures were proposed notably on the basis of mutual information. Also, using graph theory, one can study the characteristic path length L of a network and its clustering coefficient C. Small-world networks are a variety of graphs whose properties lie between the two extreme configurations (random and lattice), with small L (comparable to random networks) and great C (comparable to lattice networks) (Fig. 5). By fMRI, the brain architecture was found to exhibit small-world properties. Many other measures exist, such as alternative measures of efficiency, modularity, or motifs. For reviews of graph theory and its application to brain networks, see [6] and [42].

Reliability of graph-theoretic measures

While the test–retest reliability of graph representation of RSNs has not been investigated to the best of our knowledge, a few studies have investigated the algorithmic reliability which is influenced by two factors: how nodes are selected and what signal is assigned to each node. Using two atlases of 90 and 70 regions, Wang et al [47] showed that node definition has an influence on the quantitative measures of graph theory, but not on the qualitative nature of the graph. Hayasaka and Laurenti [21] compared voxel-based and region-based (with 90 regions) graphs and showed that even qualitative features of the graph could differ. As to the signal assigned to a node, it is usually the spatial average of the time courses of all voxels within a neighborhood. As will be discussed in the next section, removal of structured physiological processes is an important step and can modify the whole pattern of functional connectivity and, consequently, all graph-theoretic measures [9].

Neurophysiological basis of RSNs

Since the early controversy regarding the neural origin of resting-state BOLD fluctuations, a number of key aspects of the neurophysiological basis of RSNs have been established, namely: (1) RSNs are not only consequences of physiological noise; (2) RSNs reflect, to some extent, the underlying structural connections; (3) RSNs can be related to the neural activity as measured with other imaging modalities; and (4) RSNs represent sets of brain regions that are involved in similar functions.

Physiological noise

The BOLD signal only accounts for a part of a fMRI signal, which also contains artifacts from various origins. In particular, some physiological processes (e.g., cardiac, respiratory, or movement-related) induce spurious effects that contaminate the BOLD signal in the whole brain. Nonetheless, it has been shown that spatially structured physiological fluctuations and spontaneous activity as measured by low-frequency BOLD signal seem to be two distinct processes. Still, physiological noise reduction is increasingly acknowledged as a key step in connectivity analysis and is commonly applied in recent studies [4]. However, there is not yet an agreement regarding the best methodological approach to perform this noise reduction. For discussions on physiological noises, see, e.g., the articles by Marrelec et al [29], Rogers et al [34], Auer [2], or Perlbarg and Marrelec [32].

RSNs and structural connectivity

A key question for a better understanding of RSNs is the relationship between these networks and the underlying structural connectivity. For monkeys, the DMN was found to be composed of regions that are connected to one another [5]. Also, the cortical pattern of correlation with oculomotor regions was found to be consistent with the anatomical connectivity pattern obtained by retrograde tracer injection [46]. For humans, studies comparing maps of RSN and structural connectivity from diffusion-weighted imaging showed that nodes that are functionally connected can, but do not need to, be anatomically connected [18, 43]. Still, there is converging evidence that brain regions defined on the basis of either local homogeneity of functional connectivity or local homogeneity of structural connectivity are largely consistent with each other [28, 41] (see Fig. 4). To elucidate the intricate relationship between the patterns of white-matter connections and the emergence of functionally connected RSNs, a number of recent studies have implemented complex neurocomputational models whose objective is to replicate the dynamics of neuronal assemblies in a completely controlled, digital environment. The relationship between RSNs and the underlying structural connectivity is further reviewed by Damoiseaux and Greicius [11] and van den Heuvel and Hulshoff Pol [42] from an experimental perspective, while Honey et al [22] concentrate on more theoretic aspects.

RSNs and neural activity

The spontaneous fluctuations of brain activity as measured by BOLD fMRI have been compared to more direct measures of neural activity. Scalp electroencephalography (EEG), in particular, is a well-established tool to study resting-state activity. Many studies have used simultaneous EEG–fMRI acquisitions to investigate the fMRI correlates of EEG activity. The existence of a complex relationship between both types of signal was evidenced, a relationship that depends on the features extracted from the EEG signal [26]. For monkeys, cortical electrophysiological measure of the neural activity also suggested a significant relationship with the BOLD spontaneous fluctuations [37].

RSNs and the neural basis of cognitive functions

RSNs have also been found to be specifically involved in certain tasks. The DMN has consistently been reported as showinging blocked and event-related task-induced deactivation in humans [19]. For monkeys, the cortical pattern of correlation with oculomotor regions was found to be consistent with the response pattern evoked during performance of a saccadic eye movement task [46]. For humans, a meta-analysis of task-related studies showed that voxels that were consistently coactivated during specific tasks formed networks that strikingly matched RSNs [38].

Applications

A key advantage of resting-state fMRI is that the acquisition protocol is very simple and requires no particular ability from the subject, so that it can be performed even by patients with impaired functions. It is now routinely performed for cognitive purposes and is also getting increasingly used in clinical research.

Healthy brain

Resting-state connectivity has been used as a tool in many areas of basic neuroscience. We focused here on a number of results that have a direct implication for the design of resting-state protocols in relation with the following questions: (1) Can RSNs be identified during task performance? (2) Can RSNs be identified at any stage of brain development and aging?

A number of studies have looked at the modulation of RSNs by experimental tasks. For example, the spatial extent of the DMN was found to remain stable during a simple visual task [18]. Other studies have shown some modulation of the connectivity within the DMN by a task, e.g., as a function of the working-memory load [14]. This kind of approach was also applied to a variety of other RSNs and tasks. More recently, ICA was used to assess systematically the stability of all RSNs in an oddball task [7].

Most of the RSNs found in adults could also be successfully extracted by spatial ICA on resting-state acquisitions of children [15]. Using graph-theoretic measures, a similar global organization was found in children and young adults; however, development was characterized by a simultaneous reduction of short-range connectivity and a strengthening of long-range connectivity, suggesting a process of greater functional segregation in children and greater functional integration in young adults at the whole-brain level [40].

RSNs were also studied for effects of healthy aging. Graph-theoretic measures of global and local efficiency were reduced in frontal, temporal, and limbic-paralimbic cortical areas and in subcortical areas such as the pallidum and thalamus of healthy older people. However, the same cortical areas were identified as hubs in both younger and older subjects

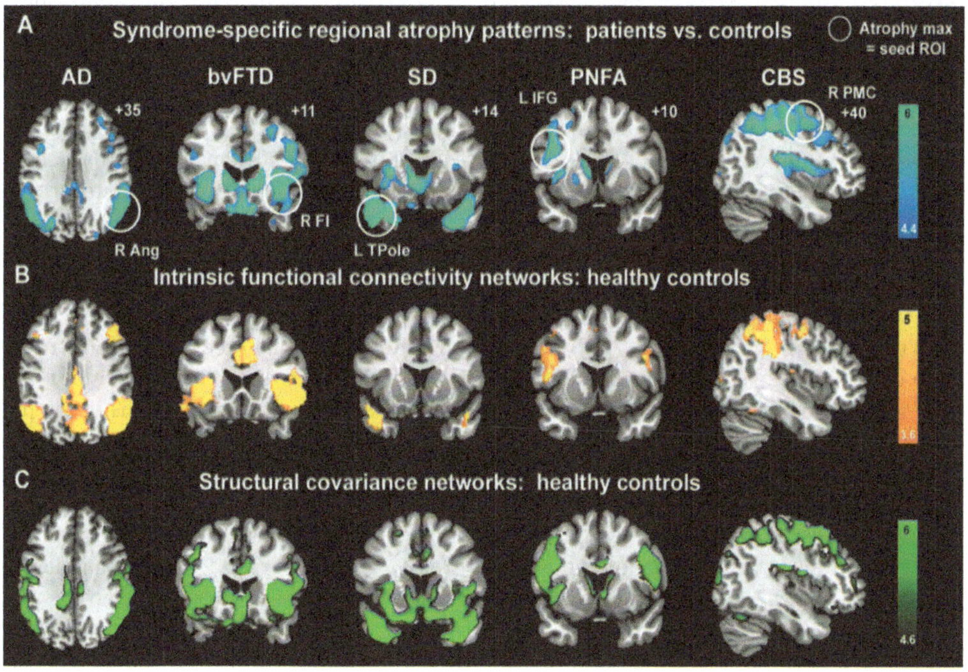

Fig. 6. (A–C). Convergent syndromic atrophy, healthy RSNs, and healthy structural covariance patterns. **(A)** Five distinct clinical syndromes showed dissociable atrophy patterns, whose cortical maxima (circled) provided seed regions of interest for RSN and structural connectivity analyses. **(B)** RSN mapping experiments identified five distinct networks anchored by the five syndromic atrophy seeds. **(C)** Healthy subjects further showed gray-matter volume covariance patterns that recapitulated results shown in panels A and B. For visualization purposes, results are shown at $p < 0.00001$ uncorrected (**A** and **C**) and $p < 0.001$ corrected height and extent thresholds (**B**). **(A–C)** Results displayed on representative sections of the Montreal Neurological Institute template brain. In coronal and axial images, the left side of the image corresponds to the left side of the brain. *Ang* angular gyrus; *FI* frontoinsula; *IFG* inferior frontal gyrus, pars opercularis; *PMC* premotor cortex; *TPole* temporal pole. From [35]. © 2009 Elsevier, reproduced with permission

[1]. This loss of efficiency was consistent with results obtained from ICA, which revealed alterations in the DMN. Decreased activity was found in this network in healthy older as compared to healthy young subjects [13, 24], with the activity in anterior regions being inversely proportional to the score of cognitive tests measuring executive functioning and processing speed.

RSNs and pathologies

RSNs have been investigated in the context of a broad range of pathologies, including Alzheimer's disease (AD), mental disorders, multiple sclerosis, and noncommunicative patients, to name a few [2, 20, 30, 42]. At least three types of benefits could be gained from these investigations. First, RSNs have been evidenced to be sensitive to the progression of certain diseases and could therefore serve as biomarkers to predict their evolution. RSNs could also provide insight into the neural correlates of a disease and thus provide a better understanding of the underlying neurophysiological process. Finally, RSNs could be used to guide an intervention. These topics will respectively be illustrated with three pathologies: AD, noncommunicative patients, and brain tumors.

A critical challenge in the treatment of AD is the identification of biomarkers for the early, prodromal diagnosis of the pathology. A seminal finding was that the regions belonging to the DMN show a loss of connectivity in AD [17]. This promising preliminary result has not been yet followed by a full realization of the potential of RSNs as an early biomarker of AD. It was recently shown that gray-matter atrophy associated with AD follows the pattern of the DMN (Fig. 6). The small-world organization of brain activity, as characterized by the clustering coefficient, also seems to be reduced by the pathology [39]. Increased correlations within prefrontal, parietal, and occipital lobes were also reported in AD patients, which could be interpreted as possible compensatory mechanisms [48].

While we are not aware of many studies using RSNs in the examination of patients with traumatic brain injury, it can be expected that such studies will gain importance in the coming years. A longitudinal follow-up of patients scanned 3 and 6 month after traumatic brain injury showed that, with recovery, graph-theoretic measures get closer to what is observed in healthy controls [31]. Interestingly, functional connectivity within the DMN in noncommunicative patients was negatively correlated with the degree of clinical consciousness impairment [45]. This finding is consistent with the putative critical role of the DMN in self-oriented mental processes.

RSNs could be used not only to provide a diagnostic but also to predict the response of a given patient to a particular treatment and to plan the treatment accordingly. A few recent studies have provided a proof of concept in the context of preoperative mapping of the sensorimotor cortex, as part of the neurosurgical planning for brain tumor resection [25, 36, 49]. Obviously, using RSNs for neurosurgical planning will require specific and thorough evaluation in terms of postoperative outcome. An important caveat is that such a use entails strong assumptions regarding the relationships between the neurophysiological basis of the pathology and BOLD RSNs. Many nonneuronal factors can have an influence on the BOLD signal, a few of which being age, sedation, sleep, and disease. In this last category, the potential effects can be as varied as the pathologies themselves, ranging from a change in the vasculature and/or blood flow alteration (as in stroke), alteration of metabolism (as in gliomas and other tumors), cortex properties (as in AD). For these reasons, there is also a lot to gain in the understanding of RSNs themselves through their use in therapy.

References

[1] Achard S, Bullmore E (2007) Efficiency and cost of economical brain functional networks. PLoS Comput Biol 3: e17

[2] Auer DP (2008) Spontaneous low-frequency blood oxygenation level-dependent fluctuations and functional connectivity analysis of the "resting" brain. Magn Reson Imaging 26: 1055–1064

[3] Biswal B, Yetkin FZ, Haughton VM, Hyde JS (1995) Functional connectivity in the motor cortex of resting human brain using echoplanar MRI. Magn Reson Med 34: 537–541

[4] Biswal BB, Mennes M, Zuo XN, Gohel S, Kelly C, Smith SM, Beckmann CF, Adelstein JS, Buckner RL, Colcombe S, Dogonowski AM, Ernst M, Fair D, Hampson M, Hoptman MJ, Hyde JS, Kiviniemi VJ, Kötter, Li SJ, Lin CP, Lowe MJ, Mackay C, Madden DJ, Madsen KH, Margulies DS, Mayberg HS, McMahon K, Monk CS, Mostofsky SH, Nagel BJ, Pekar JJ, Peltier SJ, Petersen SE, Riedl V, Rombouts SA, Rypma B, Schlaggar BL, Schmidt S, Seidler RD, Siegle GJ, Sorg C, Teng G, Veijola J, Villringer A, Walter M, Wang L, Weng XC, Whitfield-Gabrieli S, Williamson P, Windischberger C, Zang YF, Zhang HY, Castellanos FX, Milham MP (2010) Toward discovery science of human brain function. Proc Natl Acad Sci USA 107: 4734–4739

[5] Buckner RL, Andrews-Hanna JR, Schacter DL (2008) The brain's default network: anatomy, function, and relevance to disease. Ann N Y Acad Sci 1124: 1–38

[6] Bullmore E, Sporns O (2009) Complex brain networks: graph theoretical analysis of structural and functional systems. Nat Rev Neurosci 10: 186–198

[7] Calhoun VD, Kiehl KA, Pearlson GD (2008) Modulation of temporally coherent brain networks estimated using ICA at rest and during cognitive tasks. Hum Brain Mapp 29: 828–838

[8] Calhoun VD, Liu J, Adali T (2009) A review of group ICA for fMRI data and ICA for joint inference of imaging, genetic, and ERP data. Neuroimage 45 (Suppl 1): S163–S172

[9] Chang C, Glover GH (2009) Effects of model-based physiological noise correction on default mode network anti-correlations and correlations. Neuroimage 47: 1448–1459

[10] Cole DM, Smith SM, Beckmann CF (2010) Advances and pitfalls in the analysis and interpretation of resting-state FMRI data. Front Syst Neurosci 4: Article 8

[11] Damoiseaux JS, Greicius MD (2009) Greater than the sum of its parts: a review of studies combining structural connectivity and resting-state functional connectivity. Brain Struct Funct 213: 525–533

[12] Damoiseaux JS, Rombouts SA, Barkhof F, Scheutens P, Stam CJ, Smith SM, Beckmann CF (2006) Consistent resting-state networks across healthy subjects. Proc Natl Acad Sci USA 103: 13848–13853

[13] Damoiseaux JS, Beckmann CF, Sanz Arigita EJ, Barkhof F, Scheltens P, Stam CJ, Smith SM, Rombouts SARB (2008) Reduced resting-state brain activity in the "default network" in normal aging. Cereb Cortex 18: 1856–1864

[14] Esposito F, Bertolino A, Scarabino T, Latorre V, Blasi G, Popolizio T, Tedeschi G, Cirillo S, Goebel R, Di Salle F (2006) Independent component model of the default-mode brain function: assessing the impact of active thinking. Brain Res Bull 70: 263–269

[15] Fransson P, Skiöld B, Horsch S, Nordell A, Blennow M, Lagercrantz H, Åden U (2007) Resting-state networks in the infant brain. Proc Natl Acad Sci USA 104: 15531–15536

[16] Greicius MD, Krasnow B, Reiss AL, Menon V (2003) Functional connectivity in the resting brain: a network analysis of the default mode hypothesis. Proc Natl Acad Sci USA 100: 253–258

[17] Greicius MD, Srivastava G, Reiss AL, Menon V (2004) Default-mode network activity distinguishes Alzheimer's desease from healthy aging: evidence from functional MRI. Proc Natl Acad Sci USA 101: 4637–4642

[18] Greicius MD, Supekar K, Menon V, Dougherty RF (2009) Resting-state functional connectivity reflects structural connectivity in the default mode network. Cereb Cortex 19: 72–78

[19] Gusnard DA, Raichle ME (2001) Searching for a baseline: functional imaging and the resting human brain. Nat Rev Neurosci 2: 685–694

[20] Guye M, Bettus G, Bortolomei F, Cozzone PJ (2010) Graph theoretical analysis of structural and functional connecitvity MRI in normal and pathological brain networks. Magn Reson Mater Phys Biol Med 23: 409–421

[21] Hayasaka S, Laurenti PJ (2010) Comparison of characteristics between region- and voxel-based network analyses in resting-state fMRI data. Neuroimage 50: 498–508

[22] Honey CJ, Thivierge JP, Sporns O (2010) Can structure predict function in the human brain? Neuroimage 52: 766–776

[23] James GA, Kelley ME, Craddock RC, Holtzheimer PE, Dunlop BW, Nemeroff CB, Mayberg HS, Hu XP (2009) Exploratory structural equation modeling of resting-state fMRI: applicability of group models to individual subjects. Neuroimage 45: 778–787

[24] Koch W, Teipel S, Mueller S, Buerger K, Bokde ALW, Hampel H, Coates U, Reiser M, Meindl T (2010) Effects of aging on default mode network activity in resting state fMRI: does the method of analysis matter? Neuroimage 51: 280–287

[25] Kokkonen SM, Nikkinen J, Remes J, Kantola J, Starck T, Haapea M, Tuominen J, Tervonen O, Kiviniemi V (2009) Preoperative localization of the sensorimotor area using independent component analysis of resting-state fMRI. Magn Reson Imaging 27: 733–740

[26] Laufs H (2008) Endogenous brain oscillations and related networks detected by surface EEG-combined fMRI. Hum Brain Mapp 29: 762–769

[27] Li K, Guo L, Nie J, Li G, Liu T (2009) Review of methods for functional brain connectivity detection using fMRI. Comput Med Imaging Graph 33: 131–139

[28] Margulies DS, Vincent JL, Kelly C, Lohmann G, Uddin LQ, Biswal BB, Villringer A, Castellanos FX, Milham MP, Petrides M (2009) Precuneus shares intrinsic functional architecture in humans and monkeys. Proc Natl Acad Sci USA 106: 20069–20074

[29] Marrelec G, Bellec P, Benali H (2006) Exploring large-scale brain networks. J Physiol Paris 100: 171–181

[30] Mesulam M (2009) Defining neurocognitive networks in the BOLD new world of computed connectivity. Neuron 62: 1–3

[31] Nakamura T, Hillary FG, Biswal BB (2009) Resting network plasticity following brain injury. PLoS One 4: e8220

[32] Perlbarg V, Marrelec G (2008) Contribution of exploratory methods to the investigation of extended large-scale brain networks in functional MRI – methodologies, results, and challenges. Int J Biomed Imaging 2008: Article ID 218519

[33] Raichle ME, Snyder AZ (2007) A default mode of brain function: a brief history of an evolving idea. Neuroimage 37: 1083–1090

[34] Rogers BP, Morgan VL, Newtonb AT, Gore JC (2007) Assessing functional connectivity in the human brain by fMRI. Magn Reson Imaging 25: 1347–1357

[35] Seeley WW, Crawford RK, Zhou J, Miller BL, Greicius MD (2009) Neurodegenerative diseases target large-scale human brain networks. Neuron 62: 42–52

[36] Shimony JS, Zhang D, Johnston JM, Fox MD, Roy A, Leuthardt EC (2009) Resting state spontaneous fluctuations in brain activity: a new paradigm for presurgical planning using fMRI. Acad Radiol 16: 578–583

[37] Shmuel A, Leopold DA (2008) Neuronal correlates of spontaneous fluctuations in fMRI signals in monkey visual cortex: implications for functional connectivity at rest. Hum Brain Mapp 29: 751–761

[38] Smith SM, Fox PT, Miller KL, Glahn DC, Fox PM, Mackay CE, Filippini N, Watkins KE, Toro R, Laird AR, Beckmann CF (2009) Correspondence of the brain's functional architecture during activation and rest. Proc Natl Acad Sci USA 106: 13040–13045

[39] Supekar K, Menon V, Rubin D, Musen M, Greicius MD (2008) Network analysis of intrinsic functional brain connectivity in Alzheimer's disease. PLoS Comput Biol 4: e1000100

[40] Supekar K, Musen M, Menon V (2009) Development of large-scale functional brain networks in children. PLoS Biol 7: e1000157

[41] Tomassini V, Jbabdi S, Klein JC, Behrens TE, Pozzilli C, Matthews PM, Rushworth MF, Johansen-Berg H (2007) Diffusion-weighted imaging tractography-based parcellation of the human lateral premotor cortex identifies dorsal and ventral subregions with anatomical and functional specializations. J Neurosci 27: 10259–10269

[42] van den Heuvel MP, Hulshoff Pol HE (2010) Exploring the brain network: a review on resting-state fMRI functional connectivity. Eur Neuropsychopharmacol 20: 519–534

[43] van den Heuvel MP, Mandl RCW, Kahn RS, Hulshoff Pol HE (2009) Functionally linked resting-state networks reflect the underlying structural connectivity architecture of the human brain. Hum Brain Mapp 30: 3127–3141

[44] van Dijk KRA, Hedden T, Venkataraman A, Evans KC, Lazar SW, Buckner RL (2010) Intrinsic functional connectivity as a tool for human connectomics: theory, properties, and optimization. J Neurophysiol 103: 297–321

[45] Vanhaudenhuyse A, Noirhomme Q, Tshibanda LJ-F, Bruno, Boveroux P, Schnakers C, Soddu A, Perlbarg V, Ledoux D, Brichant J-F, Moonen G, Maquet P, Greicius MD, Laureys S, Boly M (2010) Default network connectivity reflects the level of consciousness in non-communicative brain-damaged patients. Brain 133: 161–171

[46] Vincent JL, Patel GH, Fox MD, Snyder AZ, Baker JT, Van Essen DC, Zempel JM, Snyder LH, Corbetta M, Raichle ME (2007) Intrinsic functional architecture in the anaesthetized monkey brain. Nature 447: 83–86

[47] Wang J, Wang L, Zang Y, Yang H, Tang H, Gong Q, Chen Z, Zhu C, He Y (2009) Parcellation-dependent small-world brain functional networks: a resting-state fMRI study. Hum Brain Mapp 30: 1511–1523

[48] Wang K, Liang M, Wang L, Tian L, Zhang X, Li K, Jiang T (2007) Altered functional connectivity in early Alzheimer's disease: a resting-state fMRI study. Hum Brain Mapp 28: 967–978

[49] Zhang D, Johnston JM, Fox MD, Leuthardt EC, Grubb RL, Chiconoine MR, Smyth MD, Snyder AZ, Raichle ME, Shimony JS (2009) Preoperative sensorimotor mapping in brain tumor patients using spontaneous fluctuations in neuronal activity imaged with functional magnetic resonance imaging: initial experience. Neurosurgery 65: 226–236

[50] Zuo X-N, Kelly C, Adelstein JS, Klein DF, Castellanos FX, Milham MP (2010) Reliable intrinsic connectivity networks: test-retest evaluation using ICA and dual regression approach. Neuroimage 49: 2163–2177

Neural network analysis and its application in neurosurgical planning

Edwin van Dellen, Linda Douw, Ingeborg Bosma, Jan J. Heimans, Cornelis J. Stam, Jaap C. Reijneveld

Introduction

The human brain is the most complex object known to man. Unraveling its wiring and functioning is one of the greatest challenges in modern science. It is now widely acknowledged that the brain should be seen as a complex network. Understanding the structure and dynamics of this network may form a crucial step towards the understanding of cognition. Moreover, it may help to explain the interaction between brain pathology and symptoms. The research field focusing on neural networks has grown rapidly over the past ten years, now combining knowledge of computational models and animal studies with advanced human neuroimaging techniques [54, 14]. Network studies focus on both structural connections and functional interactions between brain regions. Structural networks can be reconstructed by mapping cortical areas with MRI, or white matter tracts using diffusion tensor imaging (DTI). Functional interactions can be studied directly using electrophysiological registrations such as electroencephalography (EEG) and magnetoencephalography (MEG), or indirectly from blood oxygen level based fMRI.

In this chapter we will give an overview of the basic concepts of network theory especially in relation to neuroscience. Additionally we will describe clinical neurological and neurosurgical applications. In this respect, we will give our view on perspectives of network theory in the surgical treatment of patients with brain lesions and focal epilepsy.

Modern network theory

Modern network theory has its roots in both mathematics and sociology. Network theory crosses boundaries between different sciences, ranging from mathematics to language and social studies, which has led to great research advantages in recent years. Below we will discuss the basis of network theory as this is relevant for state of the art neuroscience. For a more detailed overview we recommend reviews by Bullmore and Sporns [14], Stam et al [61], and Reijneveld et al [54].

A brief history

The mathematician Leonard Euler was the first to use graph theory in 1736 when solving the problem of the bridges of Konigsberg. The question was whether it was possible to cross a river and its two islands by passing each of the seven connecting bridges only once. Euler presented the problem as an abstract network and showed that this was impossible. Since then, graph theory has become an important branch of mathematics, as it is a tool to analyze networks theoretically. It considers a given system as a collection of nodes and connections between these nodes. The nodes are referred to as vertices and the connections as edges. An important step forward in the field was the discovery of random networks, in which the vertices are connected with a fixed and equal likelihood p. Although random networks meant a boost for the science of graph theory, they were not sufficient to model real networks such as social communities or railway maps, as they are highly structured. The psychologist Stanley Milgram tried to bridge this gap, inspired by a short story called "Chains" by the Hungarian writer Frigyes Karinthy in 1929. Karinthy challenged anyone to find a person that was not connected to any other person in the world through at most five others. Milgram took up the challenge by sending letters to randomly chosen subjects in the USA. The subjects were informed to send the letter to any other person that was closer to a target person living in Boston. In 1967, Milgram postulated that every letter was only sent on average 5.5 times before reaching its destination, which is now associated with the phrase "six degrees of separation". This was the first empirical proof of the "small-world" phenomenon. Small-world properties can be found in many types of networks, including neural networks. In the next sections, we will discuss some basic concepts of graph theory and small-world networks.

Small-world networks

Watts and Strogatz were the first to define the concept of a small-world network [70]. They considered a simple model of a one-dimensional network on a ring in which the vertices were connected in a lattice configuration, i.e. each vertex is connected only to its direct neighbours and their neighbours (see Fig. 1). Two features characterize this network. The *clustering coefficient (C)*, which describes the probability that the neighbours of a node will also be connected to each other, is high. However, it takes a lot of steps to connect vertices at opposite sides of the network, which is described by the long *path length (L)* of the network. This feature is also termed low *efficiency*, which is defined as the inverse of path length. When the edges are rewired with likelihood p, the network becomes random for p=1. The path length is now short, but the system has a very low clustering coefficient. However, by only rewiring only a few edges, the network becomes small-world: high local clustering C is combined with a short path length L.

The small-world phenomenon is thought to increase the efficiency of information processing in a network to some sort of optimal state, as it combines high local clustering and high overall integration. C and L not only depend on topology but also on the size and the number of edges in the network. To correct for these factors, the measures are often divided by the C and L of a random equivalent network. The "small-worldness" or small-world index is then defined by the ratio of these two values.

Network degree

In the Watts and Strogatz model, each vertex is considered to have only small fluctuations in the number of edges connected to it. However, some networks such as the internet do not follow this rule, because the number of connections varies widely between nodes. The *degree (k)* of a vertex is the number of edges that are connected to it. The distribution of the edges over the vertices, or *degree distribution,* is an important feature of network topology. The degree distribution P(k) is defined as the probability that a randomly chosen vertex will have degree k. The key vertices of a network, that have a large number of connecting edges, are known as *hubs.* In a random network, hubs are sparsely found. A network is called *scale-free* when the degree distribution follows a power law, meaning that it has a small number of highly connected vertices and a high number of poorly connected vertices. Scale-free networks greatly depend on a few hub nodes, which take care of most of the information processing in the network. The high importance of these hub nodes also make scale-free networks vulnerable to targeted attacks, because deletion of one or more of these nodes disrupts the whole network.

The *degree correlation* is used to determine to what extent vertices with a similar degree are connected. A high correlation is present in *assortative* networks, which can be seen as the equivalent of "the rich getting richer". When vertex degrees are anticorrelated, the network is called *disassortative*, which predicts a more important role for hubs in the network. Social networks such as a network of company directors tend to be assortative, whereas technological and biological networks, including the brain, tend to be disassortative [48].

Hierarchy and modularity

Networks can contain a number of communities or *modules*. A module is a subgraph with vertices that are more connected to other vertices within the module than to vertices outside the module. Each module contains highly in-

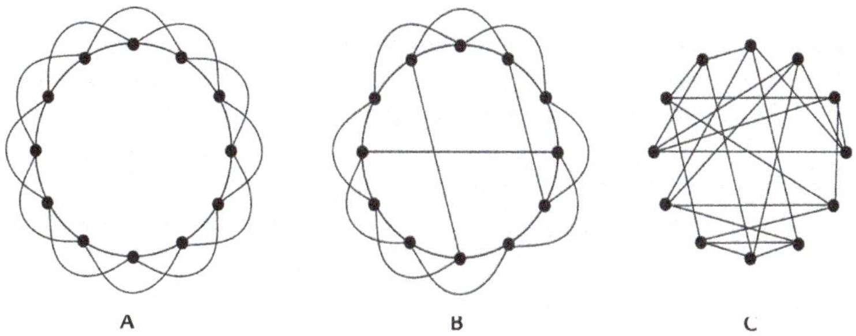

Fig. 1. Three types of networks
Three network types based on the model of Watts and Strogatz, 1998. In an ordered network the points are connected to their nearest neighbours (**A**), but there are no long-distance connections. In a random network (**C**), there is no local clustering. In a small-world network (**B**), some local connections are rewired to long distance connections, resulting in high clustering combined with a short path length

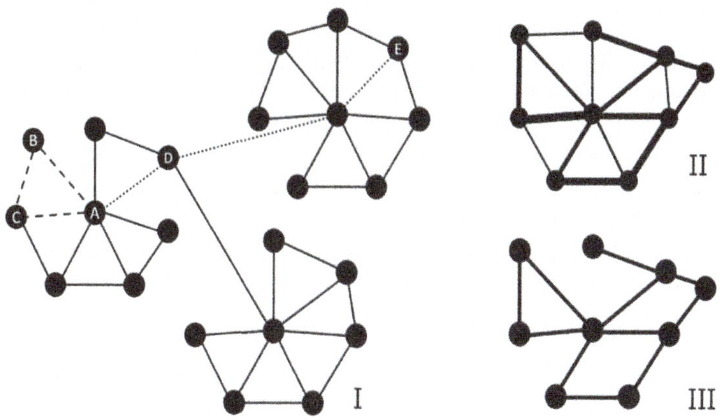

Fig. 2. Network measures
Several measures can be used to characterize network topology. **(I)** The dashed line marks a *local cluster*, as nodes A, B and C are all interconnected. The dotted line marks the *shortest path* between node A and E (L = 3). Three *modules* can be recognized in the figure. Node A is a *provincial hub* node for the left cluster, while node D functions as a *connector hub*. **(II)** A *weighted network* is constructed, for example by using functional connectivity levels between brain regions to characterize connection strengths. **(III)** An *unweighted network* can be constructed from the connectivity matrix by application of a threshold. A connection exists when the threshold value is passed

terconnected edges, and relatively few connections with other modules. Various algorithms can be used to detect modules, of which *hierarchical* clustering seems most commonly used in neuroscience [14]. Hierarchical modularity allows the description of networks at different levels or resolutions. Most global modules can be divided in a number of submodules, which can consist of even smaller submodules, etcetera. In modular networks, nodes can be described based on their function in the network. Hubs are characterized as *provincial,* when having a high degree within the module, or as *connector*, when they have a large number of connections with other modules [29]. Modular structure may be part of an optimal configuration for networks, as modules have high local clustering, while intermodular connections ensure short path lengths [44].

Weighted and unweighted graphs

In a binary or *unweighted* graph, edges either exist or do not exist. This is a convenient way to model a network as analysis of such networks is relatively easy. However, most neural network reconstructions are based on correlations in structural or functional recordings such as MRI or EEG. An unweighted graph can be constructed by application of a threshold to these data; a connection exists when correlations pass this threshold. However, the choice for a certain threshold is arbitrary. Setting it too high will exclude relevant connections and may introduce unconnected vertices, while a low threshold will introduce non-existing or unimportant edges into the network. Moreover, valuable information on the strength of connections is discarded.

In contrast, *weighted* networks include information about connection strengths in the graph. This approach is more accurate in many real networks, for instance when describing how well people know each other in a social network. Most network measures have also been described for weighted networks, such as weighted path length L, weighted local clustering C, and degree correlation [55]. Note that in a weighted network, the shortest path is not necessarily the path with the least number of edges. The path length in weighted networks is best calculated by first calculating the efficiency. In the case that a path does not exist, the path length becomes infinite and the efficiency is zero. The average shortest path length of the network is then defined as the harmonic mean of the inverse of the efficiencies [48].

Despite these efforts to improve the accuracy of network modeling, it remains difficult to compare networks of different sizes and degrees. The size or total number of vertices can differ between networks and change network characteristics. At present, there is no consensus in the field how to address these questions, and approaches differ between scientific groups.

Dynamic processes on complex networks

An interesting branch of research in modern network theory, that is highly relevant for neuroscience, is the relationship between structure and function. The relationship between network topology and synchronization dynamics on these networks is crucial in this respect, as a wide range of studies have pointed out that brain areas interact through synchronizing activity. In brief, network synchronizability characterizes the stability of synchronized state of the network. We suggest that network dynamics will be of increasing interest when studying communication patterns in brain networks. For a more detailed description of dynamic processing on complex networks we recommend reviews by Stam and Reijneveld [61], and Bocaletti et al [9].

Network and the brain

Neural network modeling and experimental studies

Watts and Strogatz were not only the first to model small-world networks, but also made the connection between graph theory and neuroscience. They showed that the clustering coefficient and average shortest path length of the C. Elegans nervous system follow the small-world principle [70]. The same was shown for cortico-cortical connection data of the cat and macaque [32]. A model study based on neuro-anatomical data found the brainstem reticular formation network best described as small-world. However, the degree distributions did not follow a power law in the models, making a scale-free architecture less probable.

Besides these studies on structural connections, some functional requirements have been formulated for neural networks. The brain must be able to perform specific tasks, which require areas of highly interconnected neurons. These processes need to be coordinated in a system that combines and plans these tasks [59]. Sporns and others defined neuronal *complexity* as a measure for the extent to which functional connectivity patterns combine local segregation with global integration. Moreover, as part of a biological system, the network has to be highly cost effective in terms of "wiring" and energy use. Computational models have been used to study network facilitation of neural processing. They showed that when networks optimize for complexity, this leads to increased small-world characteristics and decreased "wiring costs" [59].

The macaque cortex has also been used to study the correlation between anatomical and functional connectivity patterns. Functional networks proved to overlap with underlying structural networks in a computational model of macaque cortex [33]. Epileptic activity was used as a model for functional interactions. The spread of (epileptic) activity was shown to follow a small-world pattern in lesioned cortex in vivo, suggesting a correlation between anatomical and functional connectivity. Based on the epileptic activity in a large-scale cat cortex model, association fibers and their connection strengths seem useful predictors of global activation patterns [37].

Although structural networks determine functional connectivity patterns to an important extent [33], this interaction may also work the other way around. Functional connectivity seems to play a role in the development of neural networks at a cellular level. Scale-free topology was found in developing functional networks of the rat and mice hippocampus [11]. GABAergic interneurons with widespread axonal aborizations function as hub neurons, and play an important role in the shaping of the network topology and dynamics.

Human brain networks

Brain networks can be studied at various levels, ranging from structural cellular networks to functional interactions between brain areas. In this section we will focus on networks based on *in vivo* neuroimaging and neurophysiological data. Global findings seem generally in this field, but it is important to realize that differences in imaging techniques and the construction of networks may limit comparability [39]. Various measures exist for connectivity, and the method of choice when defining brain regions of interest can lead to different results [69].

Structural networks

Structural networks of the human brain have been constructed from MRI recordings. Connections between brain areas were defined as a cross-correlation in cortical thickness across individual brains [31]. This network was found to be small-world, and a modular structure was shown with resemblance to functional systems including the sensorimotor cortex and visual cortex [17]. Structural networks of grey matter areas have also been constructed using diffusion tensor imaging (DTI). Again, small-world topology was found. Also, several hub regions were identified, including the precuneus, the insula, and the superior frontal and parietal cortex, and parts of the human default mode network [30]. Interestingly, overall connectivity was higher in women, and their cortical networks were more efficiently connected than male functional networks [27]. Furthermore, local connectivity decreased with age, and older brains showed a shift of efficient connections (i.e., shorter path length) from parietal and occipital to frontal and temporal regions [27]. Changes in modular structure based on age have also been described [43]. These findings raise the question how structural neural network topology relates to functioning.

Functional networks

In this section, we will focuss on topological characteristics of the functional brain networks, how these functional structures arise during maturation, and how this correlates to brain functioning. Functional interactions between brain areas can be studied based on the concept of *functional connectivity*. Functional connectivity denotes the statistical interdependencies between elements of the system, for instance the synchronization of two simultaneously recorded EEG signals [2]. It is thought that synchronization is a key process in communication between different brain areas and cognitive functioning [68]. The ability of the network to facilitate synchronizing activity is therefore of great interest. Characteristics such as small-world index, average degree, degree distribution and degree correlation effect network synchonizability [4, 71]. For weighted graphs, random networks seem most likely to synchronize, followed by small-world networks and regular networks, respectively [15].

Functional brain networks can be constructed by studying functional connectivity (i.e., correlation) between neurophysiological time series from several brain areas. A connectivity matrix is then constructed, which can be analysed using graph theory. The connectivity matrix of electrophysiological recordings reflects a measure of synchronization or phase coupling of time series with a potentially high *temporal* resolution. Functional networks based on fMRI potentially have the highest *spatial* resolution, and are relatively easily compared with structural MRI anatomical networks. However, as with structural MRI, these images do not directly contain information about the functional coupling of brain regions. It is possible to construct a matrix of functional connectivity based on blood-oxygen level dependent (BOLD) contrast fMRI [14]. A series of recordings is made of the subject, and the correlation of BOLD responses between different regions of interest (ROIs) is calculated.

Small-world topology in the human brain was first shown with functional connectivity analysis of MEG recordings of 5 subjects [60]. The first fMRI network study on 90 ROIs of the human brain also showed small-world topology [56]. A more recent study dramatically increased network size by performing voxel-based analysis of connectivity [66]. Neural networks appear to be not only small-world, but also scale-free, which was also found earlier in the macaque cortex [25, 66]. This would implicate a major role for functional hub nodes in the human brain network. Another study showed that two types of hubs could be distinguished within the fMRI network: hubs with long-distance connections to other regions, and more "cliquishly" connected regions [1].

These findings implicate a modular structure of the functional human brain. As might be expected, modularity studies show that functionally or anatomically related brain regions are more densely interconnected [26, 44]. A hierarchical structure was found with five modules at the global level in both studies. Hub nodes were especially found in the association areas.

Brain network topology seems to have a genetically determined component. Graph analysis of EEG recordings of 574 twins and their siblings showed that clustering, path length and small-world topology are heritable characteristics [58]. Moreover, network topology changes during brain maturation [62, 10]. Boersma and colleagues followed 227 children and compared EEG recordings at the age of 7 to EEGs recorded two years earlier at 5 years of age. At age 7, the children had increased path lengths and clustering, and a decrease of weight correlation in comparison to the recording of two years earlier.

Supekar and others compared fMRI networks of children with young adults [62]. In children (ages 7–9 yrs), subcortical areas were more strongly connected with primary sensory, association and paralimbic areas. Cortico-cortical connectivity between paralimbic, limbic, and association areas was stronger in young adults (ages 19–22 yrs). Although all subjects showed small-world brain topology, hierarchy was higher in young adults, indicating a higher number of locally clustered hubs. Moreover, DTI-based wiring analysis showed that short-distance connectivity weakens during development, while long-range functional connectivity becomes stronger. This indicates a possible correlation between functional connectivity and maturation of the white matter tracts. These maturation studies indicate that the brain network evolves to a more efficient topology during maturation.

Intelligence and cognitive functioning are related to efficient functional network organization [67]. Micheloyannis and colleagues constructed unweighted graphs from EEG recordings [45]. Subjects with a few years of formal education (LE) were compared to subjects with university degrees (UE) during a working memory test. The LE group showed more prominent small-world topology during the test than the UE group, which was interpreted as a higher need for efficient network processing in this group. Van den Heuvel and colleagues analysed fMRI BOLD images of 19 subjects and constructed voxel-based networks [67]. Subjects' IQs correlated with shorter path lengths. The correlation was found to be most prominent in the medial prefrontal gyrus, the precuneus, and bilateral inferior parietal regions. Moreover, efficient wiring seems to be related not only to functional, but also to structural network topology. Li and colleagues showed that structural network topology is related to intelligence in 79 subjects [41]. Subjects with high intelligence have more efficient networks, based on DTI, than subjects with average intelligence. IQ test scores are correlated with the global efficiency of their structural networks.

In summary, the human brain network is characterized by small-world and possibly scale-free topology, found in both structural and functional brain networks. Network topology has a heritable component but becomes more efficient during maturation. Efficient brain network organization is crucial for cognitive functioning, and a more efficient network topology seems to be related to higher intelligence.

Network alterations in brain disease

As described above, efficient functional network topology is thought to reflect optimal brain functioning. Application of graph theory to brain recordings may also provide better insight into underlying mechanisms as well as diagnostic markers for brain disease. Differences have been described for a range of neurological and psychiatric diseases, including brain tumours, Alzheimer's disease, epilepsy, and schizophrenia (for an overview see [54]).

In this section, we will focus on changes in functional connectivity and network architecture of patients with lesional epilepsy, some of whom having a brain tumour, as these pathological conditions are particularly relevant for neurosurgical practice. Brain tumours and epilepsy have high co-morbidity; the incidence of epilepsy is 30% or more in brain tumour patients, depending on tumour type [64]. Conversely, the incidence of brain tumours in epilepsy patients is 4% [64]. Temporal lobe epilepsy, which accounts for half of the medically refractory epilepsies, is characterized by sclerosis of the medial temporal lobe. A better understanding of network alterations due to lesions may help to explain the frequent occurrence of epileptic seizures in these patients. We will first describe the state of the art in graph theoretical analysis of brain tumour and epilepsy patients. We will then focus on future prospects of using functional connectivity analysis as diagnostic marker and as tool for pre-surgical planning.

Brain tumours

Since graph analysis in brain tumour patients is a very novel approach, studies are still relatively sparse. Our first study in these patients was published in 2006, comparing MEG recordings of patients with various types of brain tumours with healthy controls [5, 6]. A general decrease of broad band functional connectivity (0.5–60 Hz) was found in the patient group. Separate frequency band analysis showed increased connectivity in the delta to alpha frequency ranges in the patient group, while lower connectivity was found in the beta and gamma band. Remarkably, alterations were not restricted to the lesional area, but were also found within the contralateral hemisphere [6]. Unweighted network analysis showed more random networks in patients, as clustering was decreased in the patient group in the theta and gamma band, as well as path length in the theta, beta and gamma band [5].

Guggisberg and colleagues also analysed MEG recordings for functional connectivity patterns of 15 brain lesion patients of mixed pathology, and compared them to a healthy control group [28]. Patients with lesion-induced neurological deficits had decreased connectivity levels in the lesioned areas compared to the contralateral homologue areas. Interestingly, areas with decreased connectivity could be resected without any deficits after surgery, as long as patients had no neurological deficits prior to the resection.

In two subsequent MEG studies, we analysed a more homogenous group of 17 low-grade glioma (LGG) patients and compared them to 17 healthy controls matched for age, gender and educational level [12, 13]. Again, patients had increased connectivity levels in the lower frequency bands but also in the lower gamma band, now for long-distance connections. However, the increase in long-distance connectivity was also found in the lower gamma band, while a decrease was found in the lower alpha band. A highly interesting finding in this study was that the functional connectivity changes were related to poorer cognitive functioning in the patient group [12]. Increased connectivity levels were again found in the theta band of the patient group. Unweighted network analysis showed increased theta band and decreased beta band clustering in patients, as well as lower small-world index in the beta band and lower degree correlations in the gamma band [13]. Lower degree correlation, higher clustering and longer path lengths in the delta and lower alpha band correlated with poorer cognitive performance in the patient group.

Brain tumour patients thus seem to have widespread altered functional connectivity patterns that may correlate with their cognitive deficits. Moreover, the alterations may predict surgical outcome regarding neurological deficits. This raises the question how functional connectivity patterns relate to surgical intervention. We have studied functional connectivity in MEG recordings of 15 brain tumour patients before and after surgery [19]. Long-distance interhemispheric functional connectivity in the theta band was found to decrease after tumour resection, which was not influenced by several treatment- and tumour-related factors. We speculated that the decreased connectivity after surgery may reflect recovery to a normal state, based on abovementioned increased theta band connectivity. Indeed, patients with a larger decrease of theta connectivity were more often seizure-free after tumour resection. Changes in functional networks may thus well be related to vulnerability to tumour-related epilepsy, especially in the theta band, as we will describe in the following sections.

Summarizing these findings, we conclude that altered topology is found in the functional networks of brain tumour patients, along with widespread changes in functional connectivity. These changes seem to be related to cognitive deficits in these patients. Several clinical factors may be of influence on connectivity changes, such as tumour pathology, epilepsy and treatment, which should be clarified in future studies [65, 72]. Network analysis seems a promising tool to better predict surgical outcome, which should be subject of further research.

Epilepsy

Synchronization is crucial for information processing in the brain. However, it is also an important characteristic of epileptic seizures [63,

40]. Synchronization patterns not only differ between the ictal and interictal state, but also change during a seizure. Connectivity changes over a broad range of frequencies were found in EEG recordings prior to and during seizures in focal and absence epilepsy patients [51, 57, 53]. Graph analysis showed a regularization of network topology during seizures in the broad frequency band (0.5–70 Hz), but most explicit in low frequencies (below 13 Hz). A more detailed study at seizure onset by Kramer and others suggests that network changes have patient-specific topographical characteristics [38]. They studied intracranial EEG recordings of four pharmacoresistant epilepsy patients. Again, the average path length became longer during ictal activity, and small-worldness increased, as did *betweenness centralization* (which is a measure similar to *degree correlation* described in the section *Network degree* above). In general, the degree decreased, while clustering changes were reordered differently depending on topographical localization.

Modeling studies

These findings raise the question what triggers the transition towards a pathologically synchronized state, as this might explain how seizures arise. Several model studies of seizure dynamics have suggested that epilepsy patients have altered network topology not only during seizures, but also in the interictal state. Netoff and others were the first to study neural excitation activity in epileptogenic networks, using a hippocampal slice network model [47]. A small-world network was rewired with an increasing number of long-distance connections, making the network more random. Results showed that excitation levels changed from normal to seizure-like activity to bursting activity. Bursting activity was also found in interictal electrophysiology recordings, indicating that epilepsy patients may have more random interictal networks compared to healthy individuals. Dhyrfield-Johnsen and colleagues simulated medial temporal sclerosis in a computational model network by removing long-distance connections [24]. This initially increased the local clustering of the network as well as hyperexcitability. Local clustering decreased when the number of long distance connections was further decreased. Morgan and Soltesz hypothesize that networks of epilepsy patients are characterized by "pathological hubs" that increase epileptic activity [46]. They showed that the incorporation of a small number of highly interconnected hub cells greatly increases synchronizing network activity in a rat dentate gyrus model. The network became hyperexcitable and possibly seizure-prone.

These modeling studies together with mathematical network modeling studies suggest that several factors of network topology contribute to the vulnerability for seizures, including path length, clustering and the number of hubs. Further research is needed to elucidate the interactions between measures such as overall connectivity levels, degree distribution, and synchronizability measures such as the eigenvalue ratio. Studying these interactions should lead to a integrated model for epilepsy and network topology.

Epilepsy patient studies

Several imaging techniques have been used to study alterations in functional brain networks of epilepsy patients during the interictal phase. Bettus and colleagues found a general decrease of connectivity in an fMRI study comparing mesial temporal lobe epilepsy (MTLE) patients to healthy controls [7]. Temporal connectivity on the contralateral side of the lesion seemed to be increased, which was interpreted as a possible compensation mechanism. Liao and others showed increased connectivity in the lesioned area but decreased connectivity between inferior frontal triangular and opercular regions correlated with epilepsy duration in fMRI recordings of MTLE patients [42]. Frontal and parietal within region connectivity was decreased as well as connectivity between frontal and parietal regions. The default mode network seemed to have less connectivity with other regions in patients. Network analysis showed lower path length and clustering in patients compared to healthy controls, indicating network randomization in MTLE epilepsy patients.

We recently studied the feasibility of functional connectivity of interictal EEG recordings as a diagnostic tool for epilepsy after a first seizure-like collapse [21]. We found that theta band connectivity of interictal EEG recordings was already increased in epilepsy patients after their first seizure. Moreover, increased theta band connectivity was predictive for a second seizure while visual inspection of the EEG showed no abnormalities. Horstmann and others compared interictal EEG recordings of anti-epileptic drug (AED) resistant epilepsy patients to healthy controls. Increased functional connectivity levels were found in patients in the delta, theta, and beta frequency bands of EEG recordings [35]. Weighted networks of the EEG recordings showed increased path lengths and clustering in patients in several frequency bands.

In two recent studies we analysed MEG recordings of focal epilepsy patients with different lesion types, and compared them to healthy controls [23, 65]. Functional connectivity levels showed no alterations in patients compared to controls, but theta band PLI was correlated with seizure frequency in the patient group. Clustering was higher in the theta and lower alpha band, while theta band path lengths tended to be longer in the patient group. Moreover, theta band path length was positively correlated with time since first seizure and seizure frequency. The correlation between epilepsy duration, functional connectivity and network topology was also found using fMRI and electrocorticography (ECoG) recordings of the lesioned area of MTLE patients [65, 42]. Our MEG and ECoG studies also showed differences in network topology between patients on AED polytherapy compared to patients on monotherapy. Finally, lesion type also had some effect on network topology in MEG recordings: networks showed more alterations in LGG and non-glioma patients than HGG patients [65].

Summarizing these findings, research has shown alterations of connectivity levels and network topology in lesional epilepsy patients. Altered connectivity patterns correlate with epilepsy duration in several studies. Increased theta band path lengths show correlations with

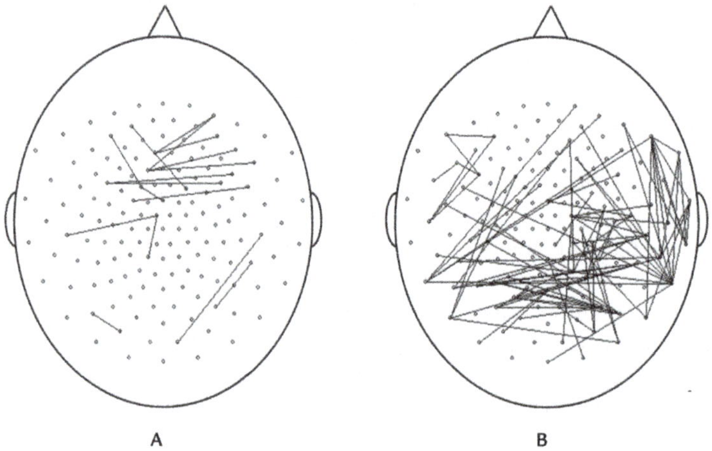

Fig. 3. Altered functional connectivity in lesional epilepsy patients
Theta band (4–8 Hz) functional connectivity differences ($p < 0.01$) between lesional epilepsy patients and healthy controls derived from MEG recordings [65, 72]. Connections that are decreased in patients are shown in the left figure, connections with increased strength in patients are shown on the right. In general, connections in patients seem increased, especially in the central, occipital and parietal regions of the right hemisphere

several factors that determine epilepsy burden, such as epilepsy duration, seizure frequency, and AED monotherapy or polytherapy. Theta band connectivity analysis of EEG and MEG recordings seems promising as diagnostic tool for epilepsy patients.

Localisation of the epileptic zone

Several studies have focused on more specific spatial alterations in functional networks of epilepsy patients in order to find markers of the *epileptic zone* from which seizures arise. Bettus and others compared functional connectivity patterns in EEG recordings of mesial temporal lobe epilepsy (MTLE) patients to those of patients with epilepsy of other origin (non-MTLE) [8]. A general increase of connectivity was found in the epileptic zone of the MTLE patients compared to the same areas in the non-MTLE group in the 3.4–97 Hz frequency range.

Ortega and colleagues used connectivity analysis of ECoG recordings to predict effectiveness of surgical intervention [49| When sharply defined connectivity peaks were found in the preoperative recordings, resection of these peak areas was positively related to surgery outcome. More equally distributed functional connectivity patterns before surgery were related to poorer outcome. In a second study on five patients with similar pathology, the authors suggested that removal of hub nodes may be crucial for successful treatment, as these hub nodes seemed involved in the development of seizures [50].

A promising approach for the identification of seizure onset zones, even without the need for invasive electrophysiological recordings, is modularity analysis. Chavez and others studied modular structures in MEG recordings of five absence epilepsy patients and five healthy controls [16]. Networks were constructed in the 5–14 Hz frequency range, combining theta and alpha band ranges. Patients were shown to have less modules than healthy controls, and the connections between modules were stronger in the patient group. Connectivity levels within the module showed only minor differences between groups. This study indicates that especially intermodular (hub) connections may play an important role in absence epilepsy. Future research is needed

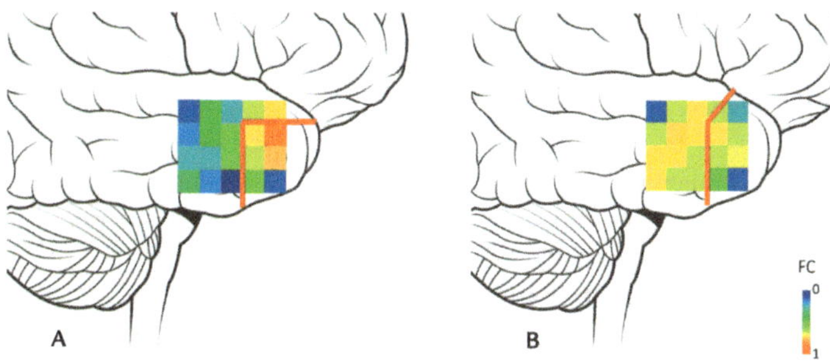

Fig. 4. Functional connectivity and surgical outcome

Functional connectivity as measured on the temporal lobe with ECoG recordings during repilepsy surgery, based on Ortega and colleagues [49]. Resection borders are marked by red lines, the colored squares indicate functional connectivity levels of 4 x 5 grid electrodes. **(A)** Functional connectivity is shown to have a peak value in the resected area which is surrounded by lower connectivity levels. Resection of such a sharply defined connectivity peak is associated with good surgical outcome regarding seizure burden. **(B)** When no clear synchronization cluster is present or the synchronization cluster is not resected, the patient is less likely to be seizure-free after surgery to elucidate whether altered modular structures are a general feature of changing networks across epilepsy types. Moreover, it would be highly interesting to see whether provincial or interregional connector hub nodes can be used to identify epileptic zones in MEG recordings.

(Impact of) surgery

Neurosurgical resection can be seen as the infliction of an acute brain lesion. It might be expected that lesions with a sudden occurrence in the brain have a different impact on brain networks than slowly developing lesions, such as tumours. How acute brain lesions directly affect functional connectivity and networks is difficult to study, as ideally one would like to have imaging of the same brain in a healthy and a lesioned state. We already mentioned our study of connectivity before and after tumour resection, in which networks seemed to return to a "healthy" topology [19]. However, this study focused on patients in which the brain was already damaged due to a tumour. In this section, we will discuss some modelling and clinical studies that indicate how neurosurgical intervention affects network topology.

Model studies

Modelling studies on the impact of brain lesions on connectivity and network topology have shown that neural networks are most vulnerable for targeted attacks on hubs. It was found in a cat and macaque cortex connectivity model that random attacks have only minor effects on path lengths, while targeted attacks on important nodes have a major impact on the global integration of the network [36]. Scale-free networks had most similarities regarding robustness against lesions with the brain networks. A model based on human diffusion MRI maps with known structural connections showed that targeted attacks also cause great disturbance on functional interactions, whereas random attacks had relatively little impact. [3]. Moreover, some of the most disruptive lesion sites were found to be central nodes that corresponded to regions where lesions are known to produce complex cognitive disturbances, and were located within the "default mode network". Another modelling study on the effects of lesions found that the most densely connected neural regions synchronize most easily [34]. Again, lesioning the nodes with most connections, especially connector hubs, had the largest impact on cortico-cortical interactions. Given the importance of provincial and connector hubs in this model, findings also indicate that modular structure might affect robustness against targeted lesions.

A stroke is an acute brain lesion of the brain that may have similar impact on brain dynamics. Only one report describes a comparison of a (right capsula interna) stroke patient with 8 healthy controls before and during a motor task [18]. EEG recordings were used to construct unweighted networks, and the patient's network was compared to that of healthy controls. Unweighted networks were reconstructed from this functional connectivity data, and the patient's network was compared to that of healthy controls. Decreased local and long-distance network efficiency was found in the beta and gamma frequency range for the patient. This indicates that less optimal network topology may be related to neurological deficits in stroke patients. Further research in a larger population, possibly with more sophisticated network analysis would be of great interest in these patients.

Clinical studies

We used the Wada test as a model for acute brain lesions in patients in [20, 22]. The Wada test or intra-arterial amobarbital procedure (IAP) is used to assess functioning of the brain after "shutting down" one hemisphere with amobarbital as part of pre-surgical screening in pharmaco-resistant epilepsy patients. Functioning of the non-anaesthetised hemisphere can temporarily be assessed by means of neuropsychological testing, while the EEG is recorded. A consistent increase of connectivity was found in the injected hemisphere after injection. Contralateral and interhemispherical functional connectivity decreased in the delta and theta band after injection, while connectivity increased in the beta band. We recently performed graph analysis on the same data set [22]. Weighted clustering decreased after injection in the theta, alpha, and beta bands. There was also a decrease of the path length in the theta and alpha bands. Moreover, edge weight correlation decreased in the theta and beta band after injection. These alterations all indicate that the whole neural network becomes more random after sedation of one hemisphere, replicating the pattern observed in brain tumour patients. We also analysed the correlation between memory performance during the Wada test with network topology after injection. Increased small-world topology in the theta band, as well as longer path lengths in the alpha band, were correlated with better memory scores, indicating direct associations between network topology and cognitive functioning.

In the previous sections, we have shown that brain tumours and epilepsy are related to altered functional networks. Resective surgery may also lead to global changes in functional network topology, and should be seen in this respect as a targeted attack of brain networks. We will give our vision on the future use of network theory in surgical planning in the final section of this chapter.

Conclusions and future prospects

In this chapter, we have shown that efficient brain network organization is crucial for optimal functioning. Moreover, network topology is disturbed in a wide range of neurological patients. Particularly brain tumour and lesional epilepsy patients may benefit from presurgical mapping of functional and structural networks. A better identification of the epileptic zone and target areas for surgical resection based on network analysis could in the future be used to guide surgical procedures and improve surgical outcome. It may also help to improve prediction cognitive deficits due to surgery, as well as post-surgical seizure burden in epilepsy patients. We consider the application of network analysis in the clinical setting an ultimate and feasible goal of this research.

The rising field of network analysis of the human brain brings a rapidly increasing understanding of neural network functioning, but improved methodology is still needed for clinical application. Several issues need to be

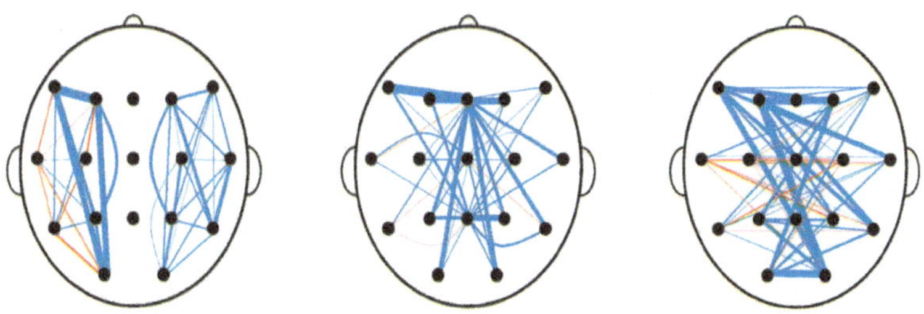

Fig. 5. Altered functional connectivity during Wada test
Example of alterations in one patients in EEG functional connectivity during the Wada test, while the left hemisphere is temporarily "shut down". The level of altered connectivity is visualized by the thickness of the connection; blue lines mark a decrease in connectivity, red lines mark an increase. (**A**) connectivity is increased in the injected hemisphere, but decreased in the contralateral hemisphere. In general, connections to the midline (**B**) and connections between both hemispheres (**C**) are also decreased

addressed before findings in different studies can be properly compared. The size of the constructed networks is still limited by technical and computation power. Moreover, no paradigm exists to properly compare networks of different size. Finally, it remains to be studied how functional networks relate to structural networks, or even how networks constructed from neurophysiological recordings such as EEG and MEG relate to those based on fMRI [14, 15].

Connectivity characteristics may provide markers for specific cognitive tasks. A lot of research in large populations is needed to clarify these interactions and to make it possible to take inter-subject variations into account. However, findings in both structural and functional imaging studies seem promising [13, 41, 67]. We speculate that advanced network analysis of functional modularity and path length and centrality per node may play an important role in the characterization of these markers.

The number of studies on the use of functional connectivity for the identification of epileptic onset zones and crucial areas for seizure spread has increased dramatically over the past years. Results of this increasing number of studies are not consistent yet, but begin to shape the prospects for clinical application. Theta band connectivity and networks based on MEG, EEG, and ECoG recordings seem to be essential in this respect, as it correlates with several factors determining epilepsy burden, including seizure frequency, duration and AED use [19, 13, 65, 35]. We have already demonstrated the possible value of theta band functional connectivity for diagnosing epilepsy after a first collapse in patients without visually detectable disturbances on EEG recordings [21]. Moreover, ECoG connectivity seems to have predictive value regarding seizure outcome [49, 50]. A properly funded definition of epileptic hub nodes seems to be a pivotal step towards clinical application of network analysis in the neurosurgical treatment of epilepsy.

References

[1] Achar, S, Salvador R et al (2006) A resilient, low-frequency, small-world human brain functional network with highly connected association cortical hubs. J Neurosci 26(1): 63–72

[2] Aertsen AM, Gerstein GL et al (1989) Dynamics of neuronal firing correlation: modulation of "effective connectivity". J Neurophysiol 61(5): 900–917

[3] Alstott J, Breakspear M et al (2009) Modeling the impact of lesions in the human brain. PLoS Comput Biol 5(6): e1000408

[4] Barahona M and Pecora LM (2002) Synchronization in small-world systems. Phys Rev Lett 89(5): 054101

[5] Bartolomei F, Bosma I et al (2006a) Disturbed functional connectivity in brain tumour patients: evaluation by graph analysis of synchronization matrices. Clin Neurophysiol 117(9): 2039–2049

[6] Bartolomei F, Bosma I et al (2006b) How do brain tumors alter functional connectivity? A magnetoencephalography study. Ann Neurol 59(1): 128–138

[7] Bettus G, Guedj E et al (2008a) Decreased basal fMRI functional connectivity in epileptogenic networks and contralateral compensatory mechanisms. Hum Brain Mapp 30(5): 1580–1591

[8] Bettus G, Wendling F et al (2008b) Enhanced EEG functional connectivity in mesial temporal lobe epilepsy. Epilepsy Res 81(1): 58–68

[9] Boccaletti S, Latora V et al (2006) Complex Networks: Structure and dynamics. Phys Reports (424): 175–308

[10] Boersma M, Smit DJ et al (2010) Network analysis of resting state EEG in the developing young brain: Structure comes with maturation. Hum Brain Mapp 32(3): 413–425

[11] Bonifazi P, Goldin M et al (2009) GABAergic hub neurons orchestrate synchrony in developing hippocampal networks. Science 326(5958): 1419–1424

[12] Bosma I, Douw L et al (2008) Synchronized brain activity and neurocognitive function in patients with low-grade glioma: a magnetoencephalography study. Neuro Oncol 10(5): 734–744

[13] Bosma I, Reijneveld JC et al (2009) Disturbed functional brain networks and neurocognitive function in low-grade glioma patients: a graph theoretical analysis of resting-state MEG. Nonlinear Biomed Phys 3(1): 9

[14] Bullmore E and Sporns O (2009) Complex brain networks: graph theoretical analysis of structural and functional systems. Nat Rev Neurosci 10(3): 186–198

[15] Chavez M, DU Hwang et al (2005) Synchronization is enhanced in weighted complex networks. Phys Rev Lett 94(21): 218701

[16] Chavez M, Valencia M et al (2009) Functional modularity of background activities in normal and epileptic brain networks. *http: //arxiv.org/abs/0811.3131v3*

[17] Chen ZJ, He Y et al (2008) Revealing modular architecture of human brain structural networks by using cortical thickness from MRI. Cereb Cortex 18(10): 2374–2381

[18] de Vico Fallani F, Astolfi L et al (2009) Evaluation of the brain network organization from EEG signals: a preliminary evidence in stroke patient. Anat Rec (Hoboken) 292(12): 2023–2031

[19] Douw L, Baayen H et al (2008) Treatment-related changes in functional connectivity in brain tumor patients: A magnetoencephalography study. Exp.Neurol. 212(2): 285–290

[20] Douw L, Baayen JC et al (2009) Functional connectivity in the brain before and during intra-arterial amobarbital injection (Wada test). Neuroimage 46(3): 584–588

[21] Douw L, de Groot M et al (2010a) Functional connectivity is a sensitive predictor of epilepsy diagnosis after the first seizure. PLoS One 5(5): e10839

[22] Douw L, van Dellen E et al (2010b) The lesioned brain: still a small world? Front Hum Neurosci 4: 12

[23] Douw L, van Dellen E et al (2010c) Epilepsy is related to theta band brain connectivity and network topology in brain tumor patients. BMC Neurosci 11(11): 103

[24] Dyhrfjeld-Johnsen J, Santhakumar V et al (2007) Topological determinants of epileptogenesis in large-scale structural and functional models of the dentate gyrus derived from experimental data. J Neurophysiol 97(2): 1566–1587

[25] Eguiluz VM, Chialvo DR et al (2005) Scale-free brain functional networks. Phys Rev Lett 94(1): 018102

[26] Ferrarini L, Veer IM et al (2009) Hierarchical functional modularity in the resting-state human brain. Hum Brain Mapp 30(7): 2220–2231

[27] Gong G, Rosa-Neto P et al (2009) Age- and gender-related differences in the cortical anatomical network. J Neurosci 29(50): 15684–15693

[28] Guggisberg AG, Honma SM et al (2008) Mapping functional connectivity in patients with brain lesions. Ann Neurol 63(2): 193–203

[29] Guimera R, Mossa S et al (2005) The worldwide air transportation network: Anomalous centrality, community structure, and cities' global roles. Proc Natl Acad Sci USA 102(22): 7794–7799

[30] Hagmann P, Cammoun L et al (2008) Mapping the structural core of human cerebral cortex. PLoS Biol 6(7): e159

[31] He Y, Chen ZJ et al (2007) Small-world anatomical networks in the human brain revealed by cortical thickness from MRI. Cereb Cortex 17(10): 2407–19

[32] Hilgetag CC, Burns GA et al (2000) Anatomical connectivity defines the organization of clusters of cortical areas in the macaque monkey and the cat. Philos Trans R Soc Lond B Biol Sci 355(1393): 91–110

[33] Honey CJ, Kotter R et al (2007) Network structure of cerebral cortex shapes functional connectivity on multiple time scales. Proc Natl Acad Sci USA 104(24): 10240–5

[34] Honey CJ and Sporns O (2008) Dynamical consequences of lesions in cortical networks. Hum Brain Mapp 29(7): 802–809

[35] Horstmann MT, Bialonski, S et al (2010) State dependent properties of epileptic brain networks: Comparative graph-theoretical analyses of simultaneously recorded EEG and MEG. Clin Neurophysiol 121(2): 172–185

[36] Kaiser M, Martin R et al (2007) Simulation of robustness against lesions of cortical networks. Eur J Neurosci 25(10): 3185–3192

[37] Kotter R and Sommer FT (2000) Global relationship between anatomical connectivity and activity propagation in the cerebral cortex. Philos Trans R Soc Lond B Biol Sci 355(1393): 127–34

[38] Kramer MA, Kolaczyk ED et al (2008) Emergent network topology at seizure onset in humans. Epilepsy Res 79(2–3): 173–186

[39] Laufs H (2008) Endogenous brain oscillations and related networks detected by surface EEG-combined fMRI. Hum Brain Mapp 29(7): 762–769

[40] Lehnertz K, Bialonski S et al (2009) Synchronization phenomena in human epileptic brain networks. J Neurosci Meth 183(1): 42–48

[41] Li Y, Liu Y et al (2009) Brain anatomical network and intelligence. PLoS Comput Biol 5(5): e1000395

[42] Liao W, Zhang Z et al (2010) Altered functional connectivity and small-world in mesial temporal lobe epilepsy. PLoS One 5(1): e8525

[43] Meunier D, Achard S et al (2009a) Age-related changes in modular organization of human brain functional networks. Neuroimage 44(3): 715–723

[44] Meunier D, Lambiotte R et al (2009b) Hierarchical modularity in human brain functional networks. Front Neuroinformatics 3: 37

[45] Micheloyannis S, Pachou E et al (2006) Using graph theoretical analysis of multi channel EEG to evaluate the neural efficiency hypothesis. Neurosci Lett 402(3): 273–277

[46] Morgan RJ and Soltesz I (2008) Nonrandom connectivity of the epileptic dentate gyrus predicts a major role for neuronal hubs in seizures. Proc Natl Acad Sci USA 105(16): 6179–6184

[47] Netoff TI, Clewley R et al (2004) Epilepsy in small-world networks. J Neurosci 24(37): 8075–8083

[48] Newman (2003) The structure and function of complex networks. SIAM Rev 45(2): 167–256

[49] Ortega GJ, de la Menendez PL et al (2008a) Synchronization clusters of interictal activity in the lateral temporal cortex of epileptic patients: intraoperative electrocorticographic analysis. Epilepsia 49(2): 269–280

[50] Ortega GJ, Sola RG et al (2008b) Complex network analysis of human ECoG data. Neurosci Lett 447(2–3): 129–133

[51] Ponten SC, Bartolomei F et al (2007) Small-world networks and epilepsy: graph theoretical analysis of intracerebrally recorded mesial temporal lobe seizures. Clin Neurophysiol 118(4): 918–927

[52] Ponten SC, Daffertshofer A et al (2009a) The relationship between structural and functional connectivity: graph theoretical analysis of an EEG neural mass model. Neuroimage 552(3): 985–994

[53] Ponten SC, Douw L et al (2009b) Indications for network regularization during absence seizures: Weighted and unweighted graph theoretical analyses. Exp Neurol 217(1): 197–204

[54] Reijneveld JC, Ponten SC et al (2007) The application of graph theoretical analysis to complex networks in the brain. Clin Neurophysiol 118(11): 2317–2331

[55] Rubinov M and Sporns O (2009) Complex network measures of brain connectivity: Uses and interpretations. Neuroimage 52(3): 1095–1069

[56] Salvador R, Suckling J et al (2005) Neurophysiological architecture of functional magnetic resonance images of human brain. Cereb Cortex 15(9): 1332–1342

[57] Schindler KA, Bialonski S et al (2008) Evolving functional network properties and synchronizability during human epileptic seizures. Chaos 18(3): 033119

[58] Smit DJ, Boersma M et al (2009) Endophenotypes in a dynamically connected brain. Behav Genet 40(2): 167–177

[59] Sporns O, Tononi G et al (2000) Theoretical neuroanatomy: relating anatomical and functional connectivity in graphs and cortical connection matrices. Cereb Cortex 10(2): 127–141

[60] Stam CJ (2004) Functional connectivity patterns of human magnetoencephalographic recordings: a "small-world" network? Neurosci Lett 355(1–2): 25–28

[61] Stam CJ and Reijneveld JC (2007) Graph theoretical analysis of complex networks in the brain. Nonlinear Biomed Phys 1(1): 3

[62] Supekar K, Musen M et al (2009) Development of large-scale functional brain networks in children. PLoS Biol 7(7): e1000157

[63] Uhlhaas PJ and Singer W (2006) Neural synchrony in brain disorders: relevance for cognitive dysfunctions and pathophysiology. Neuron 52(1): 155–168

[64] van Breemen MS, Wilms EB et al (2007) Epilepsy in patients with brain tumours: epidemiology, mechanisms, and management. Lancet Neurol 6(5): 421–430

[65] van Dellen E, Douw L et al (2009) Long-term effects of temporal lobe epilepsy on local neural networks: a graph theoretical analysis of corticography recordings. PLoS One 4(11): e8081

[66] van den Heuvel MP, Stam CJ et al (2008) Small-world and scale-free organization of voxel-based resting-state functional connectivity in the human brain. Neuroimage 43(3): 528–539

[67] van den Heuvel MP, Stam CJ et al (2009) Efficiency of functional brain networks and intellectual performance. J Neurosci 29(23): 7619–7624

[68] Varela F, Lachaux JP et al (2001) The brainweb: phase synchronization and large-scale integration. Nat Rev Neurosci 2(4): 229–239

[69] Wang J, Wang L et al (2009) Parcellation-dependent small-world brain functional networks: a resting-state fMRI study. Hum Brain Mapp 30(5): 1511–1523

[70] Watts DJ and Strogatz SH (1998) Collective dynamics of "small-world" networks. Nature 393(6684): 440–442

[71] Zhou C, Motter, AE et al (2006) Universality in the synchronization of weighted random networks. Phys Rev Lett 96(3): 034101

[72] van Dellen E, Douw L, Hillebrand A, Heimans J, Ris-Hilgersom I, Schoonheim M, Baayen J, De Witt Hamer P, Stam C and Reijneveld J. MEG functional connectivity and complex networks relate to clinical characteristics of lesional epilepsy patients [abstract]. In: AES Ann Meeting, Dec 3–7, 2010, San Antonio, USA. Poster session 2, Abstr. 2085

Futile circuits in functional neurosurgery

Alim L. Benabid

Introduction

It is not rare for a neurosurgeon to be surprised when he observes that making lesions or a resection in a patient's brain does not induce necessarily a deficit, or at least an observable deficit. This observation is common and without it neurosurgery in general, and functional neurosurgery in particular, would not be possible to perform. Trying to understand how this could happen leads quickly to the conclusion that those circuits are not fully necessary and can be deactivated or even destroyed without a significant damage to the functionality of the brain. Several questions are raised by this observation. Are these lesions or deactivations really without any effect? Why does this happen? Could it be that those circuits are embryological remnants, without real utility? Could it be that they have been added by evolution to improve the sophistication of actions, particularly of movement? Evolution does not know how to remove previously added structures. Could it be that their ablation is just equivalent to the suppression of a circuit totally useless, which would be only responsible for a higher probability of dysfunction leading to symptoms or even to a disease? When they are involved in the pathogenesis of diseases, then their ablation is acceptable as it allows the treatment of the symptoms without inducing too important defects. Are they useful but not requisite? In other words, are they futile? The term futile has been chosen because it alludes to some items used in everyone's daily life. For instance, jewels are not useless because possessing or wearing them provide some pleasure, which is not useless. But when they are removed or not worn, the essentials of life are still present. In this chapter I try to go a little bit further in this concept. This clearly might be unpleasant at the first glance, particularly for those who do spend their life working on those structures and providing valuable information about their anatomy and physiology. However, bringing this concept to reality might shed some light on the complexity of brain functions.

Where does the concept of futility come from?

During surgical procedures, it is observed that the alteration of some structures does not induce deficits or any observable postoperative change. This is particularly true after removal of rather large parts of the frontal or temporal lobes, usually on the nondominant side. Traumatic, hemorrhagic or surgical lesions of various parts of deeper structures, such as some basal ganglia (such as the ventral intermediate nucleus of the thalamus, the internal pallidum, the subthalamic nucleus, may be the nucleus accumbens), might also be clinically silent. This means that the functions born by these structures are not so important, if not even unnecessary. But most probably they may contribute to a certain level of sophistication of functions executed by more primordial systems, to which they have been added along evolution. This actually happens at the molecular level also, namely, in the metabolic futile cycles of Mitchell. Although the circumstances are different, **futile cycles** occur when two metabolic pathways run simultaneously in opposite directions and have no overall effect. For most of them, it is still debated whether they are useless or serve some nonevident purposes. In contrast to neuronal functionally futile systems, they are defined as futile because two actions, each fulfilling a purpose, run in opposite directions so that the metabolic result is nil. But each opposite way may have achieved a nonfutile goal. Here, we are considering systems as potentially futile because they seem not to achieve any necessary result.

How futile is a futile system?

The futility concept states that the absence of the system does not create any postoperative change. It can be easily argued that our postoperative evaluation of the patient is not sufficient, or not sophisticated enough, to insure that there is absolutely no change induced. There is obviously no way to refute this argument and, by default, one might think that futile systems do not exist. However, we have to cope with the observation that, as mentioned above, some structures could be altered "safely" by surgery or by accident, hemorrhage, etc. One will define a futile neuronal system as a system which can be ablated or inhibited without any consequence detectable by the usual clinical examinations or paraclinical tests evaluating the performances of the operated patient. When new methods are able to detect these changes in performance, then the structure will be withdrawn from the list of futile systems. In some way, one may find similarities with the post hoc validation hypothesis of Jean-Pierre Changeux, who assumes that after embryogenesis the neurons are interlinked by a large number of synaptic connections, possibly excessive or even redundant. Exposing the newborn brain to the reality of experience, in interaction with the external world, will validate parts of this connectivity, which will be reinforced, while the rest may be left as is or even undergoes a process of degenerative atrophy. This functionally driven atrophy may in fact compensate within an individual the inability of evolution to suppress, between subsequent generations within the same species, the useless, or unused, items.

How to recognize a nonfutile system?

For a large number, if not the majority, of the anatomical central nervous system structures, futility cannot even be discussed. These nuclei, bundles, or cortices do not tolerate alterations, which always cause, even temporarily, evident deficits. A lesion of the optic nerve, the chiasma, the optical tract, the lateral geniculate bodies, the optical radiations, and the calcarine cortex will always produce a campimetric visual deficit, the importance of which is closely related to the extent of the lesion. Many other examples can be given: Clear motor deficits will follow lesions of the primary motor cortex, the pyramidal tract, the anterior horn of the spinal cord, as well as of the peripheral motor nerves. Sensory deficits will follow lesions of the peripheral sensory nerves, the spinal cord posterior columns, the medial lemniscus, the somatosensory nuclei (VPL and VPM) of the thalamus, the thalamo-parietal projections and the primary sensory cortex. Many other structures could be mentioned which constitute the long list of nonfutile structures or systems.

When the observed deficit is transitory, the structure is nonetheless not futile as its alteration has induced a change in performances which demonstrates its clear usefulness. The transitory nature of the impaired performance might be due to temporary suffering of neighboring structures exhibiting transient edema, to neuroplasticity, to compensation by other structures (often symmetrical), or within a network. This does not make the system futile. Similarly, the work of H. Duffau has emphasized the “resectability” of tumor-invaded areas of the brain, taking into account the already lost function of the pathological area but also stressing the importance of intraoperative electrophysiological and functional exploration, in order to absolutely avoid harming important nonfutile areas and to possibly recognize, by their lack of intraoperative responses, the futile nature of other zones, which can be resected if needed. The purpose of this chapter is not to establish a hierarchy among useful structures, which might be an interesting approach. However, some examples can be provided.

The amygdalo-hippocampal formation might appear to be futile, as its total resection on the right side for treatment of some forms of epilepsy does not produce detectable cognitive deficits. However, on the dominant side this may be associated with verbal memory impairment. Moreover, its bilateral ablation creates a severe anterograde amnesia known since the famously disastrous HM case. The amygdalo-hippocampal formation could be therefore considered rather as a structure so important that its function can be achieved by only one structure, but it has been duplicated bilaterally. It is much more a very important structure, made bilateral in a design of security, than a futile system.

The ventral posterolateral nucleus of the thalamus is definitely not a futile system as it is part of a primary pathway for the acquisition of somatosensory inputs from the periphery. Does this mean that in the same nucleus or formation such as the thalamus there might be systems which are futile and systems which are not? This is probable, and this might be the case of the ventromedial intralaminar nucleus and of the ventral intermediate nucleus. The question is whether they appear at the same

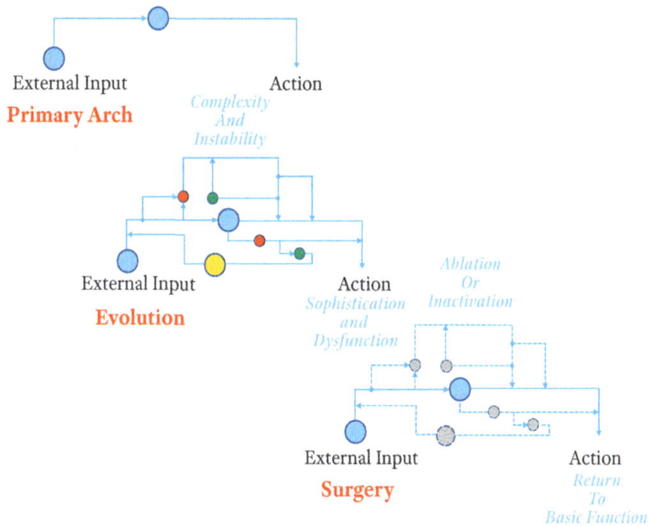

Fig. 1. The primary arch links the individual with the external world, from which it receives informations through the sensory inputs and to which it responds by action through motor outputs. Evolution achieves a progressive complexification, which produces sophistication of the function of the arch. At the same time, the risks of instability increase and create dysfunctions of the action. When surgery suppresses the networks (dotted lines and grey structures) added by evolution, it suppresses sophistication but also the pathological symptoms. The basic function of the arch is still preserved

time during the ontology of the brain or they are subsequent, the primary systems coming first, followed by the futile systems? This is also probable, and again within the thalamus its components vary strongly from rodents to nonhuman primates and finally to human primates. It cannot be excluded that during the process of maturation of the brain, some systems are added which at that time have the properties of the futile systems.

If futile systems exist, why are they here? How are they generated?

The beginning is the primary reflex arch, made of a sensory input monosynaptically connected to a motor output. To improve the function of this very basic system one might imagine, which is what happens in fact, that extracircuits impinge on the system as derivation, feedback loops, which are supposed to treat information and provide the elementary system with additional degrees of sophistication (Fig. 1). This process can be repeated several times creating a local network whose performances are more complex than that of the initial primary reflex arch but make the system progressively more complicated. Increasing the complication of a system, particularly in the nervous system, increases the probability, and then the risk, of a disturbance or of malfunctions, which ends up in abnormal systems, in abnormal symptoms, or in an abnormal functional behavior. By association with other dysfunctions, this may even create what is called a syndrome or even a disease. The question is to know if some futile systems are useless or simply futile as described which means that they are not totally useless but can be forgone leaving the system still correctly functioning. One may assume that evolution does not waste efforts to build up a network which in fact could be useless. One cannot however state for sure that this never happens, and therefore a futile system would have been generated. It may also happen that networks created to improve a given preexisting functionality become obsolete when the level of improvement they were supposed to provide is not anymore useful or even functioning, in the context of the subsequent steps of architectural complexification, brought up by the continuing process of evolution. In the same line of phylogenetic evolution, evolution, in contrast to electronic engineers, does not have the skill of redrawing a scheme or an algorithm to a simpler and still more efficient one. The only possibility evolution has shown is either to induce a mutation which would engage the genus, before the appearance of these futile networks, in a new way where these networks would be missing or to continue in the same way by adding additional loops or stages. At this moment, it is not possible to state that such useless systems exist.

Conclusion

Why are we disserting on these futile systems? Is it a pure academic brain storming, or does it have important consequences in terms of neurosurgical practice and understanding? The fact is simply that the hypothetical concept of futile systems is generated by the pragmatical observation of ablative or inactivating surgical procedures which are free of consequences and do not generate deficits. To prove this hypothesis necessitates that there is actually no induced deficit. This depends on the resolution of the methods, clinical as well as instrumental, used for the evaluation of the subject after the surgical procedure. Increasing this resolution power of the methods would probably decrease the number of systems or networks still considered as "deficit-free". For those which will remain "futile", two questions have been raised. Why are they here? And how have they been generated? The important fact, for us neurosurgeons, is to realize that, if they did not exist, functional neurosurgery as we understand it would not be feasible.

The manufacturer's authorised representative in the EU is Springer Nature Customer Service Centre GmbH, Europaplatz 3, 69115 Heidelberg, Germany. If you have any concerns regarding our products, please contact ProductSafety@springernature.com

Printed and bound by CPI Group (UK) Ltd, Croydon, CR0 4YY
22/09/2026
02221048-0020